HOMOTOPY THEORY

An Introduction to Algebraic Topology

Pure and Applied Mathematics

A Series of Monographs and Textbooks

RECENT TITLES

XIA DAO-XING. Measure and Integration Theory of Infinite-Dimensional Spaces: Abstract Harmonic Analysis

RONALD G. DOUGLAS. Banach Algebra Techniques in Operator Theory

WILLARD MILLER, JR. Symmetry Groups and Their Applications

ARTHUR A. SAGLE AND RALPH E. WALDE. Introduction to Lie Groups and Lie Algebras

T. BENNY RUSHING. Topological Embeddings

JAMES W. VICK. Homology Theory: An Introduction to Algebraic Topology

E. R. KOLCHIN. Differential Algebra and Algebraic Groups

GERALD J. JANUSZ. Algebraic Number Fields

A. S. B. HOLLAND. Introduction to the Theory of Entire Functions

WAYNE ROBERTS AND DALE VARBERG. Convex Functions

A. M. OSTROWSKI. Solution of Equations in Euclidean and Banach Spaces, Third Edition of Solution of Equations and Systems of Equations

H. M. EDWARDS. Riemann's Zeta Function

SAMUEL EILENBERG. Automata, Languages, and Machines: Volume A. *In preparation:* Volume B

MORRIS HIRSCH AND STEPHEN SMALE. Differential Equations, Dynamical Systems, and Linear Algebra

WILHELM MAGNUS. Noneuclidean Tesselations and Their Groups

J. DIEUDONNÉ. Treatise on Analysis, Volume IV

FRANÇOIS TREVES. Basic Linear Partial Differential Equations

WILLIAM M. BOOTHBY. An Introduction to Differentiable Manifolds and Riemannian Geometry

BRAYTON GRAY. Homotopy Theory: An Introduction to Algebraic Topology

ROBERT A. ADAMS. Sobolev Spaces

In preparation

D. V. WIDDER. The Heat Equation

IRVING E. SEGAL. Mathematical Cosmology and Extragalactic Astronomy

WERNER GREUB, STEPHEN HALPERIN, AND RAY VANSTONE. Connections, Curvature, and Cohomology: Volume III, Cohomology of Principal Bundles and Homogeneous Spaces

J. DIEUDONNÉ. Treatise on Analysis, Volume II, enlarged and corrected printing

JOHN J. BENEDETTO. Spectral Synthesis

I. MARTIN ISAACS. Character Theory of Finite Groups

HOMOTOPY THEORY

An Introduction to Algebraic Topology

BRAYTON GRAY
Department of Mathematics
University of Illinois at Chicago Circle
Chicago, Illinois

ACADEMIC PRESS New York San Francisco London 1975
A Subsidiary of Harcourt Brace Jovanovich, Publishers

ACADEMIC PRESS, INC.
111 Fifth Avenue, New York, New York 10003

United Kingdom Edition published by
ACADEMIC PRESS, INC. (LONDON) LTD.
24/28 Oval Road, London NW1

Library of Congress Cataloging in Publication Data

Gray, Brayton.
Homotopy theory.

(Pure and applied mathematics series, 64)
Bibliography: p.
1. Homotopy theory. I. Title. II. Series: Pure and applied mathematics; a series of monographs and textbooks.
QA3.P8 [QA612.7] 510'.8s [514'.24] 74-5686
ISBN 0-12-296050-5

PRINTED IN THE UNITED STATES OF AMERICA

Contents

Preface vii
List of Symbols xi

0 Preliminaries 1
1 Some Simple Topological Spaces 4
2 Some Simple Topological Problems 9
3 Homotopy Theory 13
4 Category Theory 17
5 The Fundamental Group 22
6 More on the Fundamental Group 29
7 Calculating the Fundamental Group 34
8 A Convenient Category of Topological Spaces 50
9 Track Groups and Homotopy Groups 62
10 Relative Homotopy Groups 70
11 Locally Trivial Bundles 77
12 Simplicial Complexes and Linearity 90
13 Calculating Homotopy Groups: The Blakers–Massey Theorem 99
14 The Topology of CW Complexes 113
15 Limits 122
16 The Homotopy Theory of CW Complexes 130
17 $K(\pi, n)$'s and Postnikov Systems 157
18 Spectral Reduced Homology and Cohomology Theories 168
19 Spectral Unreduced Homology and Cohomology Theories 183
20 Ordinary Homology of CW Complexes 187
21 Homology and Cohomology Groups of More General Spaces 200
22 The Relation between Homotopy and Ordinary Homology 215

23 Multiplicative Structure 221
24 Relations between Chain Complexes 245
25 Homological Algebra over a Principal Ideal Domain: Künneth and Universal Coefficient Theorems 256
26 Orientation and Duality 269
27 Cohomology Operations 294
28 Adem Relations 310
29 *K*-Theories 324
30 Cobordism 342

Table 1 Adem Relations $Sq^a\ Sq^b$ *for* $a < 2b \leq 12$ 358
References 360
Index 363

Preface

This book is an exposition of elementary algebraic topology from the point of view of a homotopy theorist. The only prerequisite is a good foundation in point set topology. In particular, homology theory is not assumed. Both homology and cohomology are developed as examples of (generalized) homology and cohomology theories. The idea of developing algebraic topology in this fashion is not new, but to my knowledge this is the first detailed exposition from this viewpoint. One pedagogical advantage of a course developed in this way is that it may be studied before or after a course in classical homology theory (e.g. [71] or [28]); alternatively homology and cohomology could first be introduced, as they are here, as examples of a more general theory.

The philosophical emphasis here is: to solve a geometrical problem of a global nature, one first reduces it to a homotopy theory problem; this is in turn reduced to an algebraic problem and is solved as such. This path has historically been the most fruitful one in algebraic topology.

The first few sections are introductory in nature. These are followed by a discussion of the fundamental group, covering spaces, and Van Kampen's theorem. The fundamental group serves as a model of a functor that can be calculated.

In Section 8 we introduce the category of compactly generated spaces. This seems to be the most appropriate category for algebraic topology. Many results which are most often stated in the category of CW complexes are valid in this generality.

The key result we use to make calculations is the Blakers–Massey theorem. This is strong enough to imply the suspension theorem and the Serre exact sequences. The Blakers–Massey theorem is proved by linear approximation techniques in Section 13 (after J. M. Boardman) in the case of a pair of relative cells. This allows one to calculate $\pi_n(S^n)$. The more general form of the Blakers–Massey theorem is proved in Section 16.

In Section 18 reduced "spectral" homology and cohomology theories are defined from arbitrary spectra on compactly generated spaces. They satisfy the usual axioms in this generality. It is more complicated to show that unreduced theories satisfy the usual axioms; this is done in Section 21. Spectral homology theories agree with their "singular approximations." Spectral cohomology theories on paracompact spaces have all the properties usually associated with Čech theory.

Calculations in the ordinary homology of CW complexes are studied in Section 20. We develop axioms for the chain complex of a CW complex. These are strong enough to make all the usual calculations based on ad hoc decompositions. In particular a proof of the algorithm for the homology of a simplicial complex is given. (The algorithm for singular homology follows from the functorial singular complex construction which is included as an appendix to Section 16.)

The Hurewicz theorem follows quite easily from the Blakers–Massey theorem. Duality in manifolds in its full generality for an arbitrary ring spectrum follows from the usual inductive approach using "spectral" homology and cohomology.

In Section 27 we introduce Steenrod operations geometrically via the quadratic construction. We learned this approach from J. Milgram. The Adem relations are proven in Section 29 by a method due to L. Kristensen. This necessitates calculating $H^*(K(Z_2, n); Z_2)$ for which we state, without proof, the Borel transgression theorem.

Spectral sequences have been omitted for several reasons: their introduction would increase the length of the book considerably; they are more difficult to write about than to explain; there already exist expositions which we feel we cannot improve upon ([21, 41, 31]).

In the last two sections we sketch K-theory and cobordism. This serves as an introduction to some of the more powerful functors which have been utilized in algebraic topology. Applications of K-theory to the Hopf invariant and the vector field problem are discussed. In the last section $\pi_*(MO)$ is calculated.

We have used throughout the symbol ▮ whenever we have finished the proof of some theorem, proposition, corollary, or claim made earlier. Sometimes these symbols pile up—for example, if one statement is reduced to another.

The exercises are an integral part of the development. No exercise requires outside reading, or the utilization of techniques not previously developed (usually in the relevant section). A * on an exercise indicates that its understanding is necessary for further study—not that it is difficult. Often terminology used later is first introduced in such an exercise. Many exercises are used later in only a few places. In such cases it is indicated in parentheses at the end of the exercise where the result is used.

A one-quarter introductory course could easily be based on Sections 0–13. Section 8 could be skipped if an independent proof of adjointness is given. Section 8 is used essentially in Section 18 unless one restricts the homology and cohomology theories to finite complexes.

Occasionally seminar problems are given at the end of a section. These are topics peripheral to the material in the section and not used elsewhere. They are intended for students to give a report on, based on the references given. In our opinion, these problems provide an excellent way for students to get more actively involved in the subject.

I wish to thank the many mathematicians who offered suggestions and encouragement, and the typists who suffered through my handwriting. Particular thanks in this regard are due to Ms. Shirley Bachrach.

List of Symbols

An n following a page number indicates a symbol defined in a footnote.

$f\|_A$	3n
S^n	4
Δ^n	4, 175
I^n	4
B^n	4
$\|x\|$	4
R^n	4
∂I^n	5
Int I^n	5
$\tilde{I}^n$	5
$\equiv$	6
(X, A)	6
$f: (X, A) \to (Y, B)$	6
X^∞	7
X^+	7
X/A	7
p_A	7
B_n^+	8
B_n^-	8
J^n	8
I_0^n	8
Int Δ^n	8
$\partial\Delta^n$	8
$x \cdot y$	9
$f_0 \sim f_1$	13, 15, 21
$F: f_0 \sim f_1$	13, 15, 21
$\{f\}$	14, 16, 21
$[X, Y]$	14, 15, 21
P	15
hom(X, Y)	17
$\mathcal{T}$	18
$\mathcal{S}$	18
$\mathcal{G}$	18
$\mathcal{R}$	18
$\mathcal{M}_R$	18
$F: \mathcal{C}_1 \to \mathcal{C}_2$	19
$\mathcal{T}^*$	20
$\mathcal{T}^2$	20
$\mathcal{C}^*$	20
$\mathcal{C}^2$	20
$\mathcal{T}_h$	21
$\mathcal{T}_h{}^*$	21
$\mathcal{T}_h{}^2$	21
$\pi_n(X, *)$	21
$f \sim g$ (rel A)	22
$F: f \sim g$ (rel A)	22
$\pi(X; x, y)$	22
$\pi(x, y)$	22
f_*	27, 70, 170, 183
$\Pi(X)$	27
$\pi_1(X)$	29
φ_α	29
$G_1 *_G G_2$	40
$A * B$	40, 93
Π_n	44, 110
η_n	44
ν_n	44
HP^n	44
CP^n	44
RP^n	44
H	46, 277
$\bar{q}$	47
$\mathcal{C}\mathcal{G}$	50
$k(X)$	52
$X \times_c Y$	53
$X \times Y$	53
Y^X	55
$C(X, Y)$	55
$X \wedge Y$	58
$X \vee Y$	58
$(Y, B)^{(X, A)}$	58
$\mathcal{C}\mathcal{G}^*$	58
$\tilde{\psi}$	60
$X \amalg Y$	61
$\coprod X_\alpha$	61
SX	62
f^*	62, 170, 183
C^*X	63
φ	63
∇	64
ΩX	65

γ_p	68
CX	68
ΣX	69
$\Omega(X; A, B)$	70
$\pi_n(X, A, *)$	70
$\mathcal{C}^{2*}$	70
∂	71
$\pi_n(X, A)$	75
$\pi_n(X; A, B)$	88
$O(n)$	89
$Z(\pi)$	89
ξ^σ	89
$(v_0 v_1 \cdots v_n)$	90
$\tau \prec \sigma$	91
$\lvert K \rvert$	91
lin dim A	92
$b(\sigma)$	94
K'	94
$\mu(K)$	96
Int σ	98
$X \cup_\alpha e^n$	99
χ	100, 199, 306
$B^n(\rho)$	101
$e^n(\rho)$	101
E	107
e_α	113
X^n	113
$K(A)$	114
$\mathcal{K}$	115
$\mathcal{K}^*$	115
$\mathcal{K}_h$	115
$\mathcal{K}_h{}^*$	115
RP^∞	115
CP^∞	115
HP^∞	115
$Y \cup_f CX$	118
$\bar{X}^n$	120
$Y \cup_f C^*X$	121
$\varinjlim X_\alpha$	123
X_∞	127
S	139
$\pi_r{}^S(X)$	145
$\tilde{\mathcal{S}}_r(X)$	145
$S(X)$	146
∂_i	155, 189
$K(\pi, n)$	158
$M(\pi, n)$	159
$X^{[n]}$	160
$X^{(n)}$	164
$\mathcal{O}_n(f)$	166
$\tilde{H}^n(X; \pi)$	166, 170
$\{E_n, e_n\}$	169
$\underline{X}$	169
$H\pi$	169
$\{G_n\}$	169
G_*	169
$\mathcal{S}_*$	169
$\mathcal{M}_{R*}$	169
$\mathcal{M}_{Z*}$	169
$\tilde{E}^m$	170
$\tilde{E}_m$	170
σ	170
$\tilde{H}_m(X; \pi)$	170
$\tilde{H}_m(X)$	170
$\tilde{H}^m(X)$	170
T_X	172
Δ^m	175
Δ_m	175
∇_m	175
∇^m	175
$\pi_m(E)$	177
$\pi_S{}^m(X)$	179
h	180, 216
c_φ	181
EG	181
β	181
$E_m(X, A)$	183
$E^m(X, A)$	183
$\mathcal{E}$	184
$\mathcal{N}$	184
$C_n(X, A; \pi)$	189
$C^n(X, A; \pi)$	190
δ_n	190
$\mathcal{C}_\#$	191
$\mathcal{K}_R$	191
$C_\#$	191
$SE^n(X, A)$	200
$SE_n(X, A)$	200
$S\tilde{E}_n(X)$	200
$S\tilde{E}^n(X)$	200
$\mathcal{P}$	209
$\check{H}^n(X, A; G)$	212
$\tilde{H}^n(X, A; G)$	212
c	221
E_z^n	222
Σ_z^n	222
T_σ	223
$C_{m,k}$	224
$\tilde{E}_k(X, Y)$	224
M	230
$\backslash$	233, 237
$/$	233, 237
$\barwedge$	233
$\triangle$	233
$x \cup y$	236, 238
$\underline{\times}$	237
$\overline{\times}$	237
$x \cap y$	242
$\langle x, y \rangle$	242
$\Lambda_\#$	247
$\Lambda^\#$	247
d	251, 263, 264
$\mathrm{Ext}_R(M, N)$	257
$\mathrm{Tor}_R(M, N)$	257
v	261, 343
Δ	261, 264
$[X]$	271
ρ_B^A	271
ρ_x	271
$[K]_E$	277
C	278
$\xi_{V,W}$	278
D	279
$L(f)$	288
$\mathrm{Tr}(A)$	288
H^n	290

∂M	290	$\mathrm{Vect}_n(X)$	326	K	331
Int M	290	$f^*(\xi)$	326	KO	331
$\{E, m, F, n\}$	295	k^∞	326	KSp	331
$X \rtimes Y$	296	$\gamma^n(V)$	327	β_n	332
$\Gamma^n(X)$	296	$G_n(V)$	327	$X * Y$	334
$\Gamma(X)$	297	$E_n(V)$	327	$h(f)$	334
Sq^i	301, 305	$K_k(X)$	328	ψ^k	336
P^r	308	$BO(n)$	328	$\rho(n)$	337
$\mathcal{A}(2)$	310	$BSp(n)$	328	$V_k(R^n)$	337
$\Lambda(x_1, \ldots, x_n)$	311	$BU(n)$	328	$\mathfrak{N}_n$	342
$\mathrm{ex}(I)$	311	$\xi \times \eta$	328	$T(\xi)$	343
κ_I	315	$\xi \otimes \eta$	329	$MO(k)$	344
κ	315	λ^i	329	MO	344
κ'	315	$\xi \cong_s \eta$	330	r	344
$H(\alpha)$	320	BO	331	w_i	345
$\mathcal{A}(2)^*$	322	BU	331	$w_i(\xi)$	357
$\{E, \pi, B\}$	324	BSp	331		

0

Preliminaries

An area of study in mathematics consists at least of a collection of problems and usually a collection of techniques useful in solving the problems. We begin by looking at some problems typical of those considered in algebraic topology.

Let D be the unit disk in the complex plane, i.e., the set $\{z \mid |z| \leq 1\}$.

Problem 1 Does there exist a continuous (differentiable? analytic?) function $f\colon D \to D$ with no fixed points (i.e., for no z is $f(z) = z$)?

One thinks of f as a transformation from the disk to itself, and one might visualize points moving such as a rotation of the disk or a squashing of it onto a smaller subset. Can this be done so that every point is moved? It would seem much easier to solve this problem if the answer is affirmative; one only needs to write down the equations of such a function or draw it pictorially. If the answer is no, it might be very difficult to prove.

Let $S \subset D$ be the unit circle: $\{z \mid |z| = 1\}$.

Problem 2 Does there exist a continuous (differentiable?) function $f\colon D \to S$ with $f(z) = z$ whenever $z \in S$?

Imagine a rubber disk held onto a table by the rim and try to pull the disk toward the rim without tearing (a tear would represent a discontinuity since its only use would be to move nearby points away from each other). Clearly by punching a hole in the rubber, it can be pulled to the rim, but otherwise it seems intuitively clear that no such continuous map could exist. How could one prove this?

Let S^2 be the sphere in three-dimensional space, $\{V \mid |V| = 1\}$.

Problem 3 Suppose that to each point of S^2 is associated a vector whose tail is at that point and which is tangent to the sphere. (Such a situation can be thought of as a hairy ball, with all the hairs matted down.) This can be more quantitatively stated by saying that for each vector V with $|V| = 1$ we associate a W with $V \perp W$ (i.e., $V \cdot W = 0$). Let us write W as $W(V)$ and suppose W is continuous as a function of V (continuity corresponds to having the hair on the ball combed). Does there exist a continuous function $W(V)$ with $W(V) \neq 0$ for all V? (That is, can the hair be combed on a ball without any baldspots?)

Problem 4 One can easily see that there is a continuous function $\gamma: S^2 \to S^2$ with no fixed points, namely, $\gamma(x, y, z) = (-x, -y, -z)$. This moves points very far. Does there exist a function $\gamma: S^2 \to S^2$ without fixed points so that $\|\gamma(x) - x\| < \varepsilon$ for some fixed number ε? For which ε? (Is there an "infinitesimal" transformation without fixed points?)

We can see from these representative problems that the spaces we deal with are simple. They will, to a large extent, be the spaces that arise naturally in mathematics. Many of the problems arose in analysis, linear algebra, projective geometry, etc. Our first task will be to define some of the spaces and then make some general remarks on the type of problem we are considering and the type of tools we shall use to solve the problem.

Exercises

Exercises 1, 2, and 3 are useful in homotopy theory. They are commonly contained in a point set topology course. A good exercise to confirm your mastery over this prerequisite material will be to supply proofs for them. As references for point set topology we recommend [22; 29; 36].

1.* Given a topological space X and an equivalence relation $\sim$ among the points of X, one topologizes the set of equivalence classes $X/\sim$ as follows. There is a map $\Pi: X \to X/\sim$, and we say that $U \subset X/\sim$ is open iff $\Pi^{-1}(U)$ is open in X. This is a topology. Suppose we are given a continuous map $f: X \to Y$ such that for any two points $x, x' \in X$ with $x \sim x'$ we have $f(x) = f(x')$; then there is a unique continuous map $\tilde{f}: X/\sim \to Y$ so that $\tilde{f}\Pi = f$, i.e., so that the diagram

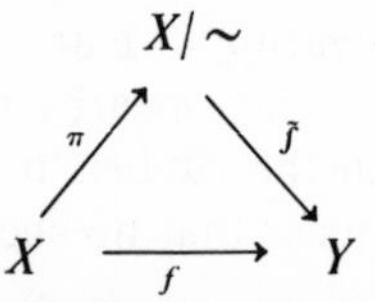

commutes.

2.* Suppose $X = F_1 \cup \cdots \cup F_k$ where each F_i is closed. Suppose $f_i\colon F_i \to Y$ is a continuous map for each i, such that[1]

$$f_i|_{F_i \cap F_j} = f_j|_{F_i \cap F_j}$$

for all i and j. Then the unique map $f\colon X \to Y$ defined by $f|_{F_i} = f_i$ is continuous.

3.* (*Lebesgue's Covering Lemma*) If X is a compact metric space and $\{\mathcal{U}_\alpha\}$ is an open cover, there exists an ε-number—i.e., a positive number ε such that if A is any set with diameter $<\varepsilon$, there exists an α such that $A \subset \mathcal{U}_\alpha$.

4. Let $f\colon X \to Y$ be a closed continuous map from X onto Y. Suppose X is Hausdorff and that either X is normal or $f^{-1}(y)$ is compact for each $y \in Y$. Prove that Y is Hausdorff. (Hint: Find open sets $U_i \supset f^{-1}(y_i)$ with $U_1 \cap U_2 = \varnothing$, and consider $W_i = Y - f(X - U_i)$.) (Exercise 8, Section 7; Exercise 13, Section 13; 16.36;27.9).

5. Let $f_1(z) = \alpha z$ for $|\alpha| < 1$, $f_2(z) = t + (1 - t)z$ for $0 < t < 1$, and $f_3(z) = z^2$. Find a fixed point for $f_3 f_2 f_1$.

[1] Given a map $f\colon X \to Y$ and a subspace $A \subset X$, we use the notation $f|_A$ for the map $A \to Y$ given by restricting f to A.

I

Some Simple Topological Spaces

Most of the topological spaces that arise in mathematical problems are subsets of n-dimensional Euclidean space, and it is natural to give preferential treatment to such spaces. We begin our study by defining some of the simpler subspaces of Euclidean space. These spaces will recur in both the theory and applications of topology, and we take some time to discuss the relationships among them.

Definition 1.1

$R^n = \{(x_1, \ldots, x_n) \mid x_i \text{ real}\}$, n-dimensional Euclidean space.

For $x \in R^n$, write $\|x\| = \sqrt{\Sigma\, x_i^2}$.

$B^n = \{(x_1, \ldots, x_n) \in R^n \mid \|x\| \le 1\}$, the n-dimensional ball.

$S^{n-1} = \{(x_1, \ldots, x_n) \in R^n \mid \|x\| = 1\}$, the $(n-1)$-dimensional sphere.

$I^n = \{(x_1, \ldots, x_n) \in R^n \mid 0 \le x_i \le 1\}$, the n-dimensional cube.

$\Delta^{n-1} = \{(x_1, \ldots, x_n) \in R^n \mid 0 \le x_i \le 1, \Sigma\, x_i = 1\}$, the n-dimensional simplex.

The last four of these are naturally imbedded in the first, via their description. They are pictured in Fig. 1.1 for $n = 2$.

The first two of these occur naturally in the four problems listed earlier ($D = B^2$, $S = S^1$). They are all closely related, however, and the rest of this section will be a technical exposition of their relation. This may be skipped

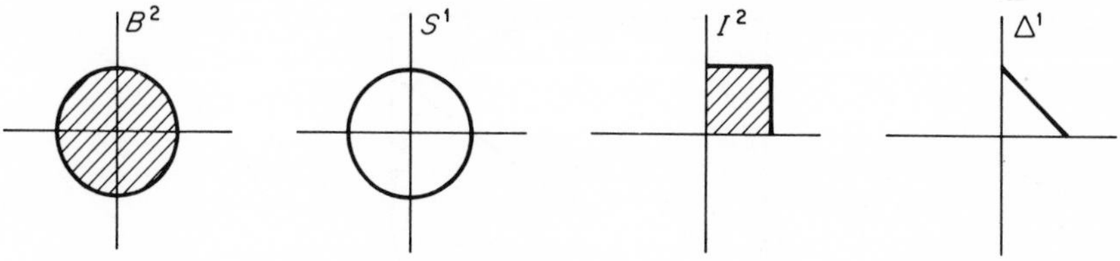

Figure 1.1

by the more impatient students, but is in fact a good introduction to "seeing in higher dimensions."

Write Int X for the interior of X considered as a subset of some larger space (given by context).

Proposition 1.2 Int $I^n = \{(x_1, \ldots, x_n) \mid 0 < x_i < 1\}$.

Proof If $0 < x_i < 1$ for each i, let $\varepsilon = \min_i(1 - x_i, x_i)$. Then a ball of radius ε about x is contained in I^n, so x is in the interior. If, for some i, $x_i = 1$ (or 0), then a ball of radius r about x will contain points with $x_i > 1$ (<0) no matter how small r is. Thus these points are not in the interior. ∎

We will write $\partial I^n = I^n - \text{Int } I^n$.

Proposition 1.3 I^n is homeomorphic to B^n. Under this homeomorphism ∂I^n corresponds to S^{n-1}.

Proof Let $\tilde{I}^n = \{(x_1, \ldots, x_n) \mid -1 \leq x_i \leq 1\}$. Clearly $\tilde{I}^n$ and I^n are homeomorphic and ∂I^n corresponds to

$$\partial \tilde{I}^n = \{(x_1, \ldots, x_n) \mid -1 \leq x_i \leq 1 \quad \text{and} \quad x_i = \pm 1 \quad \text{for some } i\}.$$

Define $\phi_1 : \tilde{I}^n \to B^n$ and $\psi_1 : B^n \to \tilde{I}^n$ by the formulas

$$\phi_1(x_1, \ldots, x_n) = \frac{\max(|x_i|)}{\|x\|}(x_1, \ldots, x_n), \qquad \phi_1(0) = 0$$

$$\psi_1(x_1, \ldots, x_n) = \frac{\|x\|}{\max(|x_i|)}(x_1, \ldots, x_n), \qquad \psi_1(0) = 0.$$

Clearly $\|\phi_1(x_1, \ldots, x_n)\| = \max(|x_i|) \leq 1$ and $\max(|\psi_{1,i}|) = \sqrt{\Sigma x_i^2} \leq 1$, so these maps are well defined and inverse to each other. The continuity at 0 follows from the inequalities

$$\max(|x_i|) \leq \|x\| \leq \sqrt{n} \max(|x_i|). \quad ∎$$

What we have done in the proof is to shrink every ray from 0 to $\partial \tilde{I}^n$ down linearly to have length 1 (Fig. 1.2). Its original length is $\|x\|/\max(|x_i|)$.

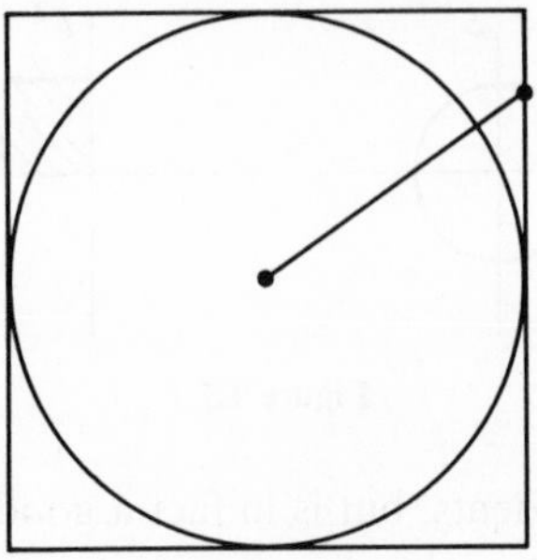

Figure 1.2

We will write $\equiv$ to mean homeomorphism. Thus we have $I^n \equiv \tilde{I}^n \equiv B^n$ and $\partial I^n \equiv \partial \tilde{I}^n \equiv S^{n-1}$ in a compatible way.

Given a space X and a subspace A, we will write (X, A) for the pair of spaces. A map from the pair (X, A) to (Y, B) is just a map $f\colon X \to Y$ with $f(A) \subset B$. A homeomorphism from (X, A) to (Y, B) is just a homeomorphism from X to Y such that A corresponds to B. Thus we have proven that $(I^n, \partial I^n)$ is homeomorphic to $(\tilde{I}^n, \partial \tilde{I}^n)$, and this is homeomorphic to (B^n, S^{n-1}). We write this as

$$(I^n, \partial I^n) \equiv (\tilde{I}^n, \partial \tilde{I}^n) \equiv (B^n, S^{n-1}).$$

Proposition 1.4 $I^n - \partial I^n \equiv R^n$.

Proof Since $I^n - \partial I^n \equiv \overbrace{(I - \partial I) \times \cdots \times (I - \partial I)}^{n}$, this follows from the fact that $I - \partial I = (0, 1) \equiv R^1$. (We use the homeomorphism $t \to \tan(\pi/2)(2t - 1)$.) ∎

Proposition 1.5 $S^n - (1, 0, 0, \ldots, 0) \equiv R^n$.

Proof Here we use the familiar stereographic projection (Fig. 1.3). Placing $R^n \subset R^{n+1}$ by making the first coordinate[2] 0 and S^n intersecting R^n in the equator, we draw a line from the north pole through a point $x \in S^n$ and record $\phi_2(x)$, its point of intersection with R^n:

$$\phi_2\colon S^n - (1, 0, \ldots, 0) \to R^n, \qquad \phi_2(x_1, \ldots, x_{n+1}) = \left(\frac{x_2}{1 - x_1}, \ldots, \frac{x_{n+1}}{1 - x_1}\right).$$

[2] This imbedding may seem strange, but it is necessary if we want $(1, 0, \ldots, 0)$ to be the missing point. This is important as it will later be chosen as a "base point" and belongs to S^n for all $n \geq 0$.

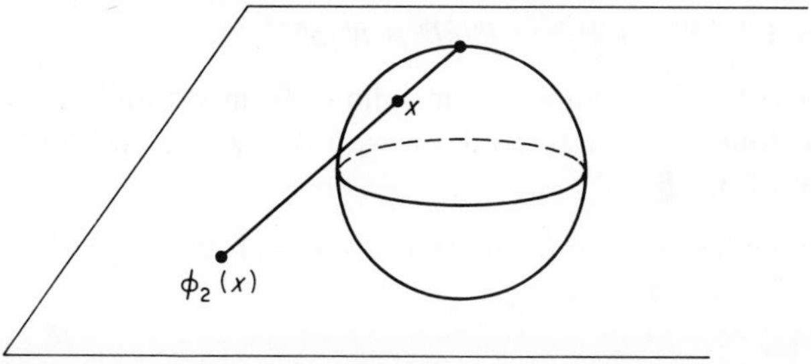

Figure 1.3

This has an inverse $\psi_2\colon R^n \to S^n - (1, 0, \ldots, 0)$ given by

$$\psi_2(x_1, \ldots, x_n) = \frac{1}{1 + \|x\|^2}(\|x\|^2 - 1, 2x_1, \ldots, 2x_n).$$

It is easy to check that $\psi_2 \phi_2 = 1$ and $\phi_2 \psi_2 = 1$ and these are clearly continuous. ∎

Let us write X/A for the quotient space of X with A identified to a point. If $A = \varnothing$, we will sometimes write X/A for X^+, the disjoint union of X and a point $+$. We will also write $p_A\colon X \to X/A$ for the quotient map. Write X^∞ for the one-point compactification of X.

Lemma 1.6 Let $U \subset X$, with X Hausdorff and regular, U open, and $\overline{U}$ compact. Then

$$U^\infty \equiv X/X - U.$$

Proof Consider the map $\gamma\colon U^\infty \to X/X - U$ given by

$$\gamma|_U = p_{X-U}|_U : U \subset X \to X/X - U$$

and $\gamma(\infty) = \{X - U\}$.

To show that this is continuous we need only consider open neighborhoods of $\{X - U\}$. If V is such a neighborhood, let $W = (p_{X-U})^{-1}(V) \subset X$. Since W is open, $\overline{U} - W$ is compact. Now $\gamma^{-1}(V) - \infty = W \cap U = U - (\overline{U} - W)$, so $\gamma^{-1}(V)$ is open. γ is clearly 1–1 and onto. Since U is open and X is regular, $X/X - U$ is Hausdorff. Hence γ is a homeomorphism. ∎

Corollary 1.7 If U is open and bounded in R^n,

$$R^n/R^n - U \equiv U^\infty. \quad ∎$$

Corollary 1.8 If X is compact Hausdorff and $x \in X$,

$$(X - x)^\infty \equiv X. \quad ∎$$

Corollary 1.9 $S^n \equiv (R^n)^\infty \equiv I^n/\partial I^n \equiv B^n/S^{n-1}$.

Proof The first homeomorphism follows from 1.5 and 1.8 with $X = S^n$. The second follows from 1.4 and 1.6 with $X = I^n$ and $U = \text{Int } I^n$. The last follows from 1.3. ▮

Note that under these homeomorphisms, $* = (1, 0, \ldots, 0) \in S^n$ corresponds to $\infty \in (R^n)^\infty$, $\{\partial I^n\}$, and $\{S^{n-1}\}$.

Definition 1.10 An n-cell is a pair (X, A) homeomorphic to (B^n, S^{n-1}).

Exercises

1. Show that $\text{Int } B^n = B^n - S^{n-1}$. (One cannot use 1.3 since the notion of interior depends on the imbedding.)

2.* Show that (X, A) is homeomorphic to (Y, B) iff there are maps $f\colon (X, A) \to (Y, B)$ and $g\colon (Y, B) \to (X, A)$ with $fg = 1_Y$ and $gf = 1_X$.

3. Let $B_n^+ = \{x \in S^n \mid x_{n+1} \geq 0\}$, and $B_n^- = \{x \in S^n \mid x_{n+1} \leq 0\}$. Show that $B_n^+ \equiv B_n^- \equiv B^n$.

4. Using the equality $I^{n-1} \times I = I^n$, define $J^{n-1} \subset I^n$ as $(\partial I^{n-1}) \times I \cup I^{n-1} \times (1)$, and $I_0^{n-1} = I^{n-1} \times (0)$. Then

$$\partial I^n = J^{n-1} \cup I_0^{n-1}, \qquad (\partial I^{n-1}) \times (0) = J^{n-1} \cap I_0^{n-1}.$$

Prove $(J^{n-1}, J^{n-1} \cap I_0^{n-1})$ is an $(n-1)$-cell. (Section 10; 11.6)

5. Show that $\text{Int } \Delta^{n-1} = \{x \in \Delta^{n-1} \mid 0 < x_i < 1\}$. Prove that $(\Delta^{n-1}, \partial\Delta^{n-1})$ is an $(n-1)$-cell where $\partial\Delta^{n-1} = \Delta^{n-1} - \text{Int } \Delta^{n-1}$. (Section 12)

6. One can easily generalize Problems 1–4 to other dimensions; they represent the case $n = 2$ of a problem for every dimension. State these problems. Solve them, if you can, for $n = 1$. Do the solutions generalize?

2

Some Simple Topological Problems

This section has two simple aims: to generalize the four problems considered in Section 0, and to study some of the relationships among them.

As generalizations, we propose the following:

Problem 1 Does there exist a continuous (differentiable?) function $f: B^n \to B^n$ without fixed points?

Problem 2 Does there exist a continuous (differentiable?) function $f: B^n \to S^{n-1}$ with $f(x) = x$ for $x \in S^{n-1}$? The condition on f is that the diagram

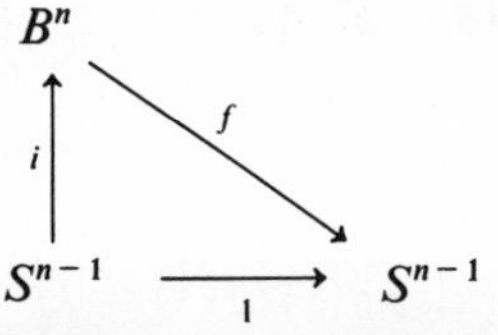

"commutes," where $i: S^{n-1} \to B^n$ is the inclusion function and 1 the identity function. For this to commute means that if you follow a point through the two paths, the result is the same.

For $x, y \in R^n$, write $x \cdot y$ for the scalar product:

$$x \cdot y = \sum_{i=1}^{n} x_i y_i .$$

Problem 3 We transfer our vectors to the origin. Does there exist a continuous (differentiable?) function

$$f: S^{n-1} \to R^n$$

satisfying $x \cdot f(x) = 0$ and $f(x) \neq 0$ for all x? Such a function is called a nonzero vector field on S^{n-1}.

Problem 4 Given $\varepsilon > 0$, does there exist a continuous (differentiable?) map $\gamma\colon S^n \to S^n$ with no fixed points and such that

$$\|\gamma(x) - x\| < \varepsilon?$$

(If $\varepsilon > 2$, the answer is yes: $\gamma(x_1, \ldots, x_{n+1}) = (-x_1, \ldots, -x_{n+1})$.)

Proposition 2.1 If S^n has a nonzero vector field, a map satisfying the conditions of Problem 4 exists for every $\varepsilon > 0$.

Proof Let $f\colon S^n \to R^{n+1}$ be a nonzero vector field. Then for $\lambda \neq 0$ a real number, λf is also a nonzero vector field. By choosing λ small we may guarantee that $\|\lambda f\| < \varepsilon$ since S^n is compact. Thus there are "small" vector fields. Suppose f satisfies $\|f\| < \varepsilon/2$.

We now get $\gamma\colon S^n \to S^n$ without fixed points by moving x in the direction of $f(x)$ (Fig. 2.1):

$$\gamma(x) = \frac{x + f(x)}{\|x + f(x)\|}.$$

Observe that

$$\|x - \gamma(x)\|^2 = [x - \gamma(x)] \cdot [x - \gamma(x)] = 2 - 2\gamma(x) \cdot x = 2\left(1 - \frac{1}{\|x + f(x)\|}\right).$$

However $\|x + f(x)\| \leq 1 + \varepsilon/2$, so

$$\|x - \gamma(x)\|^2 \leq 2\left(1 - \frac{1}{1 + \varepsilon/2}\right) \leq \varepsilon;$$

i.e., γ is close to the identity.

However, $\gamma(x) = x$ implies

$$0 = x \cdot f(x) = \gamma(x) \cdot f(x) = \frac{\|f(x)\|^2}{\|x + f(x)\|}.$$

Hence $f(x) = 0$, a contradiction. ∎

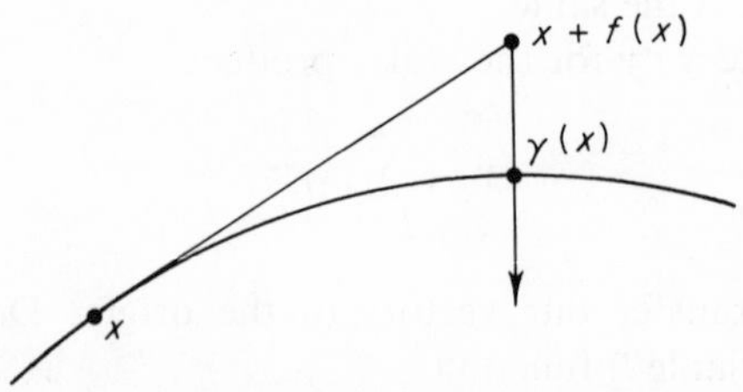

Figure 2.1

Proposition 2.2 If there exists $f\colon S^n \to S^n$ without fixed points and such that $\|f(x) - x\| < 2$, there is a nonzero vector field on S^n.

Proof $\|f(x) - x\| < 2$ is equivalent to $f(x) \neq -x$, for all $x \in S^n$.

Let $\gamma(x) = f(x) - (f(x) \cdot x)x$. Then $\gamma(x) \cdot x = 0$, so γ is a vector field and it is nonzero, for otherwise $f(x)$ and x are linearly dependent and, hence, $f(x) = \pm x$, a contradiction. ▮

Thus Problem 3 is equivalent to Problem 4. Likewise, Problem 1 is equivalent to Problem 2.

Proposition 2.3 There is a continuous (differentiable) map $\gamma\colon B^n \to S^{n-1}$ such that $\gamma i = 1$ iff there exists a continuous (differentiable) function $f\colon B^n \to B^n$ without fixed points.

The first half follows immediately from:

Lemma 2.4 Let $f\colon B^n \to B^n$ and $U = \{x \mid f(x) \neq x\}$. Then there is a map $\gamma\colon U \to S^{n-1}$ such that $\gamma|_{U \cap S^{n-1}}$ is the inclusion. If f is differentiable, so is γ.

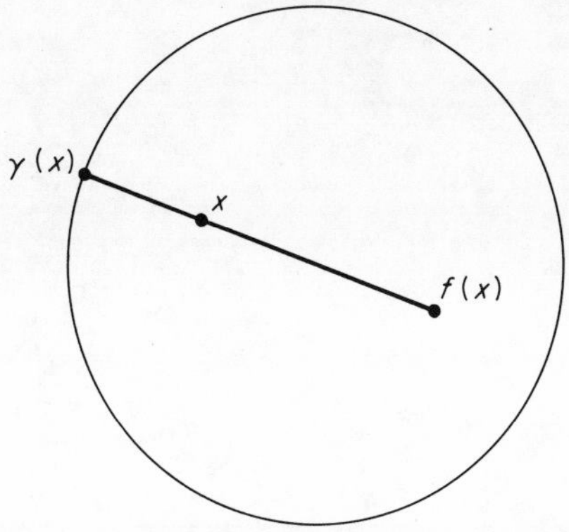

Figure 2.2

Proof Let $\gamma(x)$ be the point of intersection of S^{n-1} with the line joining x and $f(x)$ such that $\gamma(x)$, x, $f(x)$ occur in that order (Fig. 2.2). Then

$$\gamma(x) = \rho x + (1 - \rho)f(x), \qquad \rho \geq 1, \quad \|\gamma(x)\| = 1.$$

Expanding $1 = \|\gamma(x)\|$ gives a quadratic equation for ρ: $a\rho^2 + b\rho + c = 0$; one easily checks that $a > 0$, and $a + b + c \leq 0$. It follows that $b^2 - 4ac \geq (b + 2a)^2$ and hence the solution

$$\rho = \frac{-b + \sqrt{b^2 - 4ac}}{2a} \geq \frac{-b + |b + 2a|}{2a} \geq 1.$$

ρ is differentiable (and hence continuous). Now $\gamma|_{U \cap S^{n-1}}$ is the inclusion, for if $\|x\| = 1$, we get $b + 2a \geq 0$ and $a + b + c = 0$; consequently, $\rho = 1$. ∎

To prove the other half of 2.3, suppose such a $\gamma: B^n \to S^{n-1}$ exists. Define $f(x) = -\gamma(x)$. Since the image of f is $S^{n-1} \subset B^n$, any fixed point must lie in S^{n-1}. But for $x \in S^{n-1}$, $\gamma(x) = -x$. ∎

These problems are quite difficult, but will be solved in Section 13. The problem that seems most intuitively clear is Problem 2 for $n = 2$ (the original Problem 2). It seems that such a function cannot exist. At this point it is worth the reader's time to try and solve this. After consideration he will probably agree to the following:

The difficulty is that there are too many points. If the spaces involved were finite, such questions could easily be answered, but since the spaces are so big, it is not immediately clear that one can ever decide the answer in a finite number of steps.

3

Homotopy Theory

The problems that we have been considering are global problems, in the sense that if we remove one point from the spaces involved, the problem is altered. If we remove one point from S^2, it is easy to see how one could construct a nonzero vector field on the remainder (and even easier if we remove two points). If we remove a point from B^2, it is easy to see that there is a map from the remaining space to itself without a fixed point (rotate about the hole).

One of the fundamental achievements of algebraic topology is to turn global topological problems into homotopy theory problems. We will proceed to do this.

Definition 3.1 Two maps $f_0, f_1\colon X \to Y$ are said to be homotopic if there is an intermediate family of maps $f_t\colon X \to Y$ continuous jointly in x and t, i.e., if there exists

$$F\colon X \times I \to Y$$

(called a homotopy between f_0 and f_1) which is continuous and such that $F(x, 0) = f_0(x)$, and $F(x, 1) = f_1(x)$.

We write $f_0 \sim f_1$ (or $F\colon f_0 \sim f_1$) to indicate that f_0 is homotopic to f_1.

Proposition 3.2 $\sim$ is an equivalence relation.

Proof $f \sim f$ by the homotopy $F(x, t) = f(x)$. If $F\colon f \sim g$, then $G\colon g \sim f$ is given by $G(x, t) = F(x, 1 - t)$. If $F: f \sim g$ and $G: g \sim h$, then $H: f \sim h$ is given by

$$H(x, t) = \begin{cases} F(x, 2t), & 0 \le t \le \frac{1}{2} \\ G(x, 2t - 1), & \frac{1}{2} \le t \le 1. \end{cases}$$

H is continuous by Exercise 2, Section 0. ∎

Thus we speak of homotopy classes of maps. If f is a map, we denote its homotopy class by $\{f\}$. We write $[X, Y]$ for the set of homotopy classes of maps from X to Y.

Proposition 3.3 There exists $f\colon B^n \to S^{n-1}$ with $f \cdot i = 1$ iff the identity map $1\colon S^{n-1} \to S^{n-1}$ is homotopic to a constant map.

Proof If such a map f exists, we define a homotopy

$$H\colon S^{n-1} \times I \to S^{n-1}$$

by

$$H(x, t) = f(tx).$$

Clearly $H(x, 1) = f(x) = x$ since $x \in S^{n-1}$ and $H(x, 0) = f(0)$ which is independent of x.

Conversely, if H exists

$$H\colon S^{n-1} \times I \to S^{n-1}$$

with $H(x, 0) = c$ and $H(x, 1) = x$, define

$$f\colon B^n \to S^{n-1}$$

by

$$f(x) = H(x/\|x\|, \|x\|), \qquad f(0) = c.$$

Since S^{n-1} is compact, H is uniformly continuous. Thus for every $\varepsilon > 0$, there exists $\delta > 0$ depending on ε but not on x such that $\|H(x, t) - c\| < \delta$ if $t < \varepsilon$. Consequently f is continuous at 0. ∎

Proposition 3.4 Let $a_n\colon S^n \to S^n$ be the antipodal map $a_n(x_1, \ldots, x_{n+1}) = (-x_1, \ldots, -x_{n+1})$. If there is a vector field on S^n, $a_n \sim 1$.

Proof We will use a vector at $x \in S^n$ to indicate the direction of a path from x to $a_n(x)$ on S^n and hence produce a homotopy. Given $f\colon S^n \to R^{n+1}$ with $f(x) \neq 0$ and $f(x) \cdot x = 0$, we will construct a path from x to $a_n(x)$ in the plane determined by x and $f(x)$ and on the sphere:

$$H(x, t) = a(t)x + b(t)f(x), \qquad \|H(x, t)\|^2 = 1.$$

This yields the equation

$$a(t)^2 + b(t)^2 f(x) \cdot f(x) = 1;$$

we choose $a(t) = 1 - 2t$ and hence $b(t) = 2\sqrt{t - t^2}/\|f(x)\|$. These are both C^∞ for $0 \leq t \leq 1$ since $f(x) \neq 0$. Thus

$$H(x, t) = (1 - 2t)x + 2\sqrt{t - t^2} f(x)/\|f(x)\|$$

is a homotopy in S^n from 1 to a_n. ▮

Remark The converse is also true, but the technical details in proving it are harder and we will not need it to solve Problem 3. Essentially, given a homotopy from 1 to a_n, one approximates this with a differentiable homotopy. Then the tangent line to the curve $P_x(t) = H(x, t)$ at $t = 0$ contains a unit vector pointing in the direction of increasing t, which is tangent to the sphere, and nonzero.

Proposition 3.5 If n is odd, there is a nonzero vector field on S^n.

Proof We construct a linear nonsingular function $f(x)$ with $f(x) \cdot x = 0$ by

$$f(x_1, x_2, \ldots, x_{2n}) = (x_2, -x_1, x_4, -x_3, \ldots, x_{2n}, -x_{2n-1}).$$

Clearly this satisfies $f(x) \neq 0$ if $x \neq 0$ and $f(x) \cdot x = 0$. ▮

We have thus reduced our problems to homotopy theory. We will eventually show how to turn homotopy theory problems into algebraic problems. This is where algebraic topology has its strength. It transforms problems from the very complicated world of spaces and maps to the simple world of finitely generated abelian groups, or other algebraic worlds that one feels are simpler.

Exercises

1. Let P be a one-point topological space. Show that $[P, X]$ is in 1–1 correspondence with the set of arc components of X.[3]

2.* Let $f_0, f_1 \colon (X, A) \to (Y, B)$ be maps. We say f_0 is homotopic to f_1 and write $f_0 \sim f_1$ if there exists a map

$$F\colon (X \times I, A \times I) \to (Y, B)$$

[3] We use the words *arc connected* and *arc component* interchangeably with *path connected* and *path component* to refer to (not necessarily 1–1) maps $p\colon I \to X$, although there is some variety in the literature.

with $F(x, 0) = f_0(x)$, $F(x, 1) = f_1(x)$. Show that this is an equivalence relation. We write $[(X, A), (Y, B)]$ for the set of homotopy classes as before. Note that $[X, Y] = [(X, \varnothing), (Y, B)]$ for any $B \subset Y$.

3.* Let $f: (X, A) \to (Y, B)$, $g_0, g_1: (Y, B) \to (Z, C)$, and $h: (Z, C) \to (W, D)$. Suppose $g_0 \sim g_1$. Show that $g_0 f \sim g_1 f$ and $hg_0 \sim hg_1$.

4. Let $f, g: X \to S^n$ and suppose that for all $x \in X, f(x) \neq -g(x)$. Show that $f \sim g$.

5. Construct a map $\gamma: S^3 \to S^3$ satisfying Problem 4 of Section 2.

4

Category Theory

One criticism of current pedagogical methods in mathematics is that they tend to compartmentalize mathematics into subjects without emphasizing the interrelationships among subjects. Thus topology grew out of analysis, and most of modern algebra grew out of either analysis or number theory. The deeper one gets into mathematics, the closer one sees the connections.[4]

A strong connection between various fields in algebra or topology is often most conveniently expressed through the notion of categories and functors. Category theory plays somewhat the same role in algebra and topology that set theory plays in analysis. In both cases the elementary theories are a convenient language which is a bit abstract, and not very deep, but from which one obtains economy of thought. One simply has to get used to the abstraction, and this is made relatively painless by a wealth of examples.

Definition 4.1 A category consists of

(a) A class of objects.

(b) For every ordered pair of objects X and Y, a set $\hom(X, Y)$ of "morphisms" with "domain" X and "range" Y; if $f \in \hom(X, Y)$ we write $f\colon X \to Y$ or $X \xrightarrow{f} Y$. $\hom(X, Y) \cap \hom(X', Y') = \varnothing$ unless $X = X'$ and $Y = Y'$.

(c) For every ordered triple of objects X, Y, and Z, a function associating to a pair of morphisms $f\colon X \to Y$ and $g\colon Y \to Z$ their "composite"

$$g \circ f\colon X \to Z.$$

[4] In the words of the Tao Tê Ching, "enumerate the parts of a carriage and you still have not explained what a carriage is" [72, Chapter 39].

These satisfy the following two axioms:

Associativity: If $f: X \to Y$, $g: Y \to Z$, and $h: Z \to W$, then

$$h \circ (g \circ f) = (h \circ g) \circ f: X \to W.$$

Identity: For every object Y, there is a morphism $1_Y: Y \to Y$ such that if $f: X \to Y$, then $1_Y \circ f = f$, and if $h: Y \to Z$, then $h \circ 1_Y = h$.

We use the word *map* interchangeably with *morphism*.

Examples

1. $\mathfrak{T}$: As a class of objects, take all topological spaces. The set hom(X, Y) will be the set of continuous functions from X to Y. The composition rule will be composition of functions.

2. $\mathfrak{S}$: As objects, take all sets; as morphisms, take all functions.

3. $\mathfrak{G}$: As objects, take all groups; as morphisms, take all homomorphisms.

4. $\mathfrak{R}$: As objects, take all rings; as morphisms, take all ring homomorphisms.

5. $\mathcal{M}_R$: As objects, take (right) R-modules; as morphisms, take all R-module homomorphisms.

These five examples are the ones one encounters most in algebraic topology: they are all special cases of a

6. *Meta-Example* Consider as objects, sets with a given "structure." Consider as morphisms, all functions that "preserve" the "structure."

One could consider other categories, however,

7. As objects take all groups. As morphisms, take all isomorphisms, i.e., $\hom(G_1, G_2) =$ all isomorphisms $\chi: G_1 \to G_2$.

8. Let there be only two objects X_1 and X_2. Let

$$\hom(X_1, X_1) = \{1_{X_1}\}, \qquad \hom(X_2, X_2) = \{1_{X_2}\},$$

$$\hom(X_1, X_2) = \varnothing, \qquad \text{and} \qquad \hom(X_2, X_1) = \varnothing.$$

Just about anything can be considered as a category, if you try hard enough, but the important examples are 1–5.

Definition 4.2 Given two categories $\mathfrak{C}_1$ and $\mathfrak{C}_2$, a covariant functor from $\mathfrak{C}_1$ to $\mathfrak{C}_2$, F, consists of an object function which assigns to every object X of $\mathfrak{C}_1$ an object $F(X)$ of $\mathfrak{C}_2$, and a morphism function which assigns to every morphism $f: X \to Y$ of $\mathfrak{C}_1$ a morphism $F(f): F(X) \to F(Y)$ of $\mathfrak{C}_2$ such that

(a) $F(1_X) = 1_{F(X)}$;
(b) $F(g \circ f) = F(g) \circ F(f)$.

A contravariant functor from $\mathcal{C}_1$ to $\mathcal{C}_2$ consists of an object and morphism function as before except that if $f: X \to Y$, $F(f): F(Y) \to F(X)$ and instead of (b) we have

(b′) $F(g \circ f) = F(f) \circ F(g)$;

i.e., a contravariant functor reverses arrows. In either case one writes

$$F: \mathcal{C}_1 \to \mathcal{C}_2$$

to mean that F is a functor as above.

Example 1 Let $\mathcal{M}_k$ be the category of vector spaces over k and linear maps. $D: \mathcal{M}_k \to \mathcal{M}_k$ is given by $D(V) = V^*$ and $D(f) = f^*$ where V^* is the dual space and f^* the adjoint of f. D is a contravariant functor.

Example 2 From the category of R-modules and homomorphisms to itself we have, for every module M a functor T_M defined by $T_M(N) = M \otimes N$, $T_M(f) = 1 \otimes f$. T is a covariant functor. (It can also be thought of as a functor of two variables.)

Example 3 From the category $\mathcal{G}$ to itself we have the functor C defined by $C(G) =$ commutator subgroup of $G =$ subgroup generated by all $[g_1, g_2] = g_1 g_2 g_1^{-1} g_2^{-1}$. C is a covariant functor. Similarly, $A(G) = G/C(G)$ is a functor from $\mathcal{G}$ to $\mathcal{M}_Z$.

Example 4 The forgetful functor. This is a general type of covariant functor which applies in many examples. We give three examples:

(1) $\mathcal{M}_R \to \mathcal{M}_Z$,
(2) $\mathcal{M}_Z \to \mathcal{G}$,
(3) $\mathcal{G} \to \mathcal{S}$.

The functor is the identity on objects and maps, but considers them as different things. Thus, every R-module may be considered as an abelian group by forgetting the R-module structure. Every R-module homomorphism may be considered as a group homomorphism. Similarly for (2) and (3).

Example 5 The identity functor from any category to itself. It is the identity on objects and maps and is covariant.

As a method of comparing functors, we have:

Definition 4.3 A natural transformation φ from T_1 to T_2, where T_1 and T_2 are functors from a category $\mathcal{C}_1$ to a category $\mathcal{C}_2$, written

$$\varphi: T_1 \to T_2,$$

is a function from the objects of $\mathcal{C}_1$ to the morphisms of $\mathcal{C}_2$ such that for every morphism $f\colon X \to Y$ in $\mathcal{C}_1$ the (appropriate) following diagram is commutative:

$$\begin{array}{ccc} T_1(X) & \xrightarrow{\varphi(X)} & T_2(X) \\ \Big\downarrow{\scriptstyle T_1(f)} & & \Big\downarrow{\scriptstyle T_2(f)} \\ T_1(Y) & \xrightarrow{\varphi(Y)} & T_2(Y) \end{array} \qquad \begin{array}{ccc} T_1(X) & \xrightarrow{\varphi(X)} & T_2(X) \\ \Big\uparrow{\scriptstyle T_1(f)} & & \Big\uparrow{\scriptstyle T_2(f)} \\ T_1(Y) & \xrightarrow{\varphi(Y)} & T_2(Y) \end{array}$$

The best known example of this is as follows:

$\mathcal{C}_1 = \mathcal{C}_2 =$ finite-dimensional vector spaces and linear maps;
$T_1 =$ identity functor;
$T_2(V) = V^{**}$;
$\varphi(V)\colon V \to V^{**}$ is the "natural" isomorphism.

That the isomorphism $V \simeq V^{**}$ is natural means precisely that it is a natural transformation in this sense.

We are mainly concerned with "topological categories." For example:

$\mathcal{T}$ is the category of topological spaces and continuous maps.

$\mathcal{T}^*$ is the category whose objects are topological spaces with a distinguished point (called the base point and usually written $*$) and whose maps are continuous functions which preserve the base point (i.e., $f\colon X \to Y$ and $f(*) = *$, where we use $*$ ambiguously to denote the base point of any space). (The $*$ here has nothing to do with duality.)

$\mathcal{T}^2$ is the category whose objects are pairs (X, A) of topological spaces and whose morphisms are maps of pairs (see Section 1).

If $\mathcal{C}$ is any category of topological spaces and continuous maps, we will use the notation $\mathcal{C}^*$ and $\mathcal{C}^2$ with the obvious interpretation.

We will be considering functors defined on "topological categories" and taking values in some "algebraic category." The utility of such functors is that they take diagrams to diagrams, and many problems can be stated in terms of diagrams.

Exercises

1. Find several examples of categories and functors implicitly or explicitly in the most recent algebra course you have taken.

2. Prove $\varphi(V)\colon V \to V^{**}$ is a natural transformation.

3. In Section 1 we defined pairs (X, A) and maps between pairs. Show that given any category $\mathcal{C}$ in which there is a well-defined notion of subobject (certain morphisms are called inclusions), one can describe a category of pairs $\mathcal{C}^2$ from this.

4. Two objects A and B in a category $\mathcal{C}$ are called isomorphic or equivalent if there are maps $f: A \to B$ and $g: B \to A$ in $\mathcal{C}$ with $fg = 1_B$ and $gf = 1_A$. Interpret this in the examples given.

5.* Each of the categories $\mathcal{T}$, $\mathcal{T}^*$, and $\mathcal{T}^2$ has an equivalence relation called homotopy. We now define, from these, new categories $\mathcal{T}_h$, $\mathcal{T}_h{}^*$, and $\mathcal{T}_h{}^2$. These new categories will have the same objects as the old ones; the morphism sets, however, will be the set of homotopy classes of maps in the old category. Thus, in $\mathcal{T}_h$,

$$\hom(X, Y) = [X, Y];$$

in $\mathcal{T}_h{}^*$,

$$\hom((X, *), (Y, *)) = [(X, *), (Y, *)];$$

and in $\mathcal{T}_h{}^2$,

$$\hom((X, A), (Y, B)) = [(X, A), (Y, B)].$$

Show that $\mathcal{T}_h$, $\mathcal{T}_h{}^*$, and $\mathcal{T}_h{}^2$ are categories. (See Exercise 2, Section 3.)

6. Show that, fixing (X, A), $[(X, A), (Y, B)]$ is a covariant functor from $\mathcal{T}^2$ or $\mathcal{T}_h{}^2$ to the category of sets and functions. Similarly, with (Y, B) fixed, $[(X, A), (Y, B)]$ is a contravariant functor. (See Exercise 5.)

If $n \geq 0$, we will write

$$\pi_n(X, *) = [(I^n, \partial I^n), (X, *)].$$

($(I^0, \partial I^0)$ is defined as $(*, \varnothing)$.)

7.* Show that there is a natural 1–1 correspondence

$$[(X, A), (Y, *)] \leftrightarrow [(X/A, \{A\}), (Y, *)],$$

where $A \neq \varnothing$. By applying 1.9 conclude that there is a natural 1–1 correspondence

$$\pi_n(X, *) \overset{c}{\longleftrightarrow} [(S^n, *), (X, *)],$$

where $(1, 0, \ldots, 0) = * \in S^n$. (See Exercise 2, Section 3.) This correspondence will be called c.

8. Show that if (E, S) is an n-cell, there is a natural 1–1 correspondence

$$\pi_n(X, *) \leftrightarrow [(E, S), (X, *)].$$

9. For $(X, A), (Y, B) \in \mathcal{T}^2$ define $(X, A) \times (Y, B) = (X \times Y, X \times B \cup A \times Y)$. Show that this is a covariant functor in two variables. Observe that two mappings $f_0, f_1: (X, A) \to (Y, B)$ are homotopic in $\mathcal{T}^2$ iff there is a map $H: (X, A) \times (I, \varnothing) \to (Y, B)$ such that $H(x, 0) = f_0(x)$ and $H(x, 1) = f_1(x)$.

5

The Fundamental Group

The transition from homotopy theory to algebra is most often accomplished by putting an algebraic structure on sets of homotopy classes of maps. The simplest and most fundamental example of this is $\pi_1(X, *) = [(I, \{0, 1\}), (X, *)]$. In this section we will put a group structure on $\pi_1(X, *)$ in a functorial way.

We shall use the word *path* to refer to any map $p: I \to X$. If in addition $p(0) = p(1) = *$, we will call the path a based path or a loop. Thus the elements of $\pi_1(X, *)$ are homotopy classes of based paths in X. The only homotopies allowed are those that keep the end points fixed throughout the homotopy.

Definition 5.1 A homotopy $H: X \times I \to Y$ is called a homotopy relative to A, for $A \subset X$, if $H(a, t)$ does not depend on t for $a \in A$. If $H(x, 0) = f(x)$ and $H(x, 1) = g(x)$, we write $H: f \sim g$ (rel A).

Thus the homotopies involved in $\pi_1(X, *)$ are homotopies of I relative to the end points. One can generalize the construction $\pi_1(X, *)$ as follows. Choose $x, y \in X$ and consider all paths $p: I \to X$ with $p(0) = x$ and $p(1) = y$ (Fig. 5.1). Write $\pi(X; x, y)$ for the set of all homotopy classes of such paths relative to the end points. (One abbreviates this to $\pi(x, y)$ if the space X is fixed.) Thus $\pi(X; x, x) = \pi_1(X, x)$.

Our understanding of $\pi_1(X, *)$ is greatly facilitated by an organization of its elements into a group, which we now describe. We will define the composition of two based paths. This will induce a composition among the path classes. More generally, suppose we are given two paths p_1 and p_2 subject only to the requirement that $p_2(0) = p_1(1)$. We will form a new path traversing

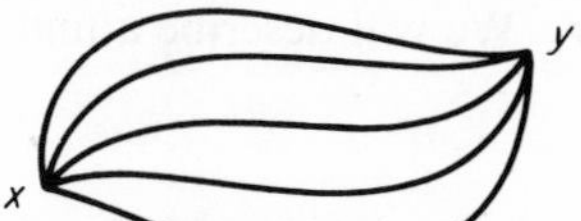

Figure 5.1

through p_1 at double speed (from $s = 0$ to $s = \frac{1}{2}$), and then through p_2 at double speed (Fig. 5.2).

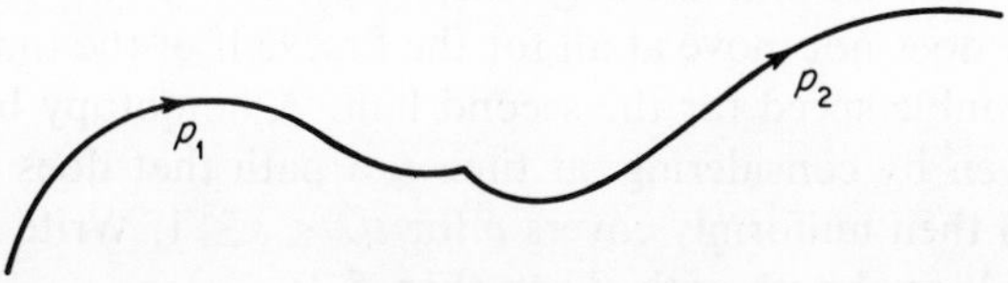

Figure 5.2

This path p_3 is defined as follows:

$$p_3(s) = \begin{cases} p_1(2s), & \frac{1}{2} \geq s \geq 0 \\ p_2(2s - 1), & 1 \geq s \geq \frac{1}{2}. \end{cases}$$

We think of p_3 as the product of p_1 and p_2:

$$p_3 = p_2 \cdot p_1.$$

If $p_1(0) = x$, $p_1(1) = p_2(0) = y$, and $p_2(1) = z$, this product defines a transformation $\pi(x, y) \times \pi(y, z) \to \pi(x, z)$. To check that this composition respects the equivalence relation, suppose

$$P_1 : p_1 \sim p_1', \qquad P_2 : p_2 \sim p_2'.$$

We must find

$$P_3 : p_3 \sim p_3'.$$

The formula is easy:

$$P_3(s, t) = \begin{cases} P_1(2s, t) & 0 \leq s \leq \frac{1}{2} \\ P_2(2s - 1, t), & \frac{1}{2} \leq s \leq 1. \end{cases}$$

In particular we have defined a composition operation $\cdot$, in $\pi_1(X, *)$.

Theorem 5.2 $\pi_1(X, *)$ together with $\cdot$ is a group, called the fundamental group of X (at $*$).

Proof We will describe a unit $1_x \in \pi(x, x)$:

$$1_x = \{u_x\},$$

where u_x is the constant path $u_x(s) = x$ for all $s \in I$.

If $\{p\} \in \pi(x, y)$ is another element, $u_y \cdot p$ and $p \cdot u_x$ are given by the formulas

$$(p \cdot u_x)(s) = \begin{cases} x, & 0 \le s \le \frac{1}{2} \\ p(2s - 1), & \frac{1}{2} \le s \le 1; \end{cases}$$

$$(u_y \cdot p)(s) = \begin{cases} p(2s), & 0 \le s \le \frac{1}{2} \\ y, & \frac{1}{2} \le s \le 1. \end{cases}$$

Thus $p \cdot u_x \neq p$ but we will show $\{p \cdot u_x\} = \{p\}$, i.e., $p \cdot u_x \sim p$. Now $p \cdot u_x$ is the path that does not move at all for the first half of the time and hurries through p at double speed for the second half. A homotopy between p and this path is given by considering, at time t, a path that does not move for $0 \le s \le t/2$ and then uniformly covers p for $t/2 \le s \le 1$. Write $P(s, t)$ for the image of $s \in I$ along the tth path. Then (Fig. 5.3)

$$P(s, t) = \begin{cases} x, & 0 \le s \le t/2 \\ p\left(\dfrac{2s - t}{2 - t}\right), & t/2 \le s \le 1. \end{cases}$$

Clearly $P(s, 0) = p(s)$, $P(s, 1) = p \cdot u_x(s)$. We note that P is well defined since

$$0 \le \frac{2s - t}{2 - t} \le \frac{2 - t}{2 - t} = 1, \qquad \text{if} \quad t/2 \le s,$$

$$p\left(\frac{2(t/2) - t}{2 - t}\right) = x, \qquad P(0, t) = x, \qquad \text{and} \qquad P(1, t) = y.$$

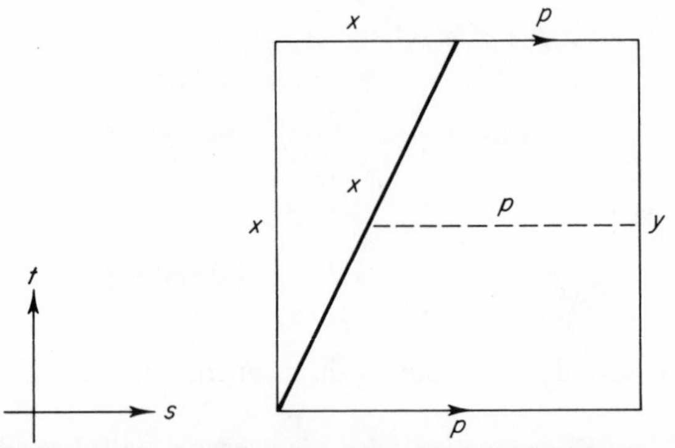

Figure 5.3

Similarly, one can prove $u_y \cdot p \sim p$; however, if we prove associativity and the existence of a right inverse, a proof of this is not needed.

Given $\{p\} \in \pi(x, y)$, we define $\{p\}^{-1} = \{q\} \in \pi(y, x)$ where $q(s) = p(1 - s)$. If: $P\ p_0 \sim p_1$, we define $Q: q_0 \sim q_1$ by

$$Q(s, t) = P(1 - s, t).$$

Since

(a) $Q(0, t) = y$, $Q(1, t) = x$,
(b) $Q(s, 0) = q_0(s)$, $Q(s, 1) = q_1(s)$,

this operation is well defined on equivalence classes.

We must show that $p \cdot p^{-1} \sim u_y$. Now $p \cdot p^{-1}$ goes through the path p twice, first backwards and then forwards:

$$(p \cdot p^{-1})(s) = \begin{cases} p(1 - 2s), & 0 \le s \le \frac{1}{2} \\ p(2s - 1), & \frac{1}{2} \le s \le 1. \end{cases}$$

There is no reason why the middle of $p \cdot p^{-1}$ must be x. We may take a homotopy from $p \cdot p^{-1}$ to u_y that at time t moves through part of p (from 0 to t) and then back again:

$$P(s, t) = \begin{cases} p(1 - 2st), & 0 \le s \le \frac{1}{2} \\ p((2s - 1)t + 1 - t), & \frac{1}{2} \le s \le 1. \end{cases}$$

Clearly:

(a) $P(0, t) = P(1, t) = y$,
(b) $p(1 - st) = p((2s - 1)t + 1 - t)$ if $s = \frac{1}{2}$,
(c) $P(s, 0) = y$, $P(s, 1) = p \cdot p^{-1}(s)$.

It remains to show that $\cdot$ is associative, let $\{p_1\}$, $\{p_2\}$, and $\{p_3\} \in \pi_1(X, *)$. Let us compute $(p_1 \cdot p_2) \cdot p_3$ and $p_1 \cdot (p_2 \cdot p_3)$:

$$(p_1 \cdot p_2) \cdot p_3(s) = \begin{cases} p_3(2s), & 0 \le s \le \frac{1}{2} \\ p_2(4s - 2), & \frac{1}{2} \le s \le \frac{3}{4} \\ p_1(4s - 3), & \frac{3}{4} \le s \le 1; \end{cases}$$

$$p_1 \cdot (p_2 \cdot p_3)(s) = \begin{cases} p_3(4s), & 0 \le s \le \frac{1}{4} \\ p_2(4s - 1), & \frac{1}{4} \le s \le \frac{1}{2} \\ p_1(2s - 1), & \frac{1}{2} \le s \le 1. \end{cases}$$

It can be seen that the only difference between these paths is the speed.

$(p_1 \cdot p_2) \cdot p_3$:

$p_1 \cdot (p_2 \cdot p_3)$:

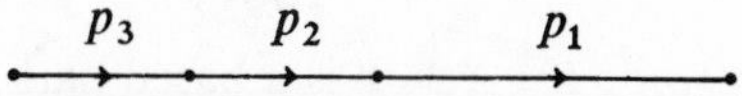

We could slide one into the other by choosing intermediate speeds; in Fig. 5.4

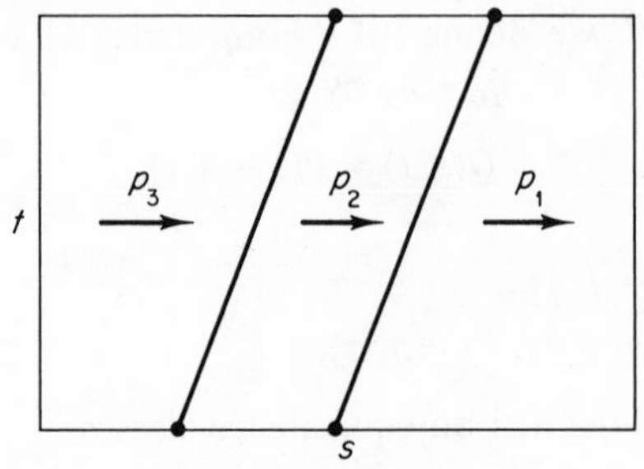

Figure 5.4

the two slanted lines are given by the equations $t = 4s - 1$ and $t = 4s - 2$. At level t one then travels the paths

$$\begin{array}{llll} p_3 & \text{from} & s = 0 & \text{to} \quad s = (t+1)/4, \\ p_2 & \text{from} & s = (t+1)/4 & \text{to} \quad s = (t+2)/4, \\ p_1 & \text{from} & s = (t+2)/4 & \text{to} \quad s = 1. \end{array}$$

This is given by the equation

$$P(s, t) = \begin{cases} p_3(4s/(t+1)), & 0 \le s \le (t+1)/4 \\ p_2(4s - t - 1), & (t+1)/4 \le s \le (t+2)/4 \\ p_1((4s - t - 2)/(2 - t)), & (t+2)/4 \le s \le 1. \end{cases}$$

This is well defined since $4s/(t+1)$, $4s - t - 1$, and $(4s - t - 2)/(2 - t)$ are between 0 and 1 in the appropriate range of s and t, and the definition is consistent for $s = (t+1)/4$ and $s = (t+2)/4$. It is easy to see that

$$\begin{array}{ll} P(0, t) = p_3(0), & P(1, t) = p_1(1), \\ P(s, 0) = p_1 \cdot (p_2 \cdot p_3)(s), & P(s, 1) = (p_1 \cdot p_2) \cdot p_3(s). \end{array}$$

This completes the proof. ▌

Theorem 5.3 π_1 is a functor from the category $\mathcal{C}^*$ to $\mathcal{G}$.

Proof Given f: $(X, *) \to (Y, *)$ we will define a homomorphism

$$\pi_1(f) \colon \pi_1(X, *) \to \pi_1(Y, *).$$

Let $\{p\} \in \pi_1(X, *)$. Define

$$\pi_1(f)(\{p\}) = \{fp\}.$$

Since $f(*) = *$, fp: $I \to X \to Y$ is a based path.

Now $\{fp\}$ depends only on $\{p\}$, for suppose $P\colon p_0 \sim p_1$. Then we define

$$Q\colon fp_0 \sim fp_1$$

by

$$Q(s, t) = f \cdot P(s, t).$$

Clearly $Q\colon fp_0 \sim fp_1$, so $\pi_1(f)$ is well defined. To see that $\pi_1(f)$ is a homomorphism, we check that

$$\pi_1(f)(\{p\} \cdot \{p'\}) = \pi_1(f)\{p\} \cdot \pi_1(f)\{p'\}.$$

This holds since both are represented by the class $\{q\}$ given by

$$q(s) = \begin{cases} f(p'(2s)), & 0 \le s \le \frac{1}{2} \\ f(p(2s-1)), & \frac{1}{2} \le s \le 1. \end{cases}$$

For reasons of tradition, we always write f_* for $\pi_1(f)$. We must show $1_* = 1$ and $(f \cdot g)_* = f_* \cdot g_*$. These are both obvious from the definitions. ∎

Exercises

1. Show that $\pi_1(X, *) = 0$ if X is a finite topological space with the discrete topology.

2. Why is it not possible to describe $\pi_1(X, *)$, as in this section, without reference to the base point?

3. Let $I \in \pi_1(S^1, (1, 0))$ be the class of the identity map. Show that nI is the class of the map $f_n\colon S^1 \to S^1$ given by $f_n(z) = z^n$. (Exercise 22, Section 7)

4. Besides using categories to discuss objects that we study, the theory of categories has another use. This is to discuss sets with a multiplication that is not always defined. Given a space X the fundamental category of X written $\Pi(X)$ is defined as follows. For objects of $\Pi(X)$ we take the points of X. We define $\hom(x, y) = \pi(x, y)$. According to the proof of 5.2, this is a category. Show that the mapping $p \to p^{-1}$ defines a transformation $r\colon \pi(x, y) \to \pi(y, x)$ satisfying:

(1) $r^2 = 1$;
(2) $r(\alpha) \cdot \alpha = 1 = \alpha \cdot r(\alpha)$;
(3) $r(\alpha \cdot \beta) = r(\beta) \cdot r(\alpha)$.

Such a category is sometimes called a groupoid. (Sections 6 and 7)

5. Let $0 < s < 1$. Given paths p and q with $p(1) = q(0)$, define h by the formula

$$h(t) = \begin{cases} p(t/s), & 0 \le t \le s \\ q((t-s)/(1-s)), & s \le t \le 1. \end{cases}$$

Prove that $\{h\} = \{q\} \cdot \{p\} \in \pi(p(0), q(1))$. State a similar result for arbitrary products and prove it by induction. (7.12)

6. Show that homotopy relative to a fixed subset is an equivalence relation.

6

More on the Fundamental Group

In this section we develop a few elementary facts about $\pi_1(X, *)$. We prove that it does not depend in an essential way on $*$, provided X is arcwise connected, and show that it is a homotopy type invariant.

On the surface it appears that $\pi_1(X, *)$ depends both on the space X and the chosen point $*$. The following theorem dispenses with the dependency on $*$.

Theorem 6.1 Let $*_1, *_2 \in X$ and suppose they belong to the same arc component. Then $\pi_1(X, *_1) \cong \pi_1(X, *_2)$.

However, there is no natural isomorphism. In any case, the isomorphism type of $\pi_1(X, *)$ as an abstract group, does not depend on the choice of $*$, only on the arc component. (A little thought shows that it could not be affected by other arc components.) If X is arc connected, one writes this isomorphism type as $\pi_1(X)$. One should be very careful here. There is no category of isomorphism types of groups and homomorphisms. Thus there is no way to make $\pi_1(X)$ a functor. Whenever one deals with induced homomorphisms, one must, at least implicitly, deal with base points.

Proof of Theorem 6.1 Since $*_1$ and $*_2$ belong to the same arc component, $\pi(*_1, *_2) \neq \varnothing$. Pick $\alpha \in \pi(*_1, *_2)$. Define (Fig. 6.1)

$$\varphi_\alpha : \pi_1(X, *_1) \to \pi_1(X, *_2)$$

by

$$\varphi_\alpha(\beta) = \alpha \cdot \beta \cdot r(\alpha)$$

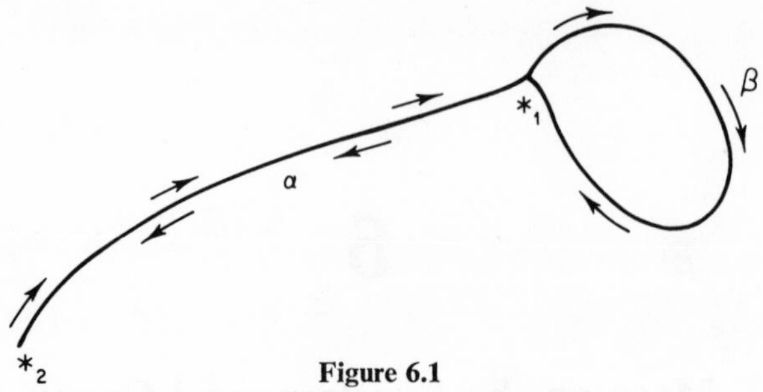

Figure 6.1

(see Exercise 4, Section 5). We also define

$$\varphi_{r(\alpha)} : \pi_1(X, *_2) \to \pi_1(X, *_1),$$

and

$$\varphi_{r(\alpha)}(\varphi_\alpha(\beta)) = r(\alpha)\alpha\beta r(\alpha)\alpha = \beta,$$
$$\varphi_\alpha(\varphi_{r(\alpha)}(\beta)) = \alpha r(\alpha)\beta\alpha r(\alpha) = \alpha.$$

Thus φ_α is 1–1 and onto.

To see that φ_α is a homomorphism, note that

$$\varphi_\alpha(\beta\gamma) = \alpha\beta\gamma r(\alpha) = \alpha\beta r(\alpha)\alpha\gamma r(\alpha) = \varphi_\alpha(\beta)\varphi_\alpha(\gamma). \quad ∎$$

Proposition 6.2 If $f_0 \sim f_1 : (X, *) \to (Y, *)$, then

$$f_{0*} = f_{1*} : \pi_1(X, *) \to \pi_1(Y, *).$$

Remark These maps of pairs must be homotopic as maps of pairs, i.e., there exists a map

$$H: X \times I \to Y$$

with

$$H(*, t) = * \qquad \text{for all} \qquad t,$$

and

$$H(x, 0) = f_0(x), \qquad H(x, 1) = f_1(x).$$

Proof $f_{0*}(\{p\}) = \{f_0 p\} = \{f_1 p\} = f_{1*}(\{p\})$ since $f_0 p \sim f_1 p$. ∎

Definition 6.3 A map $f: (X, A) \to (Y, B)$ is called a homotopy equivalence, and (X, A) and (Y, B) are said to be of the same homotopy type if there exists a map $g: (Y, B) \to (X, A)$ such that $g \circ f \sim 1$ and $f \circ g \sim 1$ (these homotopies being homotopies of pairs). In this case, we write $(X, A) \simeq (Y, B)$.

Theorem 6.4 If $f: (X, *) \to (Y, *)$ is a homotopy equivalence, $f_*: \pi_1(X, *) \to \pi_1(Y, *)$ is an isomorphism.

Proof $f_* g_* = (fg)_* = 1_*$ if $fg \sim 1$. Similarly $g_* f_* = 1$. Thus $\pi_1(X)$ depends only on the homotopy type of $(X, *)$. ∎

The first problem we will consider is how to calculate $\pi_1(X)$. For example, one would like to determine $\pi_1(B^n)$ or $\pi_1(S^n)$ for $n \geq 1$.

Definition 6.5 We say that $(X, *)$ is contractible in $\mathcal{C}^*$ if $(X, *) \simeq (*, *)$.

This means that there is a homotopy

$$H: X \times I \to X$$

satisfying:

(a) $H(x, 0) = *$;
(b) $H(x, 1) = x$;
(c) $H(*, t) = *$.

We say that X is contractible in $\mathcal{C}$ if there is a map H satisfying (a) and (b) for some point $* \in X$. Thus to be contractible means that the identity map is homotopic to a constant map in the appropriate category ($\mathcal{C}$ or $\mathcal{C}^*$).

Proposition 6.6 Let $* \in B^n$ be any point. Then $(B^n, *)$ is contractible. Hence $\pi_1(B^n, *) = 0$.

Proof A homotopy is given by $H(x, t) = tx + (1 - t)*$. $H(B^n \times I) \subset B^n$ by the Cauchy–Schwarz inequality. H clearly satisfies a, b, and c. Since there is only one path $I \to *$, $\pi_1(*, *) = 0$. Thus $\pi_1(B^n, *) = 0$ by 6.4. ∎

Let us return to Problem 1. Suppose there is a map

$$f: B^n \to S^{n-1}$$

such that the diagram

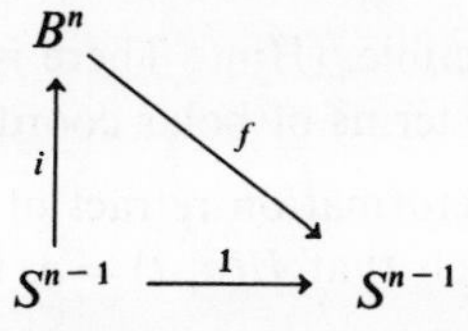

commutes. Since $* = (1, 0, \ldots, 0) \in S^{n-1}$, we have

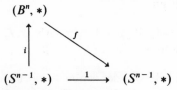

$$\begin{array}{ccc} (B^n, *) & & \\ \uparrow i & \searrow f & \\ (S^{n-1}, *) & \xrightarrow{1} & (S^{n-1}, *) \end{array}$$

Applying π_1 we get a commutative diagram

$$\begin{array}{ccc} \pi_1(B_n, *) & & \\ \uparrow i_* & \searrow f_* & \\ \pi_1(S^{n-1}, *) & \xrightarrow{1_*} & \pi_1(S^{n-1}, *) \end{array}$$

Since $\pi_1(B^n, *) = 0$, this could only happen if $\pi_1(S^{n-1}, *) = 0$. We will eventually show $\pi_1(S^1, *) \neq 0$, solving the problem for $n = 2$. Other functors will be needed for $n > 2$ since we will show that $\pi_1(S^n, *) = 0$ iff $n \neq 1$.

Definition 6.7 X is called simply connected if it is arcwise connected, and $\pi_1(X, *) = 0$.

Exercises

1. Let Q be the rational numbers. Calculate $\pi_1(Q, 0)$.

2.* Show that homotopy equivalence is an equivalence relation.

3. Show that $(B^n, (0, \ldots, 0))$ is contractible.

4. If $* \in B^n$ is any point, show directly that $(B^n, *)$ is contractible

5. Show that if $* \in R^n$ is any point, $(R^n, *)$ is contractible.

6. Show that $f: B^n \to S^{n-1}$ exists such that

$$\begin{array}{ccc} B^n & & \\ \uparrow i & \searrow f & \\ S^{n-1} & \xrightarrow{1} & S^{n-1} \end{array}$$

commutes iff S^{n-1} is contractible. (Hint: There is a natural map $\gamma: S^{n-1} \times I \to B^n$, expressing a point in terms of polar coordinates.) (13.16)

7.* A is called a strong deformation retract of X if $A \subset X$ and there is a homotopy $H: X \times I \to X$ such that $H(a, t) = a$, for $a \in A$, $H(x, 1) = x$ and

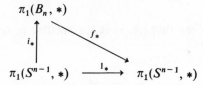

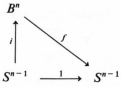

$H(x, 0) \in A$ for $x \in X$. Show that if A is a strong deformation retract of X and $B \subset A$, $(A, B) \simeq (X, B)$. Show that S^{n-1} is a strong deformation retract $R^n - 0$. (13.2, 13.7)

8. Show that there is a 1–1 correspondence $\pi(x, y) \leftrightarrow \pi(x, x)$ iff $\pi(x, y) \neq \varnothing$. (7.12)

9. Show that if $f\colon X \to X$ and $f \sim 1$ (in $\mathcal{C}$), then $f_* : \pi_1(X, *) \to \pi_1(X, f(*))$ is an isomorphism for each point $* \in X$. (Hint: Consider $\varphi_p \circ f_* : \pi_1(X, *) \to \pi_1(X, *)$ where p is the path from $f(*)$ to $*$ given by the homotopy.)

10. Using Exercise 9, show that if $X \simeq Y$ in $\mathcal{C}$, $\pi_1(X, x_0) \cong \pi_1(Y, f(x_0))$ where $f\colon X \to Y$ is a homotopy equivalence. (Hint: First show that $\pi_1(X, g(f(x_0))) \cong \pi_1(Y, f(x_0))$ where g is a homotopy inverse to f.)

11. Show that if X is connected and $X \simeq Y$, then Y is connected.

12. Generalize 6.1 as follows. In each groupoid, Hom(X, X) is a group and for each X, Y with Hom$(X, Y) \neq \varnothing$, Hom$(X, X) \simeq$ Hom(Y, Y). (See Exercise 4, Section 5.)

7

Calculating the Fundamental Group

We have done nothing, so far, to calculate $\pi_1(X)$ except in the most trivial cases. In this section we shall consider two methods of calculating π_1 and give some applications. The first method (covering spaces) is quite geometric and allows one to work from a conjecture based on intuition to the answer. It is absolutely useless in proving that a space is simply connected. The second method (the Van Kampen theorem) is analytical and somewhat more complicated, but can be easily used to show that spaces (such as S^n for $n \geq 1$) are simply connected.

We begin by defining a covering space and show how the structure of a covering space gives information about π_1. The simplest example of a covering space is the map $e: R^1 \to S^1$ given by $e(t) = e^{2\pi it}$. Thus e is periodic of period 1. We think of this as a spiral projected down onto a circle (Fig. 7.1).

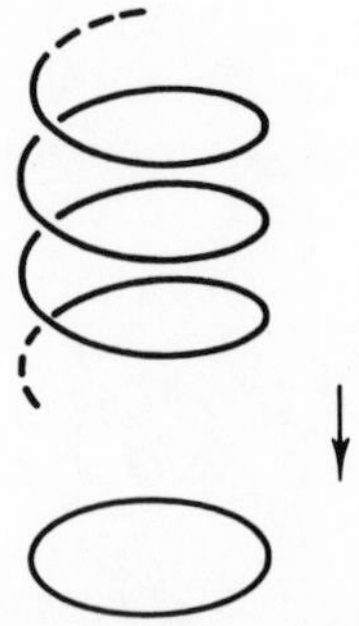

Figure 7.1

Definition 7.1 Given a space X, a covering space is a space $\tilde{X}$ and a map $\Pi: \tilde{X} \to X$ such that:

(a) Π is onto;

(b) for all $x \in X$, there is a neighborhood V of x (called a coordinate neighborhood) such that $\Pi^{-1}(V)$ is the disjoint union of open sets each of which is mapped homeomorphically onto V by Π.

We now prove that the map $e: R^1 \to S^1$ mentioned above is a covering space. Clearly e is onto. Choose open sets $V = S^1 - (1, 0)$ and $W = S^1 - (-1, 0)$. Then $V \cup W = S^1$ and $e^{-1}(V)$ has components $(n - \frac{1}{2}, n + \frac{1}{2})$ while $e^{-1}(W)$ has components $(n, n + 1)$. These are clearly mapped homeomorphically onto V and W respectively by e since e is open.

In order to relate the structure of a covering space to the fundamental group of X, we prove two useful results.

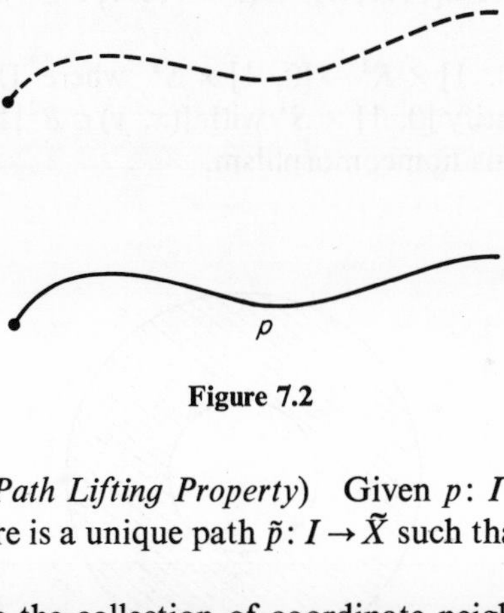

Figure 7.2

Proposition 7.2 (*Path Lifting Property*) Given $p: I \to X$ and $a \in \tilde{X}$ such that $\Pi(a) = p(0)$, there is a unique path $\tilde{p}: I \to \tilde{X}$ such that $\Pi\tilde{p} = p$ and $\tilde{p}(0) = a$. (See Fig. 7.2.)

Proof Let $\{V_\alpha\}$ be the collection of coordinate neighborhoods. $\{p^{-1}(V_\alpha)\}$ is an open cover of I. By Exercise 3, Section 0, choose $\varepsilon > 0$ such that if diam $A < \varepsilon$, $A \subset p^{-1}(V_\alpha)$ for some α. Now choose n such that $1/n < \varepsilon$, and let $t_k = k/n$. Then $p([t_{i-1}, t_i]) \subset V_{\alpha_i}$ for some α_i.

We define unique liftings $\tilde{p}_i$ over the intervals $[0, t_i]$ such that $\tilde{p}_i(0) = a$ by induction on i. For $i = 0$, this is trivial. Suppose $\tilde{p}_k : [0, t_k] \to \tilde{X}$ is defined and unique. We will show that it has a unique extension $\tilde{p}_{k+1}: [0, t_{k+1}] \to \tilde{X}$. Let W be the component of $\Pi^{-1}(V_{\alpha_{k+1}})$ containing $\tilde{p}_k(t_k)$. Any extension $\tilde{p}_{k+1}$ must map $[t_k, t_{k+1}]$ into W since $[t_k, t_{k+1}]$ is arc connected. But $\Pi|_W$ is a homeomorphism; hence there is a unique map

$$\rho: [t_k, t_{k+1}] \to W$$

with $\Pi\rho = p$. We now define

$$\tilde{p}_{k+1}(s) = \begin{cases} \tilde{p}_k(s), & 0 \le s \le t_k \\ \rho(s), & t_k \le s \le t_{k+1}. \end{cases}$$

This completes the induction and hence the theorem. ▌

We give three more examples. That they are covering spaces is left as an exercise.

Example 1 $\Pi_n : S^1 \to S^1$ given by $\Pi_n(z) = z^n$. Here every point of the base (= image space) is covered n times. (Such a covering space is called an n-fold covering.)

Example 2 $\Pi\colon R^n \to \overbrace{S^1 \times \cdots \times S^1}^{n}$ given by

$$\Pi(x_1, \ldots, x_n) = (e^{2\pi i x_1}, \ldots, e^{2\pi i x_n}).$$

Example 3 $\Pi\colon [0, 1] \times R^1 \to [0, 1] \times S^1$ where $\Pi(s, t) = (s, e^{2\pi i t})$. In this example we identify $[0, 1] \times S^1$ with $\{(x, y) \in R^2 \mid 1 \le x^2 + y^2 \le 4\}$ (Fig. 7.3) under the obvious homeomorphism.

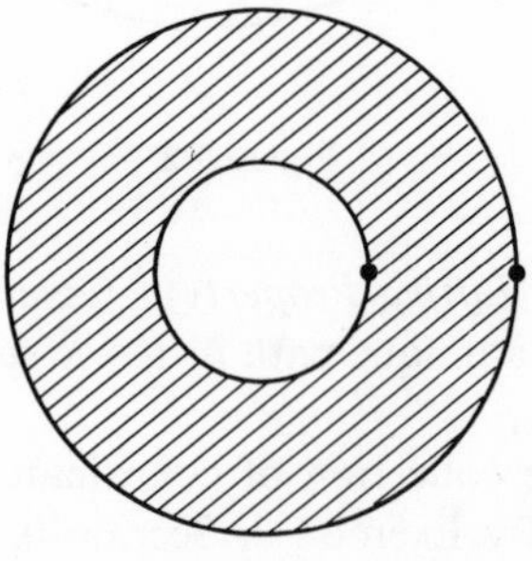

Figure 7.3

Example 3 is a good one to keep in mind for the next theorem. Fix $a \in \tilde{X}$.

Theorem 7.3 (*Monodromy Theorem*) Suppose p and p' are paths in X beginning at a and ending at b. Suppose

$$\{p\} = \{p'\} \in \pi(X; a, b).$$

Then $\tilde{p}'(1) = \tilde{p}(1)$.

Proof Let $H\colon p \sim p'$ be a homotopy (Fig. 7.4), and using Lebesgue's covering lemma as in 7.2, choose

$$0 = s_0 \leq s_1 \leq \cdots \leq s_n = 1, \qquad \text{and} \qquad 0 = t_0 \leq t_1 \leq \cdots \leq t_m = 1$$

so that

$$H([s_i, s_{i+1}] \times [t_j, t_{j+1}]) \subset V_{i,j}$$

where V_{ij} is a coordinate neighborhood.

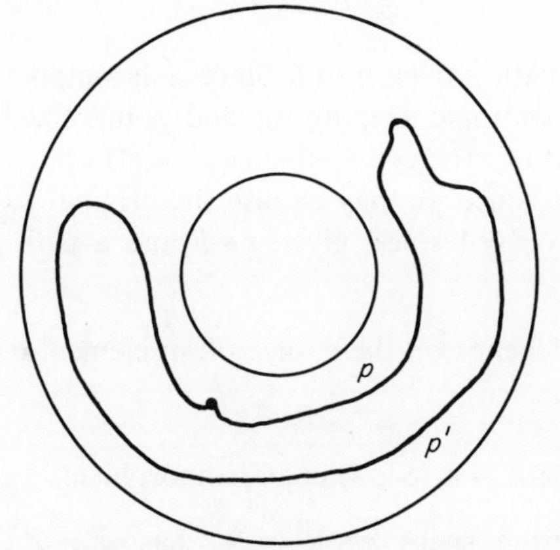

Figure 7.4

We now find a lifting $\tilde{H}$ of H; that is, we find $\tilde{H}\colon I \times I \to \tilde{X}$ such that

$$\tilde{H}(s, 0) = \tilde{p}(s), \qquad \tilde{H}(s, 1) = \tilde{p}'(s), \qquad \tilde{H}(0, t) = a.$$

The proof that this can be done is similar to the proof in 7.2: Given $\tilde{H}$ on any connected union of the rectangles $[s_i, s_{i+1}] \times [t_j, t_{j+1}]$, it can be extended over any adjacent rectangle since each rectangle is mapped into a coordinate neighborhood. We can thus proceed from the edges across the square inductively.

Now such an $\tilde{H}$ provides a path from $\tilde{p}(1)$ to $\tilde{p}'(1)$ lying in $\Pi^{-1}(p(1))$; namely

$$\gamma(t) = \tilde{H}(1, t).$$

Let V be a coordinate neighborhood with $p(1) \in V$. Each point of $\Pi^{-1}(p(1))$ is in a different component of $\Pi^{-1}(V)$. Thus $\Pi^{-1}(p(1))$ has the discrete topology. Every path in a space with the discrete topology is constant. Thus $\gamma(t) = \gamma(0) = \gamma(1)$, i.e., $\tilde{p}'(1) = \tilde{p}'(1)$. ∎

Now let $\Pi: \tilde{X} \to X$ be a covering space and $* \in \tilde{X}$. We write $\Pi(*) = * \in X$. Write $F = \Pi^{-1}(*)$. We now produce a function

$$\phi: \pi_1(X, *) \to F$$

given by $\phi(\{p\}) = \tilde{p}(1)$. By the previous two theorems, this is well defined.

Theorem 7.4 If $\tilde{X}$ is simply connected, ϕ is a 1–1 correspondence.

Proof If $\tilde{X}$ is simply connected, we will produce an inverse

$$\varphi: F \to \pi_1(X, *).$$

For $f \in F$ choose a path p from a to f. Since $\tilde{X}$ is simply connected, any two such choices are homotopic keeping the end points fixed. Thus $\{\Pi \circ p\}$ is a well-defined element of $\pi_1(X, *)$. Define $\varphi(f) = \{\Pi \circ p\}$.

Clearly $\varphi \circ \phi = 1$, since we may choose the original path to define φ. On the other hand, $\phi \circ \varphi = 1$ since, given $f \in F$ and a path p, p is a lifting of $\pi \circ p$. ▌

The function φ depends on the choice of an element $a \in \tilde{X}$. We will write this function as φ_a.

Theorem 7.5 $\varphi_0 : Z \to \pi_1(S^1, *)$ is an isomorphism.

Proof The covering space e: $R^1 \to S^1$ has $F = Z$. Now $\varphi_n(m)(s) = e^{2\pi i(m-n)s}$. The homomorphic property of φ_0 now follows from

Lemma 7.6 $\varphi_a(b) \circ \varphi_b(c) = \varphi_a(c)$.

Proof Apply the uniqueness assertion in 7.2. ▌ ▌

Corollary 7.7 There is no map f: $B^2 \to S^1$ with $f(x) = x$ for $x \in S^1$. ▌

Corollary 7.8 Every map γ: $B^2 \to B^2$ has a fixed point. ▌

Theorem 7.9 (*Fundamental Theorem of Algebra*) Every nonconstant complex polynomial has a root.

Proof Consider a polynomial $p(z) = z^n + a_{n-1}z^{n-1} + \cdots + a_0$ with no roots. Then $H(z, t) = p(trz)$ defines a homotopy

$$H: S^1 \times I \to R^2 - 0$$

for any $r > 0$. We now suppose that $r > \sum |a_i|$. Then there is a homotopy

$$\bar{H}: S^1 \times I \to R^2 - 0$$

given by $\bar{H}(z, t) = tp(rz) + (1 - t)z^n r^n$. $\bar{H}(z, t) \neq 0$; for otherwise $z^n r^n = t(z^n r^n - p(rz))$, hence

$$\begin{aligned} r^n &\leq |p(rz) - (rz)^n| \\ &= |a_{n-1}(rz)^{n-1} + \cdots + a_0| \\ &\leq |a_{n-1}| r^{n-1} + \cdots + |a_0| \leq (\sum |a_i|) r^{n-1} < r^n. \end{aligned}$$

Combining these homotopies we see $z \to r^n z^n$ is homotopic to a constant map $z \to a_0$ as maps $S^1 \to R^2 - 0$. But $R^2 - 0 \simeq S^1$, and the equivalence takes the map $z \to r^n z^n$ to the map $\varphi(n)$. This contradicts 7.5 (unless $n = 0$). ∎

This theorem is usually proved via Cauchy's theorem in complex analysis. There is in fact a relation between $\pi_1(S^1)$ and Cauchy's theorem. Given $p: I \to R^2 - 0$, $p(0) = p(1) = *$, one can consider this as a contour Γ. Then

$$\frac{1}{2\pi i} \int_\Gamma \frac{dz}{z}$$

can be calculated, and it is known that this is an integer (usually called the winding number). One has

$$\phi(\{p\}) = \frac{1}{2\pi i} \int_\Gamma \frac{dz}{z},$$

as students of complex analysis will realize.

Theorem 7.4 gives us a reasonable method for calculating $\pi_1(X)$. First one guesses the answer. This is possibly the hardest part. Having guessed the answer, it is not usually hard to see what a simply connected covering space must look like. One then defines a space $\tilde{X}$ and a map $\Pi: \tilde{X} \to X$, and proves that this is a covering space. It remains to show that $\tilde{X}$ is simply connected. This can be difficult. We now describe another useful tool for calculating π_1 which is often convenient for showing that a space is simply connected.

Suppose $X = X_1 \cup X_2$ with $X_1 \cap X_2 \neq \emptyset$. Choose $* \in X_1 \cap X_2$. We then have homomorphisms $i_{1*}: \pi_1(X_1 \cap X_2, *) \to \pi_1(X_1, *)$ and $i_{2*}: \pi_1(X_1 \cap X_2, *) \to \pi_1(X_2, *)$. In this situation one can make a general group theoretic construction. Let G, G_1, and G_2 be groups, and suppose we have homomorphisms f_1: $G \to G_1$ and f_2 : $G \to G_2$. We will define the amalgamated product of G_1 and G_2 over G. Essentially it is the smallest group generated by G_1 and G_2 with $f_1(x) = f_2(x)$ for $x \in G$. Specifically, let F be the free group generated by the set $G_1 \cup G_2$. We will write $x \cdot y$ for the product in F. Thus every element of F is of the form $x_1^{\varepsilon_1} \cdot \cdots \cdot x_k^{\varepsilon_k}$ where $\varepsilon_i = \pm 1$ and $x_i \in G_1 \cup G_2$. Consider the words $(xy)^1 \cdot y^{-1} \cdot x^{-1}$ defined if both x and y belong to either G_1 or G_2, and $f_1(g)^1 \cdot (f_2(g))^{-1}$ for $g \in G$. Let R be the normal subgroup generated by these words.

Definition 7.10 The amalgamated product of G_1 and G_2 over G, written $G_1 *_G G_2$ is the quotient group F/R.

Observe that there are homomorphisms $g_i : G_i \to F/R$ obtained as compositions $G_i \to F \to F/R$, and $g_1 f_1 = g_2 f_2$.

We now suppose that we are given a space X which is the union of two subspaces X_1 and X_2. The Van Kampen theorem allows one to calculate $\pi_1(X)$ provided we know $\pi_1(X_1)$, $\pi_1(X_2)$, and $\pi_1(X_1 \cap X_2)$. We must make an assumption about the relationship of these subspaces.

Definition 7.11 A pair of subspaces (X_1, X_2) of X is said to be excisive if $X = (\text{Int } X_1) \cup (\text{Int } X_2)$.

Let $j_1 : X_1 \to X$ and $j_2 : X_2 \to X$ be the inclusions.

Theorem 7.12 (*Van Kampen Theorem*) Suppose (X_1, X_2) is excisive, X, X_1, X_2, and $X_1 \cap X_2$ are arcwise connected, and $* \in X_1 \cap X_2$. Then there is an isomorphism

$$\pi_1(X, *) \cong \pi_1(X_1, *) *_{\pi_1(X_1 \cap X_2, *)} \pi_1(X_2, *)$$

in which $(j_1)_*$ and $(j_2)_*$ correspond to g_1 and g_2.[5]

To prove this we need some results about the amalgamated product.

Proposition 7.13 (a) Suppose $h_i : G_i \to H$ are homomorphisms such that $h_1 f_1 = h_2 f_2$. Then there is a unique homomorphism $h\colon G_1 *_G G_2 \to H$ with $hg_i = h_i$.

(b) If every element $x \in H$ can be written $x = x_1 \cdots x_k$ with $x_s = h_i(a_s)$ for some i, h is onto.

Proof (a) One defines $h' : F \to H$ by $h'|_{G_s} = h_s$ since F is free on $G_1 \cup G_2$.

$$\begin{aligned} h'((x_i x_j)^1 x_j^{-1} x_i^{-1}) &= h'((x_i x_j)^1) h'(x_j)^{-1} h'(x_i)^{-1} \\ &= g_s(x_i x_j) g_s(x_j)^{-1} g_s(x_i)^{-1}; \end{aligned}$$

$$h'(1_s^{-1}) = h'(1_s)^{-1} = g_s(1)^{-1} = 1;$$

$$h'(f_1(g)(f_2(g))^{-1}) = h_1(f_1(g)) h_2(f_2(g)^{-1}) = 1;$$

hence $h'(R) = 1$ and h' determines a homomorphism $h\colon G_1 *_G G_2 \to G$. By construction, $hg_i = h'|_{G_i} = h_i$. The uniqueness is clear.

(b) If x is of the form mentioned,

$$h(a_1 \cdots a_k) = h(g_{i_1}(a_1) \cdots g_{i_k}(a_k)) = h_{i_1}(a_1) \cdots h_{i_k}(a_k) = x_1 \cdots x_k = x. \quad \blacksquare$$

[5] Hopefully there will be no confusion between the various meanings of $*$ such as in $\pi_1(X, *)$, $G_1 *_G G_3$, $j*$, etc.

Proof of 7.12 The mappings $(j_i)_* : \pi_1(X_i, *) \to \pi_1(X, *)$ combine by 7.11(a) to give a homomorphism

$$h: \pi_1(X_1, *) *_{\pi_1(X_1 \cap X_2, *)} \pi_1(X_2, *) \to \pi_1(X, *).$$

We will show that h is an isomorphism.

Given $\{p\} \in \pi_1(X, *)$, cover I by $p^{-1}(X_1)$ and $p^{-1}(X_2)$ and choose an ε-number for this cover by Lebesgue's covering lemma. We can thus find $0 = t_0 \leq t_1 \leq \cdots \leq t_n = 1$ so that $t_i - t_{i-1} < \varepsilon$ and hence $p([t_{i-1}, t_i]) \subset X_1$ or X_2. Suppose that these are chosen so that $p(t_i) \in X_1 \cap X_2$; see Fig. 7.5.

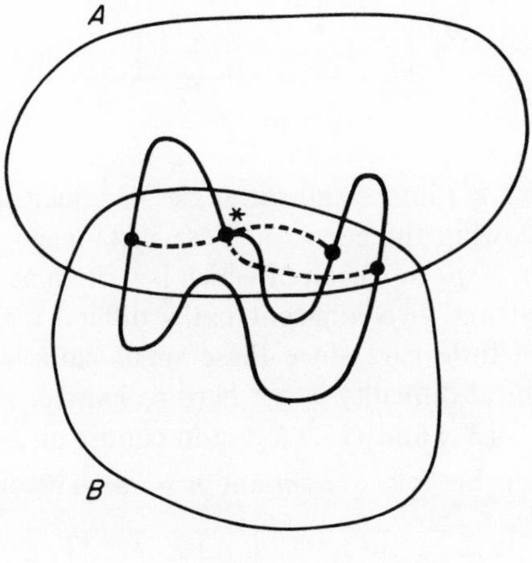

Figure 7.5

(If not, $[t_{i-1}, t_i]$ and $[t_i, t_{i+1}]$ could be combined into one interval.) Choose paths $q_i : I \to X_1 \cap X_2$ with $q_i(0) = *$, $q_i(1) = p(t_i)$ for $0 \leq i \leq n$ with $q_0 = q_n = *$. We now write $p_i : I \to X$ for the path $p|_{[t_i, t_{i+1}]}$. By Exercise 5, Section 5,

$$p = p_{n-1} \cdot \cdots \cdot p_0 \sim q_n^{-1} \cdot p_{n-1} \cdot q_{n-1} \cdot q_{n-1}^{-1} \cdot p_{n-2} \cdot q_{n-2} \cdot q_{n-2}^{-1} \cdot \cdots \cdot q_2^{-1} \cdot p_1 \cdot q_1 \cdot q_1^{-1} \cdot p_0 \cdot q_0 .$$

Now each of the paths $q_k^{-1} \cdot p_{k-1} \cdot q_{k-1}$ belongs to either $\pi_1(X_1, *)$ or $\pi_1(X_2, *)$ and hence h is onto by 7.13(b).

We speak of paths and homotopies as being small if their image lies in either X_1 or X_2. Thus the fact that H is onto can be restated by saying that every based path is the product of small based paths. To prove that ker $h = 1$, suppose we have small based paths $p_1, \ldots, p_m$ such that $h(\{p_m\} \cdots \{p_1\}) = 1$. Then there is a homotopy H in X from their product $p_m \cdot p_{m-1} \cdots p_1$ to the trivial path $*$. We would like to show that $\{p_m\} \cdots \{p_1\} = 1$ in the amalgamated product.

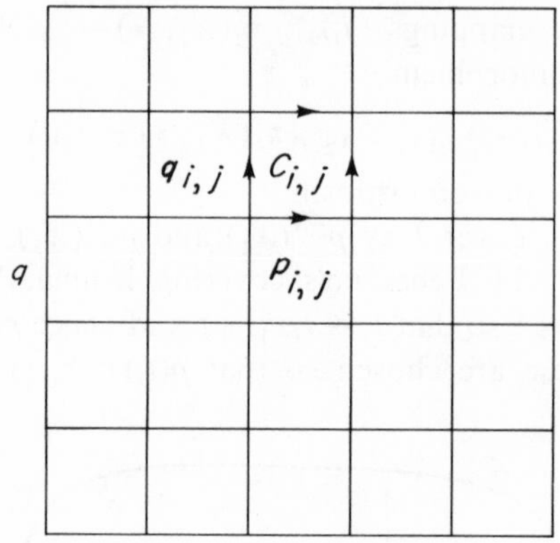

Figure 7.6

We will subdivide $I \times I$ into small rectangles $C_{i,j}$ such that $H|_{C_{i,j}}$ is small (Fig. 7.6). Then following the edges of these rectangles, we get a sequence of paths in X from $p_n \cdots p_1$ to $*$ each of which is written as a product of small paths, and such that any two adjacent paths differ by a small homotopy. One must exercise a little care since these small paths are not necessarily based, but this technical difficulty is not hard to handle.

Cover $I \times I$ by $H^{-1}(X_1)$ and $H^{-1}(X_2)$, and choose an ε-number by Lebesgue's covering lemma. Let $k > \sqrt{2}/\varepsilon m$ and $n = km$. The cubes

$$C_{i,j} = \left[\frac{i}{n}, \frac{i+1}{n}\right] \times \left[\frac{j}{n}, \frac{j+1}{n}\right]$$

are consequently mapped by H into X_1 or X_2. Let

$$p_{i,j} = H|_{[i/n,\,(i+1)/n] \times j/n} \qquad \text{and} \qquad q_{i,j} = H|_{i/n \times [j/n,\,(j+1)/n]}.$$

$p_{i,j}$ and $q_{i,j}$ are thus small paths but are not in general based paths. For each vertex $(i/n, j/n)$, choose a path $r_{i,j}$ from $H(i/n, j/n)$ to $*$ which lies in X_1, X_2, or both if $H(i/n, j/n)$ lies in X_1, X_2, or both. This is possible since X_1, X_2, and $X_1 \cap X_2$ are arcwise connected. In case $H(i/n, j/n) = *$, choose $r_{i,j}$ to be the constant path at $*$. Thus by conjugation

$$\tilde{p}_{i,j} = r_{i+1,j}\, p_{i,j} r_{i,j}^{-1} \qquad \text{and} \qquad \tilde{q}_{i,j} = r_{i,j+1} q_{i,j} r_{i,j}^{-1}$$

are small based paths. Consequently any word in $\{\tilde{p}_{i,j}\}$ and $\{\tilde{q}_{i,j}\}$ represents an element in the amalgamated product. There is a small homotopy $q_{i+1,j} p_{i,j} \sim p_{i,j+1} q_{i,j}$ relative to the end points, for both of these paths are in the image of

$$H_* : \pi\left(C_{i,j}; \left(\frac{i}{n}, \frac{j}{n}\right), \left(\frac{i+1}{n}, \frac{j+1}{n}\right)\right) \to \pi\left(X_\varepsilon; H\left(\frac{i}{n}, \frac{j}{n}\right), H\left(\frac{i+1}{n}, \frac{j+1}{n}\right)\right),$$

and by Exercise 8, Section 6

$$\pi\left(C_{i,j};\left(\frac{i}{n},\frac{j}{n}\right),\left(\frac{i+1}{n},\frac{j+1}{n}\right)\right)$$

has only one element. From the above homotopy it is easy to produce a small based homotopy

$$r_{i+1,j+1}\cdot q_{i+1,j}\cdot p_{i,j}\cdot r_{i,j}^{-1}\sim r_{i+1,j+1}\cdot p_{i,j+1}\cdot q_{i,j}\cdot r_{i,j}^{-1}$$

and hence a relation

$$\{\tilde{q}_{i+1,j}\}\cdot\{\tilde{p}_{i,j}\}=\{\tilde{p}_{i,j+1}\}\cdot\{\tilde{q}_{i,j}\}$$

in the amalgamated product. One can thus conclude that

$$\{\tilde{p}_{n-1,0}\}\cdots\{\tilde{p}_{0,0}\}=[\{\tilde{q}_{n,0}^{-1}\}\cdots\{\tilde{q}_{n,n-1}^{-1}\}][\{\tilde{p}_{n-1,n}\}\cdots\{\tilde{p}_{0,n}\}][\{\tilde{q}_{0,n-1}\}\cdots\{\tilde{q}_{0,0}\}].$$

The right-hand product consists entirely of paths constant at $*$ and hence represents $*$. We will be finished if we show that

$$\{p_m\}\cdots\{p_1\}=\{\tilde{p}_{n-1,0}\}\cdots\{\tilde{p}_{0,0}\}.$$

Choose a with $0\le a\le m-1$. Since $n=km$, $\{\tilde{p}_{s,0}\}$ are all contained in either X_1 or X_2 for $ak\le s\le(a+1)k-1$. Thus the word $\{\tilde{p}_{(a+1)k-1,0}\}\cdots\{\tilde{p}_{ak,0}\}$ is equivalent to the single-letter word $\{\tilde{p}_{(a+1)k-1,0}\cdots\tilde{p}_{ak,0}\}$ in the amalgamated product. However

$$p_{a+1}\sim p_{(a+1)k-1,0}\cdots p_{ak,0}\sim\tilde{p}_{(a+1)k-1,0}\cdots\tilde{p}_{ak,0}$$

so $\{p_{a+1}\}=\{\tilde{p}_{(a+1)k-1,0}\}\cdots\{\tilde{p}_{ak,0}\}$. ∎

Theorem 7.14 $\pi_1(S^n)=0$ for $n>1$.

Proof We write $S^n=X_1\cup X_2$ where

$$X_1=\{(x_1,\ldots,x_{n+1})\in S^n\,|\,x_{n+1}<1\},$$
$$X_2=\{(x_1,\ldots,x_{n+1})\in S^n\,|\,x_{n+1}>-1\}.$$

Since $X_1\equiv X_2\equiv R^n$, $\pi_1(X_1)=\pi_1(X_2)=0$; both are open in S^n, we need only show that $X_1\cap X_2$ is arcwise connected to apply 7.12. This is left as an exercise. (It is true only for $n>1$.) ∎

Lemma 7.15 Suppose $f_1\colon G\to G_1$, $f_2\colon G\to G_2$, and G_1 is defined by generators $x_1,\ldots,x_n$ and relations $r_1(x_1,\ldots,x_n)=1,\ldots,r_k(x_1,\ldots,x_n)=1$, and G_2 is defined by generators $y_1,\ldots,y_m$ and relations $s_1(y_1,\ldots,y_m)=1,\ldots,s_l(y_1,\ldots,y_m)=1$. Suppose finally that G is generated by $z_1,\ldots,z_j$. Then $G_1 *_G G_2$ has as generators

$$x_1,\ldots,x_n,y_1,\ldots,y_m,$$

and as relations $r_1,\ldots,r_k,s_1,\ldots,s_l$, and $f_1(z_i)f_2(z_i)^{-1}$ for $1\le i\le j$.

Proof Let $\bar{G}$ be the group defined by these generators and relations above. One can clearly find a map $\varphi: \bar{G} \to G_1 *_G G_2$ such that $\varphi(x_i) = x_i$, and $\varphi(y_i) = y_i$ since these relations hold in $G_1 *_G G_2$. On the other hand, there are maps $h_i : G_i \to \bar{G}$ with $h_1 f_1 = h_2 f_2$ given by $h_1(x_i) = x_i$, $h_2(y_i) = y_i$. By 7.13(a) there is a map $h: G_1 *_G G_2 \to \bar{G}$ with $h(x_i) = x_i$ and $h(y_i) = y_i$. Clearly $h\varphi = 1$ and $\varphi h = 1$ since these composites are the identity on a set of generators. ▮

Corollary 7.16 Let X be the union of two circles in the plane with one point in common. Then $\pi_1(X)$ is the free group on two generators.

Proof Let p be the common point and choose points p_1 and p_2 on each of the circles and not equal to p. Then $(X - p_1, p) \simeq (S^1, *)$, $(X - p_2, p) \simeq (S^1, *)$, and $((X - p_1) \cap (X - p_2), p) \simeq (*, *)$. Thus by 7.12, $\pi_1(X, p) \simeq Z *_{\{1\}} Z$. By 7.15, this is the free group on two generators. ▮

This will be generalized in Exercise 7.

As a further example of the above techniques we, will discuss some spaces that arise in projective geometry. They will be important later.

Let F be one of the division rings, R the real numbers, C the complex numbers, and H the quaternions.[6] FP^n will be thought of as the set of all lines through the origin in

$$F^{n+1} = \underbrace{F \oplus \cdots \oplus F}_{n+1}.$$

RP^n, CP^n, and HP^n are called n-dimensional real, complex, and quaternionic projective spaces. We topologize FP^n by considering it as a quotient space of $F^{n+1} - \{0\}$. Every point of $F^{n+1} - \{0\}$ determines a line through 0. Thus consider

$$\{(\xi_0, \ldots, \xi_n) \mid \xi_i \in F \quad \text{not all} \quad \xi_i = 0\}.$$

Define $(\xi_0, \ldots, \xi_n) \sim (\lambda\xi_0, \ldots, \lambda\xi_n)$, $\lambda \in F$, $\lambda \neq 0$. This is an equivalence relation. Write $[\xi_0 \mid \cdots \mid \xi_n]$ for an equivalence class, and define FP^n to be the set of equivalence classes with the quotient topology.

There is a natural map

$$F^{n+1} - \{0\} \to FP^n$$

which is continuous, and yields, on restriction to the unit sphere of F^n, maps

$$\Pi_n : S^n \to RP^n, \qquad \eta_n : S^{2n+1} \to CP^n, \qquad v_n : S^{4n+3} \to HP^n.$$

Theorem 7.17 $\Pi_n : S^n \to RP^n$ is a covering space. $\pi_1(RP^n) = Z_2$, for $n > 1$. $RP^1 \equiv S^1$.

[6] See the Appendix to this section.

Proof The sets $D_i^+ = \{(x_1, \ldots, x_{n+1}) \in S^n | x_i > 0\}$ and $D_i^- = \{(x_1, \ldots, x_{n+1}) \in S^n | x_i < 0\}$ are open and cover S^n. $\Pi | D_i^+$ is 1–1, continuous, and open. Thus if $V_i = \Pi(D_i^+) = \Pi(D_i^-)$, $\Pi^{-1}(V_i) = D_i^+ \cup D_i^-$; the sets D_i^+ and D_i^- are disjoint, open, and homeomorphic to V_i. Thus Π is a covering space. $\Pi^{-1}(*)$ contains two points. Therefore, $\pi_1(X, *)$ has two elements and must be Z_2. ∎

The maps η_n and v_n are not covering maps, but they will qualify for a generalization of a covering map—a locally trivial bundle; in Section 11 we shall make homotopy calculations using this notion in analogy with the use of covering spaces in this section.

The fundamental group has been a key tool in low-dimensional topology. Without giving details, we will indicate two applications.

A surface is a separable metric space such that every point has a neighborhood homeomorphic to the plane R^2. Given two surfaces S_1 and S_2 their connected sum $S_1 \# S_2$ is defined by removing a disk D from each of them and connecting them together by a tube $S^1 \times D^1$; see Fig. 7.7.

In surface theory one can prove (see [45, 1.5]) that any surface is either

(a) a sphere S^2,
(b) a connected sum of tori $(S^1 \times S^1)$,
(c) a connected sum of projective planes (RP^2).

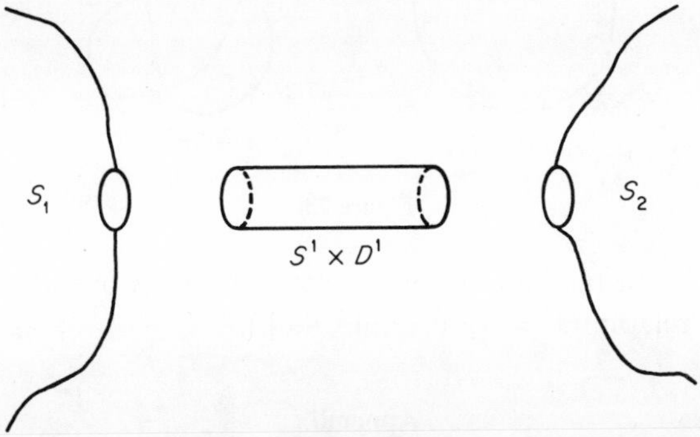

Figure 7.7

It remains to discover which of these surfaces are distinct (not homeomorphic). This is accomplished by calculating the fundamental group. The facts are (see [45, 4.5]):

(a) $\pi_1(S^2) = 0$;

(b) $\pi_1(\overbrace{T \# \cdots \# T}^{n})$ is generated by elements $a_1, \ldots, a_n, b_1, \ldots, b_n$ subject to the single relation

$$1 = (a_1 b_1 a_1^{-1} b_1^{-1})(a_2 b_2 a_2^{-1} b_2^{-1}) \cdots (a_n b_n a_n^{-1} b_n^{-1});$$

(c) $\pi_1(\overbrace{RP^2 \# \cdots \# RP^2}^{n})$ is generated by elements $\alpha_1, \ldots, a_n$ subject to the single relation

$$1 = \alpha_1{}^2\alpha_2{}^2 \cdots \alpha_n{}^2.$$

These groups are all distinct as one can see by calculating their abelianizations. Thus the fundamental group distinguishes among them and they are not homeomorphic.

The second application is to knot theory. A knot is an imbedding of S^1 in R^3. Two knots are called equivalent if there is a homeomorphism $h: R^3 \to R^3$ such that

(a) $hk_1 = k_2$;
(b) there is an integer n such that if $\|x\| > n$, $h(x) = x$.

If $k: S^1 \to R^3$ is a knot, we define the group of the knot to be $\pi_1(R^3 - k(S^1))$. It is easy to see that equivalent knots have isomorphic groups. Thus two knots with different groups are not equivalent. One can distinguish between the trivial knot and the trefoil (Fig. 7.8), for example, since the knot group of

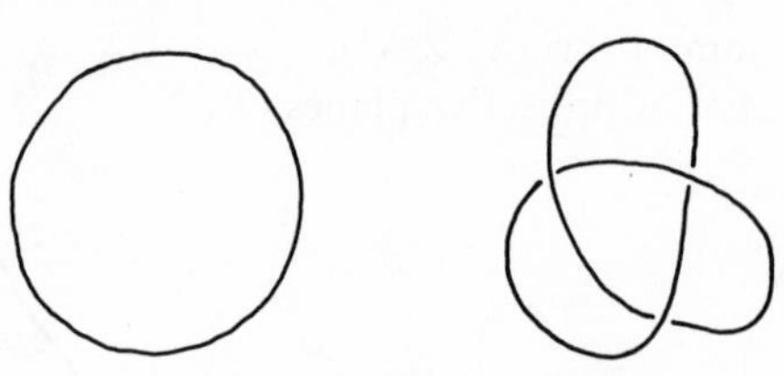

Figure 7.8

the former is Z but the knot group of the latter has two generators α and β subject to the relation $\alpha^2 = \beta^3$. (See [45, 4.6].)

Appendix

The algebra of quaternions H is a four-dimensional real vector space with basis 1, i, j, k and multiplication given by

$$ij = k \qquad jk = i \qquad ki = j$$
$$ji = -k \qquad kj = -i \qquad ik = -j$$
$$i^2 = j^2 = k^2 = -1.$$

(Observe that H is not commutative.) 1 is a unit, and multiplication is extended by linearity. Let $q = a + bi + cj + dk$ be a general quaternion for a, b, c, d

real numbers. Then define $\bar{q} = a - bi - cj - dk$ (called the conjugate of q). $q \to \bar{q}$ is linear, and one may check that $\overline{q_1 q_2} = \bar{q}_2 \cdot \bar{q}_1$, $\bar{q}q = a^2 + b^2 + c^2 + d^2 = \|q\|^2$. Hence $q^{-1} = \dfrac{\bar{q}}{\|q\|}$. The quaternions are associative.

Seminar Problems

A. We have said nothing about the existence of covering spaces. [31, Section 17] has a very readable account of this. We need some definitions:

1. X is semi-locally-simply connected if for all $b \in X$, there is a neighborhood U of b such that any two paths in U with the same end points are homotopic in X keeping the end points fixed. (This is the same as saying that the homomorphism $\pi_1(U, x_0) \to \pi_1(X, x_0)$ is 0.)

2. X is locally pathwise connected if any point has arbitrarily small pathwise connected neighborhoods.

If X is semi-locally-simply connected, locally pathwise connected, and connected, simply connected covering spaces $\tilde{X}$ exist. (See also [64, 2.5].)

B. Calculate $\pi_1(T \# T \# \cdots \# T)$ and $\pi_1(RP^2 \# \cdots \# RP^2)$. Calculate the knot group of the trivial knot and the trefoil.

Exercises

1. Suppose $\tilde{X}$, $\Pi: \tilde{X} \to X$ and $\bar{X}$, $\Pi': \bar{X} \to X$ are covering spaces. Suppose $\tilde{X}$ is simply connected. Choose $\tilde{a} \in \tilde{X}$ and $\bar{a} \in \bar{X}$ such that $\Pi(\tilde{a}) = \Pi'(\bar{a})$. Show that there is a unique continuous map $f: \tilde{X} \to X$ such that $\Pi' \circ f = \Pi$ and $f(\tilde{a}) = \bar{a}$. (Hint: A point in $\tilde{X}$ yields a path in X which can be lifted to $\bar{X}$.) Conclude that if $\bar{X}$ is also simply connected, $\tilde{X} \equiv \bar{X}$. The existence of f is called a "universal property" and a simply connected covering space is often called a universal covering space.

2. Show that S^n is arcwise connected. Show that $\{(x_1, \ldots, x_{n+1}) \in S^n \mid -1 < x_{n+1} < 1\}$ is arcwise connected. (Hint: Use the fact that it is homeomorphic to $R^n - \{x_0\}$ for some point x_0.) (7.14)

3. Show that for any $x \in CP^n$, $\eta_n^{-1}(x) \equiv S^1$, and for any $x \in HP^n$, $\nu_n^{-1}(x) \equiv S^3$.

4. Prove that S^1 and S^n do not have the same homotopy type for $n > 1$. Conclude that R^2 and R^n are not homeomorphic for $n > 2$.

5. Show that the examples given before 7.3 are covering spaces. Calculate $\pi_1(S^1 \times S^1)$.

6. Show that $\phi\colon \pi_1(X, *) \to F$ is onto if X is arc connected. Show that if ϕ is 1–1, $\tilde{X}$ is simply connected.

7. Let X_n be the union of n circles in the plane that intersect at the origin and only at the origin. Prove (by induction) that $\pi_1(X_n, 0)$ is the free group on n generators. (17.4)

8. Prove that RP^n, CP^n, and HP^n are Hausdorff. (See Exercise 4, Section 0.) (Example 3, Section 14)

9. Show that $CP^1 \equiv S^2$ and $HP^1 \equiv S^4$. (13.14)

10. Let $U_1 = CP^n - CP^{n-1}$ and $U_2 = CP^n - [0|0\cdots 0|1]$. Show that $(CP^{n-1}, *) \subset (U_2, *)$ is a homotopy equivalence, and that U_1 is contractible. Show that $U_1 \cap U_2$ is arcwise connected. (Use the fact that $C - \{0, 1\}$ is arcwise connected.) Conclude by induction that $\pi_1(CP^n) = 0$. Does the same proof work for HP^n? Why does it not work for RP^n?

11. Let $X = \{(x, y) \in RP^n \times RP^n | x = * \text{ or } y = *\}$. ($X$ is two copies of RP^n with one point $*$ in common.) Calculate $\pi_1(X, *)$. Is this group finite?

12. Show that $\pi_1(X, *)$ acts as a group of homeomorphisms on $\tilde{X}$ by using 7.2 and 7.3. (This means that for all $\sigma \in \pi_1(X)$, there is a homeomorphism $T_\sigma\colon \tilde{X} \to \tilde{X}$ such that $T_\sigma \circ T_\rho = T_{\sigma\rho}$, $T_1 = 1$.) Prove that $\Pi T_\sigma = \Pi$, and that the action is without fixed points. (This means that if for some $x \in \tilde{X}$ and some $\sigma \in \pi_1(X, *)$, $T_\sigma(x) = x$, then $\sigma = 1$.) (Exercise 14, Section 11)

13. Suppose X is simply connected and $\Pi\colon \tilde{X} \to X$ is a covering space. Show that each component of $\tilde{X}$ is mapped homeomorphically by Π onto X. (26.10)

14. The Klein bottle is defined as the quotient space:

$$K = S^1 \times I/(z, 0) \sim (z^{-1}, 1).$$

Calculate $\pi_1(K)$. What is $\pi_1(K)/[\pi_1(K), \pi_1(K)]$?

15. Let $\Pi\colon E \to B$ be an n-fold covering space (i.e., $\Pi^{-1}(x)$ consists of n points, for all $x \in X$). Show that $\Pi_*\colon \pi_1(E, *) \to \pi_1(B, *)$ is the inclusion of a subgroup of index n.

16. Find a double covering $\Pi\colon S^1 \times S^1 \to K$ where X is from Exercise 14. Calculate Π_*. Is the image of Π_* normal?

17. Let X be a "sphere with two handles," pictured in Fig. 7.9. X is the quotient of two spaces homeomorphic with $S^1 \times S^1 - D$ where D is a small open disk by identifying the boundary circles. Calculate $\pi_1(X)$.

18. Let X be a Hausdorff space and π a finite group of homeomorphisms of X such that if $\sigma \in \pi$, $x \in X$, $\sigma x = x$, then $\sigma = 1$. Define X/π to be the set

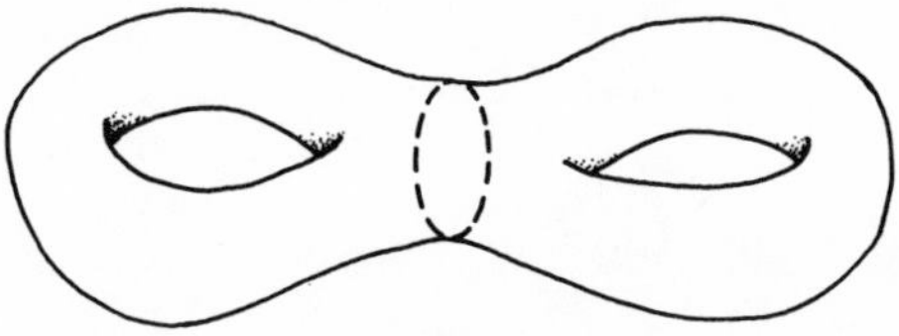

Figure 7.9

of equivalence classes where $x \sim y$ iff there exists $\sigma \in \pi$ with $x = \sigma y$. Show that the natural map $X \to X/\pi$ is a covering space. (Exercise 19)

19. In Exercise 18 let $\pi = Z_p$ and $X = S^{2n-1} = \{(z_1, \ldots, z_n) | \sum |z_i|^2 = 1)\}$. Define $\sigma(z_1, \ldots, z_n) = (\lambda z_1, \ldots, \lambda z_n)$ where $\lambda = e^{2\pi i/p}$ and σ generates Z_p. Write $L_{2n-1}(\pi) = X/\pi$. $L_{2n-1}(\pi)$ is called a Lens space. Prove that $L_{2n-1}(Z_2) \equiv RP^{2n-1}$ and $\pi_1(L_{2n-1}(Z_p)) \cong Z_p$ if $n > 1$. (Exercise 22, Section 26; appendix to Section 27)

20. The Möbius band is the space

$$M = (0, 1) \times [0, 1]/(x, 0) \sim (x, 1).$$

M can be imbedded in R^3 (i.e., M is homeomorphic to a subspace of R^3). Show that $M^\infty \equiv RP^2$

21. Let $D' = (B^2 - S^1) \cup \{(1, 0)\}$. Show that $D' \not\equiv B^2 - S^1$.

22. Prove that $\varphi \colon Z \to \pi_1(S^1, *)$ is an isomorphism by using Exercise 3, Section 5 instead of 7.6.

8

A Convenient Category of Topological Spaces

In this section we shall describe the category of compactly generated Hausdorff spaces. This contains almost all important spaces in topology, and restricting to this category and its internal operations provides stronger results with usually cleaner statements. It is our purpose to develop some of these results. This section contains excerpts from a typically well-written and definitive paper by Steenrod [68].

Definition 8.1 A compactly generated Hausdorff space is a Hausdorff space with the property that each subset which intersects every compact set in a closed set is itself closed. We denote by $\mathcal{CG}$ the category of compactly generated Hausdorff spaces and their continuous maps.

Lemma 8.2 If X is a Hausdorff space and if for each subset M and each limit point x of M there exists a compact set C in X such that x is a limit point of $M \cap C$, then $X \in \mathcal{CG}$.

Briefly, if each limit relation in X takes place in some compact subset of X, then $X \in \mathcal{CG}$.

Proof Assume M meets each compact set in a closed set, and let x be a limit point of M. By the assumption, there exists a compact C such that x is a limit point of $M \cap C$. Since $M \cap C$ is closed, we have the relation $x \in M \cap C$, hence $x \in M$. So M is closed and $X \in \mathcal{CG}$. ∎

Proposition 8.3 The category $\mathcal{CG}$ includes all locally compact spaces and all spaces satisfying the first axiom of countability (for example, metrizable spaces).

Proof In both cases we apply 8.2. If X is locally compact, we take C to be the compact closure of a neighborhood of x, and if X is first countable, C is taken to consist of x and a sequence in M converging to x. ∎

Remark The condition in the hypothesis of 8.2 is not equivalent to $X \in \mathcal{CG}$; there is an example of a space in $\mathcal{CG}$ for which the condition does not hold.

These results show that $\mathcal{CG}$ is large enough to contain most of the standard spaces. Perhaps the simplest example of a Hausdorff space not in $\mathcal{CG}$ is the following:

Example Let Y denote the ordinal numbers preceding and including the first noncountable ordinal Ω. Give to Y the topology defined by its natural order. Let X be the subspace obtained be deleting all limit ordinals except Ω. The only compact subsets of X are the finite sets, because each infinite set must contain a sequence converging to a limit ordinal of the second kind. Therefore the set $X - \Omega$ meets each compact set in a closed set, but is not closed in X because it has Ω as a limit point.

The example shows that a subspace X of a compactly generated space Y need not be compactly generated. However, the following results show that certain subspaces are in $\mathcal{CG}$.

Proposition 8.4 If X is in $\mathcal{CG}$, then every closed subset of X is also in $\mathcal{CG}$. An open set U of X is in $\mathcal{CG}$ if it is a "regular" open set, that is, if each point $x \in U$ has a neighborhood in X whose closure lies in U.

Proof Suppose A is closed in X and $B \subset A$ meets each compact subset of A in a closed set. Let C be a compact set in X. Then $A \cap C$ is a compact set of A; hence $B \cap (A \cap C) = B \cap C$ is closed in A. Since A is closed, $B \cap C$ is closed in X. Because $X \in \mathcal{CG}$, it follows that B is closed in X, hence also in A. So $A \in \mathcal{CG}$.

Let U be a regular open set in X, suppose $B \subset U$ meets each compact set of U in a closed set, and let $x \in U$ be a limit point of B. By assumption, there is a neighborhood V of x in X with closure $\overline{V} \subset U$. If C is compact in X, then $\overline{V} \cap C$ is a compact set of X in U. Since it is also compact in the relative topology of U, it follows that $B \cap \overline{V} \cap C$ is closed first in U, then in $\overline{V} \cap C$, and finally in X. Because C is any compact set in X and $X \in \mathcal{CG}$, it follows that $B \cap \overline{V}$ is closed in X. Since x is a limit point of $B \cap \overline{V}$, we see that $x \in B \cap \overline{V}$, hence $x \in B$, so B is closed in U. ∎

Proposition 8.5 If $f\colon X \to Y$ is a quotient map, $X \in \mathcal{CG}$ and Y is a Hausdorff space, then $Y \in \mathcal{CG}$.

Proof Suppose $B \subset Y$ meets each compact set of Y in a closed set. Let C be a compact set in X. Then $f(C)$ is compact, hence $B \cap f(C)$ is closed, so $f^{-1}(B \cap f(C))$ is closed, and therefore $f^{-1}(B \cap f(C)) \cap C$ is closed. Since this last set coincides with $f^{-1}(B) \cap C$, it follows that $f^{-1}(B)$ meets each compact set of X in a closed set. Because $X \in \mathcal{CG}$, this means that $f^{-1}(B)$ is closed. Since f is a quotient map, B must be closed in Y. This shows that $Y \in \mathcal{CG}$. ▌

The preceding results show that $\mathcal{CG}$ is large in the sense that it contains many spaces. By definition, it contains all continuous maps between any two of its spaces. The following proposition sometimes facilitates the recognition of the continuity of a function.

Proposition 8.6 If $X \in \mathcal{CG}$, Y is a Hausdorff space, and a function $f\colon X \to Y$ is continuous on each compact subset of X, then f is continuous.

Proof Let A be closed in Y, and let C be compact in X. Since Y is a Hausdorff space and $f|_C$ is continuous, $f(C)$ is compact, hence closed in Y. This implies that $A \cap f(C)$ is closed, hence also

$$(f|_C)^{-1}(A \cap f(C)) = (f^{-1}(A)) \cap C.$$

Because $X \in \mathcal{CG}$, it follows that $f^{-1}(A)$ is closed in X, and this shows that f is continuous. ▌

Definition 8.7 If X is a Hausdorff space, the *associated compactly generated space* $k(X)$ is the set X with the topology defined as follows: a closed set of $k(X)$ is a set that meets each compact set of X in a closed set. If $f\colon X \to Y$ is a mapping of Hausdorff spaces, $k(f)$ denotes the same function $k(X) \to k(Y)$.

Theorem 8.8

(i) The identity function $k(X) \to X$ is continuous.
(ii) $k(X)$ is a Hausdorff space.
(iii) $k(X)$ and X have the same compact sets.
(iv) $k(X) \in \mathcal{CG}$.
(v) If $X \in \mathcal{CG}$, then $I\colon k(X) \to X$ is a homeomorphism.
(vi) If $f\colon X \to Y$ is continuous on compact sets, then $k(f)$ is continuous.
(vii) $I_* : \pi_n(k(X), *) \to \pi_n(X, *)$ is a 1–1 correspondence for all n and all $*$.

The theorem can be paraphrased by saying that k is a retraction of the category $\mathcal{H}$ of Hausdorff spaces into $\mathcal{CG}$.

Proof If A is closed in X, and C is compact in X, then C is closed in X, hence also $A \cap C$. This means that A is also closed in $k(X)$, and this proves (i).

Since X is a Hausdorff space, (i) implies (ii). If a set A is compact in $k(X)$, then (i) implies that A is compact in X. Suppose now that C is compact in X, and C' denotes the set C with its relative topology from $k(X)$. By (i), the identity map $C' \to C$ is continuous; we must prove the continuity of its inverse. Let B denote a closed set of C'. By definition, B meets every compact set of X in a closed set; therefore $B \cap C = B$ is closed in C. Thus $C \to C'$ is continuous; this shows that C' is compact, and (iii) is proved. If a set A meets each compact set of $k(X)$ in a closed set, then by (iii), it meets each compact set of X in a compact (hence closed) set; therefore A is closed in $k(X)$, and (iv) is proved. (v) follows directly from (iv). To prove (vi), it suffices by 8.6 to prove that $k(f)$ is continuous on each compact set of $k(X)$. Let C' be compact in $k(X)$; by (iii), the same set with its topology in X (call it C) is compact and the identity map $C' \to C$ is a homeomorphism. Since $f|_C$ is continuous, $f(C)$ is compact, and by (iii), so is the same set $f(C')$ with its topology in $k(Y)$. Thus the function $k(f)|_{C'}: C' \to f(C)$ factors into the composition of $f|_C$ and two identity maps $C' \to C \to f(C) \to f(C')$. Hence $k(f)|_{C'}$ is continuous, and (vi) is proved. By (vi), the maps of closed cells into X coincide with those into $k(X)$; this implies (vii) because the sets in question are derived from such mappings. ∎

Given $X, Y \in \mathcal{CG}$, it may happen that $X \times_c Y$, the product space with the usual Cartesian product topology is not in $\mathcal{CG}$.

Definition 8.9 If X and Y are in $\mathcal{CG}$, their product $X \times Y$ (in $\mathcal{CG}$) is $k(X \times_c Y)$, where $\times_c$ denotes the product with the usual Cartesian topology.

Theorem 8.10 The product defined in 8.9 satisfies the universal property: There are continuous projections $\pi_1: X \times Y \to X$ and $\pi_2: X \times Y \to Y$ such that if $f: W \to X$ and $g: W \to Y$ are continuous, and W is in $\mathcal{CG}$, there is a unique map $F: W \to X \times Y$ with $f = \pi_1 F$ and $g = \pi_2 F$.

Proof Since by 8.8 the identity function $X \times Y \to X \times_c Y$ is continuous, and since the projections $X \times_c Y$ into X and Y are continuous, their compositions projecting $X \times Y$ into X and Y are continuous and, hence, belong to $\mathcal{CG}$. Let $W \in \mathcal{CG}$, and let f and g be maps $W \to X$ and $W \to Y$ in $\mathcal{CG}$. As usual, f and g are the components of a unique mapping $(f, g): W \to X \times_c Y$. Applying k and using the facts $k(W) = W$ and $k(X \times_c Y) = X \times Y$, we obtain a unique mapping $k(f, g): W \to X \times Y$ which, when composed with the projections, gives f and g. ∎

It follows from 8.10 that the product $X \times Y$ in $\mathcal{CG}$ satisfies the usual commutative and associative laws. We can extend the construction to products having any number of factors, by applying k to the usual product.

Having modified the concept of product space, we should note what effect this has on other concepts that are based on products such as topological group G ($G \times G \to G$), transformation group G of X ($G \times X \to X$), and homotopy ($I \times X \to X$). If we restrict ourselves to G and X in $\mathcal{CG}$, any multiplications $G \times_c G \to G$ or actions $G \times_c X \to X$ that are continuous in the old sense remain continuous when we apply k. Thus the effect of the new definition is to allow an increase in the number of groups and actions. The following theorem asserts that in many cases there is no change; in particular, the concept of homotopy is unaltered.

Theorem 8.11 If X is locally compact and $Y \in \mathcal{CG}$, then $X \times_c Y$ is in $\mathcal{CG}$; that is $X \times Y = X \times_c Y$.

Proof Let A be a subset of $X \times_c Y$ that meets each compact set in a closed set, and let (x_0, y_0) be a point of its complement. By local compactness, x_0 has a neighborhood whose closure N is compact. Since $N \times_c y_0$ is compact, $A \cap (N \times_c y_0)$ must be closed. It follows that x_0 has a smaller neighborhood U such that $\bar{U} \times_c y_0$ does not meet A. Let B denote the projection in Y of $A \cap (\bar{U} \times_c Y)$. If C is a compact set in Y, then $A \cap (\bar{U} \times_c C)$ is compact, and therefore $B \cap C$ is closed. Since $Y \in \mathcal{CG}$, B must be closed in Y. Since y_0 is not in B, it follows that $U \times_c (Y - B)$ is a neighborhood of (x_0, y_0) not meeting A. This proves that A is closed; hence $X \times_c Y$ is in $\mathcal{CG}$. ▮

In the category of compact spaces, it is well known that a product of decomposition spaces has the topology of the decomposition space of the product. It is not difficult to find counterexamples involving noncompact spaces. However, the following theorem asserts that each such uses either spaces not in $\mathcal{CG}$ or the wrong product.

Theorem 8.12 If $f\colon X \to X'$ and $g\colon Y \to Y'$ are quotient mappings in $\mathcal{CG}$, then $f \times g\colon X \times Y \to X' \times Y'$ is also a quotient mapping.

Proof Since $f \times g$ factors into the composition $(f \times 1)(1 \times g)$, and since a composition of quotient maps is a quotient map, it suffices to prove the special case where $Y = Y'$ and g is the identity. Suppose then that $A \subset X' \times Y$ and that $(f \times 1)^{-1}(A)$ is closed in $X \times Y$. Let C be a compact set in $X' \times Y$, and let D and E denote its projections in X' and Y, respectively. Then $D \times E$ is compact. If we can show that $A \cap (D \times E)$ is closed, it will follow that $A \cap C$ is closed, and since $X' \times Y$ is in $\mathcal{CG}$, this will show that A is closed, and the proposition will be proved. Since $(f \times 1)^{-1}(D \times E) = f^{-1}(D) \times E$ is closed in $X \times Y$, it follows that $(f \times 1)^{-1}(A \cap (D \times E))$ is closed in $f^{-1}(D) \times E$. Substituting X, X', Y for $f^{-1}(D)$, D, E, respectively, we have reduced the proof to the case where X' and Y are compact. Then, by 8.11 $X' \times Y = X' \times_c Y$ and $X \times Y = X \times_c Y$.

Suppose then that $W \subset X' \times Y$, $(f \times 1)^{-1}(W)$ is open in $X \times Y$, and $(x_0', y_0) \in W$. Choose $x_0 \in X$ so that $f(x_0) = x_0'$. Since (x_0, y_0) is in the open set $(f \times 1)^{-1}(W)$ and Y is compact, there exists a neighborhood V of y_0 such that $x_0 \times \bar{V}$ lies in $(f \times 1)^{-1}(W)$. Let U denote the set of those $x \in X$ such that $f(x) \times \bar{V} \subset W$. To see that U is open in X, let $x_1 \in U$. We can cover $x_1 \times \bar{V}$ by products of open sets contained in $(f \times 1)^{-1}(W)$, and we can select a finite subcovering; then the intersection of the X factors of these products is a neighborhood N of x_1 such that $N \times \bar{V}$ lies in $(f \times 1)^{-1}(W)$. Therefore U is open. By its definition, $U = f^{-1}(f(U))$; hence $f(U)$ is open in X', because f is a quotient map. Since (x_0', y_0) is in $f(U) \times V$, and since $f(U) \times V$ is open and contained in W, it follows that W is open. ∎

Lemma 8.13 If X and Y are Hausdorff spaces, then the two topologies $k(X) \times k(Y)$ and $k(X \times_c Y)$ on the product space coincide.

Proof Since the identity maps $k(X) \to X$ and $k(Y) \to Y$ are continuous, so also is the identity map

$$g\colon k(X) \times_c k(Y) \to X \times_c Y;$$

hence each compact set of $k(X) \times_c k(Y)$ is compact in $X \times_c Y$. Let A be a compact set of $X \times_c Y$. Since its projections B and C in X and Y, respectively, are compact, they are also compact in $k(X)$ and $k(Y)$, respectively. Therefore $B \times_c C$ is a compact set of $k(X) \times_c k(Y)$; hence $g|_{B \times_c C}$ is bicontinuous. Since $A \subset B \times_c C$, it follows that A is compact in $k(X) \times_c k(Y)$. Because $k(X) \times_c k(Y)$ and $X \times_c Y$ have the same compact sets, it follows from Definition 8.7 of k that their associated topologies in $\mathcal{CG}$ coincide. ∎

For Hausdorff spaces X, Y, let $C(X, Y)$ denote the space of continuous mappings $X \to Y$ with the compact-open topology. We recall the definition: If A is a compact set of X and U is an open set of Y, let $W(A, U)$ denote the set of $f \in C(X, Y)$ such that $f(A) \subset U$; then the family of $W(A, U)$ for all such pairs (A, U) forms a subbasis for the open sets of $C(X, Y)$. Although X and Y are in $\mathcal{CG}$, it may happen that $C(X, Y)$ is not in $\mathcal{CG}$.

Definition 8.14 For Hausdorff spaces X, Y, define $Y^X = kC(X, Y)$.

Lemma 8.15 The evaluation mapping $e\colon C(X, Y) \times_c X \to Y$, defined by $e(f, x) = f(x)$, is continuous on compact sets. If X and Y are in $\mathcal{CG}$, then e is continuous as a mapping $Y^X \times X \to Y$.

Proof Since any compact set of the product is contained in the product of its projections, it suffices to show that e is continuous on any set of the form $F \times A$, where F is compact in $C(X, Y)$, and A is compact in X. Let

$(f_0, x_0) \in F \times A$, and let U be an open set of Y containing $f_0(x_0)$. Since f_0 is continuous, there exists a neighborhood N of x_0 in A whose closure satisfies $f_0(\overline{N}) \subset U$. Therefore $(W(\overline{N}, U) \cap F) \times N$ is open in $F \times A$, it contains (f_0, x_0), and it is mapped by e into U. This shows that e is continuous on compact sets.

By 8.8(vi), if we apply k to $e\colon C(X, Y) \times_c X \to Y$, we obtain a continuous mapping. When $X \in \mathcal{CG}$, the left side gives, by 8.13,

$$k(C(X, Y) \times_c X) = k(C(X, Y)) \times k(X) = Y^X \times X;$$

and when $Y \in \mathcal{CG}$, the right-hand side becomes $k(Y) = Y$. ▮

Lemma 8.16 If X is in $\mathcal{CG}$, and Y is a Hausdorff space, then $C(X, k(Y))$ and $C(X, Y)$ are equal as sets, and the two topologies have the same compact sets, hence $k(C(X, k(Y))) = k(C(X, Y))$ as spaces in $\mathcal{CG}$.

Proof If $f\colon X \to k(Y)$ is continuous, so is its composition with $k(Y) \to Y$, and therefore $f \in C(X, Y)$. Conversely, if $f\colon X \to Y$ is continuous, then $k(f)\colon k(X) \to k(Y)$ is continuous from X to $k(Y)$. Thus $C(X, k(Y))$ and $C(X, Y)$ coincide as sets of functions. Since $k(Y) \to Y$ is continuous, it follows that the identity map $C(X, k(Y)) \to C(X, Y)$ is continuous. This implies that each compact set in $C(X, k(Y))$ is also compact in $C(X, Y)$.

Now let $F \subset C(X, Y)$ be a compact set in its relative topology in $C(X, Y)$. Let F' denote the same set with its relative topology in $C(X, k(Y))$. We wish to prove that F' is compact. It suffices to show that each open set W of $C(X, k(Y))$ meets F' in an open set of F, because this implies that the inverse correspondence $F \to F'$ is continuous, whence F is compact. It obviously suffices to prove this when W is a subbasic open set $W(C, U)$, where C is compact in X, and U is open in $k(Y)$. Suppose then that $f_0 \in W(C, U) \cap F$. Since $F \times C$ is compact, and since by 8.15 the evaluation mapping $e\colon F \times C \to Y$ is continuous, it follows from 8.8 that it is also continuous as a mapping $F \times C \to k(Y)$. Hence $e^{-1}(U)$ is an open set of $F \times C$. Since C is compact and $f_0 \times C \subset e^{-1}(U)$, there exists an open set V of F containing f_0 such that $V \times C \subset e^{-1}(U)$. It follows that $f_0 \in V \subset W(C, U)$. Hence $W(C, U) \cap F$ is open in F. This shows that F' is compact, and it completes the proof that the two topologies have the same compact sets. It follows now from Definition 8.7 of k that their associated topologies in $\mathcal{CG}$ are equal. ▮

Theorem 8.17 If X, Y, and Z are in $\mathcal{CG}$, then $Z^{Y \times X} \equiv (Z^Y)^X$.

Proof We shall first construct a natural homeomorphism

$$\text{(A)} \qquad \mu\colon C(Y \times X, Z) \xrightarrow{\equiv} C(X, C(Y, Z)).$$

Corresponding to an $f \in C(Y \times X, Z)$, define $\mu(f)\colon X \to C(Y, Z)$ by $((\mu(f))(x))(y) = f(y, x)$. To see that for each x, $(\mu(f))(x)$ is continuous from Y to Z, suppose it carries y_0 into the open set U of Z. Then $f(y_0, x) \in U$, and the continuity of f gives an open set V of Y containing y_0 such that $f(V \times x) \subset U$; therefore $(\mu(f))(x)$ maps V into U. We must now prove

(B) If $f \in C(Y \times X, Z)$, then $\mu(f)\colon X \to C(Y, Z)$ is continuous.

Let $W(B, U)$ be a subbasic open set of $C(Y, Z)$, and suppose that $(\mu(f))(x_0) \in W(B, U)$. Then $f(B \times x_0) \subset U$. Since U is open and B is compact, there is a neighborhood N of x_0 such that $f(B \times N) \subset U$. This implies $(\mu(f))(N) \subset W(B, U)$, and it proves (B).

To prove the continuity of μ, we start with the continuity of the evaluation mapping rearranged as

$$e\colon Y \times X \times C(Y \times X, Z) \to Z$$

(see 8.15). If we apply B with X replaced by $X \times C(Y \times X, Z)$, we find that

$$\mu(e)\colon X \times C(Y \times X, Z) \to C(Y, Z)$$

is continuous. Apply (B) again, with X replaced by $C(Y \times X, Z)$, Y by X, and Z by $C(Y, Z)$; then

$$\mu(\mu(e))\colon C(Y \times X, Z) \to C(X, C(Y, Z))$$

is continuous. It is readily verified that $\mu(\mu(e))$ coincides with μ of (A).

To show that μ has a continuous inverse, let

$$e\colon X \times C(X, C(Y, Z)) \to C(Y, Z), \qquad e'\colon Y \times C(Y, Z) \to Z$$

be evaluation mappings. By 8.15 the composition

$$e'(1 \times e)\colon Y \times X \times C(X, C(Y, Z)) \to Z$$

is continuous. Applying (B) with X replaced by $C(X, C(Y, Z))$, Y by $Y \times X$, and Z by Z, we see that

$$\mu(e'(1 \times e))\colon C(X, C(Y, Z)) \to C(Y \times X, Z)$$

is defined and continuous. It is readily verified that $\mu(e'(1 \times e))$ is the inverse of μ in (A).

We now apply the functor k to both sides of (A). On the right hand side we use 8.16 to obtain

$$kC(X, C(Y, Z)) \equiv kC(X, kC(Y, Z)) = (Z^Y)^X.$$

On the left side we obtain $kC(Y \times X, Z) = Z^{Y \times X}$. ∎

Theorem 8.18 For X, Y, and Z in $\mathcal{CG}$, the composition of mappings $X \to Y \to Z$ is a continuous function $Z^Y \times Y^X \to Z^X$.

Proof By 8.15, the mappings

$$Z^Y \times Y^X \times X \xrightarrow{1 \times e} Z^Y \times Y \xrightarrow{e'} Z$$

are continuous, hence $e'(1 \times e)$ is also continuous. Applying B with X replaced by $Z^Y \times Y^X$, Y by X, Z by Z, and f by $e'(1 \times e)$, we see that

$$\mu(e'(1 \times e)) \colon Z^Y \times Y^X \to C(X, Z)$$

is continuous. Then $k(\mu(e'(1 \times e)))\colon Z^Y \times Y^X \to Z^X$ is also continuous. ∎

Definition 8.19 We denote by $C((X, A), (Y, B))$ the space of continuous mappings of pairs $(X, A) \to (Y, B)$. It is the subspace of $C(X, Y)$ of maps f such that $f(A) \subset B$. We abbreviate $k(C((X, A), (Y, B)))$ by $(Y, B)^{(X,A)}$. The *smash* product[7] $X \wedge Y$ is obtained from $X \times Y$ by collapsing the *wedge* $X \vee Y = (X \times *) \cup (* \times Y)$ to a point that is the base point of $X \wedge Y$. This is Hausdorff (see Exercise 6). Define the function space of mappings of pointed spaces by

$$(X, *)^{(Y, *)} = k(C((X, *), (Y, *))),$$

where its base point is the constant map.

Our objective is to prove the analog of the exponential rule 8.17 in $\mathcal{CG}^*$; but we need a preliminary result. Let $X \in \mathcal{CG}$, and let A be a closed subspace of X such that X/A is a Hausdorff space. Recall the collapsing map $p_A \colon (X, A) \to (X/A, *)$. Let $Y \in \mathcal{CG}^*$. By composing a map $f\colon X/A \to Y$ with p_A, we obtain $fp_A \in C((X, A), (Y, *))$, and this defines a mapping of function spaces

$$(p_A)^* \colon C(X/A, Y) \to C((X, A), (Y, *)).$$

Lemma 8.20 The above mapping $(p_A)^*$ is continuous and one-to-one (bijective), and it sets up a one-to-one correspondence between compact subsets. Hence, applying the functor k, we obtain an induced natural homeomorphism

$$(Y, *)^{(X/A, *)} \equiv (Y, *)^{(X,A)}.$$

Proof The continuity and bijective properties are readily proved. The crucial point is to show that if F is a compact subset of $C((X, A), (Y, *))$, then $(p_A)^{*-1}(F)$ is compact. It suffices to show that $(p_A)^*$ is continuous on

[7] When considering functors like $X \wedge Y$ and $X \vee Y$ it is necessary to have base points in order to define them. Hence it is not necessary to include the base point in the notation (unless for some reason, more than one base point is being considered), and it is almost always suppressed.

F. Suppose $g_0 \in F$ and $W(C, U)$ is a subbasic open set of $C(X/A, Y)$ containing $(p_A)^{*-1}(g_0)$. This means that $g_0 p_A$ maps C into U. In case C does not contain the base point $*$, then $p_A^{-1}(C)$ is compact in X and $W(p_A^{-1}(C), U)$ is an open set that contains g_0, and is mapped into $W(C, U)$ by $(p_A)^{-1}$.

Suppose therefore that C contains $*$. Since F is compact, the evaluation mapping $e\colon F \times X \to Y$ is continuous, by 8.15. Since $e(F \times A) = *$ and $F \times (X/A)$ is the decomposition space of $F \times X$ obtained by collapsing $F \times A$ to $F \times *$ (see 8.12), it follows that e induces a continuous mapping $e'\colon F \times (X/A) \to Y$. Since $e'(g_0, *) \in U$, there exist a neighborhood V of g_0 in F and a neighborhood N of $*$ in X/A such that e' maps $V \times N$ into U. Set $C' = C - C \cap N$; then C' is compact and does not contain $*$. It follows that $V \cap W(p_A^{-1}(C'), U)$ is a neighborhood of g_0 in F, and any g in this neighborhood will map $(p_A)^{-1}(N)$ into U because $g \in V$, and it will map $(p_A)^{-1}(C')$ into U because $g \in W(p_A^{-1}(C'), U)$. Since $C \subset C' \cup N$, it follows that $(p_A)^{*-1}(g) \in W(C, U)$. ▮

Theorem 8.21 If X, Y, Z are in $\mathcal{CG}^*$, then

$$(Z, *)^{(Y \wedge X, *)} \equiv [(Z, *)^{(Y, *)}]^{(X, *)}.$$

Proof Abbreviate the wedge $(Y \times *) \cup (* \times X)$ by W. If in 8.20 we replace Y by Z and (X, A) by $(Y \times X, W)$, we obtain the natural homeomorphism

$$(Z, *)^{(Y \wedge X, *)} \equiv (Z, *)^{(Y \times X, W)}. \tag{C}$$

The space on the right of (C) is a subspace of $Z^{Y \times X}$ which, by 8.17 is homeomorphic to $(Z^Y)^X$. It is readily verified that, under the latter homeomorphism, $(Z, *)^{(Y \times X, W)}$ corresponds exactly to $[(Z, *)^{(Y, *)}]^{(X, *)}$. ▮

Theorem 8.22 There are natural homeomorphisms

(a) $X \wedge (Y \wedge Z) \equiv (X \wedge Y) \wedge Z$,
(b) $X \wedge Y \equiv Y \wedge X$,
(c) $(X \vee Y) \wedge Z \equiv (X \wedge Z) \vee (Y \wedge Z)$.

Proof Consider the composite

$$X \times (Y \times Z) \xrightarrow{1 \times f} X \times (Y \wedge Z) \xrightarrow{g} X \wedge (Y \wedge Z);$$

since they are both quotient maps (by 8.12), their composite is also. Hence both $X \wedge (Y \wedge Z)$ and $(X \wedge Y) \wedge Z$ are quotient spaces of $X \times Y \times Z$ under the same identifications, and are consequently homeomorphic. Similarly $X \wedge Y$ and $Y \wedge X$ are quotient spaces of $X \times Y \equiv Y \times X$. (c) is proved similarly after one establishes

Lemma 8.23 $X \vee Y$ has the quotient topology on the disjoint union $X \cup Y$.

Proof Since the inclusions $X \xrightarrow{i_1} X \cup Y$ and $Y \xrightarrow{i_2} X \cup Y$ are continuous, they determine a continuous map $X \cup Y \xrightarrow{f} X \vee Y$. Let $F \subset X \vee Y$ and suppose $f^{-1}(F)$ is closed. Then $i_1^{-1}(F)$ and $i_2^{-1}(F)$ are closed. Now $F = c_1^{-1}(F) \times * \cup * \times i_2^{-1}(F)$ is closed in $X \times_c Y$ and hence in $X \times Y$. ▌▌

Theorem 8.24 The homeomorphism of 8.21 induces a 1–1 correspondence

$$\tilde{\psi}\colon [(X, *), (Z, *)^{(Y, *)}] \to [(Y \wedge X, *), (Z, *)].$$

Thus the base point preserving homotopy classes of base point preserving maps from $Y \wedge X$ to Z are in 1–1 correspondence with the same from X to $(Z, *)^{(Y, *)}$. Sometimes we use the sloppier notation

$$[Y \wedge X, Z] \approx [X, Z^Y]$$

to represent this fact if it is understood that $X, Y, Z \in \mathcal{CG}^*$. This property is called adjointness; it will be utilized in the next two sections.

Proof Suppose $g_0 \sim g_1\colon (Y \wedge X, *) \to (Z, *)$. Let

$$\tilde{G}\colon ((Y \wedge X) \times I, * \times I) \to (Z, *)$$

be a homotopy. Consider $G\colon (Y \times X) \times I \to Z$ given by

$$Y \times X \times I \xrightarrow{\beta \times 1} (Y \wedge X) \times I \xrightarrow{\tilde{G}} Z$$

where β is the quotient map. This is continuous, hence

$$F\colon X \times I \to Z^Y$$

given by 8.17 is also continuous. It is easy to see that $F(y, t) \in (Z, *)^{(X, *)}$, $F(y, 0) = f_0(y)$, $F(y, 1) = f_1(y)$ where f_0 and f_1 correspond to g_0 and g_1 under 8.21. Finally $F(*, t) = *$, hence $F\colon f_0 \sim f_1$. The converse is similar. ▌

Exercises

1.* Show that $X \times (Y \times Z) \equiv (X \times Y) \times Z$ and $X \times Y \equiv Y \times X$, where $\times$ is the product in $\mathcal{CG}$.

2. Prove that if Y is Hausdorff, $C(X, Y)$ is Hausdorff.

3. Show that if, $X, Y, Z \in \mathcal{CG}$, $(Y \times Z)^X \equiv Y^X \times Z^X$. (Hint: First prove $C(X, Y \times_c Z) \equiv C(X, Y) \times_c C(X, Z)$.)

4. Show that if $f_0 \sim f_1\colon (Z, *) \to (Y, *)^{(X, *)}$ then $g_0 \sim g_1\colon (Z \wedge X, *) \to (Y, *)$, where g_i corresponds to f_i under 8.21. (8.24)

5. Suppose X, Y, $Z \in \mathcal{C}$ and Y is locally compact. Then $\mu\colon C(Y \times_c X, Z) \to C(X, C(Y, Z))$ is a 1–1 correspondence (compare to 8.17). Conclude that 8.24 holds with $(Z, *)^{(Y,*)}$ replaced by $C((Y, *), (Z, *))$ if X, Y, and Z are as above. This is the classical version of 8.17 and 8.24.

6. Prove that if X and Y are Hausdorff, $X \wedge Y$ is Hausdorff. (27.9)

7. If $X_\alpha \in \mathcal{CG}$ for each α, let $\amalg X_\alpha$ be the disjoint union of the X_α with a topology such that $D \subset \amalg X_\alpha$ is open iff $D \cap X_\alpha$ is open in X_α for each α. Show that $\amalg X_\alpha \in \mathcal{CG}$. In the case of two spaces this is written $X \amalg Y$. Show that $\amalg$ is commutative and associative up to homeomorphism and

$$(X \amalg Y) \times Z \equiv (X \times Z) \amalg (Y \times Z).$$

9

Track Groups and Homotopy Groups

In this section we will show that $\pi_n(X, *) = [(I^n, \partial I^n), (X, *)]$ has a natural group structure if $n \geq 1$. The composition can be constructed directly, but is most easily constructed via the results of Section 8 if $X \in \mathcal{CG}^*$. The modifications necessary to define the composition in general are indicated in the exercises.

There are two basic ways to construct a natural composition among homotopy classes. The first is to make some assumptions about the domain space.

Definition 9.1 $SX = X \wedge S^1$ is called the reduced suspension of X.

We often replace S^1 by $I/0 \sim 1$ in this definition.

Proposition 9.2 If X and Y are in $\mathcal{CG}^*$, $F(X, Y) = [(SX, *), (Y, *)]$ is a functor in two variables from $\mathcal{CG}^*$ to $\mathcal{G}$. (See [11].)

Proof

$$[(S^1 \wedge X, *), (Y, *)] \approx [(S^1, *), (Y, *)^{(X,*)}] \approx \pi_1((Y, *)^{(X,*)}, *).$$

Furthermore $(Y, *)^{(X,*)}$ is a functor covariant in $(Y, *)$ and contravariant in $(X, *)$. Hence a map

$$f\colon (X, *) \to (X', *)$$

induces

$$f'\colon (Y, *)^{(X',*)} \to (Y, *)^{(X,*)},$$

and thus $(f')_*\colon \pi_1((Y, *)^{(X',*)}, *) \to \pi_1((Y, *)^{(X,*)}, *)$; this map is usually written

$$f^*\colon [(SX', *), (Y, *)] \to [(SX, *), (Y, *)];$$

similarly if $g: (Y, *) \to (Y', *)$, we have

$$g_*: [(SX, *), (Y, *)] \to [(SX, *), (Y', *)].$$

Clearly $(f_1 f_2)^* = f_2^* f_1^*$, $(g_1 g_2)_* = (g_1)_*(g_2)_*$, $1^* = 1$, and $1_* = 1$. ▮

Proposition 9.3 Let $C^*X = (X, *) \wedge (I, 1)$. Then $X \subset C^*X$ and $C^*X/X \equiv SX$. Furthermore, $(C^*(S^{n-1}), S^{n-1}) \equiv (B^n, S^{n-1})$.

C^*X is called the reduced cone on X.

Proof Let $i: X \to C^*X$ be given by $i(x) = (x, 0)$. This is clearly an inclusion. It is also obvious that $C^*X/X \equiv SX$ by the definitions. Define $\delta: (C^*(S^{n-1}), S^{n-1}) \to (B^n, S^{n-1})$ via

$$\delta(x, t) = (1 - t)x + t* \in B^n$$

for $x, * \in S^{n-1}$. To find an inverse, apply 2.4 with $f(x) = *$. We have $\gamma(u) = \rho u + (1 - \rho)*$ and

$$* \to *, \qquad u \to \left(\gamma(u), \frac{\rho - 1}{\rho}\right) \quad (u \neq *)$$

is an inverse to δ. (It is a little difficult to show that this is continuous at $*$, but unnecessary. It is an inverse in $\mathbb{S}$ so δ is 1–1 and onto. Continuity of δ and compactness of $C^*(S^{n-1})$ finish the job.) ▮

Proposition 9.4 $S(I^{n-1}/\partial I^{n-1}) \equiv I^n/\partial I^n$. This determines a homeomorphism

$$\varphi: S(S^{n-1}) \to S^n \subset R^{n+1}$$

such that $\varphi(x, 1 - t) = (\varphi_1(x, t), \ldots, \varphi_n(x, t), -\varphi_{n+1}(x, t)) \in R^{n+1}$.

Proof Define

$$\Phi: S(I^{n-1}/\partial I^{n-1}) \to I^n/\partial I^n$$

by $\Phi(x_1, \ldots, x_{n-1}, t) = (x_1, \ldots, x_{n-1}, t)$. This is clearly a homeomorphism since $S(I^{n-1}/\partial I^{n-1})$ is compact. To define φ we must use the homeomorphism of 1.9. The homeomorphism $f_n: S^n \to I^n/\partial I^n$ is given by

$$f_n(x_1, \ldots, x_{n+1}) = \frac{1}{2}\left(1 + \frac{2}{\pi}\tan^{-1}\left(\frac{x_2}{1 - x_1}\right), \ldots, 1 + \frac{2}{\pi}\tan^{-1}\left(\frac{x_{n+1}}{1 - x_1}\right)\right)$$
$$= (\alpha_1, \ldots, \alpha_n)$$

thus

$$f_n(x_1, \ldots, -x_{n+1}) = (\alpha_1, \ldots, 1 - \alpha_n).$$

Then $\varphi = f_n^{-1} \circ \Phi \circ Sf_{n-1}$ satisfies the above formula. ▮

Corollary 9.5 $S^n \equiv S^1 \wedge \cdots \wedge S^1$.

Proof This follows by induction. ∎

Recall that $\pi_n(X, *) = [(I^n, \partial I^n), (X, *)]$. (See the remark before Exercise 7, Section 4.)

Corollary 9.6 π_n is a functor from $\mathcal{CG}^*$ to $\mathcal{G}$ for $n \geq 1$; $\pi_n(X, *)$ is called the nth homotopy group of X (at $*$).

Proof

$$\pi_n(X, *) \overset{c}{\longleftrightarrow} [(S^n, *), (X, *)] \leftrightarrow [(SS^{n-1}, *), (X, *)].$$

The result now follows from 9.2 since all of these correspondences are natural. ∎

Alternatively, we can define a composition among homotopy classes by making assumptions about the range space Y. The appropriate structure on Y is the following generalization of a topological group:

Definition 9.7 An *H*-space (X, μ) is a space X with base point e and a continuous map $\mu\colon (X \times X, e \times e) \to (X, e)$ called the multiplication such that $\mu|_{X \vee X} \sim \nabla$ in $\mathcal{T}^*$, where $\nabla\colon X \vee X \to X$ is the "folding map," $\nabla(x, e) = \nabla(e, x) = x$. (Clearly ∇ is well defined, continuous, and base point preserving.) We make no assumption about associativity or inverses. (*H* stands for Hopf who first studied such spaces [30].)

Proposition 9.8 Suppose (X, μ) is an H space. Then $[(Y, *), (X, e)]$ has a multiplication with two-sided unit, which is natural with respect to Y.

Proof Given $f, g\colon (Y, *) \to (X, e)$, we define $f \cdot g\colon (Y, *) \to (X, e)$ by

$$(f \cdot g)(y) = f(y) \cdot g(y).$$

Clearly $(f \cdot g)(*) = *$. Let $*$ be the trivial map given by $*(y) = e$ for all y. Then $* \cdot g \sim g \sim g \cdot *$. We must show that $f \cdot g$ depends only on the homotopy classes of f and g. Suppose $F, G\colon (Y \times I, * \times I) \to (X, e)$ are homotopies with $F(y, 0) = f_0$, $F(y, 1) = f_1$, $G(y, 0) = g_0$, $G(y, 1) = g_1$. Then

$$H\colon (Y \times I, * \times I) \to (X, e)$$

given by

$$H(y, t) = F(y, t) \cdot G(y, t)$$

is a homotopy from $f_0 \cdot g_0$ to $f_1 \cdot g_1$. The naturality is immediate. ∎

Proposition 9.9 If (Y, e) is an H-space, the two multiplications in $[(SX, *), (Y, e)]$ are the same, and they are commutative.

Proof Let $f, g\colon (SX, *) \to (Y, e)$. By Exercise 1, their product $f \cdot g$ in $[(SX, *), (Y, e)]$ "induced by the suspension structure" is given by

$$(f \cdot g)(x, t) = \begin{cases} g(x, 2t), & 0 \le t \le \frac{1}{2} \\ f(x, 2t - 1), & \frac{1}{2} \le t \le 1. \end{cases}$$

Their product "induced by the H-space structure" is given by

$$(f \circ g)(x, t) = \mu(f(x, t), g(x, t)).$$

These two multiplications are "independently defined," and consequently "commute" with each other. This is expressed by the formula

$$(f \circ g) \cdot (f' \circ g') = (f \cdot f') \circ (g \cdot g').$$

In fact both sides of this equation are given by the formula

$$h(x, t) = \begin{cases} \mu(f'(x, 2t), g'(x, 2t)), & 0 \le t \le \frac{1}{2} \\ \mu(f(x, 2t - 1), g(x, 2t - 1)), & \frac{1}{2} \le t \le 1. \end{cases}$$

The two multiplications are linked together by the fact that they both have a common unit, namely the trivial map. Letting $g = f' = 1$, this formula reduces to

$$f \cdot g' = f \circ g',$$

hence the multiplications are equal. Letting $f = g' = 1$, this reduces to

$$g \cdot f' = f' \circ g;$$

hence the multiplication is commutative. ∎

Definition 9.10 $\Omega X = \{\omega \in X^I \mid \omega(0) = \omega(1) = *\}$. ΩX is called the loop space on X.

As in the case of the suspension, we suppress the base point $*$ from the notation unless there will be confusion. We write $\Omega^n X = \Omega(\Omega^{n-1} X)$ and choose the constant loop at $*$ as the new base point. By 8.20 $\Omega X \equiv (X, *)^{(S^1, *)}$.

Theorem 9.11 If $n \ge 1$, $\Omega^n X$ is an H space.

Proof It is sufficient to show that ΩX is an H space. We take the constant path $*$ as a unit and path composition (as in π_1) for a multiplication

$$\mu\colon \Omega X \times \Omega X \to \Omega X.$$

It is given by

$$\mu(\omega_1, \omega_2) = \begin{cases} \omega_2(2t), & 0 \le t \le \frac{1}{2} \\ \omega_1(2t-1), & \frac{1}{2} \le t \le 1. \end{cases}$$

To show that this is continuous, observe that a map $f: Y \to X^I$ is continuous iff the corresponding map $Y \times I \to X$ given by 8.17 is continuous.

A homotopy $H: \nabla \sim \mu|_{\Omega X \vee \Omega X}$ is given by

$$H(\omega, *, s)(t) = \begin{cases} *, & 0 \le t \le \dfrac{s}{2} \\ \omega\left(\dfrac{2t-s}{2-s}\right), & \dfrac{s}{2} \le t \le 1, \end{cases}$$

and

$$H(*, \omega, s)(t) = \begin{cases} \omega\left(\dfrac{2t}{2-s}\right), & 0 \le t \le 1 - \dfrac{s}{2} \\ *, & 1 - \dfrac{s}{2} \le t \le 1. \end{cases}$$

The proof that this is continuous is as above. ∎

Proposition 9.12 The 1–1 correspondence

$$[(SX, *), (Y, *)] \leftrightarrow [(X, *), (\Omega Y, *)]$$

is an isomorphism.

Proof Let $f, g: (SX, *) \to (Y, *)$. Then both $\widetilde{f \cdot g}$ and $\tilde{f} \cdot \tilde{g}$ are given by the formula

$$h(x) = \begin{cases} \tilde{g}(x)(2t), & 0 \le t \le \frac{1}{2} \\ \tilde{f}(x)(2t-1), & \frac{1}{2} \le t \le 1. \end{cases} \quad ∎$$

Corollary 9.13 $\pi_n(X, *)$ is abelian if $n > 1$.

Proof

$$\pi_n(X, *) \leftrightarrow [(S^n, *), (X, *)] \cong [(S^1, *), (\Omega^{n-1}(X, *), *)] = \pi_1(\Omega^{n-1}(X, *), *);$$

since $\Omega^{n-1}(X, *)$ is an H-space, π_1 is abelian. ∎

Remark Since $\pi_n(X, *) \cong \pi_1(\Omega^{n-1}(X, *), *)$, one could calculate π_n if one could calculate π_1 of every space. However, not much is known, in a geometrical sense, about $\Omega(X, *)$ (e.g., what would a simply connected covering space of $\Omega(S^2, *)$ look like?).

Proposition 9.14 $[S^nX, Y]$ has n multiplications given by the n suspension coordinates

$$(f \cdot_k g)(x, s_1, \ldots, s_n) = \begin{cases} g(x, s_1, \ldots, s_{k-1}, 2s_k, s_{k+1}, \ldots, s_n), & 0 \le s_k \le \frac{1}{2} \\ f(x, s_1, \ldots, s_{k-1}, 2s_k - 1, s_{k+1}, \ldots, s_n), & \frac{1}{2} \le s_k \le 1. \end{cases}$$

They are all equal.

Proof We use induction on n. By 9.12 there is an isomorphism

$$[S^nX, Y] \cong [S^{n-1}X, \Omega Y]$$

Thus the last multiplication in $[S^nX, Y]$ corresponds to the multiplication induced by the H-space structure. The first $n - 1$ multiplications correspond to the $n - 1$ multiplications of $[S^{n-1}X, \Omega Y]$ induced by the suspension structure. By induction, these are all equal, and by 9.9 they are equal to the last one. ∎

Exercises

1. Define a multiplication in $[(SX, *), (Y, *)]$, for $X, Y \in \mathcal{C}^*$ by

$$(f \cdot g)(x, t) = \begin{cases} g(x, 2t), & 0 \le t \le \frac{1}{2} \\ f(x, 2t - 1), & \frac{1}{2} \le t \le 1. \end{cases}$$

Show that this agrees with that of 9.2 if $X, Y \in \mathcal{CG}^*$. Prove that this makes $[(SX, *), (Y, *)]$ into a group. (Think of a map $f: (SX, *) \to (Y, *)$ as a family of maps $(S^1, *) \to (Y, *)$ parametrized over X.) (9.9)

2. Show using Exercise 1 that in 9.6, 9.9, 9.12, 9.13, and 9.14, $\mathcal{CG}^*$ can be replaced by $\mathcal{C}^*$.

3. (X, μ) is said to be homotopy associative if

$$\begin{array}{ccc} X \times X \times X & \xrightarrow{\mu \times 1} & X \times X \\ \downarrow{\scriptstyle 1 \times \mu} & & \downarrow{\scriptstyle \mu} \\ X \times X & \xrightarrow{\mu} & X \end{array}$$

commutes up to homotopy. (X, μ) has a homotopy inverse if there is a map $\nu: X \to X$ such that $\nu(e) = e$ and

$$X \xrightarrow{\Delta} X \times X \xrightarrow{1 \times \nu} X \times X \xrightarrow{\mu} X$$

is homotopic to the constant one mapping all of X to e. Show that if (X, μ) is homotopy associative H-space with homotopy inverse, the set $[(Y, *), (X, e)]$ has the structure of a group. Fixing (X, e), show that $[(Y, *), (X, e)]$ is a functor from the category of spaces with base point to the category of groups and homomorphisms.

4. Give a reasonable definition of a homotopy commutative H space (X, e) and prove that if (X, e) is homotopy commutative $[(Y, *), (X, e)]$ is commutative.

5.* Show that $\pi_n(X, *) \cong \pi_{n-1}(\Omega(X, *), *)$.

6.* Show that C^*, defined in 9.3 is a functor

$$C^*: \mathcal{CG}^* \to \mathcal{CG}^*.$$

Show that if $f: (X, *) \to (Y, *)$, $f \sim *$ (in $\mathcal{CG}^*$) iff there is a map $g: C^* Y \to Y$ with

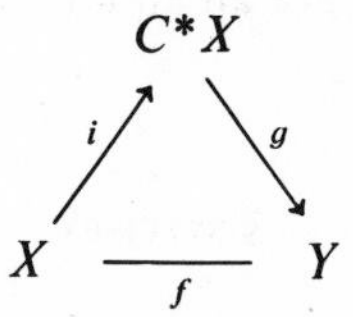

commutative, where $i(x) = (x, 0)$. Compare this with Problems 1 and 2, Section 2.

7. If $*_1$ and $*_2 \in X$ are two base points in the same arc component, a path p from $*_1$ to $*_2$ defines as isomorphism

$$\gamma_p: \pi_n(X, *_1) \cong \pi_n(X, *_2).$$

If X is simply connected, this isomorphism does not depend on the choice of paths. (Exercise 14, Section 11)

8. Let Z be the integers with the discrete topology and let $\mu: Z \times Z \to Z$ be addition. Show that $[X, Z]$ with the multiplication induced by 9.8 is isomorphic to hom(A, Z) where A is a free abelian group with a basis in 1–1 correspondence to the components of X.

9.* Let $\pi_1: X \times Y \to X$ and $\pi_2: X \times Y \to Y$ be the projections. Show that

$$F: \pi_n(X \times Y, *) \to \pi_n(X, *) \oplus \pi_n(Y, *)$$

given by $F(\{\gamma\}) = (\{\pi_1 \gamma\}, \{\pi_2 \gamma\})$ is an isomorphism $(n \geq 1)$.

10. Use Exercise 7 to show that if $X \simeq Y$ in $\mathcal{C}$, $\pi_n(X, *) \simeq \pi_n(X, f(*))$ where $f: X \to Y$ is a homotopy equivalence (compare to Exercises 9 and 10 in Section 6).

11.* Define $CX = X \times I/X \times 1$. CX is called the unreduced cone on X. Prove that $C: \mathcal{CG} \to \mathcal{CG}$ is a functor. A map $f: X \to Y$ is called nullhomotopic or inessential if f is homotopic to a constant map. Prove that $f: X \to Y$ is null-homotopic iff there is an extension $F: CX \to Y$ where $i: X \to CX$ is the inclusion given by $i(x) = (x, 0)$. Compare with Exercise 6. Prove that $CS^{n-1} \equiv B^n$. Define $\Sigma X = CX/X$. ΣX is called the unreduced suspension on X. Prove that $\Sigma S^{n-1} \equiv S^n$.

10

Relative Homotopy Groups

Given a subspace $A \subset X$, it is natural to try to relate $\pi_n(X, *)$ to $\pi_n(A, *)$. This is most easily accomplished by defining new groups $\pi_n(X, A, *)$ which measure the descrepancy. These groups can be defined in two ways. The first is to define them to be the homotopy groups of an appropriately constructed space (as we did with π_n for $n > 1$). The second is to define them as homotopy classes of maps. By appropriate use of adjointness the definitions agree and the algebraic relations between $\pi_n(A, *)$, $\pi_n(X, *)$ and $\pi_n(X, A, *)$ assume (by design) a particularly simple form.

Definition 10.1 Let $A, B \subset X$. Define

$$\Omega(X; A, B) = \{\omega \in X^I \mid \omega(0) \in A, \omega(1) \in B\} \subset X^I$$

with the induced topology; if $* \in A \subset X$, let $\pi_n(X, A, *) = \pi_{n-1}(\Omega(X; A, *), *)$. $\pi_n(X, A, *)$ is called the nth (relative) homotopy group of $(X, A, *)$.

Proposition 10.2 π_n is a covariant functor on the category of pairs with base point $(X, A, *)$, and maps of pairs preserving base point (we will call this $\mathcal{C}^{2*}$.), for $n \geq 1$. It is a group if $n \geq 2$, and is abelian if $n \geq 3$.

Proof A map f: $(X, A, *) \to (Y, B, *)$ in $\mathcal{C}^{2*}$ (i.e., a map f: $X \to Y$ with $f(A) \subset B$ and $f(*) = *$) induces a map Ωf: $\Omega(X; A, *) \to \Omega(Y; B, *)$. This is functorial and hence induces

$$f_*: \pi_n(X, A, *) \to \pi_n(Y, B, *).$$

f_* is a homomorphism if $n \geq 2$ and π_n is thus a functor. ∎

Consider now $[(I^n; \partial I^n, J^{n-1}), (X; A, *)]$, the set of homotopy classes of maps $f\colon I^n \to X$ such that $f(\partial I^n) \subset A$ and $f(J^{n-1}) = *$. The homotopies $H\colon I^n \times I \to X$ satisfy $H(\partial I^n \times I) \subset A$ and $H(J^{n-1} \times I) = *$.

Proposition 10.3 There is a natural 1–1 correspondence

$$\pi_n(X, A, *) \overset{\theta}{\longleftrightarrow} [(I^n; \partial I^n, J^{n-1}), (X; A, *)].$$

Proof A mapping

$$(I^{n-1}, \partial I^{n-1}) \xrightarrow{f} (\Omega(X; A, *), *) \subset (X^I, *)$$

determines by adjointness a mapping

$$(I^{n-1} \times I, \partial I^{n-1} \times I) \xrightarrow{\tilde{f}} (X, *).$$

One sees that $f(\partial I^n) \subset A$ and $\tilde{f}(J^{n-1}) = *$. Hence $\tilde{f}$ represents an element of $[(I^n; \partial I^n, J^{n-1}), (X; A, *)]$. Conversely, a map $g\colon (I^n; \partial I^n, J^{-n}) \to (X; A, *)$ determines a map $\bar{g}\colon I^{n-1} \to X^I$, and one observes that $\bar{g}(\partial I^{n-1}) = *$, and $\bar{g}(I^{n-1}) \in \Omega(X; A, *)$. Thus $\bar{g}$ represents an element of $\pi_n(X, A, *)$.

Homotopies are also preserved by these transformations; the proof of this is left as an exercise. ▮

There is a continuous map

$$d\colon \Omega(X; A, *) \to A, \qquad d(*) = *$$

given by $d(\omega) = \omega(0)$. This induces a homomorphism

$$\pi_n(X, A, *) = \pi_{n-1}(\Omega(X, A, *), *) \xrightarrow{d_*} \pi_{n-1}(A, *),$$

which is usually written as ∂.

Lemma 10.4 Under the correspondence θ of 10.3, ∂ is given by

$$\partial\theta(\{f\}) = \{f\,|_{I^{n-1} \times 0}\}$$

where $f\colon (I^n, \partial I^n, J^{n-1}) \to (X, A, *)$.

Proof This is easy from 10.3 and the definition of d. ▮

Note that $[(I^n, \partial I^n, J^{n-1}), (X, *, *)] = \pi_n(X, *)$.

Lemma 10.5 In the case $A = *$, the 1–1 correspondence $\theta\colon \pi_n(X, *) \to \pi_n(X, *, *)$ is an isomorphism.

Proof The composition

$$\pi_{n-1}(\Omega(X, *), *) \approx \pi_n(X, *) \xrightarrow{\theta} \pi_n(X, *, *) = \pi_{n-1}(\Omega(X, *), *)$$

is the identity. ▮

We now have constructed homomorphisms

$$i_*\colon \pi_n(A, *) \to \pi_n(X, *)$$

$$j_*\colon \pi_n(X, *) \xrightarrow{\theta} \pi_n(X, *, *) \to \pi_n(X, A, *)$$

$$\partial\colon \pi_n(X, A, *) \to \pi_{n-1}(A, *),$$

where $i\colon (A, *) \to (X, *)$ and $j\colon (X, *, *) \to (X, A, *)$ are the inclusions.

Lemma 10.6 Suppose $n \geq 1$. The compositions

$$\pi_n(X, *) \xrightarrow{j_*} \pi_n(X, A, *) \xrightarrow{\partial} \pi_{n-1}(A, *)$$

$$\pi_n(X, A, *) \xrightarrow{\partial} \pi_{n-1}(A, *) \xrightarrow{i_*} \pi_{n-1}(X, *)$$

$$\pi_n(A, *) \xrightarrow{i_*} \pi_n(X, *) \xrightarrow{j_*} \pi_n(X, A, *)$$

are all 0.

Proof Let $f\colon (I^n, \partial I^n) \to (X, *)$. Then $\partial j_*(\{f\})$ is represented by the constant map at $*$. Let $f\colon (I^n, \partial I^n, J^{n-1}) \to (X, A, *)$. Then $i_* \partial(\{f\}) = \{f|_{I^{n-1}\times 0}\}$. f is a homotopy from $f|_{I^{n-1}\times 0}$ to $f|_{I^{n-1}\times 1}$ and $f(\partial I^{n-1} \times I) = *$. Hence $i_* \partial(\{f\}) = \{f|_{I^{n-1}\times 1}\} = 0$ as elements of $\pi_n(X, *)$. To prove the last composition is zero, we invoke a lemma.

Lemma 10.7 Let $f\colon (I^n, \partial I^n, J^{n-1}) \to (X, A, *)$. Then $f \sim *$ in $\pi_n(X, A, *)$ iff there is a map $g\colon I^n \to A$ and a homotopy $H\colon f \sim g$ (rel ∂I^n). (See 5.1.)

Proof Suppose $f \sim *$ in $\pi_n(X, A, *)$ and let

$$K\colon (I^n \times I, \partial I^n \times I, J^{n-1} \times I) \to (X, A, *)$$

be a homotopy with $K(x, 0) = f(x)$, $K(x, 1) = *$, $x \in I^n$. Define H by

$$H(u, s, t) = \begin{cases} K(u, s, 2st), & 0 \leq t \leq \frac{1}{2} \\ K(u, s(2 - 2t), s), & \frac{1}{2} \leq t \leq 1 \end{cases}$$

for $u \in I^{n-1}$. The last two coordinates are those of a point pictured in Fig. 10.1, where s varies linearly from 0 to 1 as the point varies along the line. Thus we have a homotopy as t varies from $K(u, s, 0)$ to $K(u, 0, s)$. Clearly $H(u, s, 0) = f(u, s)$ and $H(u, s, 1) \in A$ while $H(u, s, t) = f(u, s)$ if $(u, s) \in \partial I^n$.

Suppose conversely that such a homotopy H exists. $H\colon f \sim g$, $g(I^n) \subset A$, and $H(x, t) = f(x) = g(x)$ for $x \in \partial I^n$. Then f and g represent the same element of $\pi_n(X; A, *)$. We must show that if $f\colon (I^n, J^{n-1}) \to (A, *)$, $\{f\} = 0$ in $\pi_n(X, A, *)$. Define

$$H\colon (I^n \times I, \partial I^n \times I, J^{n-1} \times I) \to (A, A, *) \subset (X, A, *)$$

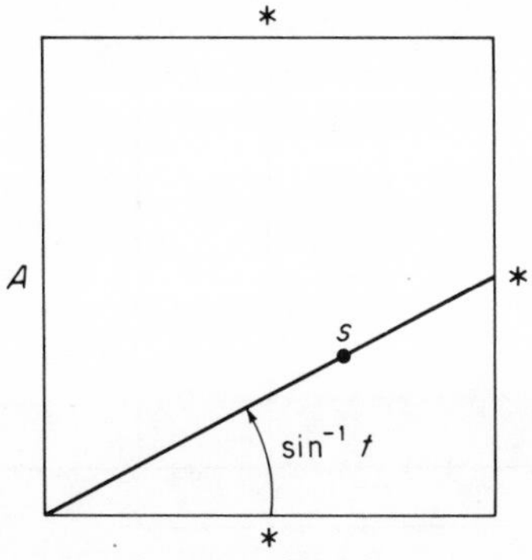

Figure 10.1

by

$$H(u, s, t) = f(u, s + t - st)$$

$$H(u, s, 0) = f(u, s), \quad H(u, s, 1) = *, \quad \text{and} \quad H(J^{n-1} \times I) = *. \quad \blacksquare\blacksquare$$

Lemma 10.8 Let $n \geq 1$. Consider the sequence

$$\pi_n(A, *) \xrightarrow{i_*} \pi_n(X, *) \xrightarrow{j_*} \pi_n(X, A, *) \xrightarrow{\partial} \pi_{n-1}(A, *) \xrightarrow{i_*} \pi_{n-1}(X, *);$$

if $j_* \beta = 0$, $\beta = i_* a$ for some $\alpha \in \pi_n(A, *)$,
if $\partial\gamma = 0$, $\gamma = j_* \beta$ for some $\beta \in \pi_n(X, *)$,
if $i_* \delta = 0$, $\delta = \partial\gamma$ for some $\gamma \in \pi_n(X, A, *)$.

This is a converse to 10.6.

Proof Let $j_*(\beta) = 0$ in $\pi_n(X, A, *)$ and let $f: (I^n, \partial I^n) \to (X, *)$ represent β. Since $j_* \beta = 0$, $f \sim g$ (rel ∂I^n) where $g(I^n) \subset A$. Thus $f \sim i_*(\{g\})$ in $\pi_n(X, *)$ and $\{g\} \in \pi_n(A, *)$.

Let $\partial\gamma = 0$ and represent γ by

$$f: (I^n, \partial I^n, J^{n-1}) \to (X, A, *).$$

Since $f|_{I^{n-1}} \sim 0$ in $\pi_{n-1}(A, *)$, there is a homotopy $H: I^n \to A$ with $H(x, 1) = f(x, 0)$ for $x \in I^{n-1}$, $H(x, 0) = *$, and $H(\partial I^{n-1} \times I) = *$.

Define $K: (I^n \times I, \partial I^n \times I, J^{n-1} \times I) \to (X, A, *)$ by (see Fig. 10.2)

$$K(x, s, t) = \begin{cases} H(x, 1 - t + s(1 + t)), & 0 \leq s \leq t/(1 + t) \\ f(x, s(1 + t) - t), & t/(1 + t) \leq s \leq 1. \end{cases}$$

Now $K(x, s, 0) = f(x, s)$ and $K(x, 0, 1) = *$, so $K(x, s, 1): (I^n, \partial I^n) \to (X, *)$. K is thus a homotopy between f and a map whose class is in $j_* \pi_n(X, *)$.

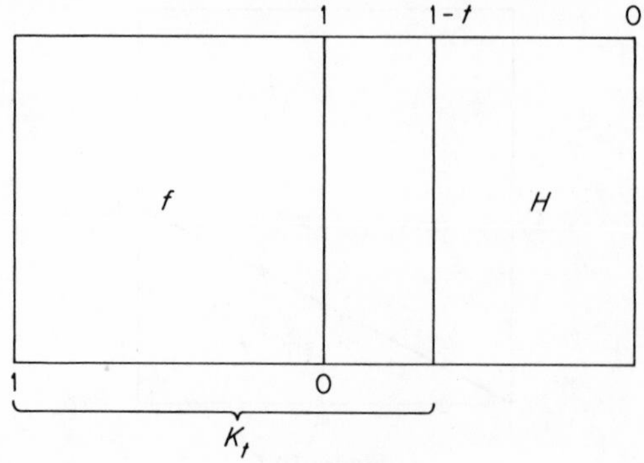

Figure 10.2

Finally, suppose $i_* \delta = 0$ and represent δ by a map

$$f: (I^{n-1}, \partial I^{n-1}) \to (A, *).$$

Let $H: I^n \to X$ be a homotopy $f \sim *$. Then $H(x, 0) = f(x)$, $H(x, 1) = *$, and $H(u, t) = *$ for $u \in \partial I^{n-1}$. Thus $H: (I^n, \partial I^n, J^{n-1}) \to (X, A, *)$ and clearly $\partial H = f$. ∎

We now have a sequence

$$\pi_n(A, *) \to \pi_n(X, *) \to \pi_n(X, A, *) \to \pi_{n-1}(A, *) \to \cdots$$
$$\cdots \to \pi_1(X, A, *) \to \pi_0(A, *) \to \pi_0(X, *)$$

such that at any point, an element maps to zero iff it is in the image of the previous map. In the case $n > 1$ this says that the kernel of a map going out of a group is equal to the image of the previous map coming in. Such a sequence is called an exact sequence. It is often very useful. For example:

Corollary 10.9 $\partial: \pi_n(B^m, S^{m-1}, *) \to \pi_{n-1}(S^{m-1}, *)$ is an isomorphism for $n > 1$.

Proof $\pi_n(B^m, *) = 0$ for all n, since $(B^n, *)$ is contractible. Thus we have the exact sequence

$$0 \to \pi_n(B^m, S^{m-1}, *) \xrightarrow{\partial} \pi_{n-1}(S^{m-1}, *) \to 0$$

Now $\ker \partial = \operatorname{Im} 0 = 0$ and $\operatorname{Im} \partial = \ker 0 = \pi_{n-1}(S^{m-1}, *)$, hence ∂ is 1–1 and onto. ∎

In the case $n = 1$, ∂ is onto. Since $\pi_0(S^{m-1}, *) = *$, if $m > 1$ we have $\pi_1(B^m, S^{m-1}, *) = *$. In the case $m = 1$ we conclude that $\pi_0(S^0, *)$ has two elements. This alone is not enough to conclude that $\pi_1(B^1, S^0, *)$ has two elements.

Exercises

1. Suppose A and X are arcwise connected, $A \subset X$, and $\pi_i(X, A, *) = 0$ for $i \leq n$. Show that the inclusion $i: A \to X$ induces an isomorphism

$$i_*: \pi_i(A, *) \to \pi_i(X, *)$$

for $i < n$ and is onto if $i = n$.

2. Suppose $\pi_2(A, *) = Z$ and $\pi_2(X, A, *) = Z_2$. What are the possibilities for $\pi_2(X, *)$?

3. Prove that $\pi_1(B^1, S^0, *)$ has two elements, where $* = -1$. They are represented by the trivial map $*: I \to * = -1$ and the homeomorphism $t: I \to B^1$ given by $t(s) = 1 - 2s$. Show that $f \sim *$ if $f(0) = -1$ and $f \sim t$ if $f(0) = 1$ for any $f: (I; 0, 1) \to (B^1; S^0, -1)$.

4. Suppose $\pi_n(A, *) = \pi_n(X, A, *) = 0$. Prove $\pi_n(X, *) = 0$.

5.* In the following diagram, three of the four sine waves are exact sequences. Use the diagram and the fact that the composite $\pi_n(B, A) \to \pi_n(X, A) \to \pi_n(X, B)$ is 0 to prove that the fourth is exact.

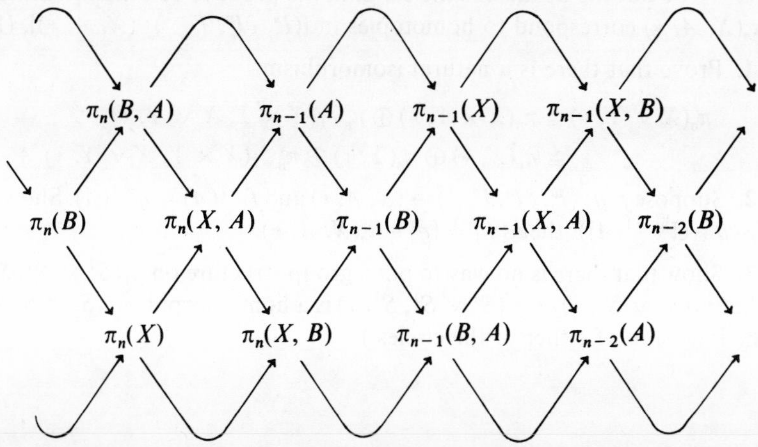

(Here, and in the future, we leave base points out of the notation for homotopy groups when the simplification will not lead to confusion.)

6.* Show that if $f, g: (X, A, *) \to (Y, B, *)$ and $f \sim g$ in $\mathscr{C}^{2*}$,

$$f_* = g_*: \pi_r(X, A, *) \to \pi_r(Y, B, *);$$

thus if $(X, A, *) \simeq (Y, B, *)$, $\pi_r(X, A, *) \simeq \pi_r(Y, B, *)$.

7. Show that the natural map

$$\tau: [(I^n, \partial I^n, J^{n-1}), (X, A, *)] \to [(I^n, \partial I^n, *), (X, A, *)]$$

is a 1–1 correspondence for $* \in J^{n-1} \subset \partial I^n \subset I^n$. (Hint: $(I^n, \partial I^n, J^{n-1}) \simeq (I^n, \partial I^n, *)$.). (16.7)

8.* Define $\partial: \mathcal{C}^{2*} \to \mathcal{C}^*$ by $\partial(X, A, *) = (A, *)$ and if $f: (X, A, *) \to (Y, B, *)$, $\partial(f) = f|_A$. Show that ∂ is a functor and

$$\partial: \pi_n \to \pi_{n-1} \circ \partial$$

is a natural transformation. Consequently a map $f: (X, A, *) \to (Y, B, *)$ induces a commutative "ladder"

$$\begin{array}{ccccccccc} \cdots \longrightarrow & \pi_n(X, *) & \longrightarrow & \pi_n(X, A, *) & \xrightarrow{\partial} & \pi_{n-1}(A, *) & \longrightarrow & \pi_{n-1}(X, *) & \longrightarrow \cdots \\ & \downarrow f_* & & \downarrow f_* & & \downarrow f_* & & \downarrow f_* & \\ \cdots \longrightarrow & \pi_n(Y, *) & \longrightarrow & \pi_n(Y, B, *) & \xrightarrow{\partial} & \pi_{n-1}(B, *) & \longrightarrow & \pi_{n-1}(Y, *) & \longrightarrow \cdots \end{array}$$

9. Use the method of Section 6 to show that if $X \supset A$ and $*_1, *_2 \in A$ are base points connected by a path in X, $\Omega(X; A, *_1) \simeq \Omega(X; A, *_2)$ (the homotopy equivalence is in $\mathcal{C}$). Hence $\pi_n(X, A, *_1) \simeq \pi_n(X, A, *_2)$ is $n \geq 2$.

10. Write out the details for the claim in the proof of 10.3 that homotopies in $\pi_n(X, A, *)$ correspond to homotopies in $[(I^n, \partial I^n, J^{n-1}); (X, A, *)]$. (10.3)

11. Prove that there is a natural isomorphism

$$\begin{aligned} \pi_n(X \vee Y, *) &\cong \pi_n(X \times Y, *) \oplus \pi_{n+1}(X \times Y, X \vee Y, *) \\ &\cong \pi_n(X, *) \oplus \pi_n(Y, *) \oplus \pi_{n+1}(X \times Y, X \vee Y, *). \end{aligned}$$

12. Suppose $f, g: (I^n, \partial I^n, J^{n-1}) \to (X, A, *)$ and $f^{-1}(A) = g^{-1}(A)$. Show that if $f \sim g$ (rel $f^{-1}(A)$), then $\{f\} = \{g\} \in \pi_n(X, A, *)$.

13. Show that there is no way to put a group structure on $\pi_1(S^1 \vee S^1, S^1, *)$ so that $\pi_1(S^1 \vee S^1, *) \to \pi_1(S^1 \vee S^1, S^1, *)$ is a homomorphism. ($S^1 \subset S^1 \vee S^1$ is the inclusion of either of the circles.)

II

Locally Trivial Bundles

The calculation of π_1 was made possible by the covering space construction and the Van Kampen theorem. The calculation of π_n for $n > 1$ is a much more difficult problem, and it is natural to try to generalize these techniques. There is no known generalization of Van Kampen's theorem, but quite an extensive generalization of covering space theory. We shall describe here a generalization of covering spaces called locally trivial bundles. We allow the inverse image of a point to be more complicated than the 0-dimensional discrete spaces that occur with covering spaces, but still demand the same uniformity of the inverse image of various points. This allows one to obtain the same homotopy information. The difficulty lies in the fact that if we attempt to construct locally trivial bundles (or more generally, fibrations), we must use very complicated spaces. The complications in the spaces are in the nature of the problem. There is, however, a wealth of locally trivial bundles that "occur in nature," and we exploit these for homotopy information.

Definition 11.1 A locally trivial bundle with fiber F is a map $\pi\colon E \to B$ such that for all $x_0 \in B$ there is a neighborhood V of x_0 and a homeomorphism

$$\varphi_V\colon V \times F \to \pi^{-1}(V)$$

such that $\pi \circ \varphi_V(x, f) = x$.

The sets V with this property are called the coordinate neighborhoods.

Examples

1. A covering space with connected base is a locally trivial bundle with discrete fiber.

2. $E = F \times B$, $\pi(f, b) = b$. This is called the trivial bundle or product bundle.

3. $\tau(M) \to M$ where M is a differential manifold and $\tau(M)$ is its tangent bundle. (See Section 29 for a definition.)

4. The maps

$$\eta_n: S^{2n+1} \to CP^n, \qquad \nu_n: S^{4n+3} \to HP^n$$

are locally trivial bundles with fibers S^1 and S^3 respectively, as we will now see. The proofs are virtually identical, and we do the case of η_n. Define

$$V_i = \{[\xi_0 | \cdots | \xi_n] \in CP^n \,|\, \xi_i \neq 0\}.$$

Then V_i for $0 \leq i \leq n$ is an open cover of CP^n. We now define a homeomorphism

$$\varphi_i: V_i \times S^1 \to \eta_n^{-1}(V_i)$$

by

$$\varphi_i([\xi_0 | \cdots | \xi_n], \lambda) = \frac{|\xi_i| \lambda}{\sqrt{\Sigma |\xi_j|^2}} (\xi_i^{-1}\xi_0, \ldots, \xi_i^{-1}\xi_k, \ldots, \xi_i^{-1}\xi_n)$$

where $\lambda \in S^1$. This is well defined since if $\xi_j = \lambda \xi_j'$, we have

$$\xi_i^{-1}\xi_j = \xi_i'^{-1}\xi_j', \qquad \text{and} \qquad \frac{|\xi_i'|}{\sqrt{\Sigma |\xi_j'|^2}} = \frac{|\xi_i|}{\sqrt{\Sigma |\xi_j|^2}}.$$

To prove it is a homeomorphism, we describe its inverse

$$\eta_n^{-1}(V_i) = \{(\xi_0, \ldots, \xi_n) | \Sigma |\xi_j|^2 = 1, \xi_i \neq 0\}.$$

Define

$$\psi_i: \eta_n^{-1}(V_i) \to V_i \times S^1$$

by

$$\psi_i(\xi_0, \ldots, \xi_n) = \left([\xi_0 | \cdots | \xi_n], \frac{\xi_i}{|\xi_i|}\right).$$

Clearly $\psi_i \varphi_i = 1$ and $\varphi_i \psi_i = 1$. ∎

We would like to have a path and homotopy lifting property for locally trivial bundles. 7.2 can be stated as follows. Let P be a one-point space. Given a commutative diagram

$$\begin{array}{ccc} P \times 0 & \xrightarrow{a} & \tilde{X} \\ \big\downarrow & & \big\downarrow \\ P \times I & \longrightarrow & X \end{array}$$

there is a unique map $\tilde{p}\colon P \times I \xrightarrow{p} \tilde{X}$ so that

$$\begin{array}{ccc} P \times 0 & \xrightarrow{a} & \tilde{X} \\ \downarrow & \nearrow_{\tilde{p}} & \downarrow \pi \\ P \times I & \xrightarrow[p]{} & X \end{array}$$

commutes.

Definition 11.2 A map $\pi\colon E \to B$ has the homotopy lifting property (HLP) with respect to X if, given a commutative diagram

$$\begin{array}{ccc} X \times 0 & \xrightarrow{f} & E \\ \downarrow & & \downarrow \pi \\ X \times I & \xrightarrow{H} & B \end{array}$$

there is a map $\theta\colon X \times I \to E$ such that

$$\begin{array}{ccc} X \times 0 & \xrightarrow{f} & E \\ \downarrow & \nearrow_{\theta} & \downarrow \pi \\ X \times I & \xrightarrow{H} & B \end{array}$$

commutes.

Definition 11.3 A map $\pi\colon E \to B$ is called a Serre fibering if it has the homotopy lifting property with respect to I^n for each $n \geq 0$.

Every locally trivial bundle is a Serre fibering. In fact

Theorem 11.4 A locally trivial bundle has the homotopy lifting property with respect to any compact Hausdorff space.

Remark The method of proof will be similar to that in 7.2 and 7.3. First we prove a lemma. Recall that $A \subset X$ is called a retract of X if there is a map $r\colon X \to A$ with $r|_A = 1$.

Lemma 11.5 Let $\pi\colon E \to B$ be a locally trivial bundle. Suppose that given any open cover $\{U_\alpha\}$ of $X \times I$ we can find $X \times I = X_n \supset X_{n-1} \supset \cdots \supset X_0 = X \times 0$ with each X_k closed such that

(a) $\overline{X_k - X_{k-1}} \subset U_\alpha$ for some α;

(b) $X_{k-1} \cap \overline{X_k - X_{k-1}}$ is a retract of $\overline{X_k - X_{k-1}}$.

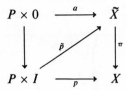

Then π has the homotopy lifting property with respect to X (Fig. 11.1).

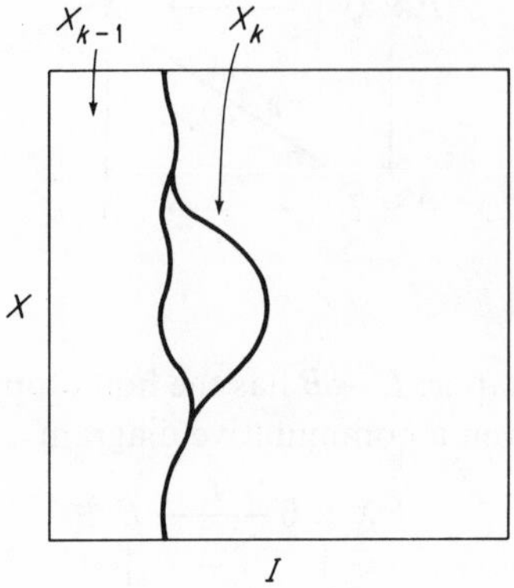

Figure 11.1

Proof Suppose we are given a commutative diagram

$$\begin{array}{ccc} X \times 0 & \xrightarrow{f} & E \\ \downarrow & & \downarrow \pi \\ X \times I & \xrightarrow{H} & B \end{array}$$

Cover $X \times I$ with $\{H^{-1}(V) | V$ is a coordinate neighborhood in $B\}$. Pick a sequence of spaces X_i as above. We will construct by induction $\theta_k: X_k \to E$ with $\pi\theta_k = H|_{X_k}$ and $\theta_k|_{X \times 0} = f$. This is trivial if $k = 0$. Suppose we have defined θ_{k-1}. To define θ_k we first define $\Gamma: \overline{X_k - X_{k-1}} \to E$. Choose a coordinate neighborhood U_α so that $H(\overline{X_k - X_{k-1}}) \subset U_\alpha$. Then define

$$\Gamma(x) = \varphi_\alpha(H(x), \pi_2\, \theta_{k-1}(r(x))),$$

where $r: \overline{X_k - X_{k-1}} \to X_{k-1} \cap \overline{X_k - X_{k-1}}$ is the retraction and $\pi_2: \pi^{-1}(U_\alpha) \equiv U_\alpha \times F \to F$ is the projection. This is continuous and well defined since $\pi\theta_{k-1}(r(x)) \in \overline{X_k - X_{k-1}} \subset U_\alpha$. Clearly $\pi\Gamma = H|_{\overline{X_k - X_{k-1}}}$. Furthermore, $\Gamma|_{\overline{X_k - X_{k-1}} \cap X_{k-1}} = \theta_{k-1}$. Γ and θ_{k-1} thus combine to define a continuous map $\theta_k: X_k = X_{k-1} \cup \overline{X_k - X_{k-1}} \to E$ satisfying $\pi\theta_k = H|_{X_k}$ and $\theta_k|_{X \times 0} = f$. This completes the inductive step. ∎

Proof of 11.4 Suppose X is compact Hausdorff and we are given an open cover $\{U_\alpha\}$ of $X \times I$. Choose a finite refinement $V_\gamma \times (a_\gamma, b_\gamma)$ and sets W_γ so that $\overline{V}_\gamma \subset W_\gamma$ and $\overline{W}_\gamma \times [a_\gamma, b_\gamma] \subset U_\alpha$ for some α. Order the indexing set $\gamma = 1, \ldots, n$, so that $i \leq j$ implies $a_i \leq a_j$. Now if $\beta \leq \gamma$, $W_\beta \times a_\beta$ does not intersect $V_\gamma \times (a_\gamma, b_\gamma)$. Since $\{V_\gamma \times (a_\gamma, b_\gamma)\}$ forms an open cover, $W_\beta \times a_\beta \subset \bigcup_{\gamma<\beta} V_\gamma \times (a_\gamma, b_\gamma)$. We now construct $t_\gamma\colon X \to I$ for $\gamma = 0, \ldots, n$ so that if $X_\gamma = \{(x, t) \mid t \le t_\gamma(x)\}$, $X_\gamma \supset \overline{V}_\beta \times [a_\beta, b_\beta]$ for $\beta \le \gamma$. Let $t_0 = 0$. Suppose we have constructed $t_0, t_1, \ldots, t_\alpha$ as above. Then

$$W_{\alpha+1} \times a_{\alpha+1} \subset \bigcup_{\gamma\le\alpha} V_\gamma \times (a_\gamma, b_\gamma) \subset X_\alpha,$$

hence $t_\alpha|_{W_{\alpha+1}} \ge a_{\alpha+1}$. By Urysohn's lemma choose $u_{\alpha+1}\colon X \to [0, b_{\alpha+1}]$ such that $u_{\alpha+1}(\overline{V}_{\alpha+1}) = b_{\alpha+1}$ and $u_{\alpha+1}(X - W_{\alpha+1}) = 0$. Define $t_{\alpha+1}(x) = \max(t_\alpha(x), u_{\alpha+1}(x))$. Now

$$X_{\alpha+1} \supset \{(x, t) \mid t \le u_{\alpha+1}(x)\} \supset \overline{V}_{\alpha+1} \times [0, b_{\alpha+1}] \supset \overline{V}_{\alpha+1} \times [a_{\alpha+1}, b_{\alpha+1}].$$

Since $X_{\alpha+1} \supset X_\alpha$ it follows that $X_{\alpha+1} \supset \overline{V}_\beta \times [a_\beta, b_\beta]$ for $\beta \le \alpha + 1$. This completes the induction. Now $X_n \supset V_\beta \times (a_\beta, b_\beta)$ for all β so $X_n = X \times I$.

$$\begin{aligned} X_\alpha - X_{\alpha-1} &= \{(x, t) \mid t_{\alpha-1}(x) < t \le t_\alpha(x)\} \\ &= \{(x, t) \mid t_{\alpha-1}(x) < t \le u_\alpha(x)\} \subset W_\alpha \times [a_\alpha, b_\alpha] \end{aligned}$$

since $u_\alpha = 0$ off W_α and $t_{\alpha-1}|_{W_\alpha} \ge a_\alpha$. Hence $\overline{X_\alpha - X_{\alpha-1}} \subset \overline{W}_\alpha \times [a_\alpha, b_\alpha] \subset U_\gamma$ for some γ. This verifies (a). To verify (b), define

$$r\colon \overline{X_\alpha - X_{\alpha-1}} \to \overline{X_\alpha - X_{\alpha-1}} \cap X_{\alpha-1}$$

by $r(x, t) = (x, t_{\alpha-1}(x))$. This clearly belongs to $X_{\alpha-1}$. If $r(x, t) \notin \overline{X_\alpha - X_{\alpha-1}}$, $t_{\alpha-1}(x) < t$. Let

$$t_n = t_{\alpha-1}(x) + \frac{t - t_{\alpha-1}(x)}{2^n}.$$

Then $(x, t_n) \in X_\alpha - X_{\alpha-1}$ and $\lim(x, t_n) = r(x, t)$. Consequently $r(x, t) \in \overline{X_\alpha - X_{\alpha-1}}$. Furthermore, if $(x, t) \in X_{\alpha-1}$, $t_{\alpha-1}(x) \ge t$; thus $t_{\alpha-1}(x) = t$ for $(x, t) \in \overline{X_\alpha - X_{\alpha-1}} \cap X_{\alpha-1}$. ▮

Lemma 11.6 $(I^n, J^{n-1}) \equiv (I^n, I^{n-1} \times 0)$.

Proof We will prove, equivalently, that

$$(B^{n-1} \times I, B^{n-1} \times 0 \cup S^{n-2} \times I) \equiv (B^{n-1} \times I, B^{n-1} \times 1).$$

Let

$$A = \{(x, t) \in B^{n-1} \times I \mid 1 - \|x\| \le \tfrac{2}{3}t\}$$

and

$$B = \{(x, t) \in B^{n-1} \times I \mid 1 - \|x\| \ge \tfrac{2}{3}t\}.$$

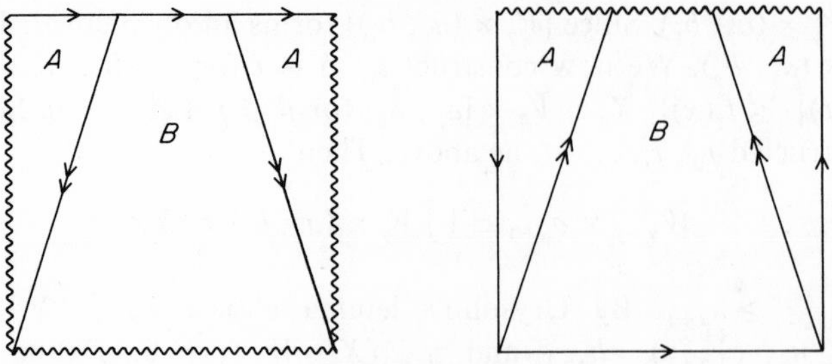

Figure 11.2

Define $\varphi: B^{n-1} \times I \to B^{n-1} \times I$ by (Fig. 11.2)

$$\varphi(x, t) = \begin{cases} \left(\dfrac{2t+1}{3\|x\|} \cdot x, \dfrac{3}{2}\|x\| - \dfrac{1}{2}\right), & (x, t) \in A \\ \left(\dfrac{1+2t}{3-2t} \cdot x, 1 - t\right), & (x, t) \in B. \end{cases}$$

If $(x, t) \in A$,

$$1 \geq \tfrac{3}{2}\|x\| - \tfrac{1}{2} \geq \tfrac{3}{2}(1 - \tfrac{2}{3}t) - \tfrac{1}{2} = 1 - t \geq 0$$

and

$$\left\| \frac{2t+1}{3x} \cdot x \right\| = \frac{2t+1}{3} < 1;$$

if $(x, t) \in B$

$$\left\| \frac{2+4t}{6-4t} \cdot x \right\| = \frac{2+4t}{6-4t} \|x\| \leq \frac{6}{6-4t} \|x\| \leq 1.$$

It is equally clear that if $(x, t) \in A \cap B$,

$$\frac{2t+1}{3\|x\|} = \frac{2+4t}{6-4t} \quad \text{and} \quad 1 - t = \frac{3}{2}\|x\| - \frac{1}{2}.$$

Hence φ is a well-defined and continuous map. Now $\varphi(A) \subset A$ and $\varphi(B) \subset B$, as can easily be seen. It is now easy to verify that $\varphi^2 = 1$, so φ is a homeomorphism. Now

$$\varphi(B^{n-1} \times 0) = \{(x, 1) \mid \|x\| \leq \tfrac{1}{3}\}$$

and

$$\varphi(S^{n-2} \times I) = \{(x, 1) \mid \tfrac{1}{3} \leq \|x\| \leq 1\}.$$

Hence

$$\varphi(B^{n-1} \times 0 \cup S^{n-2} \times I) = B^{n-1} \times 1. \quad \blacksquare$$

This lemma implies that J^{n-1} is a retract of I^n since $I^{n-1} \times 0$ is a retract of I^n.

Proposition 11.7 $\pi: E \to B$ is a Serre fibering iff given a commutative diagram

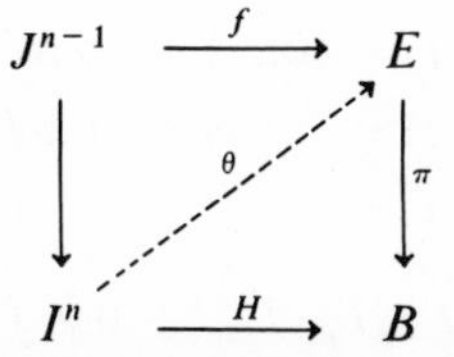

there is a lifting $\theta: I^n \to E$ making the triangles commute.

Proof Suppose $\pi: E \to B$ is a Serre fibering. Consider the composite diagram

$$I^{n-1} \times 0 \xrightarrow{\varphi} J^{n-1} \xrightarrow{f} E$$

$$I^n \xrightarrow{\varphi} I^n \xrightarrow{H} B$$

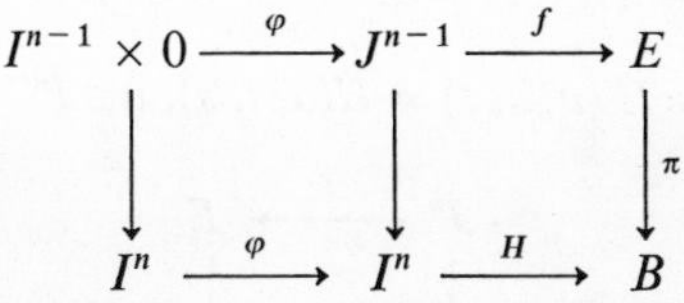

where φ is the homeomorphism from 11.6. Let $\theta: I^n \to E$ be a lifting of this diagram, i.e., $\pi\theta = H\varphi$ and $\theta|_{I^{n-1} \times 0} = f\varphi$. Then $\theta \circ \varphi^{-1}$ is a lifting for the original diagram. The converse is similar. ∎

Having a good lifting property, we would like to obtain some homotopy information. Let $\pi: E \to B$ be a Serre fibering. Choose $* \in E$ and $* = \pi(*) \in B$. Let $F = \pi^{-1}(*)$ be the fiber. Thus π induces a map $\pi: (E, F, *) \to (B, *)$.

Theorem 11.8 If π is a Serre fibering, $\pi_*: \pi_n(E, F, *) \to \pi_n(B, *)$ is a 1–1 correspondence.

Proof We first prove that π_* is onto. Let $f: (I^n, \partial I^n) \to (B, *)$. Let $g: J^{n-1} \to E$ be the trivial map $g(x) = *$ for all x. Then the diagram

$$J^{n-1} \xrightarrow{g} E$$

$$I^n \xrightarrow{f} B$$

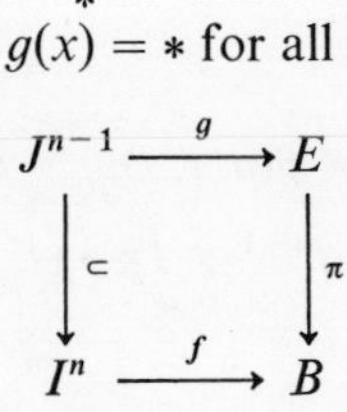

commutes. Hence there is a map $\theta: I^n \to E$ with $\theta(\partial I^n) \subseteq \pi^-(*)^1 = F$ and $\theta(J^{n-1}) = *$. Thus $\{\theta\} \in \pi_n(E, F, *)$ and $\pi_*(\{\theta\}) = \{\pi\theta\} = \{f\}$. Now suppose $\pi_*(\{f\}) = \pi_*(\{g\})$ where $f, g: (I^n, \partial I^n, J^{n-1}) \to (E, F, *)$. Define

$$L: J^n \to E$$

by

$$\begin{aligned} L(x, 1) &= *, && x \in I^n, \\ L(u, 0, t) &= f(u, t), && u \in I^{n-1}, \\ L(u, 1, t) &= g(u, t), && u \in I^{n-1}, \\ L(u, s, t) &= *, && u \in \partial I^{n-1}. \end{aligned}$$

Since $J^n = I^n \times 1 \cup \partial I^{n-1} \times I \times I \cup I^{n-1} \times 0 \times I \cup I^{n-1} \times 1 \times I$, this defines L. Let $H: I^{n+1} \to B$ be a homotopy from πf to πg:

$$\begin{aligned} H(x, 0) &= f(x), && x \in I^n, \\ H(x, 1) &= g(x), && x \in I^n, \\ H(x, t) &= *, && x \in \partial I^n. \end{aligned}$$

Define $H': I^{n+1} \to B$ by $H'(u, s, t) = H(u, t, s)$, $u \in I^{n-1}$; then

$$\begin{array}{ccc} J^n & \xrightarrow{L} & E \\ \downarrow{\scriptstyle\subset} & & \downarrow{\scriptstyle\pi} \\ I^{n+1} & \xrightarrow{H'} & B \end{array}$$

commutes. Hence there is an extension θ of L to I^{n+1}

$$\theta: I^{n+1} \to E,$$

with $\theta(x, 0) \in F$, $x \in I^n$, and of course

$$\begin{aligned} \theta(x, 1) &= *, && x \in I^n, \\ \theta(u, 0, t) &= f(u, t), && u \in I^{n-1}, \\ \theta(u, 1, t) &= g(u, t), && u \in I^{n-1}, \\ \theta(u, s, t) &= *, && u \in \partial I^{n-1}. \end{aligned}$$

Let $\theta': I^{n+1} \to E$ be given by

$$\theta'(u, s, t) = \theta(u, t, s), \qquad u \in I^{n-1};$$

then $\theta': (I^n \times I, I^{n-1} \times 0 \times I, J^{n-1} \times I) \to (E, F, *)$ and θ' is a homotopy between f and g. ∎

Corollary 11.9 If $\pi: E \to B$ is a Serre fibering with fiber F, there is an exact sequence

$$\to \pi_n(F, *) \xrightarrow{i_*} \pi_n(E, *) \xrightarrow{\pi_*} \pi_n(B, *) \xrightarrow{\partial} \pi_{n-1}(F, *) \to \cdots$$
$$\cdots \to \pi_1(B, *) \to \pi_0(F, *) \to \pi_0(E, *) \to \pi_0(B, *). \quad ∎$$

Let us consider the case of the simply connected covering space $\pi\colon \tilde{X} \to X$. Since F is discrete, $\pi_i(F, *) = 0$ for $i > 0$. Hence $\pi_i(\tilde{X}) \approx \pi_i(X)$ for $i > 1$ and

$$0 \to \pi_1(X) \xrightarrow{\partial} \pi_0(F) \to 0$$

is exact.

Corollary 11.10 $\pi_i(S^1, *) = 0$ for $i > 1$. ▮

We can also get homotopy information from the projective space fiberings. We need, first, a lemma.

Lemma 11.11 Suppose

$$0 \to A \xrightarrow{f} B \xrightarrow{g} C \to 0$$

is exact and there is a homomorphism $\gamma\colon C \to B$ with $g\gamma = 1$. Then $B \cong A \oplus C$.

Proof Define $I\colon A \oplus C \to B$ by $I(a, c) = f(a) + \gamma(c)$. I is clearly a homomorphism since f and g are homomorphisms. Suppose $I(a, c) = 0$; then $gI(a, c) = g(f(a)) + g(\gamma(c)) = g(\gamma(c)) = c$. Hence $c = 0$. Thus $f(a) = 0$, but this means $a = 0$ since we have exactness at A. We now show that I is onto. Let $b \in B$. $g(b - \gamma(g(b))) = 0$. By exactness, there is $a \in A$ with $f(a) = b - \gamma(g(b))$; thus $b = I(a, g(b))$ so I is onto. ▮

Short exact sequences with $B \cong A \oplus C$ are called split exact sequences and are said to split.

Proposition 11.12

(a) $\pi_m(RP^n, *) \cong \pi_m(S^n)$, $m > 1$;
(b) $\pi_m(CP^n, *) \cong \pi_m(S^{2n+1}, *)$, $m > 2$;
(c) $\pi_2(CP^n, *) \cong \pi_2(S^{2n+1}, *) \oplus Z$;
(d) $\pi_m(HP^n, *) \cong \pi_m(S^{4n+3}, *) \oplus \pi_{m-1}(S^3, *)$.

Proof (a) follows from the remark after 11.9. (b) follows from the exact sequence

$$\pi_m(S^1) \to \pi_m(S^{2n+1}) \to \pi_m(CP^n) \to \pi_{m-1}(S^1) \to \pi_{m-1}(S^{2n+1}),$$

using 11.10. To prove (c), consider the exact sequence

$$0 \to \pi_2(S^{2n+1}) \to \pi_2(CP^n) \to Z \to 0.$$

This decomposes into a direct sum by 11.11 since one can define $\gamma\colon Z \to \pi_2(CP^n)$ by $\gamma(n) = nx$ where $x \in \pi_2(CP^n)$ is any element with $\partial(x) = 1$.

To prove (d) we construct a homomorphism $\gamma\colon \pi_{m-1}(S^3) \to \pi_m(HP^n)$ such that $\partial\gamma = 1$ and apply 11.11 again. The existence of γ implies that ∂ is onto and $(v_n)_*$ is 1–1.

Define $F\colon (B^4, S^3) \to (S^{4n+3}, S^3)$ by

$$F(q) = (q, \sqrt{1 - q\bar{q}}, 0, \ldots, 0) \in H^{n+1};$$

clearly $\|F(q)\| = 1$ and $F|_{S^3}$ is a homeomorphism from S^3 to $v_n^{-1}(*)$ $(* = [1\,|\,0\,|\cdots|\,0])$. Let γ be the composite

$$\pi_{m-1}(S^3) \cong \pi_m(B^4, S^3, *) \xrightarrow{F_*} \pi_m(S^{4n+3}, S^3) \xrightarrow{(v_n)_*} \pi_m(HP^n).$$

Then $\partial\gamma = 1$ by the definition of ∂. ∎

We conclude this section with some general results about the construction of fiberings.

Definition 11.13 A Hurewicz fibering is a map $\pi\colon E \to B$ that has the homotopy lifting property with respect to any space X.

This definition and the following constructions can be interpreted in $\mathfrak{T}$ or in $\mathfrak{CG}$ as desired.

We now consider a construction that will "turn a map into a Hurewicz fibering." Let $f\colon X \to Y$. Define

$$E_f = \{(x, \omega) \in X \times Y^I \,|\, \omega(0) = f(x)\} \subset X \times Y^I$$

with the induced topology. There are maps

$$\pi\colon E_f \to Y, \qquad v\colon E_f \to X, \qquad \imath\colon X \to E_f,$$

given by

$$\pi(x, \omega) = \omega(1), \qquad v(x, \omega) = x, \qquad \imath(x) = (x, e_{f(x)}),$$

where $e_{f(x)}$ is the constant path at $f(x)$.

The diagram

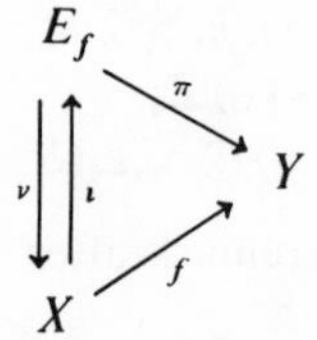

commutes, i.e., $\pi\imath = f$ and $v\imath = 1$; moreover

Proposition 11.14

(a) $\imath v \simeq 1$ so $X \simeq E_f$;
(b) π is a Hurewicz fibering;
(c) $F_f = \{(x, \omega) \in X \times Y^I \,|\, \omega(0) = f(x),\ \omega(1) = *\}$ is the fiber.

The map $j_f\colon F_f \to X$ given by $j_f = v|_{F_f}$ is given by $j_f(x, \omega) = x$.

Proof (a) As a homotopy $H\colon E_f \times I \to E_f$ we take $H(x, \omega, t) = (x, \omega_t)$ where $\omega_t(s) = \omega(st)$.

To prove (b) consider a commutative diagram

$$\begin{array}{ccc} A \times 0 & \xrightarrow{h} & E_f \\ \downarrow & & \downarrow{\scriptstyle \pi} \\ A \times I & \xrightarrow{H} & Y \end{array}$$

Let $h(a) = (h_1(a), h_2(a))$. Define $\theta\colon A \times I \to E_f \subset X \times Y^I$ by $\theta(a, t) = (\theta_1(a, t), \theta_2(a, t))$ where $\theta_1(a, t) = h_1(a)$ and

$$\theta_2(a, t)(s) = \begin{cases} h_2(a)(s(1+t)), & 0 \le s \le 1/(1+t) \\ H(a, (1+t)s - 1), & 1/(1+t) \le s \le 1. \end{cases}$$

$\theta(a, t) \in E_f$ since $f\theta_1(a, t) = \theta_2(a, t)(0)$ follows from $fh_1(a) = h_2(a)(0)$. θ is continuous since θ_1 is clearly continuous and θ_2 is continuous, by 8.17. Finally $\theta(a, 0)(s) = h(s)$ and $\pi\theta(a, t) = H(a, t)$.

(c) is trivial. ∎

Seminar Problems

1. Let $\pi\colon E \to B$ be a Hurewicz fibering and $*_1$, $*_2$ two points of B. Show that $\pi^{-1}(*_1) \simeq \pi^{-1}(*_2)$.

2. Let $\pi\colon E \to B$ be a locally trivial bundle and B a paracompact Hausdorff space (e.g., a metric space). Show that π is a Hurewicz fibering. (See [64, 2.7.14].)

Exercises

1. Let $\pi\colon \tilde{X} \to X$ be a covering space, and suppose X is arcwise connected. Then there is a 1–1 correspondence $\gamma\colon F \to \pi_0(F, *)$ such that the diagram

$$\begin{array}{ccc} \pi_1(X) & \xrightarrow{\partial} & \pi_0(F) \\ & {\scriptstyle \phi}\searrow & \downarrow{\scriptstyle \gamma} \\ & & F \end{array}$$

commutes, where ϕ is the map defined in Section 7.

2. Let $\pi\colon \Omega(X; A, B) \to A \times B$ be given by $\pi(\omega) = (\omega(0), \omega(1))$. Show that π is a Hurewicz fibering.

3. Show that the composite of two Hurewicz fiberings is a Hurewicz fibering. Is this true for locally trivial bundles?

4. Let $\iota: A \to X$ be an inclusion. Show $F_\iota \equiv \Omega(X, A, *)$. Using this, derive the exact sequence of the inclusion (10.6 and 10.8) from the exact sequence for a Serre fibering (11.9).

5. Given a locally trivial bundle $\pi: E \to B$ and $f: X \to B$, define $f^*(E) = \{(x, e) \in X \times E \mid f(x) = \pi(e)\}$. Define $f^*(\pi): f^*(E) \to X$ by $f^*(\pi)(x, e) = x$ and $f_E: f^*(E) \to E$ by $f_E(x, e) = e$. Show that these maps are continuous and $f^*(\pi)$ is a locally trivial bundle with the same fiber as π. $f^*(\pi)$ is called the induced bundle. (Section 29)

6. Let $X = A \cup B$, $* = A \cap B$. Let

$$\iota: \Omega(B; A \cap B, *) \to \Omega(X; A, *)$$

be the inclusion. Show that F_ι is homeomorphic with (Fig. 11.3)

$$\{\omega \in X^{I \times I} \mid \omega(0, t) \in A, \omega(s, 0) \in B, \omega(s, 1) = \omega(1, t) = *\}.$$

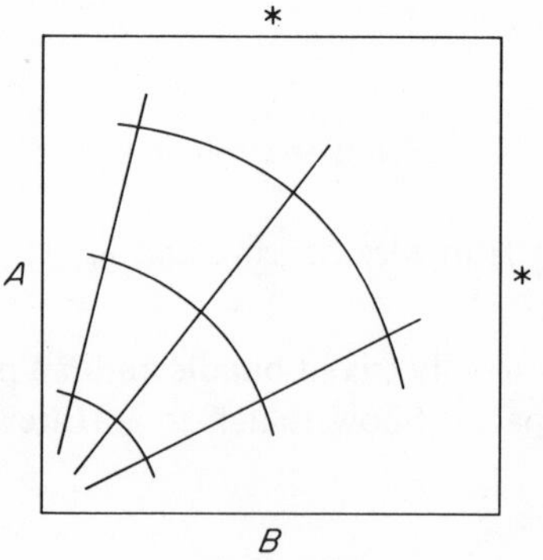

Figure 11.3

Show that this is homeomorphic to $\Omega(\Omega(X; A, B), *)$. Define $\pi_n(X; A, B) = \pi_{n-1}(\Omega(X; A, B), *)$ and conclude that there is an exact sequence

$$\cdots \to \pi_n(X, A, *) \to \pi_n(X; A, B) \xrightarrow{\partial} \pi_{n-1}(B, A \cap B, *) \to \pi_{n-1}(X, A, *) \to \cdots$$

This is called the triad exact sequence, and $(X; A, B)$ is called a triad.

7. In the notation of 11.11 observe that since g is onto, for every $c \in C$, there exists $b \in B$ with $g(b) = c$. Pick such a b and define $\gamma(c) = b$. Can we apply 11.11 and conclude that for any exact sequence

$$0 \to A \to B \to C \to 0,$$

$B \cong A \oplus C$? Explain.

8. $\pi: E \to B$ is said to be a Hurewicz fibering in $\mathcal{C}^*$ if 11.2 is satisfied for all $X \in \mathcal{C}^*$ when f and H are in $\mathcal{C}^*$ and $\theta \in \mathcal{C}^*$. Show that $\pi: E_f \to Y$ is a Hurewicz fibering in $\mathcal{C}^*$ if f is in $\mathcal{C}^*$.

9. Show that if $F \to E \to B$ is a Hurewicz fibering in $\mathcal{C}^*$ (see Exercise 8), the sequence $[X, F] \to [X, E] \to [X, B]$ of base point preserving homotopy classes is exact in $\mathcal{S}^*$. (One sometimes expresses this by saying that the sequence $F \to E \to B$ is exact in $\mathcal{C}^*$.) (17.19)

10. Let $f: X \to Y$ be a map in $\mathcal{C}^*$. Show that there is an exact sequence in $\mathcal{S}^*$

$$\begin{array}{l} [Z, \Omega X] \xrightarrow{(\Omega f)_*} [Z, \Omega Y] \to [Z, F_f] \xrightarrow{(j_f)_*} [Z, X] \xrightarrow{f_*} [Z, Y] \\ \qquad\quad \wr\wr \qquad\qquad\qquad \wr\wr \\ \cdots \to [SZ, F_f] \xrightarrow{(j_f)_*} [SZ, X] \xrightarrow{f_*} [SZ, Y] \end{array}$$

11. Let $0 \to A \xrightarrow{f} B \xrightarrow{g} C \to 0$ be an exact sequence of R-modules. Show that the following are equivalent:

(a) There exists $\gamma: C \to B$ such that $g\gamma = 1$.
(b) There exists $\varepsilon: B \to A$ such that $\varepsilon f = 1$.
(c) There exists an isomorphism $\phi: B \to A \oplus C$ such that $\pi_2 \phi = g$ and $\phi f = \iota_1$ (i.e., the sequence splits).

12. Show that the maps $R^{n+1} - \{0\} \to RP^n$, $C^{n+1} - \{0\} \to CP^n$, and $H^{n+1} - \{0\} \to HP^n$ are locally trivial bundles with fibers R, C, and H respectively. (Section 29)

13. Let $O(n)$ be the group of orthogonal $n \times n$ real matrices and $\pi: O(n) \to S^{n-1}$ be given by $\pi(A) = A(0, \ldots, 0, 1)$. Show that π is a locally trivial bundle with fiber $O(n-1)$ (Hint: Let $U = S^{n-1} - (0, \ldots, 0, 1)$ and $V = S^{n-1} - (0, \ldots, 0, -1)$. Define a map $\alpha: V \to O(n)$ by letting $\alpha(x)$ be the rotation along a great circle which takes $(0, \ldots, 0, 1)$ to x. Define coordinate functions $\phi: U \times O(n-1) \to \pi^{-1}(U)$ by $\phi(x, A) = -\alpha(-x)\binom{A\,0}{0\,1}$ and $\psi: V \times O(n-1) \to \pi^{-1}(V)$ by $\psi(x, A) = \alpha(x)\binom{A\,0}{0\,1}$). (30.7)

14. Show that $\pi_n(X, *)$ can be made into a module over $Z(\pi_1(X, *))$ for $n > 1$ by defining ξ^σ for $\xi \in \pi_n(X, *)$ and $\sigma \in \pi_1(X, *)$ to be $\Pi_* \cdot \gamma \cdot (T_\sigma)_* \Pi_*^{-1}(\xi)$ where T_σ is from Exercise 12, Section 7, γ is from Exercise 7, Section 9, and $\Pi: \tilde{X} \to X$ is a simply connected covering space. (For any group π, $Z(\pi)$ is the integral group ring. As an abelian group it is the free group generated by the elements of π. Multiplication is linear and determined on the generators by the multiplication in π.) Show that $\pi_n(X \vee S^1) \cong Z(Z) \otimes \pi_n(X)$ as modules if X is simply connected.

12

Simplicial Complexes and Linearity

Although we have several theorems relating homotopy groups, we have as yet made little progress in computing them. In this section we will discuss linearity, and in the next section use linear approximations to make some computations.

Recall that in Section 1 we defined the standard n-dimensional simplex

$$\Delta^n = \{(x_1, \ldots, x_{n+1}) \in R^{n+1} \mid 0 \leq x_i \leq 1, \sum x_i = 1\}.$$

More generally, given $n + 1$ points $v_0, \ldots, v_n \in R^n$ that are affine independent (i.e., the equations $\sum t_i v_i = 0$ and $\sum t_i = 0$ imply $t_i = 0$ for all i), one can define the n-simplex spanned by them

$$(v_0 v_1 \cdots v_n) = \left\{ x \in R^k \,\middle|\, x = \sum_{i=0}^{n} t_i v_i,\ 0 \leq t_i \leq 1, \sum t_i = 1 \right\}.$$

The v_i are called the vertices of $(v_0 v_1 \cdots v_n)$, and the t_i are called the barycentric coordinates of x.

Lemma 12.1 $(v_0 v_1 \cdots v_n) \equiv \Delta^n$.

Proof Let $\theta\colon \Delta^n \to (v_0 \cdots v_n)$ be given by

$$\theta(x_1, \ldots, x_{n+1}) = \sum_{i=0}^{n} x_{i+1} v_i.$$

This is continuous and onto. To show that it is 1–1, suppose $\theta(x_1, \ldots, x_{n+1}) = \theta(x_1', \ldots, x_{n+1}')$. Then we have equations

$$0 = \sum_{i=0}^{n} (x_{i+1} - x_{i+1}')v_i, \qquad \text{and} \qquad 0 = \sum_{i=0}^{n} (x_{i+1} - x_{i+1}')$$

Hence $x_i = x_i'$ for all i. By the compactness of Δ^n, θ is a homeomorphism. ∎

The barycentric coordinates of a simplex give it a "linear structure." We consider spaces that are the union of simplices in an appropriate sense. Observe that if $(v_0 v_1 \cdots v_n)$ is a simplex, so is $(v_{i_0} v_{i_1} \cdots v_{i_k})$ where $(i_0, \ldots, i_k)$ is a subset of $(0, \ldots, n)$. Such a simplex is called a k-face of $(v_0 \cdots v_n)$. (Note that a simplex is a face of itself, and the empty set is a face of every simplex.) Write $\tau \prec \sigma$ if τ is a face of σ.

Definition 12.2 A geometric finite simplicial complex (or complex)[8] K is a finite collection of simplices in R^m for some fixed m such that the intersection of two simplices is a face of each and each face of a simplex in K is a simplex in K. Write $|K|$ for the underlying space (i.e., the union of all simplices). A space that is homeomorphic to $|K|$ for some K is called a polyhedron. By a triangulation of a space X we mean a complex K with $|K| \equiv X$.

Examples

1. We give some examples of complexes in the plane pictorially in Fig. 12.1.

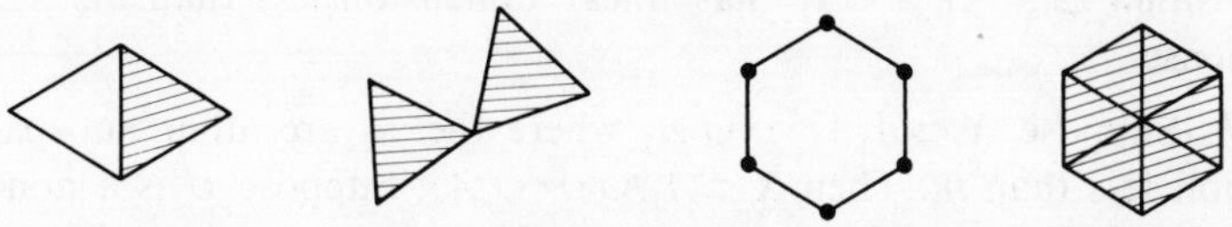

Figure 12.1

2. A simplex $(v_0 v_1 \cdots v_n)$ is a complex.

3. Define $\partial(v_0 v_1 \cdots v_n)$ as the union of the simplices:

$$(v_1 v_2 \cdots v_n), \quad (v_0 v_2 \cdots v_n), \quad (v_0 v_1 v_3 \cdots v_n), \quad \ldots, \quad (v_0 v_1 \cdots v_{n-1}).$$

This is a simplicial complex (homeomorphic to S^{n-1}).

A subcollection L of the collection of simplices in a complex K is called a subcomplex if each face of a simplex in L is a simplex in L (i.e., L itself is a complex).

If K is complex, we write K^n for the subcomplex whose simplices are the k-faces of simplices of K for $k \leq n$. K^n is called the n-skeleton of K. Thus for some r, $K = K^r$, and we have

$$\{\varnothing\} = K^{-1} \subset K^0 \subset K^1 \subset \cdots \subset K^r = K.$$

If in addition $K^{r-1} \neq K$, we say that K is r-dimensional and write $\dim K = r$.

[8] We use the word *complex* for abbreviation, although we will define more general complexes in Section 14.

Definition 12.3 Let K be a complex and $f\colon |K| \to R^n$. f is said to be linear if for each simplex $(v_0 v_1 \cdots v_s)$ of K and for each $x \in (v_0 v_1 \cdots v_s)$, we have

$$f(x) = f\left(\sum_{i=0}^{s} t_i v_i\right) = \sum_{i=0}^{s} t_i f(v_i).$$

Thus a linear map is completely determined by its value on the vertices of K. By induction on the number of simplices, one easily checks that a linear map is continuous.

Recall that a coset of a k-dimensional subspace of a vector space is called a k-dimensional affine subspace or an affine k-plane.

Definition 12.4 A set $X \subset R^m$ is said to have linear dimension $\leq k$ if there exist affine k-planes $A_1, \ldots, A_s$ with $X \subset A_1 \cup \cdots \cup A_s$. lin dim $\varnothing = -1$ where $\varnothing$ is the empty set.

Hence a k-dimensional complex has linear dimension k; for if $x \in (v_0 \cdots v_n)$,

$$x = \sum t_i(v_i - v_0) + v_0 \in V(v_1 - v_0, \ldots, v_n - v_0) + v_0$$

where $V(v_1 - v_0, \ldots, v_n - v_0)$ is the subspace spanned by $v_1 - v_0, \ldots, v_n - v_0$.

Proposition 12.5 If $X \subset R^m$ has linear dimension less than m, X is nowhere dense.

Proof Suppose $X \subset A_1 \cup \cdots \cup A_s$ where the A_i are affine subspaces of dimension less than m. Then $\bar{X} \subset A_1 \cup \cdots \cup A_s$. Suppose U is a nonempty open set and $U \subset \bar{X}$. Choose i so that $U \not\subset A_1 \cup \cdots \cup A_{i-1}$. Then $\varnothing \neq U - A_1 - \cdots - A_{i-1} \subset A_s$. Let $x \in U - A_1 - \cdots - A_{i-1}$. $A_s - x$ is a subspace of R^m of dimension less than m, so there exists a sequence $x_i \to 0$ with $x_i \notin A_s - x$. It follows that $x + x_i \to x$. Since $U - A_1 - \cdots - A_{i-1}$ is open, we must have $x + x_n \in U - A_1 - \cdots - A_{i-1} \subset A_s$ for some n, which is a contradiction. Thus $\bar{X}$ does not contain a nonempty open set. ∎

A point not in X will be said to be in "general position."

Proposition 12.6 Suppose K is a complex, $f\colon K \to R^n$ is a linear map, and $X \subset K$. Then lin dim $X \geq$ lim dim $f(X)$.

Proof Suppose $X \subset A_1 \cup \cdots \cup A_s$ with lin dim $X = \dim A_j$ for some j and $K = \bigcup \sigma_i$. $f|_{A_j \cap \sigma_i}$ extends to a linear map $\tilde{f}\colon A_j \to R^n$. Hence lin dim $f(A_j \cap \sigma_i) \leq \dim A_j$. Consequently, lin dim $f(A_j \cap K) \leq \dim A_j$. Now

$$\begin{aligned} \text{lin dim } f(X) &\leq \text{lin dim } f\left(\bigcup_{j=1}^{s} A_j \cap K\right) = \text{lin dim} \bigcup_{j=1}^{s} f(A_j \cap K) \\ &\leq \max_j \text{lin dim } f(A_j \cap K) \\ &\leq \max_j \dim A_j = \text{lin dim } X. \quad ∎ \end{aligned}$$

Let A and B be sets in R^m. Define

$$A * B = \{x \mid x = ta + (1-t)b,\ a \in A,\ b \in B\}.$$

This is called the join of A and B and is thought of as the set of all points on line segments from A to B. Define $A * \varnothing = A$ where $\varnothing$ is the empty set.

Example $(v_0 \cdots v_n) * v_{n+1} = (v_0 \cdots v_{n+1})$ if $v_0, \ldots, v_{n+1}$ are affine independent.

Lemma 12.7 Let $X \subset R^m$ and $b \in R^m$. Then lin dim $X * b \leq$ lin dim $X + 1$.

Proof If $X \subset A_1 \cup \cdots \cup A_s$, $X * b \subset A_1 * b \cup \cdots \cup A_s * b$. Let $A_s = V_s + v_s$. Then

$$A_s * b \subset \{\text{subspace spanned by the elements of } V_s \text{ and } b - v_s\} + v_s. \quad ∎$$

Definition 12.8 $X \subset R^m$ is said to be convex if $x, y \in X \Rightarrow tx + (1-t)y \in X$, when $0 \leq t \leq 1$.

Lemma 12.9 If A is an affine subspace and σ is a simplex with $A \cap \sigma \neq A \cap \partial\sigma$, we have

$$(A \cap \partial\sigma) * b = A \cap \sigma \qquad \text{for any} \qquad b \in A \cap \sigma - A \cap \partial\sigma.$$

Proof Since $A \cap \sigma$ is convex and contains both b and $A \cap \partial\sigma$, $A \cap \sigma \supset (A \cap \partial\sigma) * b$. Let $x \in A \cap \sigma$. Then x and b determine a line that must intersect $\partial\sigma$ in two points p_1 and p_2 (Exercise 1). Suppose the points occur in the order p_1, b, x, p_2. Then $x \in p_2 * b \subset (A \cap \partial\sigma) * b$, hence $A \cap \sigma \subset (A \cap \partial\sigma) * b$. ∎

Note that the lemma is true if $A \cap \partial\sigma = \varnothing$.

Proposition 12.10 Let K be an m-dimensional complex and $f\colon K \to R^n$ be linear. Then for all $\varepsilon > 0$ there is a point $a \in R^n$ with $\|a\| < \varepsilon$ such that lin dim$f^{-1}(a) \leq m - n$. ∎

In other words, if a is in general position, lin dim $f^{-1}(a) \leq m - n$.

Proof By 12.6, lin dim $f(K^{n-1}) < n$. Hence by 12.5 we may choose a with $\|a\| < \varepsilon$ such that $a \notin f(K^{n-1})$. We shall prove by induction that lin dim $f^{-1}(a) \cap K^s \leq s - n$ for $s \geq n - 1$. By choice of a, this is valid for $s = n - 1$. Supposing it to be true for $s = k - 1$, let σ be a k-simplex. Then $\sigma \cap f^{-1}(a) = \sigma \cap A$ for some affine subspace A. By induction, lin dim $\partial\sigma \cap A =$ lin dim $\partial\sigma \cap f^{-1}(a) \leq k - n - 1$. By 12.7 and 12.9, lin dim $\sigma \cap A \leq k - n$, so lin dim $K^k \cap f^{-1}(a) \leq k - n$. ∎

We will eventually approximate maps by linear maps. In order to make the approximation accurate, the domain space will have to be triangulated with small simplices. We now discuss a method of triangulating any polyhedron with simplices that are arbitrarily small. This will be achieved by interating a process called barycentric subdivision.

Given a simplex $\sigma = (v_0 \cdots v_n)$, by the barycenter of σ we mean the point $b(\sigma) = (n+1)^{-1}\ (v_0 + \cdots + v_n)$. This is the center of gravity of the vertices in the usual sense.

Definition 12.11 Let K be a complex. A barycentric subdivision of K is a complex K' such that

(a) the vertices of K' are the barycenters of simplices of K;
(b) the simplices of K' are the simplices $(b(\sigma_0) \cdots b(\sigma_m))$ where $\sigma_i < \sigma_{i+1}$ and $\sigma_i \neq \sigma_{i+1}$.

Such a complex is clearly unique if it exists, for we have specified the simplices of K' in the definition.

The barycentric subdivision of a 1-simplex and a 2-simplex are pictured in Fig. 12.2.

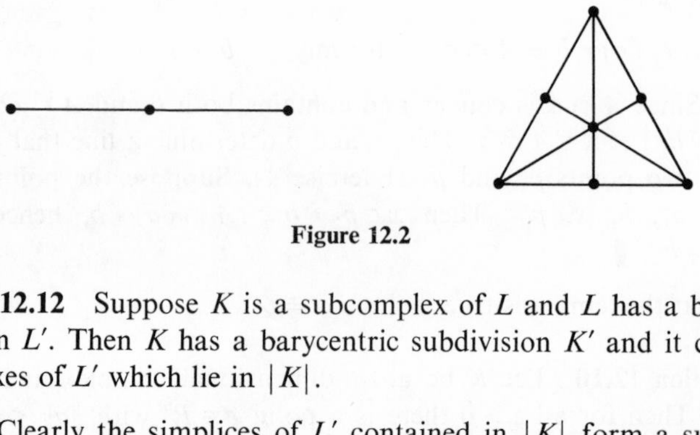

Figure 12.2

Lemma 12.12 Suppose K is a subcomplex of L and L has a barycentric subdivision L'. Then K has a barycentric subdivision K' and it consists of all simplexes of L' which lie in $|K|$.

Proof Clearly the simplices of L' contained in $|K|$ form a subcomplex of L'. Conditions (a) and (b) are immediate since if σ_i are simplices of K and $\sigma_0 < \cdots < \sigma_m$, $(b(\sigma_0) \cdots b(\sigma_m)) \subset \sigma_m \subset |K|$. ∎

Lemma 12.13 If a barycentric subdivision K' of K exists, $|K| = |K'|$.

Proof If $\tau = (b(\sigma_0) \cdots b(\sigma_m))$ is a simplex of K', $\tau \subset \sigma_m \subset |K|$. Hence $|K'| \subset |K|$. Let $x \in \sigma \subset |K|$. Order the vertices of $\sigma = (v_0 \cdots v_n)$ so that if $x = \sum t_i v_i$, $t_0 \geq t_1 \geq \cdots \geq t_n$. Let $\sigma_i = (v_0 \cdots v_i)$. Then $x = \sum s_i b(\sigma_i)$ where $s_i = (i+1)(t_i - t_{i+1})$. Now $s_i \geq 0$ and $\sum s_i = \sum t_i = 1$. Hence $x \in (b(\sigma_0) \cdots b(\sigma_m)) \in K'$ and thus $|K| = |K'|$. ∎

Lemma 12.14 Let σ be a simplex and suppose $\partial\sigma$ has a barycentric subdivision. Then σ has a barycentric subdivision.

Proof Let $\sigma = (v_0 \cdots v_n)$ and $b = (n+1)^{-1}(v_0 + \cdots + v_n)$ be the barycenter of σ. Let $\tau = (b_0 \cdots b_s)$ be a simplex of $(\partial\sigma)'$ and define $\tau b = (b_0 \cdots b_s\, b)$. We must show that this is a simplex. Suppose $\sum t_i b_i + tb = 0$, $\sum t_i + t = 0$, and assume $t \neq 0$. τ is contained in some face σ_j of σ, so $b_i = \sum_{k \neq j} \lambda_{i,k} v_k$ with $\sum_{k\neq j} \lambda_{i,k} = 1$. Hence

$$b = -\frac{1}{t}\sum t_i b_i = -\frac{1}{t}\sum_{k\neq j} t_i \lambda_{i,k} v_k = -\sum_{k\neq j} s_k v_k$$

where $\sum_{k\neq j} s_k = (-1/t)\sum_{k\neq j} t_i \lambda_{i,k} = -1$.

Thus

$$0 = \frac{1}{n+1}(v_0 + \cdots + v_n) + \sum_{k\neq j} s_k v_k = \sum r_k v_k$$

with $\sum r_k = 0$. Since σ is a simplex, $r_k = 0$ for all k, but $r_j = (n+1)^{-1}$ so we have a contradiction. Thus $t = 0$. Hence $\sum t_i b_i = 0$ and $\sum t_i = 0$. Since τ is a simplex, we must have $t_i = 0$ for each i. Consequently τb is a simplex. The vertices of σ' are the vertices of $(\partial\sigma)'$ and b. The simplices of σ' are the simplices b and τb for each simplex τ in $(\partial\sigma)'$. To show that σ' is a complex we need only show that the intersection of two simplices is a face of each. But $\tau b \cap \tau' = \tau \cap \tau'$ and $\tau b \cap \tau' b = (\tau \cap \tau')b$ so σ' is indeed a complex. ∎

Let K and L be complexes in R^m. We will write $K \cap L$ for the set of simplices in both K and L and $K \cup L$ for the set of simplices in either K or L. $K \cap L$ is a subcomplex of K and L, but it is not true in general that $K \cup L$ is a complex.

Lemma 12.15 If $|K \cap L| = |K| \cap |L|$, $K \cup L$ is a complex.

Proof We must show that if $\sigma \in K$ and $\tau \in L$, $\sigma \cap \tau$ is a face of both σ and τ. Suppose A and B are subcomplexes of K and L respectively. We claim that $|A \cap B| = |A| \cap |B| \cap |K \cap L|$. Clearly $|A \cap B| \subset |A| \cap |B| \cap |K \cap L|$. Suppose $x \in \sigma \in A$, $x \in \tau \in B$, and $x \in \rho \in K \cap L$. Since τ and ρ are simplices in L, $\tau \cap \rho < \rho$. Similarly $\sigma \cap \rho < \sigma$, hence $\sigma \cap \tau \cap \rho = (\sigma \cap \rho) \cap (\tau \cap \rho) < \sigma \cap \rho < \sigma$. By symmetry, $\sigma \cap \tau \cap \rho < \tau$. Hence $x \in \sigma \cap \tau \cap \rho \in A \cap B$ and $x \in |A \cap B|$. Now let $A(B)$ be the complex consisting of $\sigma(\tau)$ and all of its faces. Then

$$\begin{aligned}|A \cap B| &= \sigma \cap \tau \cap |K \cap L| = \sigma \cap \tau \cap |K| \cap |L| \\ &= (\sigma \cap |K|) \cap (\tau \cap |L|) = \sigma \cap \tau.\end{aligned}$$

Since $A \cap B$ is a subcomplex of A and B, $\sigma \cap \tau$ is a subcomplex of σ and τ. Since $\sigma \cap \tau$ is convex, it is a face of σ and τ. Thus $K \cup L$ is a complex. ∎

Theorem 12.16 Every complex has a barycentric subdivision.

Proof We will use double induction, first on the dimension of the complex and second on the number of simplices. Suppose the result is true for every complex of dimension less than n and every complex of dimension n with fewer than s n-simplices for $s \geq 1$. Let K be an n-dimensional complex with s n-simplices and let σ be an n-simplex. Then there is a subcomplex L of K with fewer than s simplices such that $K = L \cup \sigma$. By induction L' is a complex and $(\partial\sigma)'$ is a complex so by 12.14, σ' is a complex. Clearly $K' = L' \cup \sigma'$, so it is sufficient to show that $|K' \cap \sigma'| = |K'| \cap |\sigma'|$ by 12.15. Clearly $K' \cap \sigma' = (K \cap \sigma)'$ which is a complex by 12.12. By 12.13 and Exercise 10,

$$|K' \cap \sigma'| = |(K \cap \sigma)'| = |K \cap \sigma| = |K| \cap |\sigma| = |K'| \cap |\sigma'|. \quad ∎$$

Definition 12.17 If $K \subset R^m$ is a complex, the mesh of K (written $\mu(K)$) is the maximal diameter of the simplices.

Proposition 12.18 If $\dim K = n$,

$$\mu(K') \leq \frac{n}{n+1}\,\mu(K).$$

Proof By Exercise 6, we need only measure the length of the 1-simplices of K'. Let (b^0, b^1) be such a 1-simplex with $b^0 < b^1$. Then b^1 is the barycenter of a k-simplex $\tau = (v_0 \cdots v_k)$ in K. Now given vectors $w_0, \ldots, w_n$, ω and numbers t_i with $\sum t_i = 1$ we have

$$\|\omega - \textstyle\sum t_i w_i\| = \|\sum t_i(\omega - w_i)\| \leq \sum t_i \|\omega - w_i\|.$$

Since $b^0 \in (v_0 \cdots v_k)$,

$$\|b^1 - b^0\| = \|b^1 - \textstyle\sum t_i v_i\| \leq \sum t_i \|b^1 - v_i\|.$$

Applying the inequality again we have

$$\|v_i - b^1\| = \left\| v_i - \frac{v_0 + \cdots + v_k}{k+1} \right\| \leq \frac{1}{k+1} \sum_j \|v_i - v_j\| \leq \frac{k}{k+1}\,\mu(\tau).$$

Hence $\|b^1 - b^0\| \leq [k/(k+1)]\mu(\tau)$. Since $k \leq n$ implies that $k/(k+1) \leq n/(n+1)$, we have $\|b^1 - b^0\| \leq [n/(n+1)]\mu(K)$. ∎

Corollary 12.19 If X is a polyhedron and $\varepsilon > 0$, there is a complex K with $|K| = X$ and $\mu(K) < \varepsilon$.

Proof Let $K = X^{(r)}$. If $n = \dim X$, $\mu(K) \leq [n/(n+1)]^r \mu(X)$, and by choosing r large this can be made less than ε. ▮

We finish this section with a famous approximation theorem of the type we will consider in the next section.

Definition 12.20 Let K and L be complexes. A map $f\colon K \to L$ will be called simplicial if $f(v)$ is a vertex of L for each vertex v of K, and f is linear. If $g\colon K \to L$ is an arbitrary map, f will be called a simplicial approximation to g if for every $x \in |K|$ and every simplex $\sigma \in L$, $g(x) \in \sigma$ implies $f(x) \in \sigma$.

Proposition 12.21 Let $K_0 \subset K$ be a subcomplex and suppose $g\colon K \to L$ is a map such that $g|_{K_0}$ is simplicial. Let f be a simplicial approximation to g. Then $f \sim g$ (rel K_0).

Proof Let $L \subset R^m$ and define $f_t : K \to R^m$ by

$$f_t(x) = tf(x) + (1 - t)g(x).$$

Since both $f(x)$ and $g(x)$ belong to some simplex σ, the sum lies in σ as well. Hence $f_t(K) \subset L$. Clearly $f_t : f \sim g$ (rel K_0). ▮

Theorem 12.22 (*Simplicial Approximation Theorem*) Let $g\colon K \to L$ be continuous. Then for some $r > 0$ there is a simplicial approximation $f\colon K^{(r)} \to L$ of g.

Proof For each vertex $v \in L$, let $\mathrm{st}(v) = \bigcup_{v \in \sigma} \mathrm{Int}\, \sigma$; $\mathrm{st}(v)$ is open, contains v, and the sets $\{\mathrm{st}(v)\}$ for various v forms an open cover of L (see Exercise 11). Choose an ε-number for the covering $g^{-1}(\mathrm{st}(v))$ of $|K|$ and subdivide $|K|$ to a complex $K^{(r)}$ with $\mu(K^{(r)}) < \varepsilon/2$. Now for each vertex $v \in K^{(r)}$, diam st $v < \varepsilon$, so st $v \subset g^{-1}(\mathrm{st}\ w)$ for some vertex $w \in L$. Choose such a vertex w for each vertex $v \in K^{(r)}$ and write $f(v) = w$. This defines f on vertices, and we extend $f\colon K^{(r)} \to R^m$ linearly. We must show that $f(K^{(r)}) \subset L$. If σ is a simplex in $K^{(r)}$, $b(\sigma) \in \mathrm{st}\ v$ for each vertex v of σ. Hence $g(b(\sigma)) \in \mathrm{st}\, f(v)$. Thus $\bigcap_{v \in \sigma} \mathrm{st}\, f(v)$ is nonempty and hence the vertices $f(v)$ form a simplex, and $f(K^{(r)}) \subset L$. Clearly f is a simplicial approximation to g. ▮

Corollary 12.23 $\pi_r(S^n) = 0$ for $r < n$.

Proof Let $f\colon (S, *) \to (S^n, *)$ and assume $*$ is a vertex of triangulations of S^r and S^n. Choose a simplicial approximation $g\colon (S^r)^{(s)} \to S^n$ to f by 12.22. Since $g|_*$ is simplicial, $g \sim f$ (rel $*$) by 12.21. Now lin dim $g(S^r) \leq r$, so $g(S^r) \neq S^n$. Choosing $a \in S^n - g(S^r)$, we conclude that $g \sim *$ (rel $*$) since $\pi_r(S^n - a, *) = 0$. ▮

Exercises

1. If σ is a simplex and l is a line containing an interior point of σ, then $l \cap \sigma$ is a closed interval and $l \cap \partial\sigma$ consists of the two end points. (12.9)

2. Show that if $f\colon |\sigma| \to R^n$ is linear and 1–1, and σ is a simplex, $f(|\sigma|)$ is a simplex.

3. Show that a simplex is convex.

4. The convex hull of a set $X \subset R^n$ is the intersection of all convex sets that contain X. Show that the convex hull of a set X is convex.

5. Show that the diameter of X is the same as the diameter of its convex hull.

6. Show that a simplex is the convex hull of its vertices and that its diameter is the length of the longest 1-face. (12.18)

7. Construct complexes homeomorphic to:

(a) the annulus $\{(x, y) \in R^2 \mid 1 \leq x^2 + y^2 \leq 4\}$;
(b) the torus $S^1 \times S^1$ (use (a));
(c) the projective plane RP^2 (hard).

8. Show that lin dim $A * B \leq$ lim dim $A +$ lim dim $B + 1$.

9. Show that the intersection of two subcomplexes is a subcomplex of each, and that a subcomplex of a subcomplex is a subcomplex.

10. Show that if K_0 and K_1 are subcomplexes of L, $|K_0 \cap K_1| = |K_0| \cap |K_1|$. (12.16)

11. For a simplex σ write Int $\sigma = \sigma - \partial\sigma$. (This notion of interior is not in general the same as the interior of σ as a subspace of R^m. The notions coincide when $m = \dim \sigma$ by 26.30.) Show that for any complex K, $K = \bigcup \text{Int}\, \sigma_\alpha$ where the union is disjoint and runs over all simplices of the complex. (12.22)

13

Calculating Homotopy Groups: The Blakers–Massey Theorem

In the previous section we proved that maps between simplicial complexes are homotopic, after suitable subdivision, to simplicial maps. The basic concept involved here is that of approximating an arbitrary, and possibly highly pathological map (e.g., a map from S^1 onto S^2) by a less wild map. In certain contexts it is possible to approximate by differentiable maps—in others by linear maps. Often one approximates an arbitrary map by a differentiable map and then this differentiable map by a linear map (its derivative). We shall pursue the techniques of linear approximation in order to prove some fairly strong deformation theorems. The first such theorem will be a direct generalization of 12.23 to a relative n-cell.

Definition 13.1 A relative n-cell is a pair (Y, X) such that Y is the quotient space of $X \amalg B^n$ (see Exercise 7, Section 8) under the identifications given by a map $\alpha\colon S^{n-1} \to X$, namely, $x \sim \alpha(x)$ for $x \in S^{n-1} \subset B^n$. This identifies $S^{n-1} \subset B^n$ with a subset of X (Fig. 13.1). One often writes $X \cup_\alpha e^n$ for this space and $e^n = Y - X$. α is called the attaching map. If $X \in \mathcal{CG}$, $X \cup_\alpha e^n \in \mathcal{CG}$ (see Exercise 6).

Observe that if $X = *$,

$$(Y, X) = (X \cup_\alpha e^n, X) \equiv (B^n/S^{n-1}, *)$$
$$\equiv (S^n, *),$$

hence a relative n-cell is a generalization of a sphere with base point.

Lemma 13.2 If $p \in e^n$, then $(X \cup_\alpha e^n - p, X, *) \simeq (X, X, *)$. Consequently, $\pi_i(X \cup_\alpha e^n - p, X, *) = 0$.

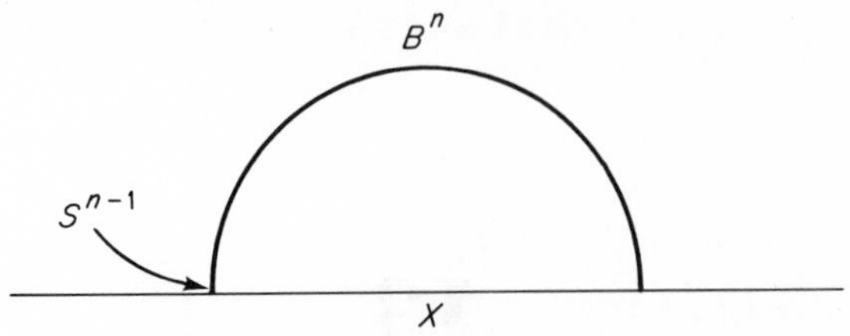

Figure 13.1

Remark In general, $\pi_i(X \cup_\alpha e^n, X, *) \neq 0$, e.g., if $X = *$, $n = 1$, and $i = 1$.

Proof Let ι: $(X, X, *) \to (X \cup_\alpha e^n - p, X, *)$ be the inclusion. We define a strong deformation retraction of $X \cup_\alpha e^n - p$ onto X. This is sufficient by Exercise 7, Section 6.

$$H: (X \cup e^n - p) \times I \to X \cup e^n - p.$$

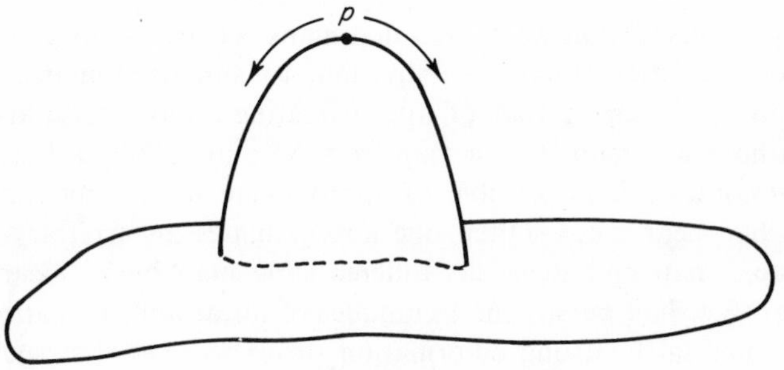

Figure 13.2

Let χ: $(B^n, S^{n-1}) \to (X \cup_\alpha e^n, X)$ be the "characteristic map" of the relative n-cell. This means χ is the restriction to B^n of the quotient map $X \amalg B^n \to X \cup_\alpha e^n$. Hence $\chi|_{S^{n-1}} = \alpha$. Let $q = \chi^{-1}(p)$. We will produce a homotopy

$$K: ((B^n - q) \times I, S^{n-1} \times I) \to (B^n - q, S^{n-1})$$

satisfying $K(y, 0) = y$, $K(y, 1) \in S^{n-1}$ and $K(y, t) = y$ if $y \in S^{n-1}$. Now if f: $X \to Y$ is a quotient map, and B is an open subset of Y, $f|_{f^{-1}(B)}: f^{-1}(B) \to B$ is a quotient map. Hence $X \cup e^n - p$ has the quotient topology on $X \amalg (B^n - q)$ and thus $(X \cup_\alpha e^n - p) \times I$ has the quotient topology on $X \times I \amalg (B^n - q) \times I$. We define H by using K on $(B^n - q) \times I$ and the constant homotopy on X. (See Fig. 13.2.)

To construct K, first apply 2.4 with $f(x) = q$ to produce γ: $B^n - q \to S^{n-1}$ with $\gamma|_{S^{n-1}} = 1$. Now define $K(y, t) = (1 - t)y + t\gamma(y)$. $\|K(y, t)\| \leq (1 - t) + t \leq 1$. If $y \in S^{n-1}$, $\gamma(y) = y$ hence $K(y, t) = y$. Clearly $K(y, 0) = y$

and $K(y, 1) = \gamma(y) \in S^{n-1}$. Finally, $K(y, t) \neq q$, for otherwise

$$q = (1 - t)y + t\rho y + t(1 - \rho)q$$

where $\gamma(y) = \rho y + (1 - \rho)q$ (see 2.4); hence

$$(1 - t + t\rho)q = (1 - t + t\rho)y;$$

since $y \neq q$, $1 = t(1 - \rho)$. This is impossible since $1 - \rho \leq 0$ and $t \geq 0$. ∎

Let us write $B^n(\rho) = \{(x_1, \ldots, x_n) \in R^n \,|\, \|x\| \leq \rho\}$. If χ: $(B^n, S^{n-1}) \to (X \cup_\alpha e^n, X)$ is the characteristic map, write $e^n(\rho) = \chi(B^n(\rho))$ if $\rho < 1$. A map φ: $K \to e^n$ will be called linear if for some ρ, $\varphi(K) \subset e^n(\rho)$ and $\chi^{-1}\varphi$: $K \to B^n(\rho) \subset R^n$ is linear.

Lemma 13.3 Let f: $I^r \to X \cup_\alpha e^n$. Then there are complexes N and N' in I^r satisfying (see Fig. 13.3):

(a) $f^{-1}(e^n(\frac{3}{4})) \subset N \subset \text{Int } N' \subset N' \subset f^{-1}(e^n(\frac{7}{8}))$;
(b) if $\sigma \in N'$, diam $\chi^{-1}(f(\sigma)) < \frac{1}{16}$.

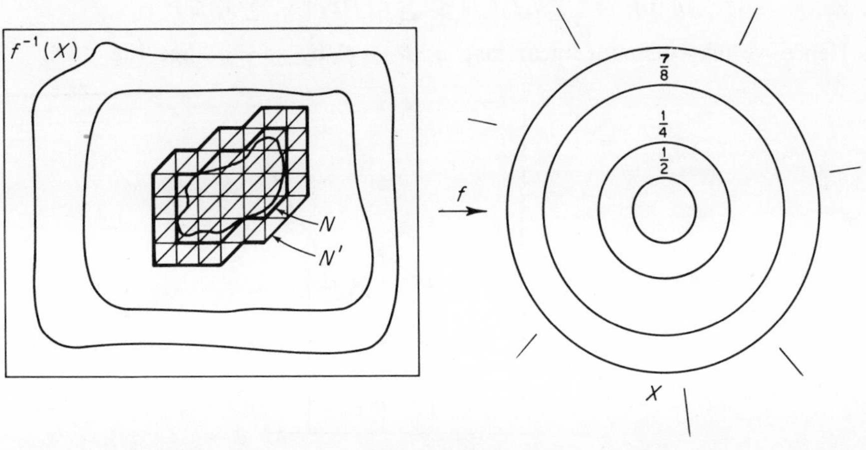

Figure 13.3

Proof By 12.16 we subdivide $I^r \equiv \Delta^r$ into a simplicial complex with mesh less than the distance between $f^{-1}(e^n(\frac{7}{8}))$ and $f^{-1}(X)$. (We take the distance between two sets to be ∞ if either of them is empty.) Thus any closed simplex meeting $f^{-1}(e^n(\frac{7}{8}))$ is mapped into e^n. We now transfer the usual metric on $B^n - S^{n-1}$ to e^n via the homeomorphism $\chi|_{B^n - S^{n-1}}$. The sets $f^{-1}(U)$ for $U \subset e^n$, U open, and diam $U < \frac{1}{16}$ form an open cover of $\{\sigma \,|\, \sigma \text{ meets } f^{-1}(e^n(\frac{7}{8}))\}$, which is compact. Choose an ε-number and further subdivide the cube I^r so that the mesh is less than this ε-number. Then if σ meets $f^{-1}(e^n(\frac{7}{8}))$, diam $f(\sigma) < \frac{1}{16}$. Let N be the union of all closed simplices of I^r meeting

$f^{-1}(e^n(\frac{3}{4}))$. Then $f(N) \subset e^n(\frac{3}{4} + \frac{1}{16}) = e^n(\frac{13}{16})$. Let N' be the union of all closed simplices meeting N. Similarly, $f(N') \subset e^n(\frac{13}{16} + \frac{1}{16}) = e^n(\frac{7}{8})$. We only need to show that $N \subset \text{Int } N'$. Suppose $\{x_i\} \to x \in N$. We must show that eventually $x_i \in N'$. Each simplex that contains infinitely many x_i must intersect N in a set that contains x, so the simplex is contained in N'. There are only a finite number of x_i which do not belong to such a simplex, so for i large enough, $x_i \in N'$. ∎

Lemma 13.4 Let $f\colon I^r \to X \cup_\alpha e^n$. Then there is an open set $U \subset I^r$ and a homotopy $h_t : \bar{U} \to e^n$ (rel ∂U) such that:

(1) $h_0 = f|_{\bar{U}}$;
(2) there is a complex $N \subset U$ such that $h_1|_N$ is linear;
(3) $\text{Int } N \supset h_1^{-1}(e^n(\frac{1}{2}))$.

Proof Choose N and N' via 13.3. Define a linear map $g'\colon N' \to E^n$ by $\chi g'(v) = f(v)$ for v a vertex N'. If $u \in N'$,

$$\|g'(u)\| = \|\sum t_i f(v_i)\| \leq \sum t_i \|f(v_i)\| \leq \sum t_i \leq 1.$$

Hence we may define a linear map $g\colon N' \to e^n$ by $\chi g = g'$. See Fig. 13.4.

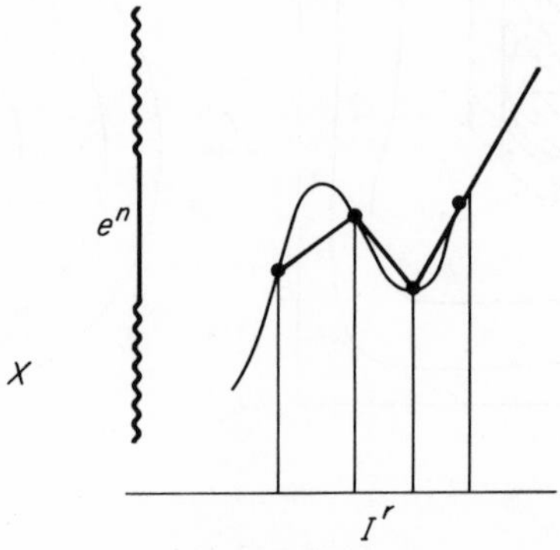

Figure 13.4

Let $U = \text{Int } N'$. Then $\bar{U} \subset N'$. Since $N \subset \text{Int } N'$, $N \cap \partial N' = \varnothing$. Choose $\varphi\colon N' \to I$ such that $\varphi(N) = 1$ and $\varphi(\partial N') = 0$. Define $h_t : \bar{U} \to e^n$ by

$$\chi^{-1}h_t(u) = \{(1-t) + t(1-\varphi(u))\}\chi^{-1}f(u) + t\varphi(u)\chi^{-1}g(u).$$

This is well defined since $(1-t) + t(1-\varphi(u)) + t\varphi(u) = 1$. Since $\partial U \subset \partial N'$, $h_t|_{\partial U} = f|_{\partial U}$. Conditions (1) and (2) are clear. To prove (3), suppose $\sigma = (v_0, \ldots, v_s)$ is a simplex of N. Then $h_1(v_i) = f(v_i)$. Thus diam $h_1(\sigma) =$ diam $(h_1(v_0), \ldots, h_1(v_s)) =$ diam $(f(v_0), \ldots, f(v_s)) \leq$ diam $f(\sigma) < \frac{1}{16}$. For

$x \in \sigma$, we have $|f(x) - f(v)| < \frac{1}{16}$ and $|h_1(x) - h_1(v)| < \frac{1}{16}$. Thus $|f(x) - h_1(x)| < \frac{1}{8}$. Consequently if $h_1(x) \in e^n(\frac{1}{2})$, $f(x) \in e^n(\frac{5}{8}) \subset \operatorname{Int} e^n(\frac{3}{4})$. Thus $x \in \operatorname{Int} f^{-1}(e^n(\frac{3}{4})) \subset \operatorname{Int} N$. ∎

Corollary 13.5 $\pi_r(X \cup_\alpha e^n, X, *) = 0$ for $r < n$.

Proof Let $f: (I^r, \partial I^r, J^{r-1}) \to (X \cup_\alpha e^n, X, *)$ represent a homotopy element, and apply 13.4. $\overline{U} \cap \partial I^r = \varnothing$ since $f(\overline{U}) \subset e^n$ and $f(\partial I^r) \subset X$. Hence h_t can be extended to a homotopy $H_t : I^r \to X \cup_\alpha e^n$ (rel ∂I^r) by

$$H_t(u) = \begin{cases} h_t(u), & u \in \overline{U} \\ f(u), & u \in I^r - U. \end{cases}$$

Now $H_0 = f$, so $\{f\} = \{H_1\}$. Choose $p \in e^n(\frac{1}{2})$ such that $p \notin H_1(N)$ by 12.5 and 12.6. $H_1^{-1}(p) \subset H_1^{-1}(e^n(\frac{1}{2})) \subset N$, so $H_1^{-1}(p) = \varnothing$. Thus H_1 is in the image of the homomorphism

$$\pi_r(X \cup_\alpha e^n - p, X, *) \to \pi_r(X \cup_\alpha e^n, X, *).$$

By 13.2, $\{H_1\} = 0$. ∎

Our second linearization theorem is more complicated. We set up our notation as follows. Let $X_2 = A \cup_\alpha e^n$ and $X_1 = A \cup_\beta e^m$. Let $X = X_1 \cup X_2$ so that $A = X_1 \cap X_2$. Let $\iota: (X_1, A) \to (X, X_2)$ be the inclusion. Note that $X_1 - A \equiv X - X_2 \equiv R^m$. See Fig. 13.5.

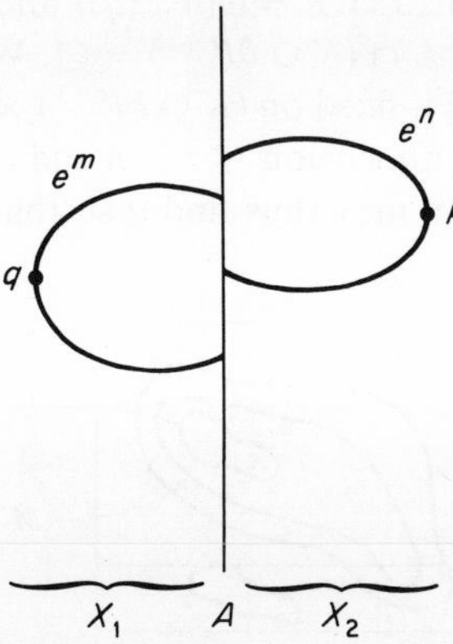

Figure 13.5

Theorem 13.6 (*Blakers–Massey Theorem—First Form*) $\iota_*: \pi_r(X_1, A, *) \to \pi_r(X, X_2, *)$ is an isomorphism if $r < m + n - 2$ and is onto if $r = m + n - 2$. [13].

We will use several lemmas in the proof.[9]

[9] The proof we give is due to J. M. Boardman.

Lemma 13.7 Let $p \in e^n$ and $q \in e^m$. Then we have a commutative diagram in which the vertical maps are isomorphisms and all maps are induced by inclusions

$$\begin{array}{ccc} \pi_r(X_1, A, *) & \xrightarrow{\iota_*} & \pi_r(X, X_2, *) \\ \downarrow \cong & & \downarrow \cong \\ \pi_r(X - p, X - p - q, *) & \longrightarrow & \pi_r(X, X - q, *) \end{array}$$

Proof By 13.2, $\pi_r(X - q, X_2, *) = 0$. By Exercise 5, Section 10 applied to the triple $X \supset X - q \supset X_2$, we have an exact sequence

$$\pi_r(X - q, X_2, *) \to \pi_r(X, X_2, *) \to \pi_r(X, X - q, *) \xrightarrow{\partial} \pi_{r-1}(X - q, X_2, *).$$

Since the end groups are zero, the middle map is an isomorphism. Similarly, $\pi_r(X_1, A, *) \to \pi_r(X - p, A, *)$ is an isomorphism since $\pi_r(X - p, X_1, *) = 0$ for all r, and $\pi_r(X - p, A, *) \to \pi_r(X - p, X - p - q, *)$ is an isomorphism since one can easily argue (as in 13.2) that $\pi_r(X - p - q, A, *) = 0$ for all r. Since all maps involved are inclusions, the diagram commutes. ∎

The idea behind the proof of 13.6 is this: We must push maps and homotopies off some point $p \in e^n$, and throughout the motion the image of ∂I^r must miss some fixed point $q \in e^m$, and J^{r-1} must stay at $*$. Let $\omega: I^r = I^{r-1} \times I \to I^{r-1}$ be the projection. Let $K = \omega(h^{-1}(q))$ and $L = \omega(h^{-1}(p))$. We wish to choose p and q so that $L \cap (K \cup \partial I^{r-1}) = \emptyset$. We can then deform I^r into $I^{r-1} \times 1$ on $L \times I$ keeping it fixed on $(K \cup \partial I^{r-1}) \times I$. If p is chosen in general position, $h^{-1}(p)$ will have dimension $\leq r - n$ and thus $L \times I$ will have dimension $\leq r - n + 1 < m$. We may thus find q so that $h^{-1}(q)$ is separated from $L \times I$. See Fig. 13.6.

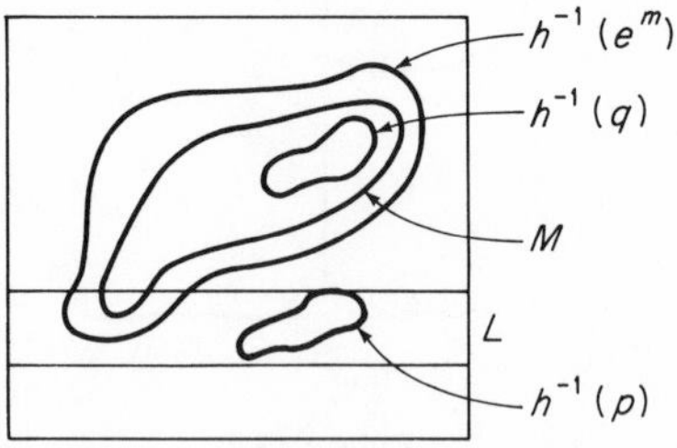

Figure 13.6

Lemma 13.8 Suppose $h: I^r \to X$ and there are complexes M and N in I^r with $h|_M : M \to e^m$ and $h|_N : N \to e^n$ linear. Suppose $r \leq m + n - 2$ and

(a) Int $N \supset h^{-1}(e^n(\frac{1}{2}))$
(b) Int $M \supset h^{-1}(e^m(\frac{1}{2}))$

Then there exist points $p \in e^n$ and $q \in e^m$ such that if $\omega: I^r = I^{r-1} \times I \to I^{r-1}$ is the projection, $K = \omega(h^{-1}(q))$ and $L = \omega(h^{-1}(p))$ are disjoint.

Proof Apply 12.10 to $h|_N : N \to e^n$. Choose $p \in e^n(\frac{1}{2})$ such that lin dim $(h|_N)^{-1}(p) \leq r - n$. By (a), $h^{-1}(p) = (h|_N)^{-1}(p)$. By 12.6 lin dim $L \leq r - n$. Thus lin dim $L \times I \leq r - n + 1 < m$. By 12.6 again lin dim $h(M \cap (L \times I)) < m$ so we may choose $q \in e^m(\frac{1}{2})$ such that $q \notin h(M \cap (L \times I))$ by 12.5. By (b), $h^{-1}(q) \subset M$, so $h^{-1}(q) \cap L \times I = \varnothing$. Now if $u \in K \cap L$, $(u, t) \in h^{-1}(q) \cap (L \times I) = \varnothing$ for some t. Consequently $K \cap L = \varnothing$. ∎

Lemma 13.9 Suppose $h: I^r \to X$, $p \in e^n$ and $q \in e^m$ are chosen in accordance with 13.8. Suppose further that

(a) $h(\partial I^{r-1} \times I) \subset X_1$;
(b) $h(I^{r-1} \times 0) \subset X_2$;
(c) $h(I^{r-1} \times 1) = *$.

Then there is a homotopy $H_t : I^r \to X$ such that

(1) $H_0 = h$;
(2) $p \notin H_1(I^r)$;
(3) $q \notin H_t(I^{r-1} \times 0)$;
(4) $H_t|_{J^{r-1}} = h|_{J^{r-1}}$.

Proof Suppose $u \in L \cap \partial I^{r-1}$. Then $h^{-1}(p)$ meets $\partial I^{r-1} \times I$. Thus $p \in h(\partial I^{r-1} \times I) \subset X_1$ by (a). This is impossible since $p \in e^n = X - X_1$. Hence $L \cap (K \cup \partial I^{r-1}) = \varnothing$. Let $\varphi: I^{r-1} \to I$ satisfy $\varphi(L) = 1$ and $\varphi(K \cup \partial I^{r-1}) = 0$. Define $H_t : I^r \to X$ by

$$H_t(x_1, \ldots, x_r) = h(x_1, \ldots, x_{r-1}, 1 - (1 - x_r)(1 - t\varphi(x_1, \ldots, x_{r-1}))).$$

Clearly $H_0 = h$. If $p = H_1(x_1, \ldots, x_r)$, $(x_1, \ldots, x_{r-1}) \in L$. Thus

$$p = H_1(x_1, \ldots, x_r) = h(x_1, \ldots, x_{r-1}, 1) \subset h(I^{r-1} \times 1) = *$$

by (c). This proves (2). If $q = H_t(x_1, \ldots, x_{r-1}, 0)$, $(x_1, \ldots, x_{r-1}) \in K$. Hence

$$q = h(x_1, \ldots, x_{r-1}, 0)\, h \subset (I^{r-1} \times 0) \subset X_2$$

by (b). This proves (3). Finally, since $J^{r-1} = \partial I^{r-1} \times I \cup I^{r-1} \times 1$, $H_t|_{J^{r-1}} = h|_{J^{r-1}}$, for either $\varphi = 0$ or $(1 - x_r) = 0$ on this set. ∎

Proof of 13.6 We first show that if $r \leq m + n - 2$, ι_* is onto. Let $f: (I^r, \partial I^r, J^{r-1}) \to (X, X_2, *)$. Choose open sets U and $V \subset I^r$ and deformations

$$h_t : \overline{U} \to e^m, \qquad k_t : \overline{V} \to e^n$$

according to 13.4. We construct $f_t : (I^r, \partial I^r, J^{r-1}) \to (X, X_2, *)$ by

$$f_t(u) = \begin{cases} h_t(u), & u \in \bar{U} \\ k_t(u), & u \in \bar{V} \\ f(u), & u \in X - U - V. \end{cases}$$

This is well defined since h_t and k_t are homotopies relative to ∂U and ∂V. $f_t(\partial I^r) \subset X_2$ and $f_t|_{J^{r-1}} = *$ since the deformations h_t and k_t remain within the cells e^m and e^n. Thus $\{f\} = \{f_0\} = \{f_1\}$. We apply 13.8 and 13.9 to f_1. Conditions (a), (b), and (c) of 13.9 hold since

$$\partial I^{r-1} \times I \cup I^{r-1} \times 1 = J^{r-1} \quad \text{and} \quad I^{r-1} \times 0 \subset \partial I^r.$$

$H_t : (I^r, \partial I^r, J^{r-1}) \to (X, X - q, *)$ since $h(J^{r-1}) = *$. Thus $f_1 = H_0 \sim H_1$ in $\pi_r(X, X - q, *)$ but $H_1(I^r) \subset X - p$. Thus $\{f_1\}$ is in the image of the homomorphism

$$\pi_r(X - p, X - p - q, *) \to \pi_r(X, X - q, *)$$

and ι_* is onto by 13.7.

Suppose now that $r + 1 \leq m + n - 2$, and $f_s : (I^r, \partial I^r, J^{r-1}) \to (X, X_2, *)$ is a homotopy with $f_0(I^r) \cup f_1(I^r) \subset X_1$. Thus $\{f_0\}, \{f_1\} \in \pi_r(X_1, A, *)$. We will prove that $\{f_0\} = \{f_1\}$. Let $F: I \times I^r \to X$ be defined by $F(s, u) = f_s(u)$. Apply 13.4 to F, once for each cell, to produce $U, V \subset I^{r+1}$ and homotopies $\alpha_t : \bar{U} \to e^m$ and $\beta_t : \bar{V} \to e^n$. As before, define

$$F_t(u) = \begin{cases} \alpha_t(u), & u \in \bar{U} \\ \beta_t(u), & u \in \bar{V} \\ F(u), & u \in I^{r+1} - U - V. \end{cases}$$

Now if $F(u) \in e^m$, $F_t(u) \in e^m$, and if $F(u) \in e^n$, $F_t(u) \in e^n$. Furthermore, if $F(u) \in A$, $F_t(u) = F(u)$. In particular $F_t(0 \times I^r \cup 1 \times I^r) \subset X_1$. Let $\mu_t = F_t|_{0 \times I^r}$ and $\nu_t = F_t|_{1 \times I^r}$. Then μ_t and ν_t are homotopies $(I^r, \partial I^r, J^{r-1}) \to (X_1, A, *)$. Since $\mu_0 = F|_{0 \times I^r} = f_0$ and $\nu_0 = F|_{1 \times I^r} = f_1$, $\{f_0\} = \{\mu_1\}$ and $\{f_1\} = \{\nu_1\}$ in $\pi_r(X_1, A, *)$. F_1 is a "linearized" homotopy from μ_1 to ν_1 (see Fig. 13.7). We will deform this homotopy so that it misses some point $p \in e^n$. We apply 13.8 to F_1 and choose points p and q accordingly. Now

$$F(\partial I^{r-1} \times I) \subset F(0 \times I^{r-1}) \cup F(1 \times I^{r-1}) \cup F(I \times J^{r-1}) \subset X_1.$$

Hence $F_1(\partial I^{r-1} \times I) \subset X_1$. Similarly, $F_1(I^{r-1} \times 0) \subset X_2$ and $F_1(I^{r-1} \times 1) = *$. Hence we may apply 13.9. Consider the homotopy H_1 constructed by 13.9. Now

$$H_1 : (I \times I^r, I \times \partial I^r, I \times J^{r-1}) \to (X - p, X - p - q, *),$$

since $F(I \times J^{r-1}) = *$ implies that $H_1(I \times J^{r-1}) = F_1(I \times J^{r-1}) = *$. But

$$H_1|_{0 \times I^r} = F_1|_{0 \times I^r} = \mu_1 \quad \text{and} \quad H_1|_{1 \times I^r} = F_1|_{1 \times I^r} = \nu_1,$$

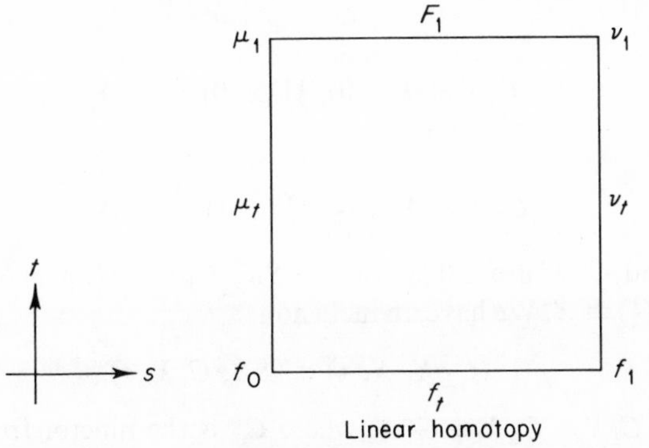

Figure 13.7

hence H_1: $\mu_1 \sim \nu_1$ in $\pi_r(X - p, X - p - q, *)$. It follows from 13.7 that $\{f_0\} = \{f_1\}$. ∎

We are nearly ready to reap the benefits of this deep theorem.

Proposition 13.10 The suspension functor S defines a transformation

$$E\colon [(X, *), (Y, *)] \to [(SX, *), (SY, *)].$$

In particular, we have $E\colon \pi_r(Y, *) \to \pi_{r+1}(SY, *)$.

Proof If $H(X \times I, * \times I) \to (Y, *)$ is a homotopy, we easily define $K\colon (SX \times I, * \times I) \to (Y, *)$ by $K((x, s), t) = (H(x, t), s)$. If $H\colon f \sim g, K\colon Sf \sim Sg$. The second part follows since $S(S^r, *) \equiv (S^{r-1}, *)$.

Proposition 13.11 The diagram

$$\begin{array}{ccc} \pi_r(X) & \xrightarrow{\;E\;} & \pi_{r+1}(SX) \\ \uparrow{\scriptstyle \partial\,\cong} & \nearrow{\scriptstyle (p_X)_*} & \\ \pi_{r+1}(C^*X, X) & & \end{array}$$

commutes. In particular, E is a homomorphism.

Proof Let f: $(I^r, \partial I^r) \to (X, *)$. We will find F: $(I^{r+1}, \partial I^{r+1}, J^r) \to (C^*X, X, *)$ such that $\partial\{F\} = \{f\}$ and $\{p_X \circ F\} = \{Sf\} = E\{f\}$. Define F by

$$F(s_1, \ldots, s_{r+1}) = (f(s_1, \ldots, s_r), s_{r+1}).$$

By 10.4, $\partial\{F\} = f$. Now $S(f)\colon I^{r+1}/\partial I^{r+1} \equiv S(I^r/\partial I^r) \xrightarrow{S(f)} SX$ is also given by this formula, according to 9.4 ∎

Let

$$C_1 X = X \times [0, \tfrac{1}{2}]/(x, 0) \sim (*, t)$$

and

$$C_2 X = X \times [\tfrac{1}{2}, 1]/(x, 1) \sim (*, t).$$

Then $C_1 X$ and $C_2 X$ are subspaces of SX, $C_1 X \cup C_2 X = SX$ and $C_1 X \cap C_2 X = X \times (\tfrac{1}{2}) \equiv X$. We have an inclusion

$$\imath\colon (C_2 X, X, *) \subset (SX, C_1 X, *).$$

Observe that $C_1 X \equiv C_2 X \equiv C^* X$, where C^* is the functor from 9.3.

Proposition 13.12 The diagram

$$\begin{array}{ccc} \pi_r(C_2 X, X, *) & \xrightarrow{(p_X)_*} & \pi_r(C_2 X/X, *) \\ \Big\downarrow{\scriptstyle i_*} & & \cong \Big\downarrow{\scriptstyle \alpha_*} \\ \pi_r(SX, C_1 X, *) & \xrightarrow{\cong} & \pi_r(SX, *) \end{array}$$

commutes where α is the natural homeomorphism $C_2 X/X \equiv SX$ given by $\alpha(x, t) = (x, 2t - 1)$.

Proof $\alpha p_X \sim i\colon (C_2 X, X, *) \to (SX, C_1 X, *)$ where the homotopy is given by

$$H(x, t, s) = (x, (2t - 1)(1 - \tfrac{1}{2}s) + \tfrac{1}{2}s). \quad \blacksquare$$

Corollary 13.13 (*Freudenthal Suspension Theorem—First Form*) $E\colon \pi_r(S^n) \to \pi_{r+1}(S^{n+1})$ is an isomorphism if $r < 2n - 1$ and onto if $r = 2n - 1$.

Proof Combine 13.11, 13.12, and 13.6, observing that, in the notation of 13.7, $X = C_2(S^n)$, $B = C_1(S^n)$. Hence $\imath_*$ is an isomorphism if $r + 1 < (n + 1) + (n + 1) - 2$, and onto if $r + 1 \leq (n + 1) + (n + 1) - 2$. $\blacksquare$

Corollary 13.14 $\pi_n(S^n) \cong Z$ generated by the identity map.

Proof By 11.12c, 12.23, and Exercise 9, Section 7, $\pi_2(S^2) \cong Z$. It now follows from 13.13 that $E\colon \pi_n(S^n) \to \pi_{n+1}(S^{n+1})$ is an isomorphism for $n \geq 1$. $\blacksquare$

Corollary 13.15 If $f\colon B^n \to B^n$, there exists x such that $f(x) = x$ (Problem 1).

Corollary 13.16 There does not exist $\gamma: B^n \to S^{n-1}$ with $\gamma|_{S^{n-1}} = 1$.

Proof This follows by applying the functor π_n to the diagram

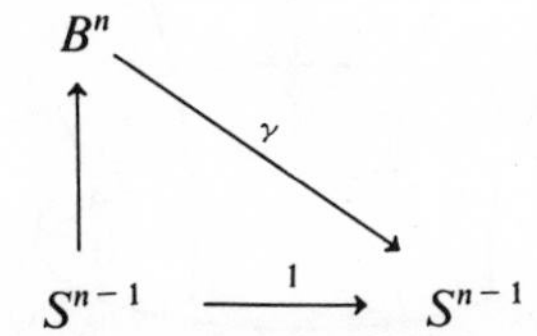

(see the end of Section 6). ∎

Corollary 13.17 If $n \neq m$, $R^n \not\equiv R^m$.

Proof If so, we would have $R^n - \{0\} \equiv R^m - \{a\} \equiv R^m - \{0\}$. But by Exercise 7, Section 6, $R^m - \{0\} \simeq S^{m-1}$ and $R^n - \{0\} \simeq S^{n-1}$, hence $S^{n-1} \simeq S^{m-1}$. If $n < m$, $\pi_{n-1}(S^{m-1}) = 0$ and $\pi_{n-1}(S^{n-1}) \cong Z$ so we have a contradiction. ∎

As an immediate corollary to 13.14 we have:

Proposition 13.18 $\pi_3(S^2) \cong Z$.

Proof Apply 13.14 and 11.12b. ∎

$\pi_3(S^2)$ is generated by the map $\eta: S_3 \to S^2$. η is the first example of an essential map between spheres of different dimensions and as such deserves a little attention. We will illuminate this map in two ways. We give an explicit formula for η (it is a polynomial), and we give a geometric description of η. If we identify S^2 with $(C)^\infty$, and S^3 as the sphere in C^2, η is given by $\eta(z_0, z_1) = z_1/z_0$. The stereographic projection map (see 1.5) $\psi_2 : C^\infty \to S^2$ is given by

$$\psi_2(z) = \frac{1}{1 + |z|^2}(|z|^2 - 1, 2x, 2y).$$

If $z_0 = x_0 + y_0 i$ and $z_1 = x_1 + y_1 i$, the composite is given by

$$\eta(x_0, y_0, x_1, y_1) = ({x_1}^2 + {y_1}^2 - {x_0}^2 - {y_0}^2, 2(x_0 x_1 + y_0 y_1), 2(x_0 y_1 - x_1 y_0)).$$

It is more illuminating to think of S^3 as the union of two solid tori $S^1 \times D^2$ and $D^2 \times S^1$ along their common boundary $S^1 \times S^1$. In fact

$$S^3 = \partial B^4 \equiv \partial(D^2 \times D^2) = \partial D^2 \times D^2 \cup D^2 \times \partial D^2 = S^1 \times D^2 \cup D^2 \times S^1.$$

To see this, picture S^3 as $(R^3)^\infty$ and a solid torus $D^2 \times S^1 \subset R^3$. The exterior of $D^2 \times S^1$ in S^3 is then a solid torus; see Fig. 13.8. $S^1 \times S^1$ is mapped onto $S^1 \subset S^2$ by the quotient map $(z_0, z_1) \to z_1/z_0$. This map is extended radially

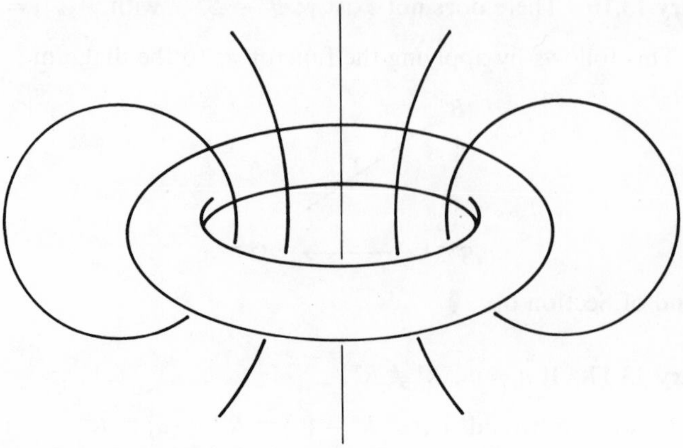

Figure 13.8

into each of the solid tori by moving through the upper and lower hemispheres toward the north and south poles respectively.

Theorem 13.13 (and hence 13.7) is best possible, for

$$E\colon \pi_2(S^1) \to \pi_3(S^2)$$

is not an isomorphism since $\pi_2(S^1) = 0$, $\pi_3(S^2) = Z$.

In fact, $E\colon \pi_3(S^2) \to \pi_4(S^3)$ is onto but not 1–1 (see 27.19). To calculate $\pi_i(S^n)$ for all i and n is an interesting and difficult unsolved problem.

Theorem 13.19 There is a continuous nonzero vector field on S^n iff n is odd (Problems 3–4).

Proof Half of this is 3.5. We will prove that if n is even, $a_n \not\sim 1$ and apply 3.4 to conclude the other half of 13.19.

For $X \in \mathcal{C}$ consider ΣX as defined in Exercise 11, Section 9, with base point $(x, 0)$. Define $\Pi_n(X) = \pi_{n+1}(\Sigma X, *)$.

Lemma 13.20 $\Pi_n : \mathcal{C} \to \mathcal{M}_Z$ is a functor for $n \geq 1$. $\Pi_n(S^n) \cong Z$.

Proof $\Sigma : \mathcal{C} \to \mathcal{C}^*$ is clearly a functor so the first part is trivial. The second part follows since ΣS^n is homeomorphic with S^{n+1}, by Exercise 11, Section 9. ∎

We will finish 13.19 by showing that $(-1)^{n+1} = \Pi_n(a_n)\colon \Pi_n(S^n) \to \Pi_n(S^n)$. Let $f_i : S^n \to S^n$ be given by

$$f_i(x_1, \ldots, x_{n+1}) = (x_1, \ldots, x_{i-1}, -x_i, x_{i+1}, \ldots, x_{n+1}).$$

Now $a_n = f_1 \circ f_2 \circ \cdots \circ f_{n+1}$. It is thus sufficient to show that $\Pi_n(f_i) = -1$. There is a homeomorphism $h_{i,j}: S^n \to S^n$ interchanging the ith and jth coordinates, and $f_i h_{ij} = h_{ij} f_j$. Hence, it is sufficient to show that $\Pi_n(f_{n+1}) = -1$. There is a natural map $\Sigma: \pi_n(X, *) \to \Pi_n(X)$, given by

$$\{f\} \to \{S^{n+1} \equiv \Sigma S^n \xrightarrow{\Sigma f} \Sigma X\},$$

where a base point of S^{n+1} is chosen corresponding to $* \in \Sigma S^n$ under the homeomorphism. This is clearly an isomorphism if $X = S^n$ so it is sufficient to show that

$$-1 = (f_{n+1})_* : \pi_n(S^n, *) \to \pi_n(S^n, *).$$

But under the homeomorphisms of 9.4

$$S^n \equiv S(S^{n-1}),$$

f_{n+1} corresponds to the inverse map. ∎

Seminar Problem

13.6 can be proven by differentiable approximation, instead of linear approximation. One first proves that one can find disjoint closed subsets M and N of I^r and a homotopy $h_t: I^r \to X$ with $h_0 = f$ and such that h_1 is C^∞ on both M and N [see (13.8)]. This follows from the fact that any map from a compact subset of R^r to R^s is close to a differentiable map. One then uses Sards theorem to pick the points $p \in e^n$ and $q \in e^m$ as in 13.9. (See [49; 53].)

Exercises

1. Let A be a closed subset of X and suppose there is a homotopy $H: X \times I \to X$ such that

(a) $H(x, 0) = x$;
(b) $H(A \times I) \subset A$;
(c) $H(A \times 1) = *$.

Prove that the collapsing map $p_A: X \to X/A$ is a homotopy equivalence. (16.33)

2. Find $X \cup e^n$ with $\pi_n(X \cup e^n, X, *) \not\cong Z$. (Hint: Let $X \cup e^n = S^1 \vee S^n$. Consider a covering space of this.) (By 16.30, it is necessary that $\pi_1(X) \neq 0$ in this example.)

3. Consider the map $\iota\colon X \to \Omega SX$ given by $\iota(x)(t) = (x, t)$. Prove that this is continuous. Show that the diagram

$$\begin{array}{ccc} \pi_n(X, *) & \longrightarrow & \pi_{n+1}(SX, *) \\ & \searrow^{\iota_*} & \downarrow \cong \\ & & \pi_n(\Omega SX, *) \end{array}$$

commutes, where the equivalence is that of Exercise 5, Section 9.

4. Show that the suspension

$$E\colon [(SX, *), (SY, *)] \to [(S^2X, *), (S^2Y, *)]$$

is a homomorphism where $S^nX = S(S^{n-1}X)$ is the iterated suspension.

5. Prove that $\pi_7(S^4)$ has an element ν of infinite order.

6. Show that if X is Hausdorff and $\alpha\colon S^{n-1} \to X$, $X \cup_\alpha e^n$ is Hausdorff. (13.1, 14.6)

7. Let $f\colon S^n \to X$ and $k\iota_n\colon S^n \to S^n$ be a map of degree k. Show that $\{f \circ k\iota_n\} = k\{f\}$. If $X = S^m$, it is not in general true that $\{k\iota_m \circ f\} = k\{f\}$. Using $(1 \wedge k\iota_1) \circ (f \wedge 1) = f \wedge k\iota_1 = (f \wedge 1) \circ (1 \wedge k\iota_1)$ and Exercise 4, show that $E\{k\iota_m \circ f\} = E\{f \circ k\iota_n\} = kE\{f\}$. (Exercise 21, Section 26)

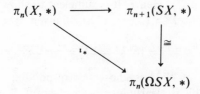

14

The Topology of CW Complexes

We now discuss a generalization of the notion of a simplicial complex which for many purposes is easier to handle.

Definition 14.1 A cell complex X is a Hausdorff space which is the union of disjoint subspaces e_α ($\alpha \in \mathcal{A}$) called cells satisfying:

(a) To each cell we associate an integer $n \geq 0$ called its dimension. If e_α has dimension n we often use the notation $e_\alpha{}^n$ for this cell. We write X^n for the union of all cells $e_\alpha{}^k$ with $k \leq n$. X^n is called the n-skeleton.

(b) If $e_\alpha{}^n$ is an n-cell, there is a "characteristic map" $\chi_\alpha : (B^n, S^{n-1}) \to (X, X^{n-1})$ such that $\chi_\alpha|_{B^n - S^{n-1}}$ is a homeomorphism from $B^n - S^{n-1}$ onto $e_\alpha{}^n$.

Examples

1. Any finite geometric simplicial complex, as described in Section 12, is a cell complex. Each open n-simplex is an n-cell, and, in this case, the maps χ_α are all homeomorphisms.

2. The n-sphere is a cell complex with two cells e^0, e^n where $e^0 = \{(1, 0, \ldots, 0)\}$ and $e^n = S^n - e^0$. Note that if we wish to write S^n as a simplicial complex, we need $\binom{n+2}{k+1}$ simplicies of dimension k; hence cell complexes are more efficient than simplicial complexes.

3. RP^n, CP^n, and HP^n are cell complexes with one cell of dimension k, $2k$, and $4k$, respectively for each $k \leq n$.

Proof: By Exercise 8, Section 7, these spaces are all Hausdorff. We will write the details out in the case of CP^n. The others are similar. CP^n has $(n + 1)$ cells e^{2k} for $0 \leq k \leq n$, given by

$$e^{2k} = \{[z_0 | \cdots | z_n] | z_k \neq 0, z_{k+1} = \cdots = z_n = 0\}.$$

Since $B^{2k} = \{z = (z_1, \ldots, z_k) \in C^k \mid \|z\| \leq 1\}$, we define $\chi_k : B^{2k} \to CP^n$ by

$$\chi_k(z_1, \ldots, z_k) = [z_1 | \cdots | z_k | \sqrt{1 - \|z\|^2}\ |0| \cdots |0].$$

We write this symbolically as $CP^n = e^0 \cup e^4 \cup \cdots \cup e^{2n}$. Similarly $RP^n = e^0 \cup e^1 \cup \cdots \cup e^n$ and $HP^n = e^0 \cup e^4 \cup \cdots \cup e^{4n}$. ∎

4. Every compact differentiable manifold can be proven to be a finite cell complex via Morse Theory [52].

We usually use a 0-cell for a base point in a cell complex (see Exercise 11).

Observe that a finite cell complex is compact since it may be covered by a finite number of compact sets $\{\chi_\alpha(B^n)\}$.

If X is a cell complex and $A \subset X$, we say A is a subcomplex if A is a union of cells e_α and $\bar{e}_\alpha \subset A$ if $e_\alpha \subset A$. Since $\bar{e}_\alpha{}^n = \chi_\alpha(B^n)$, we see that X^n is a subcomplex for every $n \geq 0$.

Suppose X and Y are cell complexes. Then $X \times Y$ can be made a cell complex by choosing as cells $e_\alpha{}^n \times e_\beta{}^m$ where $\{e_\alpha{}^n\}$ are the cells of X and $\{e_\beta{}^m\}$ are the cells of Y. We assign the dimension $n + m$ to $e_\alpha{}^n \times e_\beta{}^m$. A characteristic map $\chi_{\alpha,\beta} : B^{n+m} \to X \times Y$ is given by

$$B^{n+m} \equiv I^{n+m} = I^n \times I^m \equiv B^n \times B^m \xrightarrow{\chi_\alpha \times \chi_\beta} X \times Y.$$

This clearly satisfies 14.1.

Lemma 14.2 If X is a finite cell complex, $X = X_m \supset X_{m-1} \supset \cdots \supset X_0$, where X_0 is one point and (X_k, X_{k-1}) is a relative n_k-cell for $0 \leq n_1 \leq n_2 \leq \cdots \leq n_m$. Symbolically,

$$X = e^0 \cup e^{n_1} \cup e^{n_2} \cup \cdots \cup e^{n_m}.$$

Proof We will show that for each n-cell $e_\alpha{}^n$, $X^n \equiv (X^n - e_\alpha{}^n) \cup_f e^n$. Since $X^n - e_\alpha{}^n$ is a cell complex with one less cell, the result follows by induction. Let $f: S^{n-1} \to X^n - e_\alpha{}^n$ be $\chi_\alpha|_{S^{n-1}}$. One may construct a map $h: (X^n - e_\alpha{}^n) \cup_f e^n \to X^n$ by $h|_{X^n - e_\alpha{}^n} =$ inclusion, and $h|_{B^n} = \chi$. h is clearly well defined and continuous. Moreover, it is 1–1 and onto, so it is a homeomorphism by compactness. ∎

The structure of finite cell complexes is determined by 14.2. Infinite cell complexes do not behave as well. Any Hausdorff space is an infinite cell complex with each point as a 0-cell. We must make some restrictions on the relationship of the cellular structure to the topology if we wish to have a structure theorem like 14.2 for infinite cell complexes. We consider such restrictions now.

If $A \subset X$ is a subset of X, we define $K(A)$ to be the intersection of all subcomplexes containing A. If $A \subset B$, $K(A) \subset K(B)$. Hence if $p \in e$, $K(p) = K(e) = K(\bar{e})$. Thus $K(A)$ is a subcomplex.

Definition 14.3 (C) X is said to be closure finite if for each cell e_α^n, $K(e_\alpha^n)$ is a finite subcomplex. (W) X is said to have the weak topology if for each subset $F \subset X$, F is closed iff $F \cap \bar{e}_\alpha^n$ is compact for each cell e_α^n. A cell complex satisfying (C) and (W) is called a CW complex. Clearly every CW complex belongs to $\mathcal{CG}$.

Consider S^2 as a cell complex with every point a 0-cell. This does not have the weak topology although it is closure finite. On the other hand, B^3 with cells $e^3 = B^3 - S^2$ and one 0-cell for every point of S^2 has the weak topology, but is not closure finite.

A more illuminating example of the weak topology is as follows. Let $X = \bigvee_{\alpha=1}^{\infty} S_\alpha^1$, an infinite 1-point union of circles. This is a closure finite cell complex. This space with the weak topology is a CW complex. One can also give X the induced topology as a subset of $\prod_{\alpha=1}^{\infty} S_\alpha^1$ with the product topology. This space which we will call X' is compact. Both X and X' can be imbedded in R^2 as follows (see Fig. 14.1):

$$X \equiv \bigcup_{n=1}^{\infty} \{(x, y) \mid (x - n)^2 + y^2 = n^2\},$$

$$X' \equiv \bigcup_{n=1}^{\infty} \left\{(x, y) \middle| \left(x - \frac{1}{n}\right)^2 + y^2 = \frac{1}{n^2}\right\}.$$

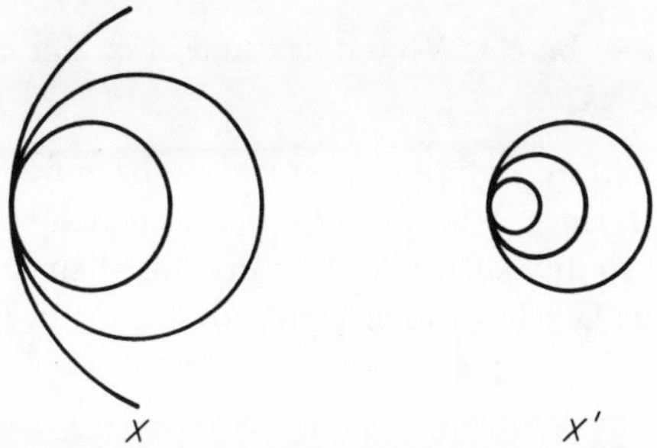

Figure 14.1

Definition 14.4 A map $f\colon X \to Y$ between two cell complexes is called cellular if $f(X^n) \subset Y^n$.

Let $\mathcal{K}$ be the category of CW complexes and cellular maps, $\mathcal{K}^*$ the corresponding pointed category where $* \in X$ is a 0-cell, and $\mathcal{K}_h^*$, $\mathcal{K}_h$ the corresponding homotopy categories.

A useful example is given as follows. $RP^n \subset RP^{n+1}$ as a subcomplex. We may define $RP^\infty = \bigcup_{n=1}^{\infty} RP^n$, and this is clearly a closure finite cell complex. We give it the weak topology so that it is a CW complex. Similarly, we define CP^∞ and HP^∞ (see Exercise 13).

Let X be a cell complex and for each cell $e_\alpha{}^n$ choose a copy $B_\alpha{}^n$ of B^n. Let $B = \amalg B_\alpha{}^n$ (see Exercise 7, Section 8) and let $\chi: B \to X$ be the characteristic map of $e_\alpha{}^n$ on $B_\alpha{}^n$.

Lemma 14.5 X has the weak topology iff $\chi: B \to X$ is a quotient map.

Proof χ is continuous since if $F \subset X$ is closed $\chi^{-1}(F) = \bigcup \chi_\alpha^{-1}(F)$. Suppose $\chi^{-1}(F)$ is closed. Then $\chi^{-1}(F) \cap B_\alpha{}^n$ is compact; by Exercise 1, $\chi_\alpha^{-1}(F \cap \bar{e}_\alpha{}^n) = \chi^{-1}(F) \cap B_\alpha{}^n$. Hence $F \cap \bar{e}_\alpha{}^n$ is compact. It follows that F is closed and thus χ is a quotient map. Suppose χ is a quotient map and $F \cap \bar{e}_\alpha{}^n$ is closed. Then $\chi^{-1}(F) \cap B_\alpha{}^n$ is closed. Hence $\chi^{-1}(F)$ is closed, and thus F is closed so X has the weak topology. ▮

Proposition 14.6 If X is a CW complex and $f: S^n \to X^n \subset X$, then $Y = X \cup_f e^{n+1}$ is a CW complex.

Proof By Exercise 6, Section 13, Y is Hausdorff. We choose as cells all cells $e_\alpha{}^m$ of X together with $e^{n+1} = Y - X$. We use the same characteristic map as before for $e_\alpha{}^m$ and $\chi: (B^{n+1}, S^n) \to (Y, X)$ as a characteristic map for e^{n+1}. Now $B_Y = B_X \amalg B^{n+1}$, hence if X has the weak topology, so does Y. To see that Y is closure finite, we use the following lemma.

Lemma 14.7 Let X be a CW complex and $A \subset X$ a compact set. Then $K(A)$ is a finite complex.

Proof Suppose $K(A)$ is infinite. There must be infinitely many cells of $K(A)$ that intersect A, since X is closure finite. For each such cell e, choose a point $x_e \in e \cap A$. Then any subset S of $\{x_e\}$ is closed since $S \cap \bar{e}$ is finite for each cell $e \subset X$. Thus $\{x_e\}$ has the discrete topology. But $\{x_e\}$ being closed is compact, a contradiction. ▮ ▮

Proposition 14.8 Let X be a CW complex and $e_\alpha{}^n$ an n-cell. Then $A = X^n - e_\alpha{}^n$ is a subcomplex and $X^n \equiv A \cup_f e^n$ for some $f: S^{n-1} \to X^{n-1}$.

Proof Clearly A is a subcomplex. As in 14.2, let $\chi_\alpha: (B^n, S^{n-1}) \to (X, X^{n-1})$ be the characteristic map for $e_\alpha{}^n$ and $f = \chi|_{S^{n-1}}$. We can then define $\phi: A \cup_f e^n \to X^n$ by $\phi|_A =$ the inclusion map, and $\phi|_{B^n} = \chi_\alpha$. This is 1–1, continuous, and onto. We will show that it is closed. In fact $A \amalg B^n \to A \cup_f e^n \to X^n$ is closed since B^n is compact and A is a closed subset of X^n. ▮

Corollary 14.9 Let $A \subset X$ and $f: A \to Y$. Suppose A is closed or open and X has the weak topology. Then f is continuous iff $f|_{A \cap \bar{e}}$ is continuous for each cell e.

Proof To show that f is continuous, it is sufficient to show that $f \circ \chi: B \cap \chi^{-1}(A) \to Y$ is continuous, by 14.5. ▮

Corollary 14.10 Let $A \subset X$ be a subcomplex and suppose X has the weak topology. Then A is closed, and the induced topology is the weak topology.

Proof Let $\chi: B \to X$. $\chi^{-1}(A)$ is clearly closed since for each α, $\chi^{-1}(A) \cap B_\alpha{}^n = \varnothing$ or $\chi^{-1}(A) \supset B_\alpha{}^n$. Thus A is closed. It follows that $\chi|_{\chi^{-1}(A)}$ is a quotient map, so A has the weak topology. ▮

Proposition 14.11 Let X and Y be CW complexes. Then $X \times Y$ (with the compactly generated topology) is a CW complex, and $X \vee Y$ is a subcomplex.

Remark The cellular structure is given above.

Proof Clearly $X \times Y$ is closure finite, since $K(e_\alpha{}^n \times e_\beta{}^m) = K(e_\alpha{}^n) \times K(e_\beta{}^m)$. We will show that $\chi_{X \times Y}: B_{X \times Y} \to X \times Y$ is a quotient map. But $B_{X \times Y} \equiv B_X \times B_Y$ and $\chi_{X \times Y}$ can be factored:

$$B_{X \times Y} \equiv B_X \times B_Y \xrightarrow{\chi_X \times \chi_Y} X \times Y.$$

8.12 implies that $\chi_{X \times Y}$ is a quotient map. ▮

We now consider an important property on a pair of spaces which is dual to the notion of the homotopy lifting property (11.2).

Definition 14.12 A pair (X, A) of spaces has the absolute homotopy extension property (AHEP) if given any space Y, any map $f: X \to Y$, and any homotopy $H: A \times I \to Y$ with $H(a, 0) = f(a)$, there is an extension $\tilde{H}: X \times I \to Y$ of H with $\tilde{H}(x, 0) = f(x)$.

The duality mentioned can be seen by comparing the diagram in 11.2 to the diagram

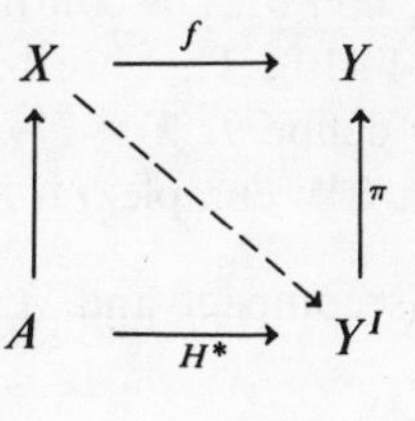

where H^* is adjoint to H and $\pi(\omega) = \omega(0)$.

This property guarantees that the extension problem in $\mathcal{C}$ is equivalent to the extension problem in $\mathcal{C}_h$ (see Exercise 15).

Proposition 14.13 If X is a CW complex and A is a subcomplex, (X, A) has the absolute homotopy extension property.

Proof Suppose we are given a map

$$\theta_{-1}: X \times 0 \cup A \times I \to Y$$

which we wish to extend to $X \times I$. Let $\bar{X}^n = X^n \cup A$. We will inductively define maps

$$\theta_n : X \times 0 \cup \bar{X}^n \times I \to Y$$

which are extensions of the previous ones. To construct θ_0, observe that the 0-skeleton of a CW complex has the discrete topology since each subset of it intersects every cell in a finite set and hence is closed. Consequently, one can extend θ_{-1} to $\theta_0 : X \times 0 \cup \bar{X}^0 \times I \to Y$ by $\theta_0(e^0, t) = \theta_{-1}(e^0, 0)$. This is clearly continuous. Suppose we have defined an extension θ_n and $n \geq 0$; see Fig. 14.2.

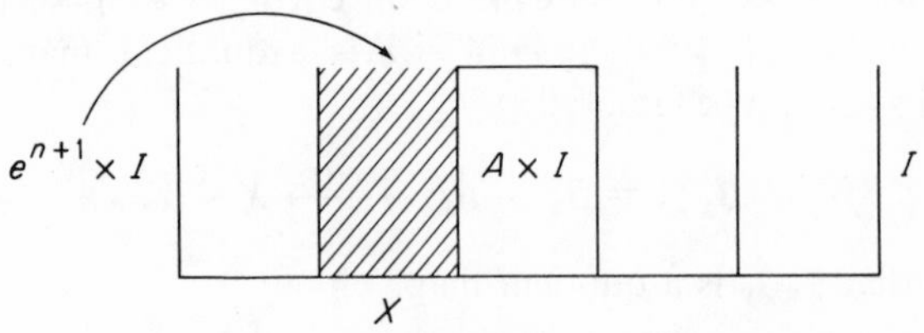

Figure 14.2

For each $(n+1)$-cell e_α^{n+1}, consider the composite

$$S_\alpha^n \times I \cup B_\alpha^{n+1} \times 0 \xrightarrow{\chi_\alpha \times 1} \bar{X}^n \times I \cup X \times 0 \xrightarrow{\theta_n} Y.$$

This has an extension $\Gamma_\alpha : B_\alpha^{n+1} \times I \to Y$ by 10.6. Define $\theta_{n+1}: \bar{X}^{n+1} \times I \cup X \times 0 \to Y$ by

$$\theta_{n+1}|_{\bar{X}^n \times I \cup X \times 0} = \theta_n \qquad \text{and} \qquad \theta_{n+1}|_{B_\alpha^{n+1} \times I} = \Gamma_\alpha .$$

These maps are compatible, and θ_{n+1} is continuous since $\bar{X}^{n+1} \times I \cup X \times 0$ has the weak topology (by 14.10). Thus we have constructed θ_{n+1}, and the induction is complete. Now define $\theta: X \times I \to Y$ by $\theta_{\bar{X}^n \times I} = \theta_n$. This is well defined, and since $X \times I$ is a CW complex, it is continuous. ∎

Definition 14.14 Let $f: X \to Y$. The mapping cone of f, written $Y \cup_f CX$ or C_f, is the quotient space

$$Y \cup CX/(x, 0) \sim f(x).$$

(See Exercise 11, Section 9.)

This construction is a generalization of the construction $X \cup_\alpha e^n$ of Section 13 and is useful because of the following property:

Proposition 14.15 Suppose we are given maps $f\colon X \to Y$ and $g\colon Y \to Z$ in $\mathcal{C}$. Then $g \circ f$ is nullhomotopic iff there exists $h\colon Y \cup_f CX \to Z$ with $h|_Y = g$.

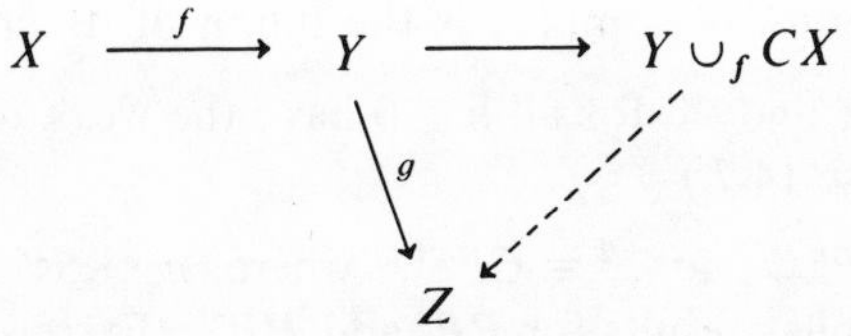

Proof By Exercise 11, Section 9, if $g \circ f$ is nullhomotopic, there is an extension $H\colon CX \to Z$ of $g \circ f$. $g \amalg H\colon Y \amalg CX \to Z$ clearly factors over $Y \cup_f CX$ and defines h. On the other hand, an extension h determines a map $CX \to Y \cup_f CX \xrightarrow{h} Z$ which extends $g \circ f$. Hence $g \circ f$ is nullhomotopic. ∎

Thus the homotopy problem: "Is $g \circ f$ nullhomotopic?" is equivalent to the extension problem: "Does g extend over $Y \cup_f CX$?"

Let X be a CW complex. Fix n and choose for each n-cell, a copy $B_\alpha{}^n$ of B^n and S_α^{n-1} of S^{n-1}. Let $B_n = \amalg B_\alpha{}^n$ and $S_{n-1} = \amalg S_\alpha^{n-1} \subset B_n$. Define $\chi_n : (B_n, S_{n-1}) \to (X^n, X^{n-1})$ by $\chi_n|_{B_\alpha{}^n} = \chi_\alpha$. Clearly $B_n = CS_{n-1}$.

Proposition 14.16 $X^n \equiv X^{n-1} \cup_f CS_{n-1} \equiv X^{n-1} \cup_f B_n$, where $f = \chi_n|_{S_{n-1}}$.

Proof Define $\theta\colon X^{n-1} \cup_f B_n \to X^n$ by letting $\theta|_{X^{n-1}}$ be the inclusion and $\theta|_{B_n} = \chi_n$. This is well defined, continuous, 1–1, and onto. To prove that it is a homeomorphism we prove that the composite $X^{n-1} \amalg B_n \xrightarrow{q} X^{n-1} \cup_f B_n \xrightarrow{\theta} X^n$ is closed. Let $A \subset X^{n-1} \amalg B_n$ be closed. Then $\theta q(A) \cap \bar{e}_\alpha{}^n = \theta q(A \cap X^{n-1} \cup A \cap B_\alpha{}^n)$. Since X^{n-1} is closed in X^n, $\theta q(A \cap X^{n-1})$ is closed and $\theta q(A \cap B_\alpha{}^n)$ is compact. Thus $\theta q(A) \cap \bar{e}_\alpha{}^n$ is closed. $\theta q(A) \cap \bar{e}_\alpha{}^m = \theta q(A \cap X^{n-1})$ for $m < n$, and consequently is closed. Thus $\theta q(A)$ is closed. ∎

The following corollary is particularly useful when $Z = *$ or $Z = I$.

Corollary 14.17 Suppose $Z \in \mathcal{CG}$. Let $f\colon X^{n-1} \times Z \to Y$ and $f_\alpha : B_\alpha{}^n \times Z \to Y$ be maps such that $f_\alpha(u, z) = f(\chi_\alpha(u), z)$ for $u \in S_\alpha^{n-1}$. Then there is a unique map $\bar{f}\colon X^n \times Z \to Y$ such that $\bar{f}|_{X^{n-1} \times Z} = f$ and $\bar{f} \circ \chi_\alpha = f_\alpha$.

Proof Apply 14.16 and 8.12. ∎

Exercises

1. Let e_α^n be a cell in a cell complex and χ_α its characteristic map. Show that $\chi_\alpha(B^n) = \bar{e}_\alpha^n$. (14.5)

2.* Show that RP^n and HP^n are CW complexes.

3. Show that every CW complex is the union of its finite subcomplexes.

4. Show that if X and X^n for all $n \geq 0$ have the weak topology, then X is a CW complex. (Use 14.7.)

5. Show that $CP^n \cup_{\eta_n} e^{2n+2} \equiv CP^{n+1}$ where $\eta_n: S^{2n+1} \to CP^n$ is from Section 7. Prove similar results for RP^n and HP^n. (Exercise 21, Section 26; 27.19; 28.19; Exercise 12, Section 16).

6. A pair (X, A) is called a relative cell complex if X is Hausdorff and $X - A$ is a union of disjoint subspaces e_α ($\alpha \in \mathfrak{U}$) called cells satisfying (a) and (b) of 14.1 except that we now define $\bar{X}^n = A \cup \bigcup \{e_\alpha^k \mid k \leq n\}$. A subcomplex B of (X, A) is a subset $B \supset A$ such that $B - A$ is a union of cells e_α and $e_\alpha \subset B$ implies $\bar{e}_\alpha \subset B$. A subcomplex is called finite if $B - A$ is a union of a finite number of cells. (X, A) is called closure finite if $K(e_\alpha^n)$ is a finite subcomplex. (X, A) is said to have the weak topology if for each subset $F \subset X$, F is closed iff $F \cap \bar{e}_\alpha^n$ is compact for each cell e_α^n and $F \cap A$ is closed. A relative CW complex is a relative cell complex which is closure finite and has the weak topology. Show that if X is a CW complex and A a subcomplex, (X, A) is a relative CW complex.

7. Show that if (X, A) is a relative CW complex, X/A is a CW complex. Hence if X and Y are CW complexes, so is $X \wedge Y$.

8. Generalize 14.13 to relative CW complexes.

9. Show that any cell complex with two cells e^0 and e^n is homeomorphic to S^n.

10. Prove that each arc component of a CW complex is a CW complex.

11.* Show that each cell complex contains a 0-cell.

12. Show that a CW complex is arcwise connected iff it is connected.

13.* Suppose $\{X_n\}$ are CW complexes and X_{n-1} is a subcomplex of X_n. Let $X = \bigcup X_n$ with the weak topology. Show that X is a CW complex and each X_n is a subcomplex (use Exercise 5, Section 0). In particular RP^∞, CP^∞, and HP^∞ are CW complexes.

14. Let $f\colon X \to Y$ be a base point preserving cellular map. Show that $Y \cup_f CX$ has the structure of a CW complex.

15. Given a pair (X, A) and a map $f\colon A \to Y$, the extension problem is that of deciding if there is a map $\tilde{f}\colon X \to Y$ such that $\tilde{f}|_A = f$. Such a problem can be formed in any category. Show that if (X, A) has the AHEP, the extension problem for (X, A) in $\mathcal{C}$ is equivalent to the extension problem for (X, A) in $\mathcal{C}_h$.

16. Define the lifting problem analogous to the extension problem in Exercise 15, and show that if $\pi\colon E \to B$ has the HLP (11.2) the lifting problem for $\pi\colon E \to B$ in $\mathcal{C}$ is equivalent to the lifting problem for $\pi\colon E \to B$ in $\mathcal{C}_h$.

17. Show that if X_α are CW complexes, $\amalg X_\alpha$ is a CW complex.

18. Suppose that (X, A) and (Y, B) have the AHEP. Show that $(X \times Y, X \times B \cup A \times Y)$ and $(X/A, *)$ have the AHEP. (16.33; Exercise 28, Section 16; 21.18)

19. Suppose X is Hausdorff and (X, A) has the AHEP. Prove that A is closed. (19.5)

20. Show that if (X, A) has the AHEP, there is a neighborhood U of A in X and a retraction $r\colon U \to A$. (21.16)

21. Given $f\colon (X, *) \to (Y, *)$, define $Y \cup_f C^*X$ as the quotient space of $Y \amalg C^*X$ given by identifying $(x, 0) \in C^*X$ with $f(x) \in Y$. Prove an analogue to 14.15. (18.4)

22. Let $A \subset X$ and $f\colon X \to Y$. Show that the natural map $I\colon Y \cup_{f|_A} CA \to Y \cup_f CX$ is an inclusion. If A is a strong deformation retract of X, show that $Y \cup_{f|_A} CA$ is a strong deformation retract of $Y \cup_f CX$. Conclude that if $f, g\colon X \to Y$ and $H\colon f \sim g$, $Y \cup_f CX \simeq Y \cup_H C(X \times I) \simeq Y \cup_g CX$. (27.19; 28.18)

15

Limits

In this section we shall discuss some algebraic and categorical notions that will be recurrent in the next few sections.

Definition 15.1 A directed set $\mathcal{A}$ is a partially ordered set $\mathcal{A}$, such that for any two elements $\alpha, \beta \in \mathcal{A}$ there is an element $\gamma \in \mathcal{A}$ with $\gamma \geq \alpha$ and $\gamma \geq \beta$. A directed system of sets (spaces, groups), directed over $\mathcal{A}$ is a collection of sets (spaces, groups) $\{X_\alpha \mid \alpha \in \mathcal{A}\}$ together with functions (maps, homomorphisms) $f_{\alpha\beta}\colon X_\alpha \to X_\beta$ defined when $\alpha \leq \beta$ such that $f_{\beta\gamma} f_{\alpha\beta} = f_{\alpha\gamma}$ and $f_{\alpha\alpha} = 1$.

The most common directed set that occurs in mathematics is the positive integers, ordered in the usual way. The skeletons of a CW complex form a directed system of spaces over this set. Here $f_{n,m}: X^n \to X^m$ is the inclusion map. Another example is given by letting $\mathcal{A}$ be the collection of finite subcomplexes of a given CW complex X, directed by inclusion, and $X_\alpha = \alpha$ considered as a space. Similarly, one could let $\mathcal{A}$ be the set of finitely generated subgroups of a group G, directed by inclusion, and $X_\alpha = \alpha$ considered as a group.

One can replace this definition with a more conceptual (although possibly more incomprehensible) one using the notions of category theory. Given a directed set $\mathcal{A}$ one can associate a category $\mathfrak{D}_\mathcal{A}$ as follows. For objects in $\mathfrak{D}_\mathcal{A}$, we take the elements of $\mathcal{A}$. We define $\hom(\alpha, \beta) = \{\delta_{\alpha\beta}\}$—a one-object set—if $\alpha \leq \beta$ and $\hom(\alpha, \beta) = \varnothing$ otherwise. We define $\delta_{\beta\gamma}\delta_{\alpha\beta} = \delta_{\alpha\gamma}$. Then a directed system of sets (spaces, groups) is simply a covariant functor $F\colon \mathfrak{D}_\mathcal{A} \to \mathcal{S}(\mathcal{T}, \mathcal{G})$. Dually one can define an inverse system by considering contravariant functors. The whole notion can be generalized by replacing $\mathfrak{D}_\mathcal{A}$ by a directed category, i.e., one in which $\hom(X, Y)$ contains at most one element. We then have sets, spaces, groups, etc. directed over a category. We will not consider this level of generality here for simplicity.

Given a directed system of objects in a category, one can sometimes associate with it an object called the direct limit. Intuitively one thinks of this as being the union (when $f_{\alpha\beta}$ are inclusions), and it is in the three examples given earlier.

Definition 15.2 Given a directed system $\{X_\alpha, f_{\alpha\beta}\}$, an object X is called the direct limit (or right limit) and is written $\varinjlim X_\alpha$, if there are maps $f_\alpha : X_\alpha \to X$ such that $f_\beta \circ f_{\alpha\beta} = f_\alpha$, and these maps satisfy the following universal property: given any space X' and system of maps $f_\alpha' : X_\alpha \to X'$ satisfying $f_\beta' \circ f_{\alpha\beta} = f_\alpha'$, there is a unique map $f: X \to X'$ so that $f_\alpha' = f \circ f_\alpha$:

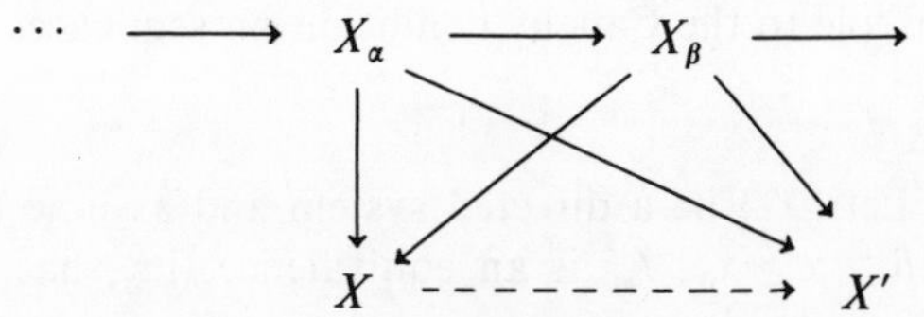

Observe that since f is unique, any two direct limits are isomorphic.

Proposition 15.3 Direct limits always exist in $\mathcal{S}, \mathcal{T}$, and $\mathcal{M}_R$.

Proof (1) In $\mathcal{S}$. Let X be the disjoint union of the X_α. Define $x \sim y$ if $x \in X_\alpha$, $y \in X_\beta$, $\alpha \le \gamma$, $\beta \le \gamma$, and $f_{\alpha\gamma}(x) = f_{\beta\gamma}(y)$. This is an equivalence relation. Let $\varinjlim X_\alpha$ be the set of equivalence classes and $f_\alpha : X_\alpha \to \varinjlim X_\alpha$ be the composite of the inclusion $X_\alpha \subset X$ and the projection $\pi : X \to \varinjlim X_\alpha$. Then $f_\beta \circ f_{\alpha\beta} = f_\alpha$. Suppose $f_\alpha' : X_\alpha \to X'$ satisfies $f_\beta' \circ f_{\alpha\beta} = f_\alpha'$; $\{f_\alpha'\}$ defines a function $F: X \to X'$ by $F|_{X_\alpha} = f_\alpha'$. Since $f_\beta' \circ f_{\alpha\beta} = f_\alpha'$, F preserves the equivalence relation, and F defines a map $f: \varinjlim X_\alpha \to X'$. Any two maps $f_1, f_2 : \varinjlim X_\alpha \to X'$ satisfying $f_1 f_\alpha = f_2 f_\alpha = f_\alpha'$ must satisfy $f_1 \pi = f_2 \pi$; since π is onto, $f_1 = f_2$.

(2) In $\mathcal{T}$. We perform the same construction as in $\mathcal{S}$ but topologize it. Let $X = \amalg X_\alpha$ and give $\varinjlim X_\alpha$ the quotient topology. Then the maps f_α and F are continuous. Uniqueness follows as before.

(3) In $\mathcal{M}_R$. Given a directed system of R-modules $\{X_\alpha\}$ let $X = \bigoplus_{\alpha \in A} X_\alpha$; i.e., the elements of X are functions $f: A \to \bigcup X_\alpha$ with $f(\alpha) \in X_\alpha$ and $f(\alpha) = 0$ for all but a finite number of elements α of A. Define $(f+g)(\alpha) = f(\alpha) + g(\alpha) \in X_\alpha$, $(-f)(\alpha) = -f(\alpha)$, $(rf)(\alpha) = rf(\alpha)$ and $O(\alpha) = 0$. This makes X into an R-module. Define $\bar{f}_\alpha : X_\alpha \to X$ by

$$\bar{f}_\alpha(x)(\beta) = \begin{cases} 0, & \beta \ne \alpha \\ x, & \beta = \alpha. \end{cases}$$

This is a homomorphism, and X is generated by the elements of the form $\bar{f}_\alpha(x)$. Let M be the submodule generated by all $\bar{f}_\alpha(x) - \bar{f}_\beta(f_{\alpha\beta}(x))$. Define $\varinjlim X_\alpha$ to be the quotient of X by M, and let $\pi\colon X \to \varinjlim X_\alpha$ be the quotient map. We then have $\pi\bar{f}_\alpha = f_\alpha : X_\alpha \to \varinjlim X_\alpha$. Now given $f_\alpha'\colon X_\alpha \to X'$ satisfying $f_\beta' f_{\alpha\beta} = f_\alpha'$, one can define $F\colon X \to X'$ by $F(f) = \sum_\alpha f_\alpha'(f(\alpha))$;

$$F(\bar{f}_\alpha(x)) = f_\alpha'(x) = f_\beta' f_{\alpha\beta}(x) = F(\bar{f}_\beta(f_{\alpha\beta}(x)),$$

so F defines a homomorphism $f\colon \varinjlim X_\alpha \to X'$ with $f\pi = F$. As before it is clear that f is unique. ∎

We think of the objects X_α as approximations to $\varinjlim X_\alpha$ in much the same way that the elements of a sequence of numbers are approximations to their limit. The condition that for all α, $\beta \in \mathcal{A}$ there exists $\gamma \in \mathcal{A}$ with $\alpha \le \gamma$ and $\beta \le \gamma$ plays a similar role to the Cauchy condition on sequences, in assuring that limits are unique.

Proposition 15.4 Let $\{X_\alpha\}$ be a directed system and assume that there is $\alpha_0 \in \mathcal{A}$ such that if $\beta \ge \alpha \ge \alpha_0$, $f_{\alpha\beta}$ is an equivalence (i.e., has a two-sided inverse). Then $\varinjlim X_\alpha \cong X_{\alpha_0}$.

Proof Define $f_\alpha : X_\alpha \to X_{\alpha_0}$ by

$$f_\alpha = \begin{cases} f_{\alpha_0\alpha}^{-1} & \text{if } \alpha \ge \alpha_0 \\ f_{\alpha\alpha_0} & \text{if } \alpha_0 \ge \alpha. \end{cases}$$

Then if $\alpha \le \beta$, $f_\beta f_{\alpha\beta} = f_\alpha$, so $\{X_{\alpha_0}, f_\alpha\}$ is a candidate for the direct limit. Suppose $f_\alpha'\colon X_\alpha \to X'$ is defined and satisfies $f_\beta' f_{\alpha\beta} = f_\alpha'$. If there exists $f\colon X_{\alpha_0} \to X'$ with $ff_\alpha = f_\alpha'$, we must have $f = f_{\alpha_0}'$ since $f_{\alpha_0} = 1$. On the other hand, $f_{\alpha_0}' f_\alpha = f_\alpha'$ so this also defines a map $f\colon X_{\alpha_0} \to X'$. ∎

Proposition 15.5 Suppose X is a space and $X = \bigcup_{\alpha\in\mathcal{A}} X_\alpha$, where X_α are subspaces and suppose X has the weak topology on the X_α(i.e., F is closed iff $F \cap X_\alpha$ is closed in X_α for all α). Then $X \equiv \varinjlim X_\alpha$, where the inclusion maps are used to form the directed system.

Remark To be consistent, it is necessary to assume that for all α, β there is a γ such that $X_\gamma \supset X_\alpha \cup X_\beta$. This can always be arranged by replacing $\{X_\alpha\}$ by the set of all finite unions $X_{\alpha_1} \cup \cdots \cup X_{\alpha_n}$.

Proof The inclusions provide maps $i_\alpha : X_\alpha \to X$ compatible with the inclusions $i_{\alpha\beta} : X_\alpha \subset X_\beta$. To show that the universal property is satisfied, suppose we are given $f_\alpha : X_\alpha \to Y$ satisfying $f_\alpha |_{X_\beta} = f_\beta$ when $\beta \le \alpha$. We are then forced to define $f\colon X \to Y$ by $f|_{X_\alpha} = f_\alpha$. These definitions are compatible, and f is continuous since if F is closed in Y, $f^{-1}(F) \cap X_\alpha = f_\alpha^{-1}(F)$, which is a closed subset of X_α for all α. ∎

Corollary 15.6 Let X be a CW complex. Then $X \equiv \varinjlim X^n$ where X^n is the n-skeleton. ▮

Corollary 15.7 Let X be a CW complex and let $\{X_\alpha\}$ be the set of finite subcomplexes. Then $X \equiv \varinjlim X_\alpha$. ▮

Proposition 15.8 Let $\{X_\alpha\}$ be a directed system in $\mathcal{M}_R$, and $X = \varinjlim X_\alpha$. Then:

(a) for each $x \in X$, there is an α and $x_\alpha \in X_\alpha$ with $f_\alpha(x_\alpha) = x$;
(b) if $x_\alpha \in X_\alpha$ and $f_\alpha(x_\alpha) = 0$, there is a $\beta \geq \alpha$ with $f_{\alpha\beta}(x_\alpha) = 0 \in X_\beta$.

Proof To prove (a), let $x = \pi\{\bar{f}_{\alpha_1}(x_1) + \cdots + f_{\alpha_n}(x_n)\}$. Pick $\alpha \geq \alpha_i$ for each i, $1 \leq i \leq n$. Then $f_{\alpha_i}(x_i) = \bar{f}_\alpha(f_{\alpha_i\alpha}(x_i))$, so

$$x = \pi \bar{f}_\alpha(f_{\alpha_1\alpha}(x_1) + \cdots + f_{\alpha_n\alpha}(x_n)).$$

To prove (b), define $F_\beta = \bigoplus_{\alpha \leq \beta} X_\alpha$, and define $\theta_\beta\colon F_\beta \to X_\beta$ by $\theta_\beta(f) = \sum f_{\beta\alpha}(f(\alpha))$. If $\alpha \leq \beta$, $\bar{f}_\alpha(X_\alpha) \subset F_\beta$ and $\theta_\beta \bar{f}_\alpha = f_{\beta\alpha}$. Suppose now that $f_\alpha(x_\alpha) = 0$. Then $\bar{f}_\alpha(x_\alpha) \in M$, so

$$\bar{f}_\alpha(x_\alpha) = \sum_{i=1}^{n} \{\bar{f}_{\alpha_i}(x_i) - \bar{f}_{\alpha_i}(f_{\alpha_i\beta_i}(x_i))\}.$$

Choose $\beta \geq \beta_i \geq \alpha_i$. Then all terms of this equation belong to F_β. By applying θ_β to this equation, we get $f_{\beta\alpha}(x_\alpha) = 0$, since $f_{\alpha_i\beta}(x_i) = f_{\beta_i\beta}(f_{\alpha_i\beta_i}(x_i))$. ▮

Proposition 15.9 Let $X = \bigcup X_\alpha$ be Hausdorff and have the weak topology as above and assume:

(a) For all α, $\beta \in A$ there exists $\delta \in A$ with $X_\alpha \cap X_\beta = X_\delta$.
(b) For all $\alpha \in A$, $\{\beta \in A \mid \beta \leq \alpha\}$ is finite ($\beta \leq \alpha$ iff $X_\beta \subset X_\alpha$).

Then $\pi_i(\varinjlim X_\alpha) = \varinjlim \pi_i(X_\alpha)$.

The proof will rely on:

Lemma 15.10 Under the hypothesis of 15.9, given a compact set $K \subset X$, there are $\alpha_i \in A$ with $K \subset X_{\alpha_1} \cup \cdots \cup X_{\alpha_n}$

Proof of 15.10 Let $e_\alpha = X_\alpha - \bigcup\{X_\beta \mid \beta < \alpha\}$. Suppose $K \not\subset X_{\alpha_1} \cup \cdots \cup X_{\alpha_n}$ for any choice of $\alpha_1, \ldots, \alpha_n$; choose inductively distinct points $x_{\alpha_i} \in e_{\alpha_i}$ as follows: Having chosen $x_{\alpha_1}, \ldots, x_{\alpha_n}$, note that $K \not\subset e_{\alpha_1} \cup \cdots \cup e_{\alpha_n}$ since $e_{\alpha_1} \cup \cdots \cup e_{\alpha_n} \subset X_{\alpha_1} \cup \cdots \cup X_{\alpha_n}$. Hence there exists $x \in K \cap X_\alpha$ for some α

and $x \notin \{x_{\alpha_1}, \ldots, x_{\alpha_n}\}$. By (b), there is $\alpha_{n+1} \leq \alpha$ with $x \in e_{\alpha_{n+1}}$. Let $x_{\alpha_{n+1}} = x$. Now let $V \subset \{x_{\alpha_i}\}$. We prove that V is closed. Suppose $x_\beta \in V \cap X_\alpha$. Now $x_\beta \in X_\beta \cap X_\alpha = X_\delta$ by (a). Since $x_\beta \in e_\beta$, $\delta \geq \beta$ and hence $\beta \leq \alpha$. Thus (b) implies that $V \cap X_\alpha$ is finite and hence closed (since X is Hausdorff). It follows that $\{x_{\alpha_i}\}$ is an infinite set with the discrete topology. On the other hand, $\{x_{\alpha_i}\}$ is a closed subset of K and hence is compact, a contradiction. ▮

Proof of 15.9 The maps $i_\alpha : X_\alpha \to X$ induce maps $(i_\alpha)_* : \pi_i(X_\alpha) \to \pi_i(X)$. Since these are consistent with the inclusions $(i_{\alpha\beta})_* : \pi_i(X_\alpha) \to \pi_i(X_\beta)$, they define a map $I: \varinjlim \pi_i(X_\alpha) \to \pi_i(X)$. Suppose $\{f\} \in \pi_i(X)$. Since $f(S^i)$ is compact, $f(S^i) \subset X_\alpha \subset X$; hence $If_\alpha(\{f\}) = \{f\}$ where $f_\alpha : \pi_i(X_\alpha) \to \varinjlim \pi_i(X_\alpha)$. Suppose $I(x) = 0$. Let $x = f_\alpha(\{f\})$ where $f: S^i \to X_\alpha$. Then $I(x) = \{i_\alpha f\} = 0$. Choose a homotopy $H: B^{i+1} \to X$. Since $H(B^{i+1})$ is compact, there exists β with $H(B^{i+1}) \subset X_\beta$. Now clearly $H: i_{\alpha\beta} f \sim 0$ in $\pi_i(X_\beta)$ so $x = 0$. ▮

Corollary 15.11 Let X be a CW complex. Then

(a) $\pi_i(X) \cong \varinjlim \pi_i(X^n)$

(b) $\pi_i(X) \cong \varinjlim \pi_i(X_\alpha)$, where the limit is taken over all finite subcomplexes.

Proof This follows immediately from 15.6, 15.7, and 15.9. ▮

Lemma 15.12 Let $\{A_\alpha, f_{\alpha\beta}\}$ and $\{B_\alpha, f'_{\alpha\beta}\}$ be two systems of abelian groups directed over the same index set A, and suppose that for each $\alpha \in A$ there is given a homomorphism $g_\alpha : A_\alpha \to B_\alpha$ such that if $\alpha \leq \beta$, the diagram

$$\begin{array}{ccc} A_\alpha & \xrightarrow{f_{\alpha\beta}} & A_\beta \\ \downarrow{\scriptstyle g_\alpha} & & \downarrow{\scriptstyle g_\beta} \\ B_\alpha & \xrightarrow{f'_{\alpha\beta}} & B_\beta \end{array}$$

commutes. Then there is a unique homomorphism $g: \varinjlim A_\alpha \to \varinjlim B_\alpha$ such that the diagram

$$\begin{array}{ccc} A_\alpha & \xrightarrow{f_\alpha} & \varinjlim A_\alpha \\ \downarrow{\scriptstyle g_\alpha} & & \downarrow{\scriptstyle g} \\ B_\alpha & \xrightarrow{f'_\alpha} & \varinjlim B_\alpha \end{array}$$

commutes.

Proof $f_\alpha' g_\alpha : A_\alpha \to \varinjlim B_\alpha$ is defined for each $\alpha \in A$. By the universal property (15.2) g exists uniquely. ▮

Lemma 15.13 Let $\{A_\alpha, f_{\alpha\beta}\}$, $\{B_\alpha, f'_{\alpha\beta}\}$, and $\{C_\alpha, f''_{\alpha\beta}\}$ be three systems directed over A and suppose we have maps $g_\alpha : A_\alpha \to B_\alpha$ and $h_\alpha : B_\alpha \to C_\alpha$ as in 15.12 such that

$$A_\alpha \xrightarrow{g_\alpha} B_\alpha \xrightarrow{h_\alpha} C_\alpha$$

is exact at B_α. Then

$$\varinjlim A_\alpha \xrightarrow{g} \varinjlim B_\alpha \xrightarrow{h} \varinjlim C_\alpha$$

is exact at $\varinjlim B_\alpha$.

Proof Let $x \in \varinjlim A_\alpha$. By 15.8, choose α and $x_\alpha \in A_\alpha$ with $f_\alpha(x_\alpha) = x$. Then

$$hg(x) = h(g(f_\alpha(x_\alpha))) = h(f_\alpha'(g_\alpha(x_\alpha))) = f_\alpha''(h_\alpha(g_\alpha(x_\alpha))) = f_\alpha''(0) = 0.$$

Let $x \in \varinjlim B_\alpha$ and suppose $h(x) = 0$. Let $x_\alpha \in B_\alpha$ be such that $f_\alpha'(x_\alpha) = x$. Then $f_\alpha''(h_\alpha(x_\alpha)) = h(f_\alpha'(x_\alpha)) = h(x) = 0$. Hence by 15.8, there exists β with $f''_{\alpha\beta}(h_\alpha(x_\alpha)) = 0$. Consequently $h_\beta f'_{\alpha\beta}(x_\alpha) = f''_{\alpha\beta}(h_\alpha(x_\alpha)) = 0$ and thus $f'_{\alpha\beta}(x_\alpha) = g_\beta(x_\beta)$. Now

$$gf_\beta(x_\beta) = f_\beta' g_\beta(x_\beta) = f_\beta' f'_{\alpha\beta}(x_\alpha) = f_\alpha'(x_\alpha) = x,$$

so ker $h = \operatorname{Im} g$. ∎

Proposition 15.14 $(\varinjlim X_\alpha) \wedge Y \equiv \varinjlim (X_\alpha \wedge Y)$, if $X_\alpha, Y \in \mathcal{CG}^*$.

Proof We produce continuous maps going both ways that are inverse to one another. The inclusion $i_\alpha : X_\alpha \to \varinjlim X_\alpha$ induces a map.

$$i_\alpha \wedge 1 : X_\alpha \wedge Y \to (\varinjlim X_\alpha) \wedge Y;$$

since these maps are compatible, they induce a map

$$I: \varinjlim(X_\alpha \wedge Y) \to (\varinjlim X_\alpha) \wedge Y.$$

Let $k_\alpha : X_\alpha \to [\lim (X_\alpha \wedge Y)]^Y$ be the adjoint of the inclusion

$$X_\alpha \wedge Y \to \varinjlim (X_\alpha \wedge Y).$$

The maps k_α are compatible and hence induce a map

$$K: \varinjlim X_\alpha \to (\varinjlim X_\alpha \wedge Y)^Y;$$

the adjoint of this is the inverse to I. ∎

Exercises

1. Show that limits directed over the positive integers exist in $\mathcal{T}_h{}^*$. (Hint: Given X_n for each $n > 0$ and $f_n : (X_n, *) \to (X_{n+1}, *)$, we define

$$X_\infty = \amalg (X_n \times I)/(x_n, 1) \sim (f_n(x_n), 0); (*, t) \sim *.$$

See Fig. 15.1. This construction is sometimes called the telescope construction.) (Exercise 2; 29.14)

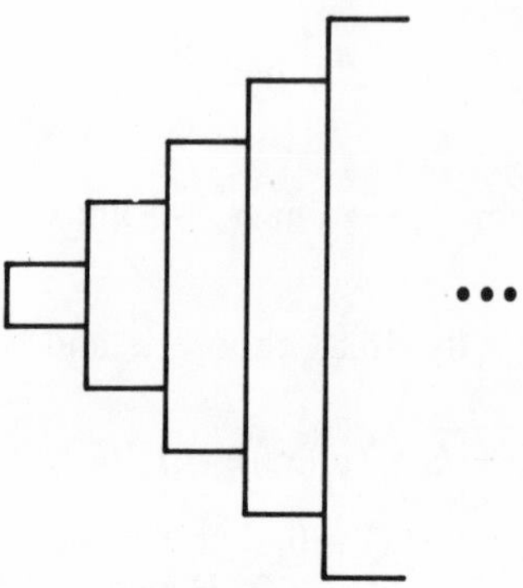

Figure 15.1

2. Show that limits directed over the positive integers exist in $\mathcal{K}_h$ (see Exercise 1).

3. Show that limits exist in $\mathcal{C}^*$ and $\mathcal{C}^2$. What about $\mathcal{C}_h$?

4. Direct limits do not exist in $\mathcal{CG}$. (Hint: Define $X_n = I$ for each $n \geq 0$ and $f_n : X_n \to X_{n+1}$ by $f_n(x) = \min(2x, 1)$.)

5. Suppose $X_n \subset X_{n+1}$ and $X = \bigcup X_n$ with X a CW complex and X_i a subcomplex. Show that $X_\infty \simeq X$ (see Exercise 1). (Hint: Define f: $X_\infty \to \bigcup X_n$ and show that it induces isomorphisms in homotopy. See 16.22.)

6. Consider the directed sequence of abelian groups $X_n = Z$ for $n \geq 0$ and $f_{n,n+1}(x) = nx$. Show that $\varinjlim X_n \cong Q$ (the rational numbers). (Exercise 1, Section 24)

7. Generalizing the ingredients of Exercise 6, show that every abelian group is isomorphic to the direct limit of its finitely generated subgroups, directed by inclusion.

8. Let $\{X_\alpha\}$ be a directed system of spaces and inclusion maps where the indexing set A satisfies (a) and (b) of 15.9. Suppose $\{B_\alpha\}$ is another such system with $B_\alpha \subset X_\alpha$. Show that $\varinjlim \pi_i(X_\alpha, B_\alpha, *) = \pi_i(X, B, *)$ where $X = \bigcup X_\alpha$, $B = \bigcup B_\alpha$, and $* \in \bigcap_{\alpha \in A} B_\alpha$. (16.4)

9. Given an inverse system $\{X_\alpha, f_{\alpha\alpha'}\}$ in $\mathcal{S}$ where $f_{\alpha\alpha'} : X_{\alpha'} \to X_\alpha$ for $\alpha' \geq \alpha$, define $\varprojlim\{X_\alpha, f_{\alpha\alpha'}\}$ as the subset of $\prod X_\alpha$ of those functions $\{x_\alpha\}$ with $f_{\alpha\alpha'}(x_{\alpha'}) = x_\alpha$. Show that this is an inverse limit in $\mathcal{S}$. If for each α, X_α is a topological space and $f_{\alpha\alpha'}$ is continuous, show that this is an inverse limit in $\mathcal{C}$ if we use the subspace topology on $\varprojlim\{X_\alpha, f_{\alpha\alpha'}\}$ and the product topology on ΠX_α.

If for each α, X_α is an R-module, show that $\varprojlim X_\alpha$ with coordinatewise addition is an inverse limit in $\mathcal{M}_R$. (Exercise 2, Section 17; 21.20; 27.4)

10. The conditions of 15.9 are necessary, for let $\{A_\alpha\}$ be the collection of countable compact subsets of S^1. Then S^1 has the weak topology on $\{A_\alpha\}$, but every map $S^1 \to A_\alpha$ is constant. Hence $\pi_1(S^1) = Z$ and $\varinjlim \pi_1(A_\alpha) = 0$.

11. Show that if all spaces belong to $\mathcal{CG}$ (or $\mathcal{CG}^*$),

$$(\varinjlim X_\alpha) \times Y \equiv \varinjlim (X_\alpha \times Y) \qquad \text{and} \qquad (\varinjlim X_\alpha) \vee Y \equiv \varinjlim (X_\alpha \vee Y). \text{ (16.6)}$$

12. Prove $(\bigvee_{\alpha \in \mathcal{A}} X_\alpha) \wedge Y \equiv \bigvee_{\alpha \in \mathcal{A}} (X_\alpha \wedge Y)$ (Hint: Use 15.14.) (24.1)

13. Let R be a commutative ring. Let $\{X_\alpha\}$ be a direct system in $\mathcal{M}_R$ and $Y \in \mathcal{M}_{R'}$. Prove that $(\varinjlim X_\alpha) \otimes_R Y \cong \varinjlim (X_\alpha \otimes_R Y)$. (24.1, Section 26).

14. Given a collection of R-modules $\{X_\alpha\}$, define $\prod_{\alpha \in \mathcal{A}} X_\alpha$ as the set of all functions $f \colon A \to \bigcup_{\alpha \in \mathcal{A}} X_\alpha$ with $f(\alpha) \in X_\alpha$. This is an R-module, as in the proof of 15.3, and $\oplus_{\alpha \in \mathcal{A}} X$ is a submodule. Show that

$$\hom_R\Big(\bigoplus_{\alpha \in \mathcal{A}} X_\alpha, Y\Big) \cong \prod_{\alpha \in \mathcal{A}} \hom_R (X_\alpha, Y). \quad (24.11)$$

15. Let X_∞ be the construction from Exercise 1. Prove that if K is compact,

$$[(K, *), (X_\infty, *)] \cong \varinjlim[(K, *), (X_n, *)].$$

16. Let $A_{n,m}$ be a system of abelian groups with homomorphisms $f_{n,m} : A_{n,m} \to A_{n+1,m}$ and $g_{n,m} : A_{n,m} \to A_{n,m+1}$ such that the diagrams

$$\begin{array}{ccc} A_{n,m} & \xrightarrow{f_{n,m}} & A_{n+1,m} \\ \Big\downarrow{\scriptstyle g_{n,m}} & & \Big\downarrow{\scriptstyle g_{n,m+1}} \\ A_{n,m+1} & \xrightarrow{f_{n,m+1}} & A_{n+1,m+1} \end{array}$$

commute. Let $A_n = \varinjlim \{A_{n,m}, g_{n,m}\}$ and $B_m = \varinjlim\{A_{n,m}, f_{n,m}\}$. Prove that $\varinjlim A_n \cong \varinjlim B_m$. (Exercise 5, Section 28)

16

The Homotopy Theory of CW Complexes

This section deals with the homotopy theory of CW complexes. Most of the results are fairly technical—relating homotopy groups and cellular structures to various extension and deformation problems. We discuss the approximation of spaces by CW complexes, construct an (ad hoc) singular complex for a space and prove Whitehead's theorem. Using these techniques we prove the general version of the Blakers–Massey theorem and make applications. In an appendix we give the functorial singular complex construction. This will be used in Section 21.

Throughout this section we shall discuss CW complexes X and relative CW complexes (X, A) (as defined in Exercise 6, Section 14). If desired, one may replace the phrase "relative CW complex (X, A)" by (the more restricted notion of a) "pair (X, A) where X is a CW complex and A is a subcomplex." Such a change will not restrict most of the applications of our results.

Definition 16.1 A space X is said to be n-connected if $\pi_i(X, *) = 0$ for $i \leq n$.

Remark This is independent of the choice of $*$. 0-connected is the same as arcwise connected, and 1-connected is the same as simply connected. S^n is $(n-1)$-connected by 12.23.

Lemma 16.2 Let $g: S^{n-1} \to X$, $f: X \to Y$, and suppose Y is arcwise connected and $\pi_{n-1}(Y, *) = 0$. Then there is an extension of f to $X \cup_g e^n$:

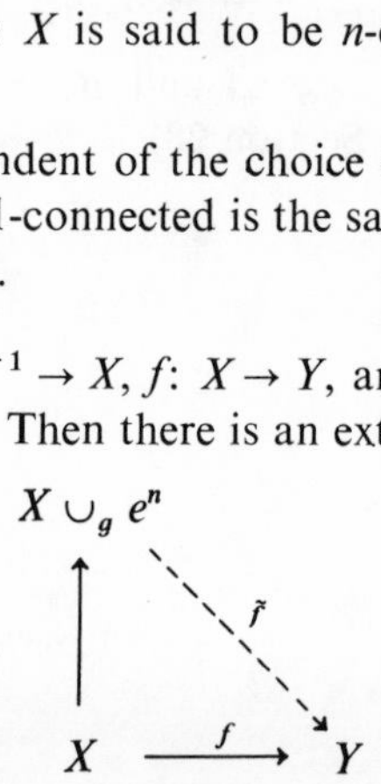

Proof This follows from Exercise 11, Section 9 and 14.15. ▮

Corollary 16.3 Suppose $\mathcal{S}$ is a set of integers and (Y, X) is a relative CW complex such that if $e_\alpha \subset Y - X$ is a cell, $\dim e_\alpha \in \mathcal{S}$. Suppose that if $n \in \mathcal{S}$, $\pi_{n-1}(Z, *) = 0$. Then any map $f\colon X \to Z$ admits an extension $\tilde{f}\colon Y \to Z$:

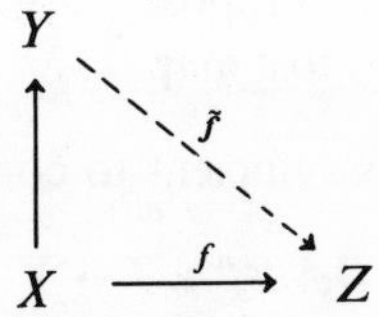

Proof Given any finite subcomplex of (Y, X), it is clear that f extends over this subcomplex, for the extension can be done a cell at a time using 16.2. In the general case we apply Zorn's lemma. Consider the set C of all pairs (X_α, f_α) where $X \subset X_\alpha \subset Y$ is a subcomplex and $f_\alpha : X_\alpha \to Y$ is a extension of f. Partially order C by inclusion and restriction: $(X_\alpha, f_\alpha) < (X_\beta, f_\beta)$ if $X_\alpha \subset X_\beta$ and $f_\beta|_{X_\alpha} = f_\alpha$. Increasing chains in C have an upper bound since if $\{X_\alpha\}$ is such a chain, $(\bigcup X_\alpha, f)$ belongs to C, where $f|_{X_\alpha} = f_\alpha$ (f is continuous by 14.9). Thus C has a maximal element (X', f'). Now $X' = Y$, for if $X' \neq Y$ choose a cell $e \subset Y - X'$ of minimal dimension. Then $\bar{e} - e \subset X'$. Apply 16.2 to this cell to obtain an extension $\bar{f}\colon X' \cup e \to Z$ of f', contradicting maximality. ▮

Proposition 16.4 Suppose (Y, X) is a relative CW complex with cells (in $Y - X$) only in dimensions $\geq n$, then $\pi_i(Y, X, *) = 0$ for $i < n$.

Proof By Exercise 8, Section 15, it is enough to show that if (X_α, X) is a finite subcomplex of (Y, X), $\pi_i(X_\alpha, X, *) = 0$ for all $i < n$. We do this by induction making key use of 13.5. Suppose (X', X) is a finite subcomplex of (Y, X) and $\overline{X} = X' \cup e^k \subset Y$. The exact sequence

$$\pi_i(X', X, *) \to \pi_i(\overline{X}, X, *) \to \pi_i(\overline{X}, X', *)$$

has first and last terms equal to 0 for $i < n$ by induction and 13.5 since $k \geq n$. Hence $\pi_i(\overline{X}, X, *) = 0$ for $i < n$. ▮

Corollary 16.5 If X is a CW complex with one 0-cell and all other cells in dimensions $\geq n$, $\pi_i(X, *) = 0$ for $i < n$.

We now prove a general deformation theorem.

Theorem 16.6 Let $\mathcal{S}$ be a set of integers. Let $f\colon (X, A) \to (Y, B)$ and suppose (X, A) is a relative CW complex with cells (in $X - A$) whose dimensions belong to $\mathcal{S}$. Suppose that for any choice of $*$, $\pi_k(Y, B, *) = 0$ if $k \in \mathcal{S}$. Then there is a map $g\colon X \to B$ with $g \sim f$ (rel A).

The inductive step will be based on:

Lemma 16.7 Let (X, A) be a relative CW complex and $\bar{X}^n = X^n \cup A$. Let $H\colon \bar{X}^{n-1} \times I \cup \bar{X}^n \times 0 \to Y$. Suppose[10] $\pi_n(Y, B, *) = 0$ for any choice of $*$ and $H(x, 1) \in B$. Then H extends to a map $\tilde{H}\colon \bar{X}^n \times I \to Y$ such that $\tilde{H}(x, 1) \in B$.

Proof of 16.7 By 14.17 it is sufficient to construct maps

$$H_\alpha\colon B_\alpha^n \times I \to Y$$

such that

$$\begin{aligned} H_\alpha(x, t) &= H(\chi_\alpha(x), t) \qquad &&\text{for } x \in S_\alpha^{n-1} \\ H_\alpha(x, 0) &= H(\chi_\alpha(x), 0), &&H_\alpha(x, 1) \in B. \end{aligned}$$

By 11.6 there is a homeomorphism

$$\varphi\colon (B^n \times I, B^n \times 0 \cup S^{n-1} \times I) \to (B^n \times I, B^n \times 1).$$

One can easily check from the definition that $\varphi|_{S^{n-1} \times 1} = 1$. Define $h_\alpha\colon (B^n, S^n) \to (Y, B)$ by

$$h_\alpha = H \circ (\chi_\alpha \times 1) \circ \varphi^{-1}|_{B^n \times 1}.$$

By Exercise 7, Section 10 there is a homotopy $K_\alpha\colon B_\alpha^n \times I \to Y$ with $K_\alpha(x, 1) = h_\alpha(x)$, $K_\alpha(x, 0) = *$ and $K_\alpha(S_\alpha^{n-1} \times I) \subset B$. Let $H_\alpha = K_\alpha \circ \varphi$. Then $\varphi(S_\alpha^{n-1} \times I) \subset B_\alpha^n \times 1$, so $K_\alpha|_{\varphi(S_\alpha^{n-1} \times I)}) = h_\alpha|_{\varphi(S_\alpha^{n-1} \times I)}$. Consequently $H_\alpha(x, t) = H(\chi_\alpha(x), t)$ for $x \in S_\alpha^{n-1}$. Similarly, $H_\alpha(x, 0) = H(\chi_\alpha(x), 0)$. Since $\varphi^2 = 1$, $\varphi(x, 1) \in B_\alpha^n \times 0 \cup S_\alpha^{n-1} \times I$. Thus $H_\alpha(x, 1) \in B$. ∎

Proof of 16.6 Using 16.7 we construct homotopies $H^n\colon \bar{X}^n \times I \to Y$ by induction on n such that

(a) $H^n|_{\bar{X}^{n-1} \times I} = H^{n-1}$;
(b) $H^n(x, 0) = f(x)$;
(c) $H^n(x, 1) \in B$;
(d) $H^n(a, t) = f(a)$.

Define $H\colon X \times I \to Y$ by $H|_{\bar{X}^n \times I} = H^n$. This is well defined by (a) and is continuous by 15.6 and Exercise 11, Section 15. Define $g(x) = H(x, 1)$. By (c), $g(x) \in B$ and by (b) and (d), $H\colon f \sim g$ (rel A). ∎

[10] We interpret the statement $\pi_0(Y, B, *) = 0$ to mean that $\pi_0(B, *) \to \pi_0(Y, *)$ is onto; i.e., every point of Y may be joined by a path to a point in B.

Definition 16.8 Let (X, A) and (Y, B) be relative CW complexes and $f: (X, A) \to (Y, B)$. f is called cellular if $f(\bar{X}^n) \subset \bar{Y}^n$.

Theorem 16.9 Let (X, A) and (Y, B) be relative CW complexes $f: (X, A) \to (Y, B)$. Then $f \sim g$ (rel A) where g is a cellular map.

Proof We construct homotopies $H^n: \bar{X}^n \times I \to Y$ inductively such that:

(a) $H^n|_{\bar{X}^{n-1} \times I} = H^{n-1}$;
(b) $H^n(x, 0) = f(x)$;
(c) $H^n(x, 1) \in \bar{Y}^n$;
(d) $H^n(a, t) = f(a)$.

by 16.7, using the fact that for any choice of $*$, $\pi_n(Y, \bar{Y}^n, *) = 0$ by 16.4. Define $H: X \times I \to Y$ by $H|_{\bar{X}^n \times I} = H^n$. As in 16.6, this completes the proof. ▌

Corollary 16.10 Let $f, g: (X, A) \to (Y, B)$ be cellular and homotopic (rel A). Then there is a cellular homotopy (rel A) between them ($H: (X, A) \times I \to (Y, B)$ is cellular if $H(\bar{X}^n \times I) \subset \bar{Y}^{n+1}$).

Proof $(X \times I, X \times 0 \cup A \times I \cup X \times 1)$ is a relative CW complex with n-skeleton $X \times 0 \cup \bar{X}^{n-1} \times I \cup X \times 1$. Apply 16.9 to the given homotopy to obtain a new one. ▌

As an important special case, we have:

Corollary 16.11 Every map between CW complexes is homotopic to a cellular map and every two homotopic cellular maps are cellularly homotopic.

Proof Apply 16.9 and 16.10 with $A = B = \varnothing$. ▌

Thus it is only necessary, from the homotopy theory point of view, to consider cellular maps and cellular homotopies. One should be careful, however, because one extra dimension is needed for a homotopy.

Example Let $X = S^n$ and $Y = B^{n+1}$ and consider the inclusion $X \to Y$. We make X into a complex with one 0-cell and one n-cell. We make Y into a complex with a 0-, n-, and $(n + 1)$-cell. Thus $X = Y^n$ and the inclusion is cellular. The inclusion is homotopic to another cellular map, namely, the map sending all of X to the 0-cell of Y. However, there is no homotopy $H: X^n \times I \to Y^n$ between these maps.

Definition 16.12 A map $f: X \to Y$ is called a weak homotopy equivalence if $f_*: \pi_n(X, x) \to \pi_n(Y, f(x))$ is 1–1 and onto for all $n \geq 0$ and all $x \in X$.

Note that this is not an equivalence relation. If X is arcwise connected, it is sufficient to consider f_* for one choice of $x \in X$.

Definition 16.13 Given a space Y, a cellular approximation to Y or resolution of Y is a pair (K, f) where K is a CW complex $f\colon K \to Y$ is a weak homotopy equivalence.

We will show that resolutions always exist and any two of them are equivalent (in a sense we will define later). However, there is considerable choice in finding a resolution (as the proof will indicate), and neither the dimensions of the cells nor the number of cells are invariants. As an example, consider the two resolutions of a one-point space: K_1 = one-point space with the point as a 0-cell; $K_2 = B^3$ with cells $e^0 = (1, 0, 0)$, $e^2 = S^2 - e^0$, and $e^3 = B^3 - S^2$.

If (K, f) is a resolution of Y, K is sometimes called a singular complex for Y.

Proposition 16.14 Given an $(n-1)$-connected space Y there is a resolution (K, f). If $n \geq 1$, we can furthermore assume that K has no cells of dimension $< n$ except for a single 0-cell $*$.

Before proving 16.14 we introduce a lemma.

Lemma 16.15 Given $f\colon X \to Y$ such that $f|_A$ is an inclusion, there is a commutative diagram

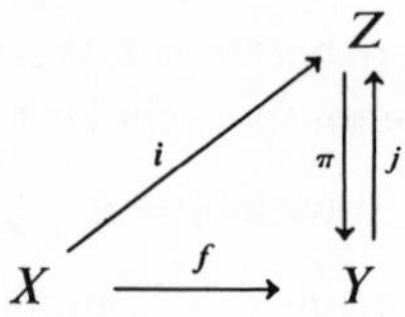

where i is an inclusion, $\pi j = 1$ and $j\pi \sim 1$ (rel $i(A)$). The space Z is called the mapping cylinder of the map f.

Remark To visualize this consider first the case $A = \varnothing$. Thus we replace Y with Z where $Z \simeq Y$ and f then corresponds to the inclusion i. If $f|_A$ is an inclusion, one can achieve the same result, but without altering Y on $f(A) \equiv A$.

Proof Define $Z = Y \cup X \times I/(x, 0) \sim f(x)$, $(a, t) \sim f(a)$. See Fig. 16.1. Define $i(x) = (x, 1)$, $\pi(y) = y$, $\pi(x, t) = f(x)$, and $j(y) = y$. All the claims are obvious. The homotopy $H\colon j\pi \sim 1$ (rel $i(A)$) is given by $H(y, t) = y$, $H(x, s, t) = (x, st)$. (Compare this construction with 11.14.) ∎

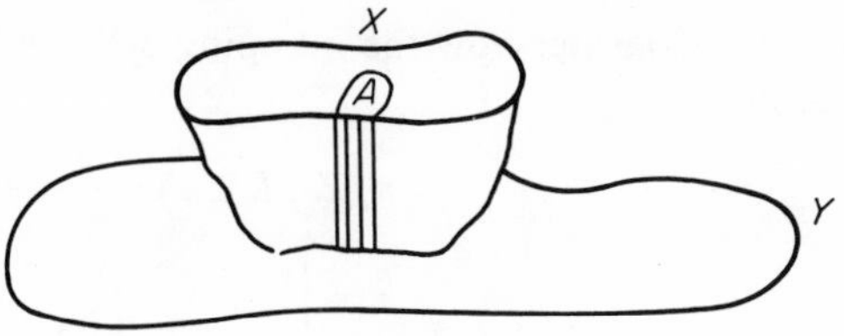

Figure 16.1

Definition 16.16 Let $f_i\colon G_i \to H_i$ be a homomorphism for each $i \in Z^+$. $\{f_i\}$ is called a k-isomorphism if f_i is an isomorphism for $i < k$ and f_k is onto. If X and Y are spaces and $f\colon X \to Y$, f is called a k-equivalence[11] if $f_*\colon \pi_i(X, x) \to \pi_i(Y, f(x))$ is a k-isomorphism for all $x \in X$.

Proof of 16.14 Assume first that $n \geq 1$. By induction we will construct an m-dimension complex K^m such that

$$K^m \supset K^{m-1} \supset \cdots \supset K^0 = *,$$

and m-equivalences $f_m\colon (K^m, *) \to (Y, *)$ such that $f_m|_{K^{m-1}} = f_{m-1}$. We begin the induction with $* = K^0 = K^{n-1}$, and $f_{n-1}(*) = *$. Suppose now that we have constructed (K^m, f_m). Let Z be the mapping cylinder of f_m $(A = *)$. Then we have a commutative diagram

$$\begin{array}{ccccccc}
 & & & & \pi_i(Z, *) & \longrightarrow & \pi_i(Z, K^m, *) \longrightarrow \cdots \\
 & & & \nearrow^{i_*} & & & \\
\cdots \longrightarrow & \pi_{i+1}(Z, K^m, *) & \longrightarrow & \pi_i(K^m, *) & \approx \downarrow \pi_* & & \\
 & & & \searrow_{(f_m)_*} & & & \\
 & & & & \pi_i(Y, *) & &
\end{array}$$

Hence $\pi_i(Z, K^m, *) = 0$ for $i \leq m$. Let $\{f_\alpha\}$ generate $\pi_{m+1}(Z, K^m, *)$, $f_\alpha\colon (B_\alpha^{m+1}, S_\alpha{}^m, *) \to (Z, K^m, *)$. We construct K^{m+1} as follows:

$$K^{m+1} = K^m \cup \amalg\, B_\alpha^{m+1}/x \sim f_\alpha(x) \qquad \text{for} \quad x \in S_\alpha{}^m \subset B_\alpha^{m+1}.$$

K^{m+1} is a closure finite cell complex, and we give it the weak topology. Hence K^{m+1} is a CW complex and K^m is a subcomplex.

Define $F\colon (K^{m+1}, K^m) \to (Z, K^m)$ extending the identity by $F|_{B_\alpha{}^{m+1}} = f_\alpha\colon (B_\alpha^{m+1}, S_\alpha{}^m, *) \to (Z, K^m, *)$. Define $f_{m+1} = \pi F$ where $\pi\colon Z \to Y$. Then

[11] The literature is not consistent on the use of the term k-equivalence. For example, Whitehead does not assume that f_k is onto [73].

$f_{m+1}|_{K^m} = \pi|_{K^m} = f_m$. Consider now the mapping cylinder Z' of F with $A = K^m$, and the diagram

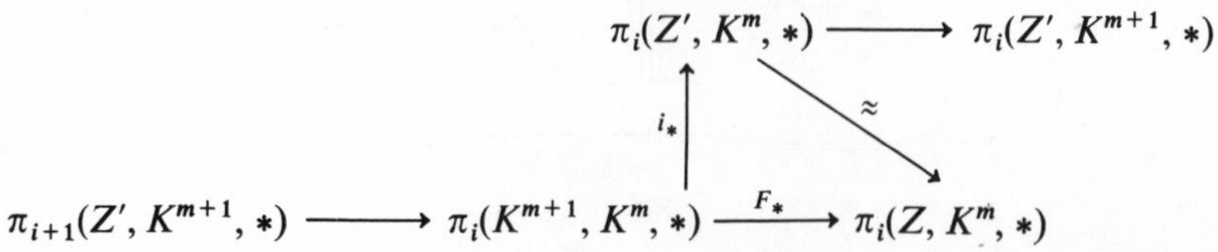

$\pi_i(Z, K^m, *) = 0$ for $i \leq m$ and F_*: $\pi_{m+1}(K^{m+1}, K^m, *) \to \pi_{m+1}(Z, K^m, *)$ is onto by construction. Hence $\pi_i(Z', K^{m+1}, *) = 0$ for $i \leq m + 1$, and i_* is an $m + 1$ isomorphism. Now consider π': $Z' \to Z$ and π: $Z \to Y$. These are homotopy equivalences and $\pi\pi'i = \pi F = f_{m+1}$. Hence $(f_{m+1})_*$ has the desired properties, and the induction is complete.

We now define $K = \bigcup K^m$ and f: $K \to Y$ by $f|_{K^m} = f_m$. If we give K the weak topology, we have $\pi_i(K, K^m, *) = 0$ for $i \leq m$. Hence in the diagram

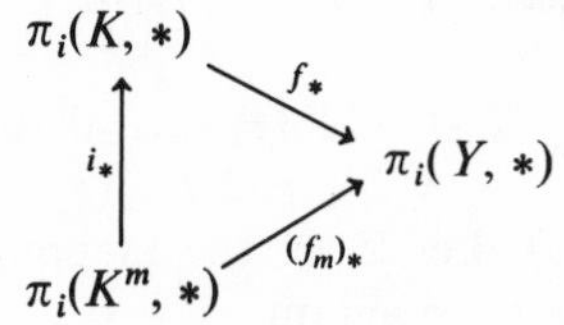

all maps are isomorphisms.

If $n = 0$, choose a resolution K_α for each arc component Y_α of Y and let $K = \amalg K_\alpha$. Define f by setting $f|_{K_\alpha}$ to be a resolution of Y_α. ∎

Lemma 16.17 Suppose f: $X \to Y$ is a base point preserving map. f is a weak homotopy equivalence iff given any CW pair (L, L_0) and maps α: $L_0 \to X$, β: $L \to Y$ with $f\alpha = \beta|_{L_0}$ there is a map g: $L \to X$ with $g|_{L_0} = \alpha$ and $fg \sim \beta$ (rel L_0)

$$
\begin{array}{ccc}
X & \xrightarrow{\ f\ } & Y \\
\uparrow{\scriptstyle\alpha} & \nwarrow{\scriptstyle g} & \uparrow{\scriptstyle\beta} \\
L_0 & \longrightarrow & L
\end{array}
$$

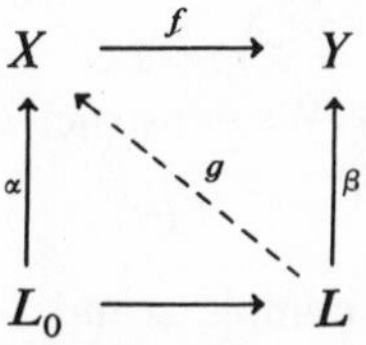

Proof If this property is satisfied, consideration of the diagrams

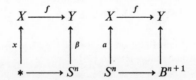

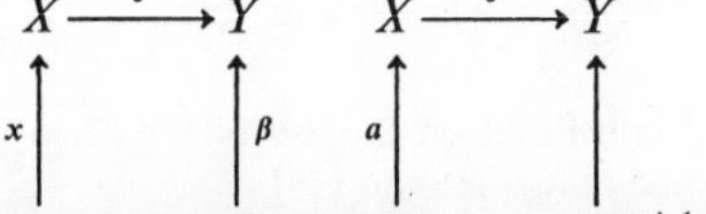

leads to the conclusion that f_* is onto and 1–1 in π_n, for any choice of $x \in X$. Suppose conversely that f induces isomorphisms in homotopy. Let Z be the mapping cylinder of f:

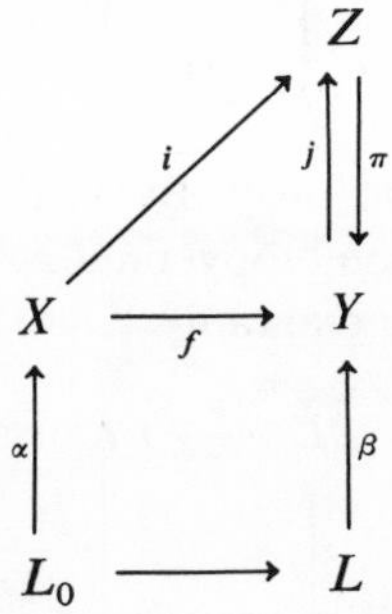

Define $\bar{F}$: $L \times 0 \cup L_0 \times I \to Z$ by $\bar{F}(l, 0) = j\beta(l)$ and $\bar{F}(l, t) = (\alpha(l), t)$ for $l \in L_0$. Extend $\bar{F}$ to F: $L \times I \to Z$ by 14.13. Let γ: $L \to Z$ be given by $\gamma(l) = F(l, 1)$. Then $\gamma(L_0) \subset X \times 1$. Apply 16.6 to produce g: $L \to X \times 1$ with $g \sim \gamma$ (rel L_0). Now $g|_{L_0} = \alpha$, and $fg = \pi g \sim \pi \gamma \sim \beta$ (rel L_0) where the last homotopy is given by $(l, t) \to \pi F(l, t)$. ∎

Definition 16.18 A map f: $(X, A) \to (Y, B)$ is called a weak homotopy equivalence if the associated maps f: $X \to Y$ and $f|_A$: $A \to B$ are weak homotopy equivalences. A resolution of a pair (X, A) is a CW pair (K, L) and a map f: $(K, L) \to (X, A)$ which is a weak homotopy equivalence.

Proposition 16.19 Any pair (X, A) has a resolution.

Proof Let f_0: $L \to A$ be a resolution. By adding cells to L we may form a complex $K \supset L$ and an extension of f_0 to a resolution f of X by Exercise 7.

Proposition 16.20 Let f: $(K, L) \to (X, A)$ and g: $(K', L') \to (Y, B)$ be resolutions and h: $(X, A) \to (Y, B)$. Then there is a map φ: $(K, L) \to (K', L')$ unique up to homotopy of pairs such that

$$
\begin{array}{ccc}
(K', L') & \xrightarrow{\ g\ } & (Y, B) \\
\uparrow{\scriptstyle \psi} & & \uparrow{\scriptstyle h} \\
(K, L) & \xrightarrow{\ f\ } & (X, A)
\end{array}
$$

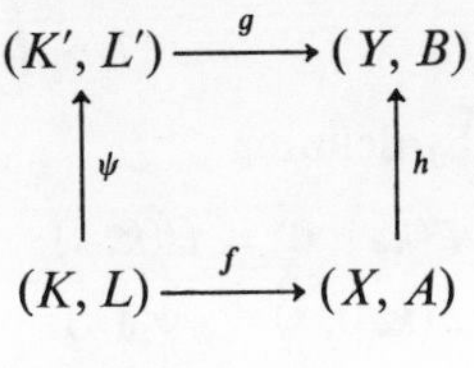

commutes up to homotopy (of pairs).

Proof Apply 16.17 to the commutative diagram

$$\begin{array}{ccc} \varnothing & \longrightarrow & L' \\ \big\downarrow & & \big\downarrow g \\ L & \xrightarrow{hf} & B \end{array}$$

to produce $\psi' \colon L \to L'$ with $g\psi' \sim hf$. By 14.13 $\gamma \colon K \times 0 \cup L \times I \to Y$ given by hf on K and the above homotopy on $L \times I$ extends to $\Gamma \colon K \times I \to Y$. Now apply 16.17 again to the diagram

$$\begin{array}{ccc} L & \xrightarrow{\psi'} & K' \\ \big\downarrow & & \big\downarrow g \\ K & \xrightarrow{\Gamma(\ ,1)} & Y \end{array}$$

to produce an extension $\psi \colon (K, L) \to (K', L')$. Then $g\psi \sim \Gamma(\ , 1) \sim hf$ and the homotopy maps L into B.

To prove uniqueness suppose ψ_0 and ψ_1 both satisfy the conclusion. Choose a homotopy $H \colon g\psi_0 \sim g\psi_1$. This is possible since both are homotopic to hf. We can thus apply 16.17 to the diagram

$$\begin{array}{ccc} L \times 0 \cup L \times 1 & \xrightarrow{\psi_0 \cup \psi_1} & L' \\ \big\downarrow \subset & & \big\downarrow g \\ L \times I & \xrightarrow{H} & B \end{array}$$

Let $J \colon L \times I \to L'$ be a homotopy between $\psi_0|_L$ and $\psi_1|_L$ such that $gJ \sim H$ (rel $L \times 0 \cup L \times 1$). Call this homotopy P. Then

$$P \colon L \times I \times I \to B$$

satisfies

$$\begin{aligned} P(l, s, 1) &= gJ(l, s) \qquad & P(l, 0, t) &= g\psi_0(l) \\ P(l, s, 0) &= H(l, s) \qquad & P(l, 1, t) &= g\psi_1(l). \end{aligned}$$

We will extend P to $K \times I \times I$. First we extend P to a map from

$$Z = (K \times I \times 0) \cup (K \times 0 \times I) \cup (L \times I \times I) \cup (K \times 1 \times I)$$

to Y. This is accomplished by defining

$$\begin{aligned} P(k, s, 0) &= H(k, s) \\ P(k, 0, t) &= g\psi_0(k) \\ P(k, 1, t) &= g\psi_1(k). \end{aligned}$$

Now $Z = (K \times I) \times 0 \cup (K \times 0 \cup L \times I \cup K \times 1) \times I$ and hence by 14.13 there is an extension to $K \times I \times I$ which we call $\bar{P}$. Define $\bar{H}$ to be the composition

$$K \times I \equiv K \times I \times 1 \subset K \times I \times I \xrightarrow{\bar{P}} Y.$$

Now

$$\bar{H}(k, 0) = \bar{P}(k, 0, 1) = g\psi_0(k)$$
$$\bar{H}(l, s) = \bar{P}(l, s, 1) = gJ(l, s)$$
$$\bar{H}(k, 1) = \bar{P}(k, 1, 1) = g\psi_1(k).$$

Again we have a commutative diagram

$$\begin{array}{ccc} K \times 0 \cup L \times I \cup K \times 1 & \xrightarrow{\psi_0 \cup J \cup \psi_1} & K' \\ \downarrow \subseteq & & \downarrow g \\ K \times I & \xrightarrow{\bar{H}} & Y \end{array}$$

and we apply 16.17 to construct $\bar{J}$: $K \times I \to K'$, a homotopy from ψ_0 to ψ_1 which extends J; thus $\psi_0 \sim \psi_1$ as maps of pairs $(K, L) \to (K', L')$. ∎

Proposition 16.21 There is a functor S: $\mathcal{C}^2 \to \mathcal{K}_h{}^2$ and a natural transformation f_X: $I \circ S(X, A) \to (X, A)$ where I: $\mathcal{K}_h{}^2 \to \mathcal{C}_h{}^2$ is the inclusion. Finally $(f_X)_*$ is a weak homotopy equivalence.

Proof We use the axiom of choice for classes. For each pair $(X, A) \in \mathcal{C}^2$ we choose a CW pair (K, L) and a map f: $(K, L) \to (X, A)$ which is a resolution. Define $S(X, A) = (K, L)$. If f: $(X, A) \to (Y, B)$, there is a unique homotopy class $S(f)$: $S(X, A) \to S(Y, B)$ by 16.20. By uniqueness, $S(1) = 1$ and $S(f \circ g) = S(f) \circ S(g)$. The natural transformation follows from the construction. ∎

There seems to be a lot of choice involved in the construction of this functor. This is more apparent than real, however. Given two resolutions (K, f) and (K', f') of X, 16.20 provides a map h: $K \to K'$ with $f'h \sim f$. Thus h induces isomorphisms in all homotopy groups. h is in fact a homotopy equivalence, as one sees from the following famous

Theorem 16.22[12] (*Whitehead Theorem*) Let X and Y be CW complexes and assume that g: $X \to Y$ is a weak homotopy equivalence. Then g is a homotopy equivalence.

[12] This theorem is usually attributed to Whitehead and appears in his classical paper [73], where CW complexes were first defined. An earlier theorem which is actually more general can be found in [33].

Proof Since $g\colon X \to Y$ is a resolution, 16.17 implies that there exists $h\colon Y \to X$ with $gh \sim 1$ in $\mathcal{C}^*$. Since h is a resolution, we can similarly find $j\colon X \to Y$ with $kj \sim 1$ in $\mathcal{C}^*$. Now $g \sim ghj \sim j$, hence $hg \sim 1$ and h is a homotopy inverse for g. ▮

Corollary 16.23 Suppose X is a connected CW complex and $\pi_i(X, *) = 0$ for $i < n$. Then X is the pointed homotopy type of a CW complex X' with no cells in dimensions $<n$ except for one 0-cell.

Remark In particular, every connected CW complex is the pointed homotopy type of a CW complex with only one 0-cell.

Proof Combine 16.22 and 16.14. ▮

Lemma 16.24[13] Suppose (X_1, X_2) is ecxisive in X and (Y_1, Y_2) is ecxisive in Y. Let $\varphi\colon Y \to X$ with $\varphi(Y_1) \subset X_1$ and $\varphi(Y_2) \subset X_2$. If $\varphi|_{Y_1}\colon Y_1 \to X_1$, $\varphi|_{Y_2}\colon Y_2 \to X_2$, and $\varphi|_{Y_1 \cap Y_2}\colon Y_1 \cap Y_2 \to X_1 \cap X_2$ are weak homotopy equivalences, so is φ.

Proof Given $f\colon \Delta^n \to X$, $g\colon \partial\Delta^n \to Y$ with $\varphi g = f|_{\partial\Delta^n}$ we will find $F\colon \Delta^n \to Y$ with $F|_{\partial\Delta^n} = g$ and $\varphi \circ F \sim f$ (rel $\partial\Delta^n$). This is clearly enough to prove that φ_* is an isomorphism:

$$\begin{array}{ccc} Y & \xrightarrow{\varphi} & X \\ {\scriptstyle g}\uparrow & \nwarrow^{F} & \uparrow{\scriptstyle f} \\ \partial\Delta^n & \longrightarrow & \Delta^n \end{array}$$

Let $A_i = g^{-1}(Y - \operatorname{Int} Y_i) \cup f^{-1}(X - \operatorname{Int} X_i)$ for $i = 1, 2$. Then A_1 and A_2 are disjoint closed sets. Subdivide Δ^n so that no simplex meets both A_1 and A_2. Define

$$K_i = \{\sigma \mid g(\sigma \cap \partial\Delta^n) \subset \operatorname{Int} Y_i, f(\sigma) \subset \operatorname{Int} X_i\}.$$

Then K_1 and K_2 are subcomplexes and $\Delta^n = K_1 \cup K_2$, for if σ is a simplex that misses A_i, $\sigma \subset K_i$. Furthermore $f(K_i) \subset \operatorname{Int} X_i$ and $g(K_i \cap \partial\Delta^n) \subset \operatorname{Int} Y_i$. By restriction we have a commutative diagram

$$\begin{array}{ccc} Y_1 \cap Y_2 & \xrightarrow{\varphi} & X_1 \cap X_2 \\ {\scriptstyle g}\uparrow & & \uparrow{\scriptstyle f} \\ \partial\Delta^n \cap K_1 \cap K_2 & \longrightarrow & K_1 \cap K_2 \end{array}$$

[13] This result is not in the standard expositions on homotopy theory. An equivalent result is stated without proof in [4, Section 10].

There exists $F\colon K_1 \cap K_2 \to Y_1 \cap Y_2$ with $F|_{K_1 \cap K_2 \cap \partial\Delta^n} = g$ and $\varphi F \sim f$ (rel $\partial\Delta^n \cap K_1 \cap K_2$). Now define $G_1\colon K_1 \cap (\partial\Delta^n \cup K_2) \to Y_1$ by $G_1|_{K_1 \cap \partial\Delta^n} = g|_{K_1 \cap \partial\Delta^n}$ and $G_1|_{K_1 \cap K_2} = F$. Then

$$\varphi G_1 \sim f|_{K_1 \cap (\partial\Delta^n \cup K_2)} \,(\text{rel } K_1 \cap \partial\Delta^n).$$

By 14.13 this homotopy may be extended to a homotopy $H\colon K_1 \times I \to X_1$ of f to a map $f_1\colon K_1 \to X_1$ with $f_1|_{K_1 \cap (\partial\Delta^n \cup K_2)} = \varphi G_1$. Thus the diagram

$$\begin{array}{ccc} Y_1 & \xrightarrow{\varphi} & X_1 \\ \uparrow{\scriptstyle G_1} & & \uparrow{\scriptstyle f_1} \\ K_1 \cap (\partial\Delta^n \cup K_2) & \longrightarrow & K_1 \end{array}$$

commutes and we may find $F_1\colon K_1 \to Y_1$ with

$$F_1|_{K_1 \cap (\partial\Delta^n \cup K_2)} = G_1 \qquad \text{and} \qquad \varphi F_1 \sim f_1 \quad (\text{rel}(K_1 \cap (\partial\Delta^n \cup K_2))).$$

This implies that $\varphi F_1 \sim f(\text{rel}(K_1 \cap \partial\Delta^n))$. Similarly we construct $F_2\colon K_2 \to Y_2$ with $F_2|_{K_2 \cap (\partial\Delta^n \cup K_1)} = G_2$ and $\varphi F_2 \sim f$ (rel$(K_2 \cap \partial\Delta^n)$). Now F_1 and F_2 agree on $K_1 \cap K_2$, so they define a map $\tilde{F}\colon I^n \to Y$ with $\tilde{F}|_{K_1} = F_1$, $\tilde{F}|_{K_2} = F_2$. Then $\tilde{F}|_{\partial\Delta^n} = g$. The homotopies $\varphi F_1 \sim f_1 \sim f$ and $\varphi F_2 \sim f_2 \sim f$ agree on $(K_1 \cap K_2) \times I$. Since $\varphi F_1 \sim f_1$ and $\varphi F_2 \sim f_2$ are homotopies rel $K_1 \cap K_2$, and the homotopies $f_1 \sim f$ and $f_2 \sim f$ when restricted to $(K_1 \cap K_2) \times I$ both yield the homotopy $\varphi F \sim f$. Hence $\varphi\tilde{F} \sim f$, and this is a homotopy rel$(K_1 \cap \partial\Delta^n) \cup (K_2 \cap \partial\Delta^n) = \partial\Delta^n$. ∎

This result will be applied to a forthcoming chain of generalizations of 13.6.

Theorem 16.25 Let X be a CW complex, X_1 and X_2 subcomplexes, and $A = X_1 \cap X_2$. Suppose (X_1, A) has cells in dimensions $\geq n$ and (X_2, A) has cells in dimensions $\geq m$. Then if $i\colon (X_1, A) \to (X, X_2)$, $i_*\colon \pi_r(X_1, A, *) \to \pi_r(X, X_2, *)$ is an $(m + n - 2)$-isomorphism.

Proof *Case 1* $X_1 = A \cup_\alpha e^n$. $X_2 - A$ consists of a finite number of cells.

Let k be the number of cells in $X_2 - A$. The result is true by 13.6 if $k = 1$. By induction suppose $X_2 - A$ has k cells and $X_2 = X_2' \cup e^t$ where $X_2' \supset A$, $X_2' - A$ has $k - 1$ cells, and $t \geq m$. Then we have a commutative diagram of inclusion maps

$$\begin{array}{ccc} (X_1, A) & \xrightarrow{i_1} & (X, X_2) \\ & {\scriptstyle i_2}\searrow \qquad \nearrow{\scriptstyle i_3} & \\ & (X_1 \cup X_2', X_2') & \end{array}$$

Both i_2 and i_3 qualify for the inductive step hence i_{2*} and i_{3*} are $(m+n-2)$-isomorphisms. Consequently $(i_1)_*$ is an $(m+n-2)$-isomorphism.

Case 2 $X_1 - A$ and $X_2 - A$ consist of a finite number of cells.

We now do induction on the number of cells in $X_1 - A$. Call this number k and note that $k = 1$ is the above case. Suppose now that $X_1 = X_1' \cup e^t$, $X_1' \supset A$, and $t \geq n$. Consider the ladder diagram

$$\begin{array}{ccccccccc} \pi_{r+1}(X_1, X_1') & \longrightarrow & \pi_r(X_1', A) & \longrightarrow & \pi_r(X_1, A) & \longrightarrow & \pi_r(X_1, X_1') & \longrightarrow & \pi_{r-1}(X_1', A) \\ \downarrow {\scriptstyle i_{1*}} & & \downarrow {\scriptstyle i_{2*}} & & \downarrow {\scriptstyle i_*} & & \downarrow {\scriptstyle i_{1*}} & & \downarrow {\scriptstyle i_{2*}} \\ \pi_{r+1}(X, X_1' \cup X_2) & \rightarrow & \pi_r(X_1' \cup X_2, X_2) & \rightarrow & \pi_r(X, X_2) & \rightarrow & \pi_r(X, X_1' \cup X_2) & \rightarrow & \pi_{r-1}(X_1' \cup X_2, X_2) \end{array}$$

(where base points are suppressed from the notation for brevity). The horizontal sequences are exact. i_{1*} and i_{2*} qualify for the inductive hypothesis and hence are $(m+n-2)$-isomorphisms (since $t \geq n$). The conclusion for i_* follows from the famous

Lemma 16.26 (*5-Lemma*) Consider a commutative diagram of abelian groups and homomorphisms

$$\begin{array}{ccccccccc} A_1 & \rightarrow & A_2 & \rightarrow & A_3 & \rightarrow & A_4 & \rightarrow & A_5 \\ \downarrow {\scriptstyle \alpha_1} & & \downarrow {\scriptstyle \alpha_2} & & \downarrow {\scriptstyle \alpha_3} & & \downarrow {\scriptstyle \alpha_4} & & \downarrow {\scriptstyle \alpha_5} \\ B_1 & \rightarrow & B_2 & \rightarrow & B_3 & \rightarrow & B_4 & \rightarrow & B_5 \end{array}$$

in which the horizontal sequences are exact. Then

(a) if α_2 and α_4 are onto and α_5 is 1–1, α_3 is onto:
(b) if α_2 and α_4 are 1–1 and α_1 is onto, α_3 is 1–1.

Remark No proof is given for this lemma because an essential element to any understandable proof is a certain amount of manual motion, and written proofs already abound in standard texts.

We return to the proof of 16.25.

Case 3 *General case* Let $\{\alpha\} \in \pi_r(X, X_2, *)$. Since $\alpha(I^r)$ is compact, it is contained in a finite subcomplex K of X. Consider the diagram

$$\begin{array}{ccc} \pi_r(X_1, A, *) & \xrightarrow{i_*} & \pi_r(X, X_2, *) \\ \uparrow & & \uparrow {\scriptstyle j_*} \\ \pi_r(K \cap X_1, K \cap A, *) & \xrightarrow{(i|_K)_*} & \pi_r(K, K \cap X_2, *) \end{array}$$

If $r \leq m+n-2$, $(i|_K)_*$ is onto and $\{\alpha\}$ is in the image of j_*. Hence $i*$ is onto. Let $\{\alpha\} \in \pi_r(X_1, A, *)$ and $i_*(\{\alpha\}) = 0$. Let $H: I^{r+1} \to X$ be a homotopy. As before $H(I^{r+1})$ is contained in a finite subcomplex K of X and hence the same diagram shows that $(i|_K)_*(\{\alpha\}) = 0$ and hence $\{\alpha\} = 0$ if $r < m + n - 2$. ▮

Corollary 16.27 (*Blakers–Massey Theorem—General Version*) Suppose (X_1, X_2) is excisive in X. Suppose $(X_1, X_1 \cap X_2)$ is $(n-1)$-connected and $(X_2, X_1 \cap X_2)$ is $(m-1)$-connected. Then $i_*\colon \pi_r(X_1, X_1 \cap X_2, *) \to \pi_r(X_1 \cup X_2, X_2, *)$ is an $(m+n-2)$-isomorphism.

Proof Let $f_{12}\colon K_{12} \to X_1 \cap X_2$ be a resolution. The composite $K_{12} \to X_1 \cap X_2 \to X_1$ is an $(n-1)$-equivalence. By Exercise 7 there is a resolution $f_1\colon K_1 \to X_1$ with $K_{12} \subset K_1$, $f_1|_{K_{12}} = f_{12}$ and all cells of $K_1 - K_{12}$ have dimension $\geq n$. Similarly there is a resolution $f_2\colon K_2 \to X_2$ with $K_{12} \subset K_2$, $f_2|_{K_{12}} = f_{12}$, and all cells of $K_2 - K_{12}$ have dimension $\geq m$.

Let $K = K_1 \cup K_2$ and define $f\colon K \to X$ by $f|_{K_1} = f_1$ and $f|_{K_2} = f_2$. Let $\pi\colon K \times I \to K$ be the projection and define $\bar{K} \subset K \times I$ by $\bar{K} = K_1 \times 0 \cup K_1 \cap K_2 \times I \cup K_2 \times 1$. Let $U_1 = \bar{K} - K_2 \times 1$ and $U_2 = \bar{K} - K_1 \times 0$. Then U_1 and U_2 are open in $\bar{K}$ and $\bar{K} = U_1 \cup U_2$. Define $\bar{f} = f\pi|_{\bar{K}}$. Then $\bar{f}(U_1) \subset X_1$ and $\bar{f}(U_2) \subset X_2$. Since $\pi\colon U_1 \simeq K_1$, $\pi\colon U_2 \simeq K_2$, and $\pi\colon U_1 \cap U_2 \simeq K_1 \cap K_2$, we may apply 16.24 and conclude that $\bar{f}$ induces isomorphisms in homotopy. To see that f induces isomorphisms, it is sufficient to prove that the inclusion $\bar{K} \subset K \times I$ induces isomorphisms. We utilize:

Lemma 16.28 Suppose (X, A) has the AHEP. Then the inclusion $X \times 0 \cup A \times I$ is a strong deformation retract of $X \times I$.

Proof Let $h\colon X \times I \to X \times 0 \cup A \times I$ be a retraction. Define $H\colon X \times I \times I \to X \times I$ by

$$H(x, t, s) = (h_1(x, t(1-s)), st + (1-s)h_2(x, t))$$

where $h(x, t) = (h_1(x, t), h_2(x, t)) \in X \times I$. This clearly satisfies the conditions. ▮

Returning to 16.27, we see that $K_1 \times 0 \cup K_1 \cap K_2 \times I$ is a strong deformation retract of $K_1 \times I$, and $K_1 \cap K_2 \times I \cup K_2 \times 1$ is a strong deformation retract of $K_2 \times I$. Hence $\bar{K}$ is a strong deformation retract of $K \times I$. It now follows that $f_*\colon \pi_r(K, K_2, *) \to \pi_r(X, X_2, *)$ and $f_*\colon \pi_r(K_1, K_{12}, *) \to \pi_r(X_1, A, *)$ induce isomorphisms in homotopy by applying the 5-lemma to the exact sequences for the pairs involved. The conclusion follows by applying 16.25 to the diagram

$$\begin{array}{ccc}
\pi_r(X_1, A, *) & \longrightarrow & \pi_r(X, X_2, *) \\
\uparrow \cong & & \uparrow \cong \\
\pi_r(K_1, K_{12}, *) & \longrightarrow & \pi_r(K, K_2, *)
\end{array}$$

since $K_1 - K_1 \cap K_2$ has cells in dimensions $\geq n$ and $K_2 - K_{12}$ has cells in dimensions $\geq m$. ∎

Theorem 16.29 Let $X = X_1 \cup X_2$, $A = X_1 \cap X_2$ and assume (X_1, A) and (X_2, A) have the AHEP. Suppose (X_1, A) is $(n-1)$-connected and (X_2, A) is $(m-1)$-connected. Then $i_*: \pi_r(X_1, A, *) \to \pi_r(X, X_2, *)$ is an $(m+n-2)$-isomorphism.

Proof Replace X by $\overline{X}$ as in the proof of 16.27. $\overline{X} = U_1 \cup U_2$ and one may apply 16.27 to deduce the conclusion since $(U_1, U_1 \cap U_2) \simeq (X_1, A)$ and $(\overline{X}, U_2) \simeq (X, A)$. ∎

Proposition 16.30 Suppose (X, A) has the AHEP and is $(n-1)$-connected. Suppose A is $(s-1)$-connected. Then $(p_A)_*: \pi_r(X, A, *) \to \pi_r(X/A, *)$ is an $(n+s-1)$-isomorphism. (Note $\pi_0(X, A, *)$ is not defined so this applies for $r > 0$ only.)

Proof Consider the inclusion $i: (X, A) \subset (X \cup CA, CA)$. Since (CA, A) has the AHEP (Exercise 4), and $X \cup CA - CA \equiv X = A$, $i_*: \pi_r(X, A, *) \to \pi_r(X \cup CA, CA, *)$ is an $(n + (s+1) - 2)$-isomorphism. Now $\pi_r(X \cup CA, CA, *) \simeq \pi_r(X \cup CA, *)$. The conclusion follows from:

Lemma 16.31 Suppose (X, A) has the AHEP. Then $p_{CA}: X \cup CA \to X/A$ is a homotopy equivalence.

Proof By 16.28, $X \cup A \times I$ is a strong deformation retract of $X \times I$. It follows easily that $X \cup CA$ is a strong deformation retract of $X \times I/A \times 1$. We now show that $X/A \times 1$ is a strong deformation retract of $X \times I/A \times 1$. A deformation is given by $H(x, t, s) = (x, t + (1-s)(1-t))$. ∎ ∎

Definition 16.32 A space will be called well pointed or will be said to have a nondegenerate base point $*$ if $(X, *)$ has the AHEP.

Lemma 16.33 If X is well pointed, $\Sigma X \simeq SX$.

Proof By Exercise 18, Section 14, $(X \times I, X \times 0 \cup * \times I \cup X \times 1)$ has the AHEP. Define a homotopy $H: (X \times 0 \cup * \times I \cup X \times 1) \times I \to \Sigma X$ by

$$H(x, 0, t) = (*, t/2), H(x, 1, t) = (*, 1 - t/2), H(*, s, t) = (*, s(1-t) + t/2).$$

Define $f: X \times I \times 0 \to \Sigma X$ by $f(x, s, 0) = (x, s)$. Since H and f agree on the intersection of their domains, there is an extension $K: X \times I \times I \to \Sigma X$ of both f and H by the AHEP. $K(x, 0, t)$ and $K(x, 1, t)$ do not depend on x, so K defines a map $\overline{K}: \Sigma X \times I \to \Sigma X$ by 8.12. Clearly $K(x, s, 0) = (x, s)$, $K(\Sigma * \times 1) = (*, \frac{1}{2})$ and $K(\Sigma * \times I) \subset \Sigma *$. By Exercise 1, Section 13, $\Sigma X \simeq \Sigma X/\Sigma * \equiv SX$. ∎

Theorem 16.34 (*Freudenthal Suspension Theorem—General Version*) Suppose X is $(n-1)$-connected with $n \geq 1$ and is well pointed. Then $E\colon \pi_r(X, *) \to \pi_{r+1}(SX, *)$ is a $(2n-1)$-isomorphism.

Proof Apply 16.30, 13.11, and 16.33 to the diagram

$$\begin{array}{ccc}
\pi_r(X, *) & \xrightarrow{E} & \pi_{r+1}(SX, *) \\
\cong \uparrow & & \cong \uparrow (p_{\Sigma *})_* \\
\pi_{r+1}(C^*X, X) & & \\
\cong \uparrow & & \\
\pi_{r+1}(CX, X) & \xrightarrow[(p_X)_*]{} & \pi_{r+1}(\Sigma X, *)
\end{array}$$ ∎

If we consider the sequence of groups $\pi_{r+n}(S^n X, *)$ for $n = 0, 1, \ldots$ and the suspension homomorphisms between them we observe that if X is a CW complex $S^n X$ has no cells in dimension $<n$ except for a 0-cell. Hence it is $(n-1)$-connected and $\pi_{r+n}(S^n X, *) \cong \pi_{r+n+1}(S^{n+1} X, *) \cong \cdots$ if $n > r+1$. Thus for n large the sequence "stabilizes" and this stable value is called the rth stable homotopy group of X and is written $\pi_r{}^S(X, *)$ or $\tilde{S}_r(X)$. The determination of these groups for simple spaces seems to be a very difficult problem. For example, there is no known space X for which $\pi_r{}^S(X, *)$ is known for all r. An interesting conjecture which also seems very hard is the Freyd conjecture: If X and Y are finite CW complexes and $f\colon X \to Y$ induces the zero homomorphism in $\pi_r{}^S$ for all r, then $S^k f \sim *$ for some k.

Appendix

We shall describe the classical construction of the singular complex of a space and show that it is a resolution. As a corollary (which we will use in Section 21) we prove the every CW complex is the homotopy type of a simplicial CW complex (defined below).

The classical construction will be called the functorial singular complex, and the construction in 16.21 will be called the ad hoc singular complex when we wish to make a distinction. The main advantage of the functorial singular complex is that it is a functor from $\mathcal{C}$ to $\mathcal{K}$, whereas the ad hoc construction is only functorial in the homotopy category $\mathcal{K}_h$. One pays for this advantage with size. The functorial construction on any finite geometric simplicial complex of positive dimension has 2^c cells in each positive dimension, where c is the cardinality of the continuum. On the other hand, the ad hoc construction of a simply connected space can be made very efficiently by Exercise 9, Section 22. We will also use the notation $S(X)$ for the functorial construction if it will not lead to confusion; we now define the functorial construction.

Let S_n be the set of continuous maps $F: \Delta^n \to X$ with the discrete topology. Let $B_n(X) = S_n \times \Delta^n$. $B_n(X)$ is a disjoint union with one copy of Δ^n for each map $f \in S_n$. Let $\iota_s: \Delta_{n-1} \to \Delta_n$ be defined for $0 \le s \le n$ by $\iota_s(t_0, \ldots, t_{n-1}) = (t_0, \ldots, t_{s-1}, 0, t_s, \ldots, t_{n-1})$. Let $B(X) = \coprod_{n=0}^{\infty} B_n(X)$ and let $\sim$ be the equivalence relation in $B(X)$ generated by

$$(f\iota_s, u) \sim (f, \iota_s(u))$$

for all $0 \le s \le n$, $u \in \Delta^{n-1}$, $f \in S_n$, and all $n \ge 0$.

Definition 16.35 $S(X) = B(X)/\sim$ with the quotient topology.

Proposition 16.36

(a) $S(X)$ is a CW complex;
(b) $S: \mathcal{C} \to \mathcal{K}$ is a functor;
(c) if $A \subset X$, $S(A) \subset S(X)$;
(d) There is a natural transformation $\pi: S(X) \to X$.

Proof We use Exercise 5, Section 0 to show that $S(X)$ is Hausdorff. $B(X)$ is normal. Let $q: B(X) \to S(X)$ be the quotient map. Then q is closed since if $A \subset B(X)$ is closed, $\{b \mid b \sim a, a \in A\}$ is closed. Thus $S(X)$ is Hausdorff. Let $S^n(X) = q(B_n(X) \cup \cdots \cup B_0(X))$. Then

$$S^n(X) - S^{n-1}(X) = q(S_n \times \text{Int } \Delta^n) \equiv S_n \times \text{Int } \Delta^n$$

since $q|_{S_n \times \text{Int } \Delta^n}$ is 1–1 and $S_n \times \text{Int } \Delta^n$ is open in $B(X)$. For each $f \in S_n$, define $\chi_f: \Delta^n \to S(X)$ by $\chi_f(u) = q(f, u)$. This is continuous and $\chi_f|_{\text{Int } \Delta^n}$ is a homeomorphism. Furthermore $\chi_f(\Delta^n, \partial\Delta^n) \subset (S^n(X), S^{n-1}(X))$. Thus choosing $e_f^n = \chi_f(\text{Int } \Delta^n)$ as n-cells, we have a cellular structure on $S(X)$. Clearly $S(X)$ is closure finite and has the weak topology by 14.5. This proves (a).

Let $h: X \to Y$ and define $B(h): B(X) \to B(Y)$ by $B(h)(f, u) = (hf, u)$. This is continuous and preserves the identifications. Hence it defines $S(h): S(X) \to S(Y)$. This is clearly functorial, so we have proven (*b*).

Suppose $A \subset X$. Then $B(A)$ is a closed subset of $B(X)$. Hence the subset $q(B(A)) \subset S(X)$ has the quotient topology and is thus equal to $S(A)$. Thus $S(A) \subset S(X)$. This proves (c).

Define $\pi: S(X) \to X$ by $\pi q(f, u) = f(u)$, It is easy to verify that this is well defined, continuous, and a natural transformation. ∎

Definition 16.37 By a semisimplicial CW complex we will mean a pair $(K, \{\chi_\alpha\})$, where K is a CW complex and $\{\chi\}$ is a collection of characteristic maps $\chi_\alpha: \Delta^n \to K$ with one for each cell e_α^n such that for each cell e_α^n and each s

with $0 \leq s \leq n$ there is a cell $e^{n-1}_{\beta(\alpha, s)}$ such that $\chi_{\beta(\alpha, s)} = \chi_\alpha \circ \iota_s$. A CW complex is called regular if there is a characteristic map for each cell that is a homeomorphism. By a simplicial CW complex we will mean a regular semisimplicial CW complex such that each cell $e_\alpha{}^n$ is determined by the set $\{\chi_\alpha(v_i)\}$ where $\Delta^n = (v_0, \ldots, v_n)$. We will write $|K|$ for the underlying topological space of a semisimplicial CW complex K.

To understand the meaning of these definitions it is helpful to notice that S^1 can be made into a semisimplicial CW complex with only one 0-cell. To make S^1 into a regular CW complex, one needs two 0-cells and to make S^1 into a simplicial CW complex one needs three 0-cells. However one cannot make S^2 into a semisimplicial CW complex without any 1-cells.

Proposition 16.38 $(S(X), \{\chi_f\})$ is a semisimplicial CW complex.

Proof $\chi_f \iota_s = \chi_{f\iota_s}$. ∎

The function $\beta(\alpha, s)$ defined in a semisimplicial CW complex K is often written $\partial_s \alpha$. Thus an operator ∂_s is defined from the set of n-cells to the set of $(n-1)$-cells for $0 \leq s \leq n$. (See Exercise 25.)

Clearly a subcomplex of a semisimplicial regular or simplicial CW complex is a CW complex of the same sort.

One can easily give a finite geometric simplicial complex K the structure of a simplicial CW complex. By ordering the vertices of K, one defines for each n-simplex a characteristic map $\chi_\sigma \colon \Delta^n \to K$ which is order preserving on the vertices. As a converse, we have:

Proposition 16.39 Every finite simplicial CW complex is homeomorphic to a finite geometric simplicial complex.

Proof Let $(K, \{\chi_\alpha\})$ be a finite simplicial CW complex. Let $V = \{a_0, \ldots, a_m\}$ be the set of 0-cells. Define $F\colon V \to \Delta^m$ by $F(a_i) = v_i$. We will extend this over K. Define $F_\alpha \colon \bar{e}_\alpha{}^n \to \Delta^m$ by $F_\alpha(\chi_\alpha(\Sigma t_i v_i)) = \Sigma t_i\, F(\chi_\alpha(v_i))$. Since K is regular this is well defined. Since $\chi_\alpha \iota_s = \chi_{\beta(\alpha, s)}$, $F_\alpha|_{\bar{e}^{n-1}_{\beta(\alpha, s)}} = F_{\beta(\alpha, s)}$. Thus if $\bar{e}_{\alpha'}{}^m \subset \bar{e}_\alpha{}^n$, $F_\alpha|_{\bar{e}_{\alpha'}{}^m} = F_{\alpha'}$ and the maps F_α therefore define a continuous map $f\colon K \to \Delta^m$. Since K is simplicial, F is 1–1 and hence is a homeomorphism from K to $F(K)$.

Now $F(K)$ is the union of the simplicies $(F(\chi_\alpha(v_0)), \ldots, F(\chi_\alpha(v_n)))$ and hence is a subcomplex of $\Delta^m \subset R^m$. ∎

The process of barycentric subdivision (12.16) can be applied to semisimplicial CW complexes and we consider this construction next.

A sequence of subcomplexs $\tau_0 \prec \tau_1 \prec \cdots \prec \tau_k \prec \Delta^{n'}$ with $\tau_i \neq \tau_{i+1}$ determines a k-simplex $(b(\tau_0), \ldots, b(\tau_k))$ in Δ^n. We take as a characteristic map for this cell the map

$$I^{\tau_0, \ldots, \tau_k} \colon \Delta^k \to \Delta^n$$

given by

$$I^{\tau_0, \ldots, \tau_k}(t_0, \ldots, t_k) = \sum t_i b(\tau_i).$$

Let $(K, \{\chi_\alpha\})$ be a semisimplicial CW complex. Define maps $\chi_\alpha^{\tau_0, \ldots, \tau_k}: \Delta^k \to K$ as the composition $\chi_\alpha \circ I^{\tau_0, \ldots, \tau_k}$. Define the redundancy of $\chi_\alpha^{\tau_0, \ldots, \tau_k}$ by $r(\chi_\alpha^{\tau_0, \ldots, \tau_k}) = n - \dim \tau_k \geq 0$.

Lemma 16.40 Each map $\chi_\alpha^{\tau_0, \ldots, \tau_k}$ is equal to a map $\chi_\beta^{\sigma_0, \ldots, \sigma_k}$ with $r(\chi_\beta^{\sigma_0, \ldots, \sigma_k}) = 0$.

Proof We will use induction. Suppose $r(\chi_\alpha^{\tau_0, \ldots, \tau_k}) > 0$. Then $\tau_k \subset \iota_s(\Delta^{n-1})$ for some s. Define $\sigma_i \subset \Delta^{n-1}$ by $\sigma_i = \iota_s^{-1}(\tau_i)$. Then $\sigma_0 \prec \sigma_1 \prec \cdots \prec \sigma_k \prec \Delta^{n-1}$ and $\sigma_i \neq \sigma_{i+1}$. Furthermore $\iota_s(b(\sigma_i)) = b(\tau_i)$. Hence $I^{\tau_0, \ldots, \tau_k} = \iota_s I^{\sigma_0, \ldots, \sigma_k}$ and consequently $\chi_\alpha^{\tau_0, \ldots, \tau_k} = \chi_{\beta(\alpha, s)}^{\sigma_0, \ldots, \sigma_k}$. But $r(\chi_{\beta(\alpha, s)}^{\sigma_0, \ldots, \sigma_k}) = r(\chi_\alpha^{\tau_0, \ldots, t_k}) - 1$. ∎

We call $\chi_\alpha^{\tau_0, \ldots, \tau_k}$ nonredundant in case $r(\chi_\alpha^{\tau_0, \ldots, \tau_k}) = 0$; i.e., $\tau_k = \Delta^n$. We will write K' for the underlying space K together with the nonredundant maps $\chi_\alpha^{\tau_0, \ldots, \tau_k}$ as characteristic maps.

Theorem 16.41 If K is a semisimplicial CW complex, K' is a regular semisimplicial CW complex. If K is also regular, K' is a simplicial CW complex.

Proof We first observe that if $0 \leq s \leq k$,

$$\chi_\alpha^{\tau_0, \ldots, \tau_k} \iota_s = \chi_\alpha^{\tau_0, \ldots, \tau_{s-1}, \tau_{s+1}, \ldots, \tau_k} = \chi_\beta^{\sigma_0, \ldots, \sigma_{k-1}}$$

for some nonredundant map $\chi_\beta^{\sigma_0, \ldots, \sigma_{k-1}}$ by 16.40. We show now that each $\chi_\alpha^{\tau_0, \ldots, \tau_k}$ is a homeomorphism by induction on k. This is trivial if $k = 0$. Choose a map $\chi_\alpha^{\tau_0, \ldots, \tau_k}$. By 16.40 we can assume that it is nonredundant. Suppose $\chi_\alpha^{\tau_0, \ldots, \tau_k}(x_0) = \chi_\alpha^{\tau_0, \ldots, \tau_k}(x_1)$. Let $u_i = I_\alpha^{\tau_0, \ldots, \tau_k}(x_i)$. Then $u_i \in \partial\Delta^n$. But $(I^{\tau_0, \ldots, \tau_k})^{-1}(\partial\Delta^n) \subset \iota_k(\Delta^{k-1})$, so $x_i = \iota_k(y_i)$. Now $\chi_\alpha^{\tau_0, \ldots, \tau_k} \iota_k = \chi_\alpha^{\tau_0, \ldots, \tau_{k-1}}$ so $\chi_\alpha^{\tau_0, \ldots, \tau_{k-1}}(y_0) = \chi_\alpha^{\tau_0, \ldots, \tau_{k-1}}(y_1)$. By induction $y_0 = y_1$ so $x_0 = x_1$ and $\chi_\alpha^{\tau_0, \ldots, \tau_k}$ is a homeomorphism.

Now define cells in K' by

$$e_\alpha^{\tau_0, \ldots, \tau_k} = \chi_\alpha(\operatorname{Int}(b(\tau_0), \ldots, b(\tau_k))) = \chi_\alpha^{\tau_0, \ldots, \tau_k}(\operatorname{Int} \Delta^k)$$

in case $\chi_\alpha^{\tau_0, \ldots, \tau_k}$ is nonredundant. Then

$$\operatorname{Int} \Delta^n = \bigcup \operatorname{Int}(b(\tau_0), \ldots, b(\tau_k)),$$

where the union is disjoint and is taken over all sequences $\tau_0 \prec \cdots \prec \tau_k = \Delta^n$ with $\tau_i \neq \tau_{i+1}$. Hence the cells $e_\alpha^{\tau_0, \ldots, \tau_k}$ are disjoint and cover $K' = K$. Now $\chi_\alpha^{\tau_0, \ldots, \tau_k}$ is a characteristic map for $e_\alpha^{\tau_0, \ldots, \tau_k}$ and since $\chi_\alpha^{\tau_0, \ldots, \tau_k} \iota_s = \chi_\beta^{\sigma_0, \ldots, \sigma_{k-1}}$, $\chi_\alpha^{\tau_0, \ldots, \tau_k}(\partial\Delta^k) \subset (K')^{k-1}$. Thus K' with these cells is a cell complex. K' is clearly closure finite and has the weak topology on the cells $\bar{e}_\alpha^{\tau_0, \ldots, \tau_k}$.

We have already proved that it is semisimplicial and regular so the first part is done.

Suppose now that K is regular. Let $\Delta^k = (v_0, \ldots, v_k)$, and suppose $\chi_\alpha^{\tau_0, \ldots, \tau_k}(v_i) = \chi_\beta^{\sigma_0, \ldots, \sigma_k}(v_i)$ for each i. Since $\chi_\alpha^{\tau_0, \ldots, \tau_k}(v_k) = \chi_\alpha(b(\tau)_k) = \chi_\alpha(b(\Delta^k)) \in e_\alpha$ and similarly $\chi_\beta^{\sigma_0, \ldots, \sigma_k}(v_k) \in e_\beta$ we must have $\alpha = \beta$. Since K is regular, we must have $I^{\tau_0, \ldots, \tau_k}(v_i) = I^{\sigma_0, \ldots, \sigma_k}(v_i)$ for each i. Hence $b(\tau_i) = b(\sigma_i)$ and this implies $\tau_i = \sigma_i$. Thus the cell $e_\alpha^{\tau_0, \ldots, \tau_k}$ is determined by the points $\chi_\alpha^{\tau_0, \ldots, \tau_k}(v_i)$ for $0 \leq i \leq k$. ▌

Corollary 16.42 (*Barratt*) If K is a semisimplicial CW complex, K'' is a simplicial CW complex. ▌

We now state the main result and prove a corollary.

Theorem 16.43 (*Giever–Whitehead* [25, 74]) $(S(X), \pi)$ is a resolution of X.

Corollary 16.44 Every CW complex is the homotopy type of a simplicial CW complex.

Proof of 16.42 By 16.22, 16.36, and 16.43, $S(X) \simeq X$. By 16.38 and 16.41, $|S(X)| \equiv |S''(X)|$ is a simplicial CW complex. Thus $X \simeq S''(X)$. ▌

The proof of 16.43 is complicated and requires some lemmas.

Let S_n be the set of characteristic maps of n-cells in a semisimplicial CW complex K, with the discrete topology. Let $B = \coprod S_n \times \Delta^n$. Let $\sim$ be the equivalence relation in B generated by

$$(\chi_\alpha, \iota_s(u)) \sim (\chi_{\beta(\alpha, s)}, u)$$

for $x_\alpha \in \Delta_n$ and $u \in \Delta^{n-1}$.

Lemma 16.45 $B/\!\sim\, \equiv K$.

Proof Define F: $B/\!\sim\, \to K$ by $F(\{(\chi_\alpha, u)\}) = \chi_\alpha(u)$. This is well defined, continuous, and onto. To see that it is 1–1 note that every point $x \in B/\!\sim$ has a representative (χ_α, u) with $u \in \operatorname{Int} \Delta^n$. Thus if $F(x) = F(x')$, $\chi_\alpha(u) = \chi_{\alpha'}(u')$. But e_α and $e_{\alpha'}$ are equal or disjoint, so $\alpha = \alpha'$ and $u = u'$. To see that F is open, note that both $B/\!\sim$ and K are quotient spaces of the disjoint union of the closed cells. Hence the topologies agree. ▌

Definition 16.46 Let $(K, \{\chi_\alpha\})$ and $(L, \{\chi_\beta\})$ be semisimplicial CW complexes. We will call a map f: $K \to L$ simplicial if for every α there is a β such that $\chi_\beta = f\chi_\alpha$.

Clearly simplicial maps are continuous.

Lemma 16.47 Let K be a semisimplicial CW complex and $f: K \to X$ be a continuous map. Then there is a unique simplicial map $S(f): K \to S(X)$ such that $\pi S(f) = f$.

Proof We use 16.45 to construct $S(f)$. Define

$$B(f): B \to S(X)$$

by $B(f)(\{\chi_\alpha, u\}) = \{(f\chi_\alpha, u)\}$. This is clearly well defined and continuous, and preserves the equivalence relation. Thus $B(f)$ determines a map $S(f): K \to S(X)$ and $\pi S(f) = f$.

Suppose $g: K \to S(X)$ is a simplicial map with $\pi g = f$. Then $g(\{(\chi_\alpha, u\}) = \{(\theta_\alpha, u)\}$ for some map $\theta_\alpha: \Delta^n \to X$. If $\pi g = f$ we must have $\theta_\alpha(u) = f\chi_\alpha(u)$ so $g = S(f)$. ▮

If $f: K \to S(X)$ is a simplicial map, there is a unique simplicial map $f': K' \to S(X)$ such that the diagram

$$\begin{array}{ccc} K' & \xrightarrow{f'} & S(X) \\ \downarrow{\scriptstyle 1} & & \downarrow{\scriptstyle \pi} \\ K \xrightarrow{f} & S(X) & \xrightarrow{\pi} X \end{array}$$

commutes.

Lemma 16.48 $f \sim f'$ (rel K^0).

This is the key to 16.43; we defer its proof temporarily.

Proof of 16.43 It is clearly sufficient to consider the case that X is arcwise connected. Choose $* \in X$ and let $e \in S(X)$ be the 0-cell with $\pi(e) = *$. Then it is sufficient to show that

$$\pi_*: \pi_i(S(X), e) \to \pi_i(X, *)$$

is an isomorphism for all i. We first show that π_* is onto. Let $\alpha: (S^n, *) \to (X, *)$. Choose a semisimplicial complex K with $|K| = S^n$ such that $*$ corresponds to a 0-cell $v \in K$. By 16.47 there is a simplicial map $S(\alpha)$: $S^n \to S(X)$, with $\pi S(\alpha) = \alpha$. Since $S(\alpha)(v) \in S^0(X)$, $S(\alpha)(v) = e$. Hence $S(\alpha): (S^n, *) \to (S(X), e)$ and $\pi_*\{S(\alpha)\} = \{\alpha\}$.

Now let $\alpha: (S^n, *) \to (S(X), e)$. We will show that there is a semisimplicial complex K with $|K| \equiv S^n$, a 0-cell $v \in K$ corresponding to $*$, and a simplicial map $f: K \to S(X)$ with $f \sim \alpha$ (rel v). Now there is a simplicial map $\gamma: S'(X) \to$

$S(X)$ with $\pi\gamma = \pi$ and a simplicial map γ': $S''(X) \to S(X)$ with $\pi\gamma' = \pi$. By 16.48 with $K = S(X)$ we see that $\gamma \sim 1$ (rel $S^0(X)$). By 16.48 with $K = S'(X)$ and $f = \gamma$ we see that $\gamma' \sim \gamma$ (rel $S'(X)^0$). Hence $\gamma' \sim 1$ (rel $S^0(X)$). Thus there is a map β: $(S^n, *) \to (S''(X), e)$ with $\gamma'\beta \sim \alpha$ (rel $*$). Now $\beta(S^n)$ is contained in a finite subcomplex of $S''(X)$. By 16.39 and the simplicial approximation theorem, there is a simplicial complex K with $|K| = S^n$ and 0-cell v corresponding to $*$ so that β is homotopic to a simplicial map δ: $K \to S''(X)$ relative to $*$. Thus $\alpha \sim \gamma'\delta$ (rel $*$) and $f = \gamma'\delta$ is simplicial for some choice of characteristic maps in K. Consequently $\{\alpha\} = \{f\}$. Suppose $\pi_*\{f\} = 0$. Then there is a map H: $B^{n+1} \to X$ with $H|_{S^n} = \pi f$. Now there is a semisimplicial complex L with $|L| = B^{n+1}$ and K a subcomplex of L (if $K = (\partial\Delta^n)^{(r)}$, let $L = (\Delta^n)^{(r)}$). Then $S(H)$: $B^{n+1} \to S(X)$ is simplicial and since K is a subcomplex of L, $S(H)|_K = f$ by uniqueness. Hence $\{f\} = 0$. ▮

It remains to prove 16.48.

Proof of 16.48 We will call a map ι: $\Delta^k \to \Delta^n$ inclusive is ι if induced by a 1–1 order preserving map of the vertices. Since every inclusive map is a composite of maps of the form ι_s, we have $(\chi_\alpha \iota, u) \sim (\chi_\alpha, \iota(u))$ in the equivalence relation of 16.45, for ι inclusive, $u \in \Delta^k$, and χ_α: $\Delta^n \to K$.

Now let λ_n be a 1–1 correspondence from the nonempty faces of Δ^n to the integers $0, \ldots, 2^{n+1} - 2$ such that if $\tau \prec \tau'$, $\lambda_n(\tau) \le \lambda_n(\tau')$ and $\lambda_n(\{v_i\}) = i$. Define linear maps

$$\iota_\lambda: \Delta^n \to \Delta^{2^{n+1}-2}, \qquad b_\lambda: (\Delta^n)' \to \Delta^{2^{n+1}-2}, \qquad \pi_\lambda: \Delta^{2^n-2} \to \Delta^n,$$

by $\iota_\lambda(v_i) = v_i$, $b_\lambda(b(\tau)) = v_{\lambda_n(\tau)}$ where $b(\tau)$ is the vertex of $(\Delta^n)'$ which is the barycenter of τ, and $\pi_\lambda(v_i) = b(\lambda_n^{-1}(i))$.

If $0 \le s \le n$, ι_s: $\Delta^{n-1} \to \Delta^n$ induces an inclusion of the faces of Δ^{n-1} into the faces of Δ^n; by the condition on λ_n, this inclusion corresponds to an inclusive map I_s: $\Delta^{2^n-2} \to \Delta^{2^{n+1}-2}$. We then have $I_s \iota_\lambda = \iota_\lambda \iota_s$, $I_s b_\lambda = b_\lambda \iota_s$, and $\pi_\lambda I_s = \iota_s \pi_\lambda$. Now define θ_t:$\Delta^n \to \Delta^{2^{n+1}-2}$ by

$$\theta_t = t b_\lambda + (1 - t)\iota_\lambda .$$

We are now prepared to define a homotopy f_t: $K' \to S(X)$. We represent a point in K' via 16.45 by a pair $(\chi_\alpha^{\tau_0, \ldots, \tau_k}, u)$ for $u \in \Delta^k$ and $\chi_\alpha^{\tau_0, \ldots, \tau_k}$ nonredundant. Recall that f: $K \to S(X)$ is a simplicial map. Define

$$f_t(\chi_\alpha^{\tau_0, \ldots, \tau_k}, u) = (\pi f \chi_\alpha \pi_\lambda, \theta_t I^{\tau_0, \ldots, \tau_k}(u)).$$

Note that $\theta_t I^{\tau_0, \ldots, \tau_k}(u) \in \Delta^{2^{n+1}-2}$ and $\pi f \chi_\alpha \pi_\lambda$: $\Delta^{2^{n+1}-2} \to X$. Thus this pair represents a point in $S(X)$.

We show that the formula for f_t is valid even if $\chi_\alpha^{\tau_0, \ldots, \tau_k}$ is redundant by induction on the redundancy. Suppose it holds for pairs $(\chi_\alpha^{\tau_0, \ldots, \tau_k}, u)$ of redundancy less than p, and $\dim \tau_k + p = n$. Let $\sigma_j = \iota_s^{-1}(\tau_k)$. Then

$$\begin{aligned} f_t(\chi_\alpha^{\tau_0, \ldots, \tau_k}, u) &= f_t(\chi_{\beta(\alpha, s)}^{\sigma_0, \ldots, \sigma_k}, u) \\ &= (\pi f \chi_\alpha \iota_s \pi_\lambda, \theta_t I^{\sigma_0, \ldots, \sigma_k}(u)) \\ &= (\pi f \chi_\alpha \pi_\lambda I_s, \theta_t I^{\sigma_0, \ldots, \sigma_k}(u)) \\ &\sim (\pi f \chi_\alpha \pi_\lambda, I_s \theta_t I^{\sigma_0, \ldots, \sigma_k}(u)) \\ &= (\pi f \chi_\alpha \pi_\lambda, \theta_t \iota_s I^{\sigma_0, \ldots, \sigma_k}(u)) \\ &= (\pi f \chi_\alpha \pi_\lambda, \theta_t I^{\tau_0, \ldots, \tau_k}(u)). \end{aligned}$$

Hence the formula holds in this case as well.

We must show that f_t preserves the equivalence relation in K (16.45). But

$$\begin{aligned} f_t(\chi_\alpha^{\tau_0, \ldots, \tau_k}, \iota_s(u)) &= (\pi f \chi_\alpha \pi_\lambda, \theta_t I^{\tau_0, \ldots, \tau_k} \iota_s(u)) \\ &= (\pi f \chi_\alpha \pi_\lambda, \theta_t I^{\tau_0, \ldots, \tau_{s-1}, \tau_{s+1}, \ldots, \tau_k}(u)) \\ &= f_t(\chi_\alpha^{\tau_0, \ldots, \tau_{s-1}, \tau_{s+1}, \ldots, \tau_k}, u) \\ &= f_t(\chi_\alpha^{\tau_0, \ldots, \tau_k} \iota_s, u). \end{aligned}$$

Note that in the case $s = k$, $\chi_\alpha^{\tau_0, \ldots, \tau_{k-1}}$ is redundant and we have used our earlier result.

Now $\theta_0 = \iota_\lambda$ which is inclusive. Hence

$$f_0(\chi_\alpha^{\tau_0, \ldots, \tau_k}, u) = (\pi f \chi_\alpha, I^{\tau_0, \ldots, \tau_k}(u)).$$

The homeomorphism $K' \equiv K$ is given by $(\chi_\alpha^{\tau_0, \ldots, \tau_k}, u) \leftrightarrow (\chi_\alpha, I^{\tau_0, \ldots, \tau_k}(u))$, so $f_0 = f$.

On the other hand, $\theta_1 = b_\lambda$, and $b_\lambda I^{\tau_0, \ldots, \tau_k}$ is the inclusive map that sends v_i to $v_{\lambda_n(\tau_i)}$. Hence

$$\begin{aligned} f_1(\chi_\alpha^{\tau_0, \ldots, \tau_k}, u) &= (\pi f \chi_\alpha \pi_\lambda, b_\lambda I^{\tau_0, \ldots, \tau_k}(u)) \\ &= (\pi f \chi_\alpha \pi_\lambda b_\lambda I^{\tau_0, \ldots, \tau_k}, u) \\ &= (\pi f \chi_\alpha^{\tau_0, \ldots, \tau_k}, u). \end{aligned}$$

Since f_1 is simplicial, $f_1 = f'$.

Finally, if $(\chi_\alpha^{\tau_0, \ldots, \tau_k}, u)$ is a vertex of K, $I^{\tau_0, \ldots, \tau_k}(u) = v_i$. Hence $f_t(\chi_\alpha^{\tau_0, \ldots, \tau_k}, u) = (\pi f \chi_\alpha \pi_\lambda, \theta_t v_i) = (\pi f \chi_\alpha \pi_\lambda, v_i)$, since $\theta_t(v_i) = v_i$. ∎

Exercises

1. Show that if $\bar{U} \subset \operatorname{Int} A$, $i_*: \pi_r(X - U, A - U) \to \pi_r(X, A)$ is an isomorphism if $r < m + n - 2$ and is onto if $r = m + n - 2$ if $(X - U, A - U)$ is $(n - 1)$-connected and $(A, A - U)$ is $(m - 1)$-connected.

2. Write down a homotopy that makes $\bar{K}$ a strong deformation retract of $K \times I$ (notation from Theorem 16.27).

3. Prove that if $f: (X, *) \to (Y, *)$ and $(X, *)$ is well pointed, $Y \cup_f C^*X \simeq Y \cup_f CX$. (23.8)

4. Show that (CX, X) has the AHEP. (16.30)

5. Suppose X and Y are arcwise connected and the homotopy type of CW complexes. Let $f: X \to Y$ induce $f_*: \pi_*(X, *) \to \pi_*(Y, *)$. Prove that if f_* is an isomorphism, f is a homotopy equivalence.

6. Suppose $f: X \to Y$ is a weak homotopy equivalence and K is a CW complex. Show that $f_*: [K, X] \to [K, Y]$ is a 1–1 correspondence. (Section 17)

7. Using the proof of 16.14, prove the following generalization: Given $f_0: A \to Y$ such that $(f_0)_*: \pi_i(A, *) \to \pi_i(Y, *)$ is an $(n-1)$-isomorphism, there is a space $X \supset A$ such that (X, A) is a relative CW complex with cells in dimension $\geq n$ and an extension $f: X \to Y$ of f_0 which is a weak homotopy equivalence. (Exercise 10; 16.19; 16.27; Exercise 6, Section 21; 22.5)

8. By considering the pair $(X \times Y, X \vee Y)$ prove that if X is $(n-1)$-connected and Y is $(m-1)$-connected, and both X and Y are CW complexes,

$$\pi_k(X \vee Y, *) \cong \pi_k(X) \oplus \pi_k(Y)$$

for $k < m + n$, the isomorphism being given by $\alpha \to ((p_1)_*(\alpha), (p_2)_*(\alpha))$. Use this, induction, and a limit argument to prove that if X_α is $(n-1)$-connected for all α,

$$\pi_k\left(\bigvee_{\alpha \in A} X_\alpha\right) \cong \bigoplus_{\alpha \in A} \pi_k(X_\alpha)$$

for $k < 2n$. (23.8)

9. Give an alternative proof of 16.4 without using Zorn's lemma based on proving that if $X \subset \bar{X} \subset Y$ and $\bar{X} = X \cup e_{\alpha_1} \cup \cdots \cup e_{\alpha_n}$, $\pi_i(\bar{X}, X, *) = 0$.

10. Let $K \to X$ be a resolution and suppose X is $(n-1)$-connected and well pointed. By applying 16.34, Exercise 3, Section 13, and Exercise 7 above conclude that there is a resolution K' of ΩSX such that $K' \supset K$ and $K' - K$ consists of cells of dimension $\geq 2n$:

$$\begin{array}{ccc} K & \longrightarrow & X \\ \downarrow{\scriptstyle \subset} & & \downarrow{\scriptstyle i} \\ K' & \longrightarrow & \Omega SX \end{array}$$

By applying (16.10) conclude that if M is a CW complex E: $[M, X] \to [SM, SX]$ is a 1–1 if dim $M < 2n - 1$ and is onto if dim $M < 2n$.

11. Show that if $A \subset X$ and A is contractible in $\mathscr{C}$, X is a strong deformation retract of $X \cup CA$. Prove a similar result in $\mathscr{C}^*$.

12. Prove that $\pi_n(RP^n, RP^{n-1}, *) \cong Z \oplus Z$ if $n > 1$. (Hint: Use Exercise 5, Section 14 to evaluate the homomorphism $\pi_r(RP^{n-1}) \to \pi_r(RP^n)$.) Note that $\pi_n(RP^n/RP^{n-1}, *) \cong Z$. Compare this to 16.30.

13. Let $S^\infty = \bigcup_{n=1}^\infty S^n$ with the weak topology. Prove that S^∞ is contractible.

14. Show that i_*: $\pi_r(X, A, *) \to \pi_r(X \cup CA, CA, *)$ is an $(n + s - 1)$-isomorphism if (X, A) is $(n - 1)$-connected and A is $(s - 1)$-connected. (Hint: Use 16.27 with $X \cup CA = (X \cup A \times [0, \frac{1}{2})) \cup (X \cup CA - X)$.)

15. Suppose (X, A) is $(n - 1)$-connected and A is $(s - 1)$-connected. Prove that there is an exact sequence

$$\pi_{n+s-2}(A) \to \pi_{n+s-2}(X) \to \pi_{n+s-2}(X \cup CA) \to \pi_{n+s-3}(A) \to \cdots$$

truncated on the left at $\pi_{n+s-2}(A)$. (Exercise 23; 23.8)

16. Let A be the graph of $\sin(1/x)$, with $x > 0$ in R^2 and $B = \{(x, y) \mid y \le 0, x \ge 0, x^2 + y^2 = 16\}$. Let $X = A \cup B \cup 0 \times [-4, 0]$. Let $* = (0, -4)$. Show that $\pi_r(X, *) = 0$ for all $r \ge 0$ but X is not contractible.

17. Using the formula for $E(\{f\})$ in 13.11 construct a homomorphism

$$E: \pi_n(X, A, *) \to \pi_{n+1}(SX, SA, *)$$

such that there is a commutative diagram

$$\begin{array}{ccccc}
\pi_n(X, *) & \longrightarrow & \pi_n(X, A, *) & \xrightarrow{\partial} & \pi_{n-1}(A, *) \\
\downarrow E & & \downarrow E & & \downarrow E \\
\pi_{n+1}(SX, *) & \longrightarrow & \pi_{n+1}(SX, SA, *) & \xrightarrow{\partial} & \pi_n(SA, *)
\end{array}$$

Prove that if A is $(m - 1)$-connected, X is $(k - 1)$-connected, $*$ is a non-degenerate base point in both A and X, and $r = \min(2k - 1, 2m)$, then E: $\pi_n(X, A, *) \to \pi_{n+1}(SX, SA, *)$ is an r-isomorphism.

18. Let X be a connected one-dimensional CW complex. Show that X has the homotopy type of a wedge of circles. (A one-dimensional CW complex is topologically the same as a graph.)

19. Show that f: $X \to Y$ is a k-equivalence iff given any CW pair (L, L_0) of dimension $\le k$ and maps α: $L_0 \to X$, β: $L \to Y$ with $f\alpha \simeq \beta|_{L_0}$, there is a map g: $L \to X$ with $g|_{L_0} = \alpha$ and $fg \sim \beta$ (rel L_0).

20. Suppose (X, A) has the AHEP. Show that $p_{C*A}\colon X \cup C^*A \to X/A$ is a homotopy equivalence in $\mathscr{C}^*$. (18.10, 18.11)

21. Use 16.24 to show that if $A \subset X$ has a neighborhood U such that A is a strong deformation retract of U, $X \cup CA \to X/A$ is a weak homotopy equivalence.

22. Let $X = \{x \in R^1 \mid x = 0 \text{ or } 1/x \in Z\}$. Show that X is not the homotopy type of a CW complex. (Hint: Calculate $\pi_*(X)$ and find a resolution $f\colon K \to X$. Show that f has no homotopy inverse.)

23. Let $F \to E \to B$ be a Serre fibration. Construct a map $\gamma\colon E \cup CF \to B$ extending π and show that $\gamma_*\colon \pi_*(E \cup CF, *) \to \pi_*(B, *)$ is an $(n+m)$-isomorphism if F is $(n-1)$-connected and B is $(m-1)$-connected. (Hint: Consider the maps $(E, F) \to (E \cup CF, CF) \to (B, *)$.) (Exercise 6, Section 21)

24. Use 16.20 to show that if K is a CW complex and there are maps $f\colon K \to X$ and $g\colon X \to K$ such that $fg \sim 1$, X is the homotopy type of a CW complex.

25. An abstract semisimplicial complex is a sequence of sets X_n for $n \geq 0$ and transformations $\partial_i\colon X_n \to X_{n-1}$ for $0 \leq i \leq n$ such that $\partial_i \partial_j = \partial_{j-1}\partial_i$ if $0 \leq i < j \leq n$. Show that there is a 1–1 correspondence between semisimplicial CW complexes and abstract semisimplicial complexes.

26. An abstract simplicial complex is a pair (V, S) where V is a set and S is a collection of nonempty finite subsets of V (called simplices) such that:

(a) if $v \in V$, $\{v\} \in S$
(b) if $\sigma \in S$ and $\tau \subset \sigma$, $\tau \in S$.

Show that every simplicial CW complex determines an abstract simplicial complex, and if two simplicial CW complexes determine the same abstract simplicial complex they are homeomorphic. (21.14, 21.15)

27. Let X be a regular CW complex (16.37). Show that there is a simplicial CW complex Y with $|X| \equiv |Y|$. (Hint: Suppose K is a simplicial CW complex with $|K| \equiv S^n$ find a simplicial CW complex L with $|L| \equiv B^{n+1}$ and K a subcomplex.)

28. Let $F\colon [(X, *), (Y, *)] \to [X, Y]$ be the transformation which ignores the basepoint. Suppose that $* \in X$ is nondegenerate and Y is arcwise connected.

(a) Prove that F is onto.
(b) Define an action of $\pi_1(Y, *)$ on $[(X, *), (Y, *)]$ as follows. Let $f\colon (X, *) \to (Y, *)$ and $\alpha\colon (I, \{0, 1\}) \to (Y, *)$ and choose $K\colon X \times I \to Y$ such that $K(x, 0) = f(x)$ and $K(*, t) = \alpha(t)$. Let $f^\alpha = K(x, 1)$. Show that the homo-

topy class of f^α depends only on that of f and α (use Exercise 18, Section 14).

(c) Show that $F(\{f^\alpha\}) = F(\{f\})$

(d) Show that if $F(\{f\}) = F(\{g\}), f = g^\alpha$

(e) Show that if $(Y, *)$ is an H-space with unit, K may be chosen so that $f^\alpha = f$.

(Compare with Exercise 14, Section 11.)

29. Suppose $f: (X, A) \to (Y, B)$ is a weak homotopy equivalence. Prove that the induced map $f: X \cup CA \to Y \cup CB$ is a weak homotopy equivalence. (21.8).

17

$K(\pi, n)$'s and Postnikov systems

At the time of writing there is no finite simply connected CW complex all of whose homotopy groups are known—with the exception of contractible complexes. In the absence of such information it is reasonable to try to turn the problem around. Given a sequence of homotopy groups can one find a space X realizing this sequence? Do any conditions have to be put on the sequence? The question then is one of constructing spaces with preassigned homotopy properties. We cannot expect our constructions to be finite cell complexes in general. In fact they will be objects somewhat beyond ordinary geometric imagination. We will think of them in terms of their categorical properties rather than their geometry and treat them with secondary concern—as tools and guideposts. Their properties will make them useful, as we will see in the sequel.

We begin by looking at a few examples.

Proposition 17.1 (a)

$$\pi_i(RP^\infty) = \begin{cases} Z_2, & i = 1 \\ 0, & i \neq 1; \end{cases}$$

(b)

$$\pi_i(CP^\infty) = \begin{cases} Z, & i = 2 \\ 0, & i \neq 2; \end{cases}$$

(c)

$$\pi_i(S^1) = \begin{cases} Z, & i = 1 \\ 0, & i \neq 1. \end{cases}$$

Proof This follows from 11.10, 11.12, 13.14, 13.5, and 15.9. ∎

The simplest constructional question is to ask if there are other spaces with one nontrivial homotopy group. The existence of such spaces could be useful. 16.6 shows, for example, that we can expect to have good control over the mappings of a CW complex into such a space.

Definition 17.2 A CW complex with a single nonvanishing homotopy group π occurring in dimension n is called an Eilenberg–MacLane space $K(\pi, n)$.

Thus RP^∞ is a $K(Z_2, 1)$, CP^∞ is a $K(Z, 2)$, and S^1 is a $K(Z, 1)$.

There are three questions which naturally arise:

1. For which π and n do $K(\pi, n)$ spaces exist? More generally, which sequences of groups $\pi_1, \pi_2, \ldots$ with π_1 abelian for $i \geqq 2$ can be realized as the homotopy groups of some space.

2. Can spaces with many nonzero homotopy groups be decomposed into spaces with fewer nonzero homotopy groups.

3. Does 16.6 give enough information to calculate $[(X, *), (K(\pi, n), *)]$ (assuming $K(\pi, n)$ exists)?

We will show:

Theorem 17.3 If π is a group that is abelian and if $n > 1$, there exists a CW complex $K(\pi, n)$.

The construction will depend on several lemmas.

Let $\bigvee S_\alpha{}^n$ be a one-point union of n-spheres $S_\alpha{}^n$ (where α runs over an indexing set $\mathcal{A}$) with the weak topology.

Lemma 17.4 If $n > 1$, $\pi_n(\bigvee S_\alpha{}^n)$ = free abelian group generated by $\{i_\alpha\}$ where $i_\alpha\colon S^n \equiv S_\alpha{}^n \subset \bigvee_\alpha S_\alpha{}^n$ is the inclusion. $\pi_1(\bigvee S_\alpha{}^1)$ is free and generated by $\{i_\alpha\}$.

Proof Consider first the case that the indexing set is finite. If $n = 1$, the result is Exercise 7, Section 7. Suppose $n > 1$. Then $S^n_{\alpha_1} \vee \cdots \vee S^n_{\alpha_k} \subset S^n_{\alpha_1} \times \cdots \times S^n_{\alpha_k}$ and is a subcomplex. The cells of $S^n_{\alpha_1} \times \cdots \times S^n_{\alpha_k} - (S^n_{\alpha_1} \vee \cdots \vee S^n_{\alpha_k})$ are in dimensions $\geq 2n$ so

$$\pi_i(S^n_{\alpha_1} \times \cdots \times S^n_{\alpha_k}, S^n_{\alpha_1} \vee \cdots \vee S^n_{\alpha_k}, *) = 0$$

for $i < 2n$ and hence

$$\begin{aligned}\pi_n(S^n_{\alpha_1} \vee \cdots \vee S^n_{\alpha_k}, *) &\approx \pi_n(S^n_{\alpha_1} \times \cdots \times S^n_{\alpha_k}) \\ &\approx \pi_n(S^n_{\alpha_1}) \oplus \cdots \oplus \pi_n(S^n_{\alpha_k}) \simeq Z \oplus \cdots \oplus Z.\end{aligned}$$

For the case of arbitrary indexing sets, apply 15.11. ∎

Let $F\mathcal{A}^n$ be the free (abelian if $n > 1$) group generated by the elements of the set $\mathcal{A}$. 17.4 implies that $F\mathcal{A}^n \cong \pi_n(\bigvee_{\alpha \in \mathcal{A}} S_\alpha{}^n)$.

Lemma 17.5 Let $F\mathcal{A}^n$ and $F\mathcal{B}^n$ be as above and $\varphi\colon F\mathcal{A}^n \to F\mathcal{B}^n$ be a homomorphism. Then there is a unique (up to homotopy) map $f\colon \bigvee_{\alpha\in\mathcal{A}} S_\alpha^n \to \bigvee_{\beta\in\mathcal{B}} S_\beta^n$ such that $\varphi = f_*\colon \pi_n(\bigvee_{\alpha\in\mathcal{A}} S_\alpha^n) \to \pi_n(\bigvee_{\beta\in\mathcal{B}} S_\beta^n)$.

Proof Let $\bar{\alpha} \in F\mathcal{A}^n$ be the generator corresponding to $\alpha \in \mathcal{A}$. Then $\varphi(\bar{\alpha}) \in F\mathcal{B}^n = \pi_n(\bigvee_{\beta\in\mathcal{B}} S_\beta^n)$. Choose $f_\alpha\colon S^n \to \bigvee_{\beta\in\mathcal{B}} S_\beta^n$ with $f_\alpha \in \varphi(\bar{\alpha})$. Define f by $f|_{S_\alpha^n} = f_\alpha$. This is well defined since f_α is base point preserving and continuous since $\bigvee_{\alpha\in\mathcal{A}} S_\alpha^n$ has the weak topology. Clearly $f_*(\bar{\alpha}) = \varphi(\bar{\alpha})$ for all $\alpha \in \mathcal{A}$ so $f_* = \varphi$. To prove uniqueness, suppose $g\colon \bigvee_{\alpha\in\mathcal{A}} S_\alpha^n \to \bigvee_{\beta\in\mathcal{B}} S_\beta^n$ has the required property. Since $i_\alpha\colon S^n \equiv S_\alpha^n \subset \bigvee_{\alpha\in\mathcal{A}} S_\alpha^n$ represents $\bar{\alpha}$, gi_α represents $\varphi(\bar{\alpha})$. Hence $g|_{S_\alpha^n} \sim f_\alpha$ (rel $*$). It follows that $g \sim f$ (rel $*$). ∎

Proposition 17.6 Let n be an integer and π be an (abelian if $n > 1$) group. Then there is a CW complex with one 0-cell and all other cells in dimensions n and $n+1$, $M(\pi, n)$, such that $\pi_n(M(\pi, n)) \cong \pi$.

Proof By 17.5 such a space exists if π is free (free abelian if $n > 1$).

Let $0 \to R \xrightarrow{\varphi} F \to \pi \to 0$ be a resolution of π, i.e., a short exact sequence with R and F free. Let $f\colon M(R, n) \to M(F, n)$ be a cellular map such that $f_* = \varphi$.

Let Z be the mapping cylinder of f (16.15) with $A = \varnothing$ and define

$$M(\pi, n) = M(F, n) \cup_f CM(R, n) \equiv Z/M(R, n)$$

If $n > 1$, we have an exact sequence:

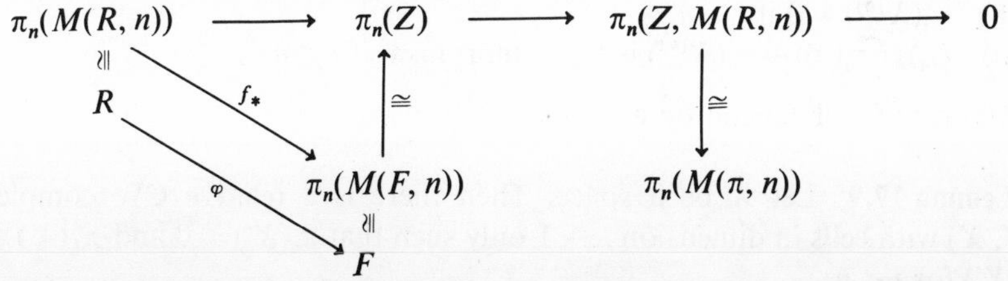

The last isomorphism follows from 16.30 since $n > 1$. Consequently $\pi \cong \pi_n(M(\pi, n))$. Suppose now that $n = 1$. Applying 7.12 with $X = M(\pi, 1)$, $X_1 = M(\pi, 1) - M(R, n) \times 1$, and $X_2 = M(\pi, 1) - M(F, n)$, we conclude that $\pi_1(M(\pi, 1)) = F *_R \{1\} \cong F/R \cong \pi$. ∎

$M(\pi, n)$ is called a Moore space for the group π.

Proposition 17.7 Let $\varphi\colon \pi \to \rho$. Then there is a map $f\colon M(\pi, n) \to M(\rho, n)$ with $f_* = \varphi$.

Proof Construct a commutative diagram of groups

$$\begin{array}{ccccccccc} 0 & \longrightarrow & R_1 & \xrightarrow{\varphi_1} & F_1 & \longrightarrow & \pi & \longrightarrow & 0 \\ & & \downarrow{\scriptstyle \alpha} & & \downarrow{\scriptstyle \beta} & & \downarrow{\scriptstyle \varphi} & & \\ 0 & \longrightarrow & R_2 & \xrightarrow{\varphi_2} & F_2 & \longrightarrow & \rho & \longrightarrow & 0 \end{array}$$

By 17.5, there are maps

$$\begin{array}{ccccc} M(R_1, n) & \xrightarrow{f_1} & M(F_1, n) & \longrightarrow & M(\pi, n) \\ \downarrow{\scriptstyle f_\alpha} & & \downarrow{\scriptstyle f_\beta} & & \vdots{\scriptstyle f} \\ M(R_2, n) & \xrightarrow{f_2} & M(F_2, n) & \longrightarrow & M(\rho, n) \end{array}$$

The diagram commutes up to homotopy by the uniqueness assertion in 17.5. We thus may construct f by 14.15. Applying π_n to the diagram, one easily concludes that $f_* = \varphi$. ∎

We now consider a construction for "killing" homotopy groups. We will apply this to $M(\pi, n)$, killing all homotopy groups above the nth to construct $K(\pi, n)$.

Proposition 17.8 Given X and n there is a space $X^{[n]}$ and inclusion $i_n: X \to X^{[n]}$ such that:

(a) $(X^{[n]}, X)$ is a relative CW complex with cells in dimensions $\geq n + 2$.
(b) $\pi_i(X^{[n]}) = 0$ if $i > n$.
(c) $(i_n)_*: \pi_i(X) \to \pi_i(X^{[n]})$ is an isomorphism if $i \leq n$.

The proof will depend on a lemma.

Lemma 17.9 Let X be a space. Then there is a relative CW complex (X', X) with cells in dimension $n + 1$ only such that $\pi_n(X') = 0$ and $\pi_i(X) \approx \pi_i(X')$ for $i < n$.

Proof of 17.9 Let $\{e_\alpha\}$ be a set of generators and $f_\alpha: S^n \to X$ represent e_α. Define

$$X' = X \cup \coprod B_\alpha^{n+1} / x \in S_\alpha^{\,n} \sim f_\alpha(x).$$

Let $i: X \to X'$ be the inclusion; then $i_*(e_\alpha) = 0$ so $i_*(\pi_n(X)) = 0$. But $\pi_i(X', X) = 0$ for $i \leq n$ by 16.9; consequently $\pi_n(X') = 0$ and $\pi_i(X) \simeq \pi_i(X')$ for $i < n$. ∎

Proof of 17.8 Apply 17.9 to produce X' with $\pi_{n+1}(X') = 0$. Apply it again to X' to produce X'' with $\pi_{n+2}(X'') = 0$, etc. Let $X^{[n]} = \bigcup X^{(r)}$ with the weak topology. (a) is clear. (b) follows since $\pi_i(X^{(r)}) = 0$ for $n < i \leq r$, and $\pi_i(X^{[n]}) \cong \varinjlim \pi_i(X^{(r)})$ by 15.9. (c) follows from (a). ∎

Proof of 17.3 Let $K(\pi, n) = M(\pi, n)^{[n]}$. $\pi_i(K(\pi, n)) = 0$ for $i > n$ by 17.8(b), $\pi_n(K(\pi, n)) = \pi$ by 17.8(c), and $\pi_i(K(\pi, n)) = 0$ for $i < n$ by 17.8(c) and 17.6. ∎

Lemma 17.10 Let $\varphi\colon \pi \to \rho$, $M(\pi, n)$ be as constructed in 17.6 and suppose $K(\rho, n)$ satisfies 17.3. Then there is a unique (up to homotopy) map $f\colon M(\pi, n) \to K(\rho, n)$ such that $\varphi = f_*\colon \pi_n(M(\pi, n)) \to \pi_n(K(\rho, n))$.

Proof We show existence by constructing a map $j\colon M(\rho, n) \to K(\rho, n)$ and combining this with 17.7. Since $M(\rho, n)^n$ has cells only in dimension n excepting for a 0-cell, $M(\rho, n)^n = \bigvee_{\alpha \in \mathcal{A}} S_\alpha^n$. Let $i_\alpha\colon S_\alpha^n \subset \bigvee_{\alpha \in \mathcal{A}} S_\alpha^n$ and $i\colon \bigvee_{\alpha \in \mathcal{A}} S_\alpha^n \to M(\rho, n)$ be the inclusions. Let $x_\alpha = \{ii_\alpha\} \in \pi_n(M(\rho, n)) \simeq \rho$. Let $f_\alpha\colon S_\alpha^n \to K(\rho, n)$ be a representative of $x_\alpha \in \rho = \pi_n(K(\rho, n))$. Define $F\colon \bigvee_{\alpha \in \mathcal{A}} S_\alpha^n \to K(\rho, n)$ by $F|_{S_\alpha^n} = f_\alpha$. Now $\ker F_* = \ker i_* \subset \pi_n(\bigvee_{\alpha \in \mathcal{A}} S_\alpha^n)$ since $F_*(\{i_\alpha\}) = x_\alpha = i_*(\{i_\alpha\})$. Hence for any $(n+1)$-cell e_β^{n+1} of $M(\rho, n)$ with attaching map f_β, we have $\{Ff_\beta\} = F_*(\{f_\beta\}) = *$. We thus construct an extension F_β of F over e_β^{n+1} and hence an extension $j\colon (M\rho, n) \to K(\rho, n)$ of F. Since the diagram

$$\begin{array}{ccc} \pi_n(M(\rho, n)) & \xrightarrow{j_*} & \pi_n(K(\rho, n)) \\ & {\scriptstyle i_*}\nwarrow \quad \nearrow{\scriptstyle F_*} & \\ & \pi_n\Big(\bigvee_{\alpha \in \mathcal{A}} S_\alpha^n\Big) & \end{array}$$

commutes and i_* is onto, j_* is an isomorphism.

Given two maps f_0 and f_1 with $(f_0)_* = (f_1)_*$, we have

$$f_0\Big|_{\bigvee_{\alpha \in \mathcal{A}} S_\alpha^n} \sim f_1\Big|_{\bigvee_{\alpha \in \mathcal{A}} S_\alpha^n}$$

since

$$\{f_0 i_\alpha\} = (f_0)_*\{ii_\alpha\} = (f_1)_*\{ii_\alpha\} = \{f_1 i_\alpha\}.$$

To construct an extension of this homotopy to $M(\pi, n) \times I$, consider the diagram

$$\begin{array}{ccc} M(\pi, n) \times I & & \\ \uparrow & \searrow{\scriptstyle (\text{dashed})} & \\ M(\pi, n) \times 0 \cup \bigvee_{\alpha \in \mathcal{A}} S_\alpha^n \times I \cup M(\pi, n) \times 1 & \longrightarrow & K(\rho, n) \end{array}$$

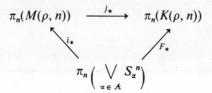

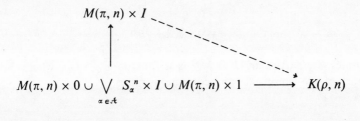

Since the cells of $M(\pi, n) \times I - \{M(\pi, n) \times 0 \cup \bigvee_{\alpha \in \mathcal{A}} S_\alpha^n \times I \cup M(\pi, n) \times 1\}$ are of dimension $n + 2$, the extension exists by 16.3. ▮

Proposition 17.11 Let $\varphi: \pi \to \rho$. Let $K(\pi, n) = M(\pi, n)^{[n]}$ and let $K(\rho, n)$ satisfy 17.2. Then there is a map $f: K(\pi, n) \to K(\rho, n)$ unique up to homotopy with $f_* = \varphi$.

Proof By 17.10 we can construct $f': M(\pi, n) \to K(\rho, n)$. f' can be extended to $f: K(\pi, n) \to K(\rho, n)$ by 16.3. To prove uniqueness, let $f_1, f_2: K(\pi, n) \to K(\rho, n)$ and suppose $(f_1)_* = (f_2)_*$; then $f_1|_{M(\pi,n)} \sim f_2|_{M(\pi,n)}$ by 17.10. We consider the extension problem

$$\begin{array}{ccc} K(\pi, n) \times I & & \\ \uparrow & \searrow & \\ K(\pi, n) \times 0 \cup M(\pi, n) \times I \cup K(\pi, n) \times 1 & \longrightarrow & K(\rho, n) \end{array}$$

and apply 16.3 again. ▮

Corollary 17.12 Any two CW complexes $K(\pi, n)$ satisfying 17.2 are of the same homotopy type.

Proof They are both the same homotopy type as $[M(\pi, n)]^{[n]}$ by 17.11 applied to $\varphi = 1$. ▮

Corollary 17.13 For each $n \geq 1$, there is a functor from $\mathcal{M}_Z$ to $\mathcal{K}_h$ (from $\mathcal{G}$ to $\mathcal{K}_h$ if $n = 1$) taking π to a $K(\pi, n)$ space.

Proof This follows immediately from 17.11 and 17.12. ▮

Proposition 17.14 Let $f: X \to Y$ and $i_n: X \to Y^{[n]}$, $i_n': Y \to Y^{[m]}$ satisfy the conclusion of 17.8 with $m \leq n$. Then there is a unique (up to homotopy) map $f_{n,m}: X^{[n]} \to Y^{[m]}$ such that the diagram

$$\begin{array}{ccc} X & \xrightarrow{f} & Y \\ \downarrow{\scriptstyle i_n} & & \downarrow{\scriptstyle i_m'} \\ X^{[n]} & \xrightarrow{f_{n,m}} & Y^{[m]} \end{array}$$

commutes. In particular, $X \to X^{[n]}$ is a functor $\tilde{\mathcal{K}}_h \to \tilde{\mathcal{K}}_h$, where $\tilde{\mathcal{K}}$ is the category of CW complexes and continuous maps.

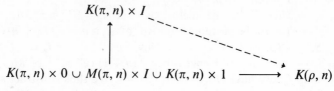

Proof We apply 16.3 to the diagram

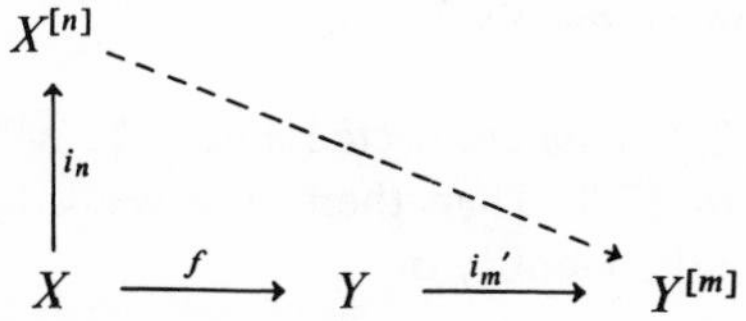

to conclude the existence of $f_{n,m}$. To prove uniqueness, suppose $f_{n,m}$ and $f'_{n,m}$ are extensions. Consider the diagram

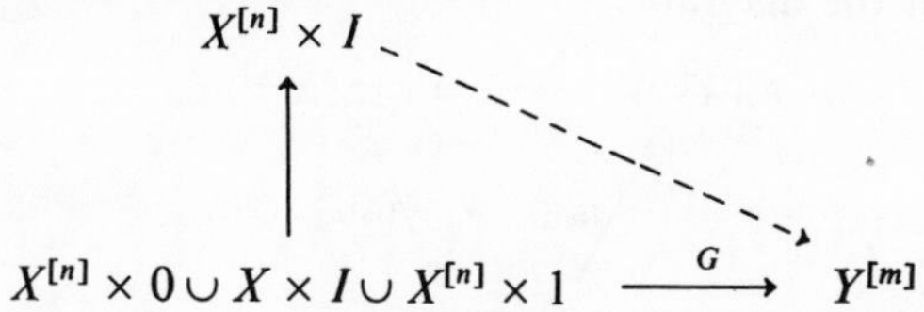

where $G(x, 0) = f_{n,m}(x)$, $G(x, 1) = f'_{n,m}(x)$ for $x \in X^{[n]}$ and $G(x, t) = i_m'(f(x))$ for $x \in X$, and apply 16.3 again. The map $f_{n,n}\colon X^{[n]} \to Y^{[n]}$ will be written $f^{[n]}$. Clearly $f^{[n]} \circ g^{[n]} = (f \circ g)^{[n]}$ by the uniqueness assertion. A functor is thus defined by choosing for each X, a space $X^{[n]}$ satisfying the conclusion of 17.8. ∎

The space $X^{[n]}$ is called the nth Postnikov section of X. These sections fit together to form a tower called the Postnikov system or Postnikov tower of X.

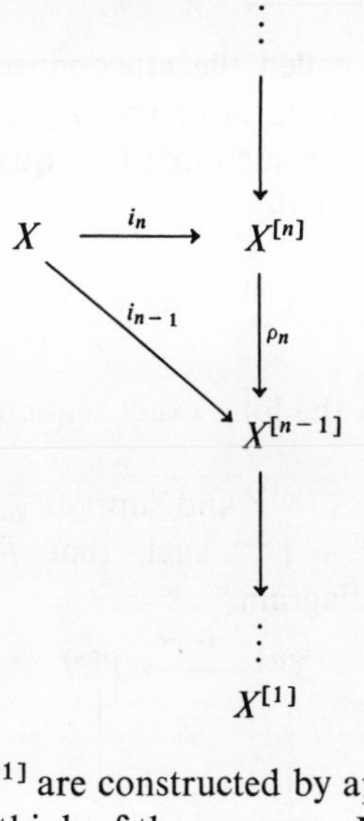

The maps $\rho_n\colon X^{[n]} \to X^{[n-1]}$ are constructed by applying 17.14 to the identity map with $m = n - 1$. We think of the sequence $X^{[n]}$ as being approximations to X (see Exercise 2).

The next result shows that if X is a CW complex, the homotopy type of $X^{[n]}$ does not depend on the choice of $X^{[n]}$.

Proposition 17.15 Let X be connected and $X^{[n]}$, $\bar{X}^{[n]}$ be two spaces satisfying the conclusion of 17.8. Then there is a weak homotopy equivalence $e\colon X^{[n]} \to \bar{X}^{[n]}$ which is the identity on X.

Proof By 17.14 there is a map $e\colon X^{[n]} \to \bar{X}^{[n]}$ which is the identity on X. It is easy to see that e induces isomorphisms in homotopy groups by the commutativity of the diagram

$$\begin{array}{ccc} \pi_i(X^{[n]}) & \xrightarrow{e_*} & \pi_i(\bar{X}^{[n]}) \\ & {\scriptstyle (i_n)_*}\nwarrow \quad \nearrow{\scriptstyle (i_n)_*} & \\ & \pi_i(X) & \end{array} \qquad \blacksquare$$

Definition 17.16 Let $X^{(n)}$ be a resolution of the fiber F of the map $i_n\colon X \to X^{[n]}$, i.e., we convert i_n to a fibering

$$\begin{array}{ccccc} F & \longrightarrow & E & \xrightarrow{\pi_n} & X^{[n]} \\ \uparrow & & \uparrow \simeq & \nearrow_{i_n} & \\ X^{(n)} & \xrightarrow{j_n} & X & & \end{array}$$

according to 11.14. $X^{(n)}$ is called the nth connective covering space of X. If $n = 1$, this is equivalent to the ordinary covering space construction (see Exercise 1). A map $j_n\colon X^{(n)} \to X$ such that the square commutes up to homotopy is induced by the construction.

Proposition 17.17 $(j_n)_*\colon \pi_i(X^{(n)}) \to \pi_i(X)$ is an isomorphism if $i > n$, and $\pi_i(X^{(n)}) = 0$ if $i \leq n$.

Proof This follows from the long exact sequence of the fibering. ∎

Proposition 17.18 Let $f\colon X \to Y$ and suppose $j_n\colon X^{(n)} \to X$. Then if $m \leq n$, there are maps $f^{n,m}\colon X^{(n)} \to Y^{(m)}$ such that $f^{n,m} \circ g^{k,n} \sim (f \circ g)^{k,m}$, and $1^{n,n} = 1$. Furthermore, the diagram

$$\begin{array}{ccc} X^{(n)} & \xrightarrow{f^{n,m}} & Y^{(m)} \\ \downarrow{\scriptstyle j_n} & & \downarrow{\scriptstyle j_m} \\ X & \xrightarrow{f} & Y \end{array}$$

commutes up to homotopy.

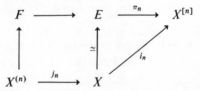

Proof In 17.16, $F = \Omega(X^{[n]}, X, *)$. We define

$$\bar{f}^{n,m}\colon \Omega(X^{[n]}, X, *) \longrightarrow \Omega(Y^{[m]}, Y, *)$$

by $\bar{f}^{n,m}(\omega)(t) = f_{n,m}(\omega(t))$. Clearly $\bar{f}^{n,m} \circ \bar{g}^{k,n} \sim (\overline{fg})^{k,m}$ and $\bar{1}^{n,n} \sim 1$. $f^{n,m}$ is constructed by applying 16.20. The diagram commutes up to homotopy since

$$\begin{array}{ccc} \Omega(X^{[n]}, X, *) & \xrightarrow{\bar{f}^{n,m}} & \Omega(Y^{[n]}, Y, *) \\ \Big\downarrow{\scriptstyle \pi_1} & & \Big\downarrow{\scriptstyle \pi_1} \\ X & \xrightarrow{\quad f \quad} & Y \end{array}$$

commutes. ∎

We describe now a technique for analyzing various homotopy theory problems, called obstruction theory. We will consider only one problem: Given $f\colon X \to Y$, is f nullhomotopic? We assume X is the homotopy type of a CW complex, and we take a Postnikov system for Y. We will say f is n-trivial if $i_n f$ is nullhomotopic:

$$X \xrightarrow{\ f\ } Y \xrightarrow{\ i_n\ } Y^{[n]}.$$

This does not depend on the choice of $Y^{[n]}$, for if we choose $Y \xrightarrow{\bar{i}_n} \bar{Y}^{[n]}$, there is a map $e\colon Y^{[n]} \to \bar{Y}^{[n]}$ inducing isomorphisms in homotopy and such that $ei_n \sim \bar{i}_n$. Since X is the homotopy type of a CW complex, there is a 1–1 correspondence $[X, Y^{[n]}] \leftrightarrow [X, \bar{Y}^{[n]}]$ so $i_n f \sim *$ iff $\bar{i}_n f \sim *$. (See Exercise 6, Section 16.)

Lemma 17.19 If X is a k-dimensional CW complex, f is k-trivial iff f is trivial.

Proof If f is k-trivial, there is a map $\tilde{f}\colon X \to Y^{(k)}$ such that $f \sim j_k \tilde{f}$ by Exercise 9, Section 11. Now apply 16.3 to the extension problem

$$\begin{array}{ccc} CX & & \\ \Big\uparrow & \searrow & \\ X & \xrightarrow{\ \tilde{f}\ } & Y^{(k)} \end{array}$$

Since $CX - X$ has all cells in dimensions $\leq k + 1$ and $\pi_i(Y^{(k)}) = 0$ for $i \leq k$, an extension exists. Thus $\tilde{f}$ and hence f is trivial. The converse is easy. ∎

If X is an infinite CW complex, there may exist maps that are k-trivial for all k but which are essential. Such maps are called phantom maps. (See [26].)

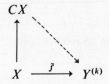

For finite CW complexes one may take an inductive approach to the problem. Let $\tilde{H}^n(X: \pi) = [(X, *), (K(\pi, n), *)]$. As we shall see in the next section, this set is an abelian group called the n-dimensional (ordinary) cohomology of X with coefficients in π, if π is abelian.

Proposition 17.20 Let $f: X \to Y$. If f is $(n-1)$-trivial, there is defined a set $\mathcal{O}_n(f) \subset \tilde{H}^n(X; \pi_n(Y))$. $0 \in \mathcal{O}_n(f)$ iff f is n-trivial.

Proof Consider the exact sequence in $\mathcal{S}^*$ (by Exercise 10, Section 11)

$$[X, K(\pi_n(Y), n)] \xrightarrow{\alpha_*} [X, Y^{[n]}] \to [X, Y^{[n-1]}];$$

let $\mathcal{O}_n(f) = \alpha_*^{-1}(i_n \cdot f)$. This is nonempty iff f is $(n-1)$-trivial and $0 \in \mathcal{O}_n(f)$ iff f is n-trivial. ▌

$\mathcal{O}_n(f)$ is called the n-dimensional obstruction set to f being essential. 17.20 is most useful if we know that $\tilde{H}^n(X, \pi_n(Y)) = 0$ for all n.

Corollary 17.21 If $0 = \tilde{H}^n(X; \pi_n(Y))$ for all n and X is an arcwise connected finite CW complex, $[X, Y] = *$. ▌

Exercises

1. Show that if X is the homotopy type of a CW complex, $X^{(1)}$ is the homotopy type of the simply connected covering space of X.

2. Show that if the maps ρ_n are converted inductively into fibrations, there is a map $i: X \to \varprojlim X^{[n]}$ such that $\pi_n i = i_n$, where the i_n are chosen so that $\rho_n i_n = i_{n-1}$. Show that i induces isomorphisms in homotopy. (Show first that $\pi_i(\varprojlim X^{[n]}) \cong \varprojlim \pi_i(X^{[n]})$.) See Exercise 9, Section 15.

3. Let $f: E \to B$ be a fibering with B connected. Show that there are fiberings $\pi_n: E^{[n]} \to E^{[n-1]}$ for all $n \geq 0$ and factorizations $f_n: E \to E^{[n]}$ with $\pi_n f_n = f_{n-1}$ such that

(a) $E^{[0]} = B, f_0 = f$;
(b) the fiber of π_n is $K(\pi_n(F), n)$ where F is the fiber of f;
(c) $(f_n)_*: \pi_i(E) \to \pi_i(E^{[n]})$ is an isomorphism for $i \leq n$;
(d) If $f^n = \pi_1 \circ \cdots \circ \pi_n: E^{[n]} \to B$, $(f^n)_*: \pi_i(E^{[n]}) \to \pi_i(B)$ is an isomorphism for $i > n$.

This generalizes the constructions $X^{[n]}$ and $X^{(n)}$.

4. Generalize 17.21 to show that if X is an arcwise connected finite CW complex and $\tilde{H}^k(X; \pi_k(Y)) = 0$ for all $k \neq n$, there is a transformation from $\tilde{H}^n(X; \pi_n(Y))$ onto $[X, Y]$. Furthermore, if $H^k(X; \pi_{k+1}(Y)) = 0$ for all $k < n - 1$, this transformation is also 1–1.

5. Deduce from Exercise 4 the famous *Hopf theorem*: If X is an arcwise connected finite CW complex of dimension n, there is a 1–1 correspondence between $\tilde{H}^n(X; Z)$ and $[X, S^n]$.

6. Show that any sequence of groups $\pi_1, \pi_2, \ldots$ with π_i abelian for $i \geq 2$ can be realized as the homotopy groups of some space. (Hint: Do it first for a finite sequence.)

7. Calculate $[RP^\infty, CP^\infty]$. That is how many homotopy classes are there? (Hint: Compare with $[RP^2, CP^\infty]$.)

8. Let K be the Klein bottle (see Exercise 14, Section 7). Show that $K = K(\pi, 1)$ where π is a group on two generators x, y with the single relation $xyx = y$.

9. Let $p\colon RP^2 \to RP^2/RP^1 \equiv S^2$, $\Pi_2\colon S^2 \to RP^2$. Let $f = \Pi_2 p$. Prove that:

(1) f is essential;
(2) $0 = f_*\colon \pi_1(RP^2, *) \to \pi_1(RP^2, *)$;
(3) $RP^2 \equiv M(Z_2, 1)$.

(Compare to 17.7 and 17.10.)

10. Let $0 \to \pi \xrightarrow{\varphi} \rho \xrightarrow{\psi} \sigma \to 0$ be a short exact sequence of abelien groups. Construct maps

$$K(\pi, n) \xrightarrow{f} K(\rho, n) \xrightarrow{g} K(\sigma, n)$$

as in 17.11. Prove that this sequence is homotopy equivalent to a fiber sequence; i.e., if g is converted into a fibering there is a homotopy commutative diagram

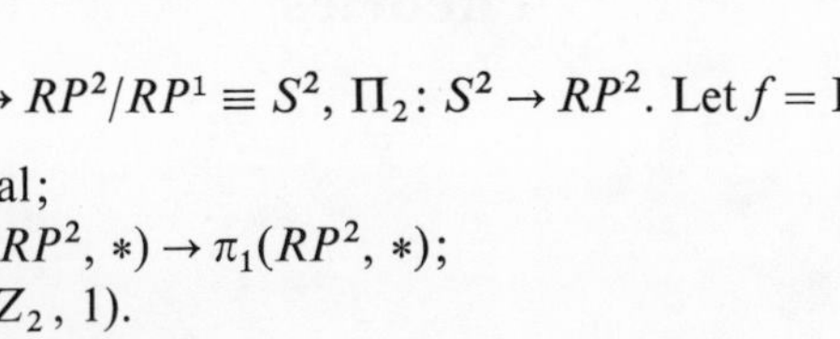

11. Show that $K(\pi, n)$ is an H-space iff π is abelian.

18

Spectral Reduced Homology and Cohomology Theories

This section is concerned with the definition of spectral homology and cohomology functors. The ordinary homology group $H_n(X)$ can be thought of as an approximation to $\pi_n(X)$.

We think of $\pi_n(X)$ as a classification of n-dimensional "elements" in X. In this case, by an n-dimensional element we mean a continuous image of S^n. The groups $\pi_n(X)$ are easy to define, and, as we have seen, hard to calculate. A different notion of element is given if we consider elements as represented by imbedded cells. Then $S^1 \times S^1$ has a 2-dimensional element even though $\pi_2(S^1 \times S^1) = 0$. The number of cells in a given dimension is not, however, a topological invariant. The ordinary homology in dimension n is designed to be the classification of certain invariant combinations of the n-dimensional cells (called cycles). The difficulty with homology theory is exactly the opposite to homotopy theory. Homology groups are easy to calculate, but hard to define (in an invariant way). In Section 20 we shall give a more detailed explanation of what we mean by a cycle and when two cycles are homologous. Our present task is to define certain general functors called homology and cohomology theories on any space in $\mathcal{CG}^*$. They will be topological (in fact homotopy) invariants. In Section 20 we shall show that they correspond to the homology classification of cycles in the case of ordinary homology theory. Our general theories[14] will include stable homotopy theory and

[14] Sometimes the theories described here are called extraordinary homology and cohomology to distinguish them from the ordinary theory described above. However, as times goes on, they become less extraordinary.

other functors which have recently become important in algebraic topology—K-theories and cobordism theories. A brief description of these is found in Sections 29 and 30.

We begin by defining the notion of a spectrum $E = \{E_n, e_n\}$ and show that each spectrum gives rise to two sequences of functors $\tilde{E}_m$ and $\tilde{E}^m$ from $\mathcal{CG}^*$. to $\mathcal{M}_Z$ the first covariant and the second contravariant. These are the spectral homology and cohomology functors.

Definition 18.1 A spectrum $E = \{E_n, e_n\}$ is a sequence of spaces E_n and maps e_n: $SE_n \to E_{n+1}$ for $n \geq 0$ (or equivalently $\bar{e}_n$: $E_n \to \Omega E_{n+1}$) in $\mathcal{CG}^*$. E is called a suspension spectrum if e_n is a weak homotopy equivalence, for all n sufficiently large and an Ω-spectrum if $\bar{e}_n$ is a weak homotopy equivalence for all n sufficiently large.

Examples

1. Let $X \in \mathcal{CG}$ and define $\underline{X}$ by $\underline{X}_n = S^n X$, and $\underline{x}_n$: $S(S^n X) \to S^{n+1} X$ to be the natural homeomorphism. Any suspension spectrum is obviously of this form "up to weak homotopy" where $X = \underline{X}_0$. This spectrum will be written $\underline{X}$ and $\underline{S}^0$ will be abbreviated $\underline{S}$.

2. $H\pi$ is given by $(H\pi)_n = K(\pi, n)$, and $(\overline{h\pi})_n$: $K(\pi, n) \to \Omega K(\pi, n+1)$ a chosen resolution.

Given a spectrum E (we often suppress the spaces E_n and maps e_n from the notation, when it will not lead to confusion), and a space X we will define groups $\tilde{E}_m(X)$ and $\tilde{E}^m(X)$ for each integer m.

Definition 18.2 A graded abelian group is a sequence $\{G_n\}$ of abelian groups, defined for each integer n. A homomorphism f: $\{G_n\} \to \{G_n'\}$ of graded groups is a sequence $\{f_n\}$ of homomorphisms f_n: $G_n \to G_n'$. One often writes G_* for $\{G_n\}$. Similar definitions may be made for graded R-modules, or graded sets.

Such objects and homomorphisms form a category written $\mathcal{M}_{Z*}$, $\mathcal{M}_{R*}$, or $\mathcal{S}_*$ in the cases of graded abelian groups, graded R-modules, and graded sets respectively.

Example The sequence $G_n = \pi_n(X, *)$ for $n \geq 1$ and $G_n = 0$ if $n \leq 0$ is a graded abelian group if π_1 is abelian (otherwise it will be called just a graded group), and the sequence of homotopy groups yields a functor π_*: $\mathcal{C}^* \to \mathcal{M}_{Z*}$. This sequence is called positively graded since $G_n = 0$ if $n < 0$.

Let $\mathcal{C}^*$ be a category of spaces with base point and base point preserving mappings.

Definition 18.3 A reduced homology (cohomology) theory on $\mathcal{C}^*$ is a covariant (contravariant) functor $\{\tilde{E}_m\}$ ($\{\tilde{E}^m\}$) from $\mathcal{C}^*$ to $\mathcal{M}_{Z*}$ satisfying:

(A) If $f: X \to Y$, write $f_*: \tilde{E}_m(X) \to \tilde{E}_m(Y)$ ($f^*: \tilde{E}^m(Y) \to \tilde{E}^m(X)$) for the induced homomorphism. Then if $f \sim g$ in $\mathcal{C}^*$, $f_* = g_*$ and $f^* = g^*$.

(B) There is a natural transformation:

$$\sigma: \tilde{E}_m(X) \to \tilde{E}_{m+1}(SX) \qquad (\sigma: \tilde{E}^m(X) \to \tilde{E}^{m+1}(SX))$$

which is an isomorphism.

(C) If $f: X \to Y$ and $i: Y \to Y \cup_f C^* X$, the sequence:

$$\tilde{E}_m(X) \xrightarrow{f_*} \tilde{E}_m(Y) \xrightarrow{i_*} \tilde{E}_m(Y \cup_f C^*X)$$

$$(\tilde{E}^m(X) \xleftarrow{f^*} \tilde{E}^m(Y) \xleftarrow{i^*} \tilde{E}^m(Y \cup_f C^*X))$$

is exact in the middle.

We now construct, for each spectrum E, functors $\{\tilde{E}_m\}$ and $\{\tilde{E}^m\}$ which are reduced homology and cohomology theories on $\mathcal{CG}^*$. These will be called spectral homology and cohomology theories (to distinguish them from theories constructed in other ways). If $E = H\pi$, these groups will be called the ordinary spectral reduced homology and cohomology theories with coefficients in π. These are classically written $\tilde{H}_m(X; \pi)$ and $\tilde{H}^m(X; \pi)$. As we shall see, ordinary cohomology agrees with the functor introduced in Section 17 with the same name. If $\pi = Z$, this is abbreviated $\tilde{H}_m(X)$ and $\tilde{H}^m(X)$.

Given $X \in \mathcal{CG}^*$, consider the directed systems

$$\cdots \to \pi_{n+m}(X \wedge E_n) \xrightarrow{\gamma_n} \pi_{n+m+1}(X \wedge E_{n+1}) \to \cdots$$

$$\cdots \to [S^{n-m}X, E_n] \xrightarrow{\lambda_n} [S^{n+1-m}X, E_{n+1}] \to \cdots \qquad (n \geq m),$$

where the homomorphisms γ_n and λ_n are the composites

$$\pi_{n+m}(X \wedge E_n) \xrightarrow{E} \pi_{n+m+1}(X \wedge E_n \wedge S^1) \xrightarrow{(1 \wedge e_n)_*} \pi_{n+m+1}(X \wedge E_{n+1})$$

$$[S^{n-m}X, E_n] \xrightarrow{E} [S^{n+1-m}X, SE_n] \xrightarrow{(e_n)_*} [S^{n+1-m}X, E_{n+1}].$$

Define

$$\tilde{E}_m(X) = \underrightarrow{\lim}\{\pi_{n+m}(X \wedge E_n), \gamma_n\} \qquad \text{and} \qquad \tilde{E}^m(X) = \underrightarrow{\lim}\{[S^{n-m}X, E_n], \lambda_n\}.$$

Theorem 18.4 $\{\tilde{E}^m\}$ is a reduced cohomology theory on $\mathcal{CG}^*$.

Proof $f: X \to Y$ induces homomorphisms

$$\begin{array}{ccccc}
[S^{n-m}Y, E_n] & \xrightarrow{E} & [S^{n+1-m}Y, SE_n] & \xrightarrow{(e_n)_*} & [S^{n+1-m}Y, E_{n+1}] \\
\downarrow{\scriptstyle (S^{n-m}f)^*} & & \downarrow{\scriptstyle (S^{n+1-m}f)^*} & & \downarrow{\scriptstyle (S^{n+1-m}f)^*} \\
[S^{n-m}X, E_n] & \xrightarrow{E} & [S^{n+1-m}X, SE_n] & \xrightarrow{(e_n)_*} & [S^{n+1-m}X, E_{n+1}]
\end{array}$$

This induces a homomorphism from the direct sequence for $\tilde{E}^m(Y)$ to the direct sequence for $\tilde{E}^m(X)$, and hence a homomorphism $f^*\colon \tilde{E}^m(Y) \to \tilde{E}^m(X)$ by 15.12. By the uniqueness assertions, $1^* = 1$, and $(fg)^* = g^*f^*$.

Proof of (*A*) Suppose $f \sim g$ in $\mathcal{CG}^*$. Then $S^{n-m}f \sim S^{n-m}g$ in $\mathcal{CG}^*$ so

$$(S^{n-m}f)^* = (S^{n-m}g)^*\colon [S^{n-m}Y, E_n] \to [S^{n-m}X, E_n].$$

Hence $f^* = g^*$. ∎

Proof of (*B*) Replace SX by $X \wedge S^1$ and note that the natural homeomorphisms $X \wedge S^k \equiv (X \wedge S^1) \wedge S^{k-1}$ induce (unlabeled) natural isomorphisms

$$\begin{array}{ccccc}
[X \wedge S^{n-m}, E_n] & \xrightarrow{E} & [X \wedge S^{n-m} \wedge S^1, E_n \wedge S^1] & \xrightarrow{(e_n)^*} & [X \wedge S^{n-m+1}, E_{n+1}] \\
\Big\| \cong & & \Big\| \cong & & \Big\| \cong \\
[(X \wedge S^1) \wedge S^{n-m-1}, E_n] & \xrightarrow{E} & [(X \wedge S^1) \wedge S^{n-m-1} \wedge S^1, E_n \wedge S^1] & \xrightarrow{(e_n)^*} & [(X \wedge S^1) \wedge S^{n-m}, E_{n+1}]
\end{array}$$

Now the diagram commutes, and both horizontal composites are λ_n, as occurring in the direct limit for $\tilde{E}^m(X)$ and $\tilde{E}^{m+1}(X \wedge S^1)$, it follows that the limits are naturally isomorphic. ∎

Proof of (*C*) By Exercise 21, Section 14, the sequence

$$\begin{array}{ccccc}
[S^{n-m}(Y \cup_f C^*X), E_n] & \xrightarrow{i^*} & [S^{n-m}Y, E_n] & \xrightarrow{f^*} & [S^{n-m}X, E_n] \\
\wr\| & & \wr\| & & \wr\| \\
[Y \cup_f C^*X, \Omega^{n-m}E_n] & \xrightarrow{i^*} & [Y, \Omega^{n-m}E_n] & \xrightarrow{f^*} & [X, \Omega^{n-m}E_n]
\end{array}$$

is exact at the middle. By 15.13 we are done. ∎

Theorem 18.5 $\{\tilde{E}_m\}$ is a reduced homology theory on $\mathcal{CG}^*$.

Proof To define f_*, consider the commutative diagram

$$\begin{array}{ccccc}
\pi_{n+m}(X \wedge E_n) & \xrightarrow{E} & \pi_{n+m+1}(X \wedge E_n \wedge S^1) & \xrightarrow{(1 \wedge e_n)_*} & \pi_{n+m+1}(X \wedge E_{n+1}) \\
\Big\downarrow (f \wedge 1)_* & & \Big\downarrow (f \wedge 1)_* & & \Big\downarrow (f \wedge 1)_* \\
\pi_{n+m}(Y \wedge E_n) & \xrightarrow{E} & \pi_{n+m+1}(Y \wedge E_n \wedge S^1) & \xrightarrow{(1 \wedge e_n)_*} & \pi_{n+m+1}(Y \wedge E_{n+1})
\end{array}$$

15.12 provides a map $f_*\colon \tilde{E}_m(X) \to \tilde{E}_m(Y)$ and the uniqueness assertions imply that $1_* = 1$ and $(fg)_* = f_*g_*$.

Proof of (*A*) This follows as before since if $f \sim g$, $f \wedge 1 \sim g \wedge 1$ and hence

$$(f \wedge 1)_* = (g \wedge 1)_*\colon \pi_{n+m}(X \wedge E_n) \to \pi_{n+m}(Y \wedge E\ ).\quad ∎$$

Proof of (*B*) We define σ as follows. For any space X, define

$$\Sigma\colon \pi_n(X) \to \pi_{n+1}(S^1 \wedge X)$$

by $\Sigma\{\theta\} = \{1 \wedge \theta\}$. The diagram

$$\begin{array}{ccc}
\pi_{n+m}(X \wedge E_n) & \xrightarrow{\Sigma} & \pi_{n+m+1}(S^1 \wedge X \wedge E_n) \\
\downarrow{\scriptstyle E} & & \downarrow{\scriptstyle E} \\
\pi_{n+m+1}(X \wedge E_n \wedge S^1) & \xrightarrow{\Sigma} & \pi_{n+m+2}(S^1 \wedge X \wedge E_n \wedge S^1) \\
\downarrow{\scriptstyle (1 \wedge e_n)_*} & & \downarrow{\scriptstyle (1 \wedge e_n)_*} \\
\pi_{n+m+1}(X \wedge E_{n+1}) & \xrightarrow{\Sigma} & \pi_{n+m+2}(S^1 \wedge X \wedge E_{n+1})
\end{array}$$

commutes, hence Σ induces a natural transformation $\sigma\colon \tilde{E}_m(X) \to \tilde{E}_{m+1}(S^1 \wedge X) \cong \tilde{E}_{m+1}(SX)$, where the isomorphism is given by the natural transposition homeomorphism $T_X\colon S^1 \wedge X \to X \wedge S^1 = SX$.

The proof that σ is an isomorphism will depend on a lemma.

Lemma 18.6 The diagram

$$\begin{array}{ccc}
\pi_n(X) & \xrightarrow{\Sigma} & \pi_n(S^1 \wedge X) \\
\downarrow{\scriptstyle E} & \swarrow{\scriptstyle (T_X)_*} & \downarrow{\scriptstyle E} \\
\pi_n(X \wedge S^1) & \xrightarrow{\Sigma} & \pi_{n+1}(S^1 \wedge X \wedge S^1)
\end{array}$$

commutes up to sign.

Proof $T_{S^n}\colon S^1 \wedge S^n \to S^n \wedge S^1$ is a homeomorphism and hence, after identification of the spaces involved with spheres, is homotopic to ± 1. Thus the upper triangle commutes up to sign since for $f\colon S^n \to X$,

$$\begin{array}{ccc}
S^1 \wedge S^n & \xrightarrow{\Sigma(f)} & S^1 \wedge X \\
\downarrow{\scriptstyle T_{S^n}} & & \downarrow{\scriptstyle T_X} \\
S^n \wedge S^1 & \xrightarrow{E(f)} & X \wedge S^1
\end{array}$$

commutes. Let $f\colon S^n \to S^1 \wedge X$. Then

$$\begin{array}{ccccc}
S^1 \wedge S^n & \xrightarrow{1 \wedge f} & S^1 \wedge S^1 \wedge X & & \\
\downarrow{\scriptstyle T_{S^n}} & & \downarrow{\scriptstyle T_{S^1 \wedge X}} & \searrow{\scriptstyle T_{S^1} \wedge 1} & \\
S^n \wedge S^1 & \xrightarrow{f \wedge 1} & S^1 \wedge X \wedge S^1 & \xrightarrow{1 \wedge T_X^{-1}} & S^1 \wedge S^1 \wedge X
\end{array}$$

commutes.

Now $T_{S^1} = E(\alpha)$ for some $\alpha\colon S^1 \to S^1$ with $\{\alpha\} = \pm 1$. Thus $(T_{S^1} \wedge 1) \circ (1 \wedge f) \sim \alpha \wedge f$. Consequently

$$(f \wedge 1) \circ (T_{S^n}) = (1 \wedge T_X) \circ (T_{S^1} \wedge 1) \circ (1 \wedge f) \sim \alpha \wedge T_X f.$$

Thus $\pm E(\{f\}) = \{(1 \wedge T_X f) \circ (\alpha \wedge 1)\} = \pm\, \Sigma((T_X)_*(\{f\}))$. ▮

We now complete the proof of B. By 18.6 the diagram

$$\begin{array}{ccc}
\pi_{n+m}(X \wedge E_n) & \xrightarrow{\Sigma} & \pi_{n+m+1}(S^1 \wedge X \wedge E_n) \\
\Big\downarrow E & \swarrow \cong\ (T_{X \wedge E_n})_* & \Big\downarrow E \\
\pi_{n+m+1}(X \wedge E_n \wedge S^1) & \xrightarrow{\Sigma} & \pi_{n+m+2}(S^1 \wedge X \wedge E_n \wedge S^1) \\
\Big\downarrow (1 \wedge e_n)_* & & \Big\downarrow (1 \wedge e_n)_* \\
\pi_{n+m+1}(X \wedge E_{n+1}) & \xrightarrow{\Sigma} & \pi_{n+m+2}(S^1 \wedge X \wedge E_{n+1})
\end{array}$$

commutes up to sign. If $\sigma(\alpha) = 0$, there is an $x \in \pi_{n+m}(X \wedge E_n)$ such that x represents α and $\Sigma x = 0$. Consequently $Ex = 0$ so $\alpha = 0$. If $\alpha \in \tilde{E}_{m+1}(S^1 \wedge X)$ there is an $x \in \pi_{n+m+1}(S^1 \wedge X \wedge E_n)$ which represents α. But $\gamma_n(x) = \Sigma((1 \wedge e_n) \cdot T_{X \wedge E_n})_*(x)$, so $\alpha \in \text{Image } \sigma$. ▮

To prove (C), we need some lemmas.

Lemma 18.7 Let $f\colon X \to Y$. Then in the diagram

$$\begin{array}{ccc}
\pi_{n+1}(Y \cup_f C^*X) & \xrightarrow{(p_Y)_*} & \pi_{n+1}(SX) \\
 & & \Big\uparrow E \\
 & & \pi_n(X) \xrightarrow{f_*} \pi_n(Y)
\end{array}$$

$E(\ker f_*) \subset \text{Im}(p_Y)_*$.

Proof Let $\alpha\colon S^n \to X$ represent an element of $\pi_n(X)$ and $H\colon C^*S^n \to Y$ be an extension of $f\alpha$. Define $\beta\colon S^{n+1} = SS^n \to Y \cup_f C^*X$ by

$$\beta(\theta, t) = \begin{cases} (\alpha(\theta), 2t - 1), & \frac{1}{2} \le t \le 1 \\ H(\theta, 1 - 2t), & 0 \le t \le \frac{1}{2} \end{cases}$$

for $\theta \in S^n$. A homotopy $\bar{H}\colon S\alpha \sim p_Y \circ \beta$ is given by

$$\bar{H}(\theta, t, s) = \begin{cases} (\alpha(\theta), (2t - s)/(2 - s)) & s/2 \le t \le 1 \\ *, & 0 \le t \le s/2. \end{cases}$$ ▮

Lemma 18.8 The diagram

$$\begin{array}{ccc} C^*Y \cup_f C^*X & \xrightarrow{p_{Y \cup_f C^*X}} & SY \\ & \searrow^{p_{C^*Y}} \quad \nearrow_{-Sf} & \\ & SX & \end{array}$$

commutes up to homotopy.

Proof At time t we will pinch a piece of size t off of C^*Y and size $1-t$ off of C^*X (see Fig. 18.1):

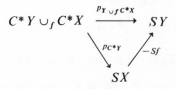

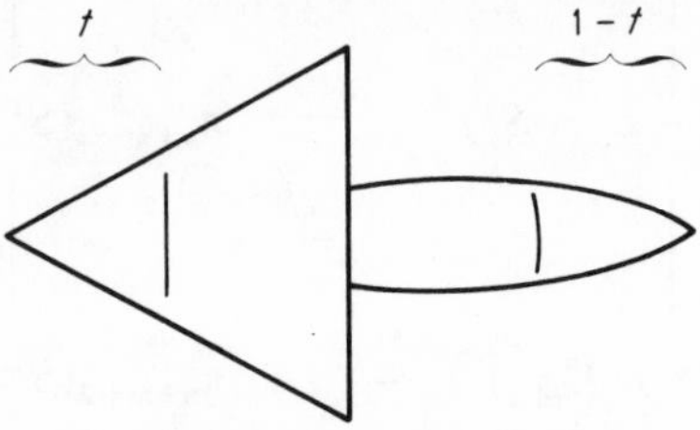

Figure 18.1

$$H(y, s, t) = \begin{cases} (y, t+s), & s \leq 1-t \\ *, & s \geq 1-t; \end{cases}$$

$$H(x, s, t) = \begin{cases} (f(x), t-s), & s \leq t \\ *, & s \geq t. \end{cases}$$

This is a well-defined homotopy $H\colon (C^*Y \cup_f C^*X) \times I \to SY$ between p_{C^*X} and $(-Sf) \circ p_{C^*Y}$. ∎

Proof of (C) Let $\alpha \in \tilde{E}_m(Y)$ with $i_*(\alpha) = 0$. There exists $x \in \pi_{m+n}(Y \wedge E_n)$ with $i_*(x) = 0 \in \pi_{m+n}((Y \cup_f C^*X) \wedge E_n)$ such that x represents α. Applying 18.7 to the diagram

$$\begin{array}{ccc} & \pi_{m+n}(Y \wedge E_n) \longrightarrow & \pi_{m+n}((Y \cup_f C^*X) \wedge E_n) \\ & & \Big\downarrow E \\ \pi_{m+n+1}(C^*(Y \wedge E_n) \cup_{f \wedge 1} C^*(X \wedge E_n)) & \longrightarrow & \pi_{m+n+1}(S(Y \wedge E_n)) \\ \Big\downarrow & \nearrow_{(-Sf)_*} & \Big\uparrow (Sf)_* \\ \pi_{m+n+1}(S(X \wedge E_n)) & \xrightarrow{(1 \wedge (-1))_*} & \pi_{n+m+1}(S(X \wedge E_n)) \end{array}$$

(which commutes by 18.8), one finds an element $y \in \pi_{n+m+1}(S(X \wedge E_n))$ such that $(Sf)_*(y) = E(x)$. Consequently, $(1 \wedge e_n)_*(y) \in \pi_{n+m+1}(X \wedge E_{n+1})$ is a representative for an element $\beta \in \tilde{E}_m(X)$ such that $f_*(\beta) = \alpha$. This proves C and completes the proof of 18.5. ▮ ▮

Theorem 18.9 Let $f\colon X \to Y$. Let $\{\tilde{E}_m\}$ and $\{\tilde{E}^m\}$ be reduced homology and cohomology theories. Then there are exact sequences

$$\cdots \to \tilde{E}_m(X) \xrightarrow{f_*} \tilde{E}_m(Y) \xrightarrow{i_*} \tilde{E}_m(Y \cup_f C^*X) \xrightarrow{\Delta_m} \tilde{E}_{m-1}(X) \to \cdots$$

$$\cdots \leftarrow \tilde{E}^m(X) \xleftarrow{f^*} \tilde{E}^m(Y) \xleftarrow{i^*} \tilde{E}^m(Y \cup_f C^*X) \xleftarrow{\Delta^{m-1}} \tilde{E}^{m-1}(X) \leftarrow \cdots$$

where $i\colon Y \to Y \cup_f C^*X$ is the inclusion, $\Delta_m = \sigma^{-1} \circ (p_Y)_*$, and $\Delta^{m-1} = (p_Y)^* \circ \sigma$, where σ is the suspension isomorphism.

Corollary 18.10 Let (X, A) have the AHEP. Then there are exact sequences

$$\cdots \to \tilde{E}_m(A) \xrightarrow{i_*} \tilde{E}_m(X) \xrightarrow{p_{A*}} \tilde{E}_m(X/A) \xrightarrow{\nabla_m} \tilde{E}_{m-1}(A) \to \cdots$$

$$\cdots \leftarrow \tilde{E}^m(A) \xleftarrow{i^*} \tilde{E}^m(X) \xleftarrow{p_A^*} \tilde{E}^m(X/A) \xleftarrow{\nabla^{m-1}} \tilde{E}^{m-1}(A) \leftarrow \cdots$$

where $i\colon A \to X$ is the inclusion, $p_A\colon X \to X/A$ is the quotient map, $\nabla_m = \Delta_m \circ (p_{C^*A})_*^{-1}$, and $\nabla^m = (p_{C^*A})^* \circ \Delta^m$.

The corollary follows from the theorem since $p_{C^*A}\colon X \cup C^*A \to X/A$ is a homotopy equivalence by Exercise 20, Section 16. ▮

The theorem follows from:

Proposition 18.11 The following diagram commutes up to homotopy, where the last two horizontal arrows are the inclusion into the mapping cone of the previous map, and the vertical arrows are homotopy equivalences:

$$\begin{array}{ccccccccc}
X & \xrightarrow{f} & Y & \longrightarrow & Y \cup_f C^*X & \xrightarrow{\pi} & C^*Y \cup_f C^*X & \longrightarrow & C^*_\pi \\
 & & & & & \searrow^{p_Y} & \downarrow{\simeq\, p_{C^*Y}} & \searrow^{p_{Y \cup_f C^*X}} & \downarrow{\simeq\, p_{C^*(Y \cup_f C^*X)}} \\
 & & & & & & SX & \xrightarrow{-Sf} & SY
\end{array}$$

Proof The statements about the right-hand triangle follow from those about the left-hand one by replacing $f\colon X \to Y$ by $i\colon Y \to Y \cup_f C^*X$. p_{C^*Y} is a homotopy equivalence by Exercise 20, Section 16. The middle triangle commutes by 18.8. ▮

18.11 implies that the sequence

$$\tilde{E}_m(X) \xrightarrow{f_*} \tilde{E}_m(Y) \xrightarrow{i_*} \tilde{E}_m(Y \cup_f C^*X) \xrightarrow{(p_Y)_*} \tilde{E}_m(SX) \xrightarrow{(-Sf)_*} \tilde{E}_m(SX)$$

is exact. $(-Sf)_*$ may be replaced by $(Sf)_*$ since $-1: SX \to SX$ is a homotopy equivalence and

$$\begin{array}{ccc} \tilde{E}_m(SX) & \xrightarrow{(Sf)_*} & \tilde{E}_m(SY) \\ \downarrow{\scriptstyle (1\wedge(-1))_*} & \searrow{\scriptstyle (-Sf)_*} & \downarrow{\scriptstyle (1\wedge(-1))_*} \\ \tilde{E}_m(SX) & \xrightarrow{(Sf)_*} & \tilde{E}_m(SY) \end{array}$$

commutes. Similarly in cohomology we have an exact sequence

$$\tilde{E}^m(X) \xleftarrow{f^*} \tilde{E}^m(Y) \xleftarrow{i^*} \tilde{E}^m(Y \cup_f C^*X) \xleftarrow{(p_Y)^*} \tilde{E}^m(SX) \xleftarrow{(Sf)^*} \tilde{E}^m(SY)$$

Piecing these together and using the natural isomorphism σ, one proves 18.9 ▮

Proposition 18.12 The sequences of 18.9 are natural, i.e., a commutative square

$$\begin{array}{ccc} X & \xrightarrow{f} & Y \\ \downarrow{\scriptstyle \alpha_1} & & \downarrow{\scriptstyle \alpha_2} \\ X' & \xrightarrow{f'} & Y' \end{array}$$

induces commutative ladders

$$\begin{array}{ccccccccc} \cdots \longrightarrow & \tilde{E}_m(X) & \xrightarrow{f_*} & \tilde{E}_m(Y) & \xrightarrow{i_*} & \tilde{E}_m(Y \cup_f C^*X) & \xrightarrow{\Delta_m} & \tilde{E}_{m-1}(X) & \longrightarrow \cdots \\ & \downarrow{\scriptstyle (\alpha_1)_*} & & \downarrow{\scriptstyle (\alpha_2)_*} & & \downarrow{\scriptstyle \alpha_*} & & \downarrow{\scriptstyle (\alpha_1)_*} & \\ \cdots \longrightarrow & \tilde{E}_m(X') & \xrightarrow{(f')_*} & \tilde{E}_m(Y') & \xrightarrow{i_*'} & \tilde{E}_m(Y' \cup_{f'} C^*X') & \xrightarrow{\Delta_m} & \tilde{E}_{m-1}(X') & \longrightarrow \cdots \end{array}$$

$$\begin{array}{ccccccccc} \cdots \longleftarrow & \tilde{E}^m(X) & \xleftarrow{f^*} & \tilde{E}^m(Y) & \xleftarrow{i^*} & \tilde{E}^m(Y \cup_f C^*X) & \xleftarrow{\Delta^{m-1}} & \tilde{E}^{m-1}(X) & \longleftarrow \cdots \\ & \uparrow{\scriptstyle (\alpha_1)^*} & & \uparrow{\scriptstyle (\alpha_2)^*} & & \uparrow{\scriptstyle (\alpha)^*} & & \uparrow{\scriptstyle (\alpha_1)^*} & \\ \cdots \longleftarrow & \tilde{E}^m(X') & \xleftarrow{f^*} & \tilde{E}^m(Y') & \xleftarrow{(i')^*} & \tilde{E}^m(Y' \cup_{f'} C^*X') & \xleftarrow{\Delta^{m-1}} & \tilde{E}^{m-1}(X') & \longleftarrow \cdots \end{array}$$

where $\alpha: Y \cup_f C^*X \to Y' \cup_{f'} C^*X'$ is constructed from α_1 and α_2 in the obvious way.

Proof This follows immediately from the various naturality results about $\tilde{E}_*$, $\tilde{E}^*$, and σ. ▮

The sequences of 18.9 are called the long exact sequences of the homology and cohomology theories. They are infinite in both directions.

Proposition 18.13 Suppose E is an Ω-spectrum and X is a CW complex.[15] There is a natural isomorphism $\tilde{E}^m(X) \cong [X, E_m]$.

Proof Consider the diagram

$$\begin{array}{ccc} [S^{n-m}X, E_n] & \xrightarrow{\lambda_n} & [S^{n-m+1}X, E_{n+1}] \\ & \searrow^{(\bar{e}_n)_*} & \updownarrow \tilde{\psi} \\ & & [S^{n-m}X, \Omega E_{n+1}] \end{array}$$

where $\bar{e}_n\colon E_n \to \Omega E_{n+1}$ is the adjoint to $e_n\colon SE_n \to E_{n+1}$ and $\tilde{\psi}$ is the 1–1 correspondence of 8.24. It is easy to verify that the diagram commutes, and since $(\bar{e}_n)_*$ and $\tilde{\psi}$ are 1–1 correspondences λ_n is a 1–1 correspondence. Hence

$$\tilde{E}^m(X) \cong [S^{n-m}X, E_n] \cong [X, \Omega^{n-m}E_n] \cong [X, E_m]. \quad \blacksquare$$

Proposition 18.14 For any spectrum E, $\tilde{E}_m(S^k) \cong \tilde{E}^k(S^m)$.

Proof By property (B), it is sufficient to show that $\tilde{E}_m(S^0) \cong \tilde{E}^{-m}(S^0)$. These are both the direct limit of the sequence

$$\begin{array}{ccccccc} \cdots & \longrightarrow & \pi_{n+m}(E_n) & & \longrightarrow & & \pi_{n+m+1}(E_{n+1}) & \longrightarrow & \cdots \\ & & & \searrow^{E} & & \nearrow^{(e_n)_*} & & & \\ & & & & \pi_{n+m+1}(SE_n) & & & & \end{array} \quad \blacksquare$$

This group is called the mth homotopy group of the spectrum E and is sometimes written $\pi_m(E)$. It is also called the group of coefficients for the theories $\tilde{E}_*$ and $\tilde{E}^*$.

Theorem 18.15 Suppose $X = \bigcup X_\alpha = \varinjlim X_\alpha$ has the weak topology and assume[16]:

(a) For all $\alpha, \beta, \in A$, there exists $\delta \in A$ such that $X_\alpha \cap X_\beta = X_\delta$.
(b) For all $\alpha \in A$, $\{\beta \in A \mid \beta \leq \alpha\}$ is finite ($\beta \leq \alpha$ iff $X_\beta \subset X_\alpha$).

Then $\tilde{E}_m(\varinjlim X_\alpha) \cong \varinjlim \tilde{E}_m(X_\alpha)$.

[15] The hypothesis that X is a CW complex is used in our proof, but may be dropped if each E_n is a CW complex since it is known [50] that in this case ΩE_n is the homotopy type of a CW complex (see the proof).

[16] Compare to 15.9.

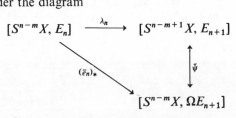

Proof The inclusions $i_\alpha: X_\alpha \to X$ induce homomorphisms $(i_\alpha)_*: \tilde{E}_m(X_\alpha) \to \tilde{E}_m(X)$. These are compatible with the homomorphisms $(i_{\alpha\beta})_*: \tilde{E}_m(X_\alpha) \to \tilde{E}_m(X_\beta)$ and hence induce a homomorphism $I: \varinjlim \tilde{E}_m(X_\alpha) \to \tilde{E}_m(X)$. Suppose $I(x) = 0$. x must have a representative $v \in \pi_{m+n}(X_\alpha \wedge E_n)$ such that $(i_\alpha)_*(v) = 0 \in \pi_{m+n}(X \wedge E_n)$. Let $f: S^{m+n} \to X_\alpha \wedge E_n$ represent v and $H: i_\alpha f \sim *$ be a homotopy. Since $X \wedge E_n = (\varinjlim X_\alpha) \wedge E_n \equiv \varinjlim(X_\alpha \wedge E_n)$, we may apply 15.10 and conclude that there exists $\beta \in A$ with $H(S^{m+n} \times I) \subset X_\beta \wedge E_n$. Thus $(i_{\alpha\beta})_*(v) = 0$ so $x = 0$. Hence I is 1–1. Let $x \in \tilde{E}_m(X)$. Choose $v \in \pi_{m+n}(X \wedge E_n)$ to represent x and $f: S^{m+n} \to X \wedge E_n$ to represent v. As before $f(S^{m+n}) \subset X_\alpha \wedge E_n$ for some α. So $v \in (i_\alpha)_*(\pi_{m+n}(X_\alpha \wedge E_n))$. Hence I is onto. ▮

Theorem 18.16

$$\tilde{E}_m(X_1 \vee \cdots \vee X_n) \cong \tilde{E}_m(X_1) \oplus \cdots \oplus \tilde{E}_m(X_n),$$
$$\tilde{E}^m(X_1 \vee \cdots \vee X_n) \cong \tilde{E}^m(X_1) \oplus \cdots \oplus \tilde{E}^m(X_n).$$

The decomposition is given by the induced homomorphisms from the maps $i_k: X_k \to X_1 \vee \cdots \vee X_k$ and $p_k: X_1 \vee \cdots \vee X_n \to X_k$ for $1 \le k \le n$.

Proof We first do the case $n = 2$. Observe that since C^*Y is contractible in $\mathcal{CG}^*$. $p_{C^*Y}: X \vee C^*Y \to X$ is a homotopy equivalence. Consequently, the sequence

$$\tilde{E}_m(Y) \xrightarrow{(i_2)_*} \tilde{E}_m(X \vee Y) \xrightarrow{(p_1)_*} \tilde{E}_m(X)$$

is exact in the middle. Since the diagram

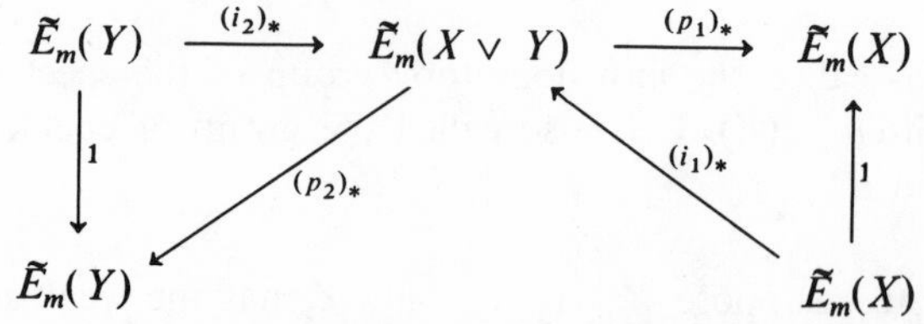

commutes, $(i_2)_*$ is a monomorphism, $(p_1)_*$ is an epimorphism, and we may apply 11.11 to prove the splitting. If $n > 2$, we apply induction, observing that $X_1 \vee \cdots \vee X_n \equiv (X_1 \vee \cdots \vee X_{n-1}) \vee X_n$. The case of cohomology is similar. ▮

Given R-modules M_α, the direct sum $\bigoplus_{\alpha \in A} M_a$ was introduced in 15.3. One similarly defines the direct product $\prod_{\alpha \in A} M_\alpha$ to be the set of all sequences $\{x_\alpha\}$ with $x_\alpha \in M_\alpha$. Addition and scalar multiplication are coordinatewise. Note that $\bigoplus_{\alpha \in A} M_\alpha \subset \prod_{\alpha \in A} M_\alpha$ and if A is finite, they are equal.

Let $(X_\alpha, *_\alpha) \in \mathcal{CG}^*$ be indexed by a set A. By $\bigvee_{\alpha \in A} X_\alpha$ we will mean the quotient space $\coprod_{\alpha \in A} X_\alpha / *_\alpha \sim *_\beta$ If A is finite, this is the one-point union $X_{\alpha_1} \vee \cdots \vee X_{\alpha_n}$.

Theorem 18.17 $\tilde{E}_m(\bigvee_{\alpha \in A} X_\alpha) \cong \oplus \tilde{E}_m(X_\alpha)$. If $(X_\alpha, *_\alpha)$ are CW complexes[17] and $\{E_n, e_n\}$ is an Ω-spectrum, $\tilde{E}^m(\bigvee_{\alpha \in A} X_\alpha) \cong \prod_{\alpha \in A} \tilde{E}^m(X_\alpha)$.

Proof Let $X = \bigvee_{\alpha \in A} X_\alpha$ and consider all subspaces $X_{\alpha_1, \dots, \alpha_n} = X_{\alpha_1} \vee \cdots \vee X_{\alpha_n}$ of X. This satisfies the hypothesis of 18.15 so

$$\begin{aligned} \tilde{E}_m(X) &\cong \varinjlim \tilde{E}_m(X_{\alpha_1} \vee \cdots \vee X_{\alpha_n}) \\ &\cong \varinjlim \tilde{E}_m(X_{\alpha_1}) \oplus \cdots \oplus \tilde{E}_m(X_{\alpha_n}) \cong \oplus_{\alpha \in A} \tilde{E}_m(X_\alpha). \end{aligned}$$

In the cohomology case, the hypothesis implies that $X = \bigvee_{\alpha \in A} X_\alpha$ is a CW complex so we may apply 18.13. Let $i_\alpha \colon X_\alpha \to Y$ be the inclusion, and consider the homomorphisms $(i_\alpha)^* \colon \tilde{E}^m(X) \to \tilde{E}^m(X_\alpha)$. These induce a homomorphism $I \colon \tilde{E}^m(X) \to \prod_{\alpha \in A} \tilde{E}^m(X_\alpha)$. Consider now the diagram

$$\begin{array}{ccc} \tilde{E}^m(X) & \xrightarrow{\ I\ } & \prod\limits_{\alpha \in A} \tilde{E}^m(X_\alpha) \\ \uparrow \cong & & \uparrow \cong \\ [X, E_m] & \xrightarrow{\ I\ } & \prod\limits_{\alpha \in A} [X_\alpha, E_m] \end{array}$$

In the bottom row I is a 1–1 correspondence, for given $\{f_\alpha\}$ where $f_\alpha \colon X_\alpha \to E_m$ one easily constructs $f \colon X \to E_m$ with $f|_{X_\alpha} = f_\alpha$. If $f|_{X_\alpha} \sim *$ for each α; the homotopies $H_\alpha \colon X_\alpha \times I \to E_m$ define a homotopy $H \colon X \times I \to E_m$ by $H|_{X_\alpha \times I} = H_\alpha$ since they are base point preserving homotopies. Hence I is an isomorphism. ∎

The conclusion of 18.17 is sometimes called the wedge axiom.

Consider now the sphere spectrum $\underline{S}$. In this case the homology group $\underline{\tilde{S}}_m(X)$ is given by the direct limit

$$\pi_{m+n}(X \wedge S^n) \xrightarrow{E} \pi_{m+n+1}(X \wedge S^{n+1}) \to \cdots.$$

This is also written $\pi_m{}^S(X)$, and is the functor defined at the end of Section 16. 18.15 implies that $\pi_m{}^S(X) = 0$ if $m < 0$ and X is a CW complex.

Similarly one defines the stable cohomotopy groups $\pi_S{}^m(X)$ as the cohomology theory associated with this spectrum, i.e., the direct limit of the sequence

$$[S^{n-m}X, S^n] \xrightarrow{E} [S^{n-m+1}X, S^{n+1}] \to \cdots.$$

These groups can be nonzero for negative values of m when applied to CW complexes. For example, $\pi_S^{-1}(S^0) \cong \pi_4(S^3) \cong Z_2$ (see 27.19).

[17] See footnote 15.

Exercises

1. Give a detailed proof that the diagrams in the conclusion of 18.12 commute.

2. Give a detailed proof of the isomorphism

$$\varinjlim \tilde{E}_m(X_{\alpha_1}) \oplus \cdots \oplus \tilde{E}_m(X_{\alpha_n}) \cong \bigoplus_{\alpha \in A} \tilde{E}_m(X_\alpha)$$

in the proof of 18.17.

3. Define $h\colon \pi_i(X) \to \pi_i^S(X)$ to be the injection homomorphism from the first group of the direct system for $\pi_i^S(X)$ into the limit. Show that if X is simply connected, the following are equivalent:

(a) $\pi_i(X) = 0$ for $i < n$;

(b) $\pi_i^S(X) = 0$ for $i < n$;

they imply that $h\colon \pi_i(X) \to \pi_i^S(X)$ is an isomorphism if $1 \leq i < 2n - 1$ and onto if $i = 2n - 1$.

4. Let $\varphi_n\colon E_n \to E_n'$ be defined for each $n \geq N_0$ such that the diagrams

$$\begin{array}{ccc} SE_n & \xrightarrow{S\varphi_n} & SE'_{n+k} \\ \Big\downarrow{\scriptstyle e_n} & & \Big\uparrow{\scriptstyle e'_{n+k}} \\ E_{n+1} & \xrightarrow{\varphi_{n+1}} & E'_{n+k+1} \end{array}$$

commute. Such a sequence $\{\varphi_n\}$ will be called a map of spectra of degree k. Show that spectra form a category with this definition of morphism. Show that if $\varphi = \{\varphi_n\}\colon E \to E'$ is a map of spectra of degree k, it induces natural homomorphisms of homology and cohomology theories

$$\varphi\colon \tilde{E}_m(X) \to \tilde{E}'_{m-k}(X), \qquad \varphi\colon \tilde{E}^m(X) \to \tilde{E}'^{m+k}(X)$$

for all m commuting with the suspension isomorphism; i.e., $\varphi(\sigma(x)) = \sigma(\varphi(x))$. Such a transformation is called a stable homology or cohomology operation. (Exercise 11; Exercise 1, Section 22; Section 27)

5. A spectrum E and the corresponding theories are called connective if E_n is $(n-1)$-connected. Show by taking connected covering spaces that given any spectrum E_n there is a connective spectrum $\bar{E}_n$ and a mapping $\varphi_n\colon \bar{E}_n \to E_n$ of spectra such that $(\varphi_m)_*$ is an isomorphism in π_i for $i \geqslant 0$.

6. Let P be a one-point topological space. Show that $\tilde{E}_m(P) = 0 = \tilde{E}^m(P)$ for any functors $\tilde{E}_m$ and $\tilde{E}^m$ satisfying axioms (A)–(D).

7. Let $f\colon X \to Y$ be a map in $\mathcal{CG}^*$ and suppose $f \sim *$ in $\mathcal{CG}^*$. Show that $f^* = 0 = f_*$.

8. Let E be a spectrum and define W_n to be the telescope construction (Exercise 1, Section 15) on the system

$$E_n \to \cdots \to \Omega^k E_{n+k} \xrightarrow{\Omega^k \bar{e}_{n+k}} \Omega^{k+1} E_{n+k+1} \to \cdots.$$

Show that $\tilde{E}^n(X) = [(X, *), (W_n, *)]$ for X a compact Hausdorff space. Show that $\{W_n, w_n\}$ is an Ω-spectrum where $\bar{w}_n \colon W_n \to \Omega W_{n+1}$ is given on the kth term of the telescope by

$$\Omega^k E_{n+k} \xrightarrow{\Omega^k \bar{e}_{n+k}} \Omega^{k+1} E_{n+k+1}.$$

9. Show that $\tilde{E}_m(X) \cong \varinjlim \tilde{E}_m(K)$ where the limit is taken over all compact subsets K of X containing $*$.

10. Let $\varphi\colon \tilde{E}_*(X) \to \tilde{E}_*'(X)$ be a natural transformation such that $\varphi\colon \tilde{E}_*(S^0) \to E_*'(S^0)$ is an isomorphism. Show that $\varphi\colon \tilde{E}_*(X) \to \tilde{E}_*'(X)$ is an isomorphism for each CW complex X. Prove an analogue for cohomology if X is a finite complex. (30.25)

11. Using Exercise 4, construct for each homomorphism $\varphi\colon \pi \to \rho$ coefficient transformations

$$c_\varphi\colon \tilde{H}_n(X; \pi) \to \tilde{H}_n(X; \rho), \qquad c_\varphi\colon \tilde{H}^n(X; \pi) \to \tilde{H}^n(X; \rho)$$

and show, using these, that homology and cohomology are covariant functions of the coefficient groups. (Exercise 18, Section 23)

12. Given a spectrum E and an abelian group G define $EG = \{EG_n, eG_n\}$ by

$$EG_n = M(G, 1) \wedge E_{n-1}$$

and

$$(eG)_n = 1 \wedge e_{n-1}\colon M(G, 1) \wedge E_{n-1} \wedge S^1 \to M(G, 1) \wedge E_n.$$

We write $\tilde{E}_*(X; G)$ for $\widetilde{EG}_*(X)$ and $\tilde{E}^*(X; G)$ for $\widetilde{EG}^*(X)$. Show that if $E = H$, these definitions agree with the ordinary definitions if X is a CW complex.[18] (Hint: Use Exercise 10.) (Exercise 13; Exercise 11, Section 22)

13. Let $0 \to G \xrightarrow{\varphi} H \xrightarrow{\psi} J \to 0$ be a short exact sequence of abelian groups. Construct natural long exact sequences

$$\cdots \to \tilde{E}^i(X; G) \xrightarrow{c} \tilde{E}^i(X; H) \xrightarrow{d} \tilde{E}^i(X; J) \xrightarrow{\beta} \tilde{E}^{i+1}(X; G) \to \cdots$$

$$\cdots \to \tilde{E}_i(X; G) \xrightarrow{c} \tilde{E}_i(X; H) \xrightarrow{d} \tilde{E}_i(X; J) \xrightarrow{\beta} \tilde{E}_{i-1}(X; G) \to \cdots.$$

[18] It is actually only necessary to assume that X is well pointed by 21.7.

This is called the Bockstein sequence and β is called the Bockstein homomorphism. Show that if $E = H$, $c = c_\varphi$, and $d = c_\psi$. (Exercise 8, Section 21; 27.15)

14. A spectrum E is called properly convergent if e_n: $SE_n \to E_{n+1}$ is a $(2n+1)$-isomorphism for each $n \geq 0$. Show that if E_n is well pointed for each n and E is properly convergent, E_n is $(n-1)$-connected. Show that if E is an Ω-spectrum and E_n is connected for each n, E is properly convergent. (Exercise 12, Section 22; 27.5)

19

Spectral Unreduced Homology and Cohomology Theories

By a simple transformation we can transfer the domain of our theories from $\mathcal{CG}^*$ to $\mathcal{CG}^2$. Homology and cohomology theories defined on pairs (X, A) are called unreduced homology and cohomology theories (sometimes the adjective *unreduced* is dropped). We define unreduced theories here and develop their properties on the category $\mathcal{E}$ of pairs in $\mathcal{CG}$ with the AHEP. In the next section we will consider unreduced theories on more general pairs.

Definition 19.1 For $(X, A) \in \mathcal{CG}^2$ we set

$$E_m(X, A) = \tilde{E}_m(X \cup CA), \qquad E^m(X, A) = \tilde{E}^m(X \cup CA),$$

where the vertex of the cone is chosen as base point. If $A = \varnothing$ we interpret $X \cup CA$ to mean X with a disjoint point added and used as base point. E_m and E^m are called the unreduced homology and cohomology theories associated with the spectrum $\{E_n\}$.

Definition 19.2 Let $\mathcal{C}$ be a category of pairs of topological spaces. An unreduced homology (cohomology) theory defined on $\mathcal{C}$ is a sequence of covariant (contravariant) functors E_m (E^m) for $m \in Z$ satisfying the axioms:

(A) (*Homotopy*) Let $f, g\colon (X, A) \to (Y, B)$. Suppose $f \sim g$ in $\mathcal{C}$ (i.e., there is a map $H\colon (X \times I, A \times I) \to (Y, B)$ in $\mathcal{C}$ with $H(x, 0) = f(x)$, $H(x, 1) = g(x)$). Then

$$f_* = g_*\colon E_m(X, A) \to E_m(Y, B) \quad \text{and} \quad f^* = g^*\colon E^m(Y, B) \to E^m(X, A).$$

(B) (*Excision*) If U is open and $\bar{U} \subset \text{Int } A$, the inclusion $e: (X - U, A - U) \to (X, A)$ induces isomorphisms in homology and cohomology.

Abbreviate $E_m(X, \varnothing)$ and $E^m(X, \varnothing)$ as $E_m(X)$ and $E^m(X)$.

(C) (*Exactness*) There are natural transformations $\partial: E_m(X, A) \to E_{m-1}(A)$ and $\delta: E^m(A) \to E^{m+1}(X, A)$ which fit into exact sequences

$$\cdots \to E_m(A) \to E_m(X) \to E_m(X, A) \xrightarrow{\partial} E_{m-1}(A) \to \cdots$$

$$\cdots \leftarrow E^m(A) \leftarrow E^m(X) \leftarrow E^m(X, A) \xleftarrow{\delta} E^{m-1}(A) \leftarrow \cdots.$$

For the homology and cohomology theories we construct, we will prove stronger excision properties than axiom (B). There are two types of strengthenings of axiom (B):

Type 1 excision If $\bar{U} \subset \text{Int } A$, e induces isomorphisms (U is not assumed to be open).

Type 2 excision If U is open and $\bar{U} \subset A$, e induces isomorphisms.

As we will see in section 21, type 1 excision is natural for homology and type 2 excision is natural for cohomology.

Lemma 19.3 Type 1 excision is equivalent to the condition that if (X_1, X_2) is excisive in X,

$$E_m(X_1 \cup X_2, X_1) \cong E_m(X_2, X_1 \cap X_2)$$

and

$$E^m(X_1 \cup X_2, X_1) \cong E^m(X_2, X_1 \cap X_2).$$

Proof To prove B, let $X_1 = A$ and $X_2 = X - U$. Then $\text{Int } X_1 \cup \text{Int } X_2 = \text{Int } A \cup X - \bar{U} = X$. Conversely, if B holds, making the same substitutions we see that $\bar{U} = X - \text{Int } X_2 \subset \text{Int } X_1 = \text{Int } A$. ∎

Definition 19.4 Let $\mathcal{E}$ be the atcegory of pairs $(X, A) \in \mathcal{CG}^2$ with the AHEP. Let $\mathcal{N} = \mathcal{E} \cap \mathcal{CG}^*$ be the category of well-pointed spaces.

Theorem 19.5 On $\mathcal{E}$, E_m and E^m are unreduced homology and cohomology theories.

Proof If $f: (X, A) \to (Y, B)$ define $\tilde{f}: X \cup CA \to Y \cup CB$ by $\tilde{f}(x) = f(x)$ and $\tilde{f}(a, t) = (f(a), t)$. $\tilde{f}$ is base point preserving. Furthermore, if $f \sim g$ in $\mathcal{E}$, $\tilde{f} \sim \tilde{g}$ in $\mathcal{CG}^*$. Since $X \cup CA \simeq X/A$, $E_m(X_1 \cup X_2, X_1) \cong E_m(X_2, X_1 \cap X_2)$ and $E^m(X_1 \cup X_2, X_1) \cong E^m(X_2, X_1 \cap X_2)$ with no hypothesis, for $X_1 \cup X_2/X_1$

$\equiv X_2/X_1 \cap X_2$. (By Exercise 19, Section 14, X_1 is closed.) Exactness follows immediately from the homeomorphism

$$X^+ \cup C^*A^+ \equiv X \cup CA$$

and 18.9. ∎

Corollary 19.6

$$E_m(X) \cong E_m(X, *) \oplus E_m(*), \qquad E^m(X) \cong E^m(X, *) \oplus E^m(*).$$

Proof The exact sequences

$$\to E_m(*) \to E_m(X) \to E_m(X, *) \to \cdots$$
$$\cdots \leftarrow E^m(*) \leftarrow E^m(X) \leftarrow E^m(X, *) \leftarrow$$

split since there is a map $p_X\colon X \to *$ with $p_X i = 1_*$ by Exercise 11, Section 11 and 11.11. ∎

Proposition 19.7 If $X \in \mathcal{N}$, $E_m(X, *) \cong \tilde{E}_m(X)$ and $E^m(X, *) \cong \tilde{E}^m(X)$. Hence

$$E_m(X) \cong \tilde{E}_m(X) \oplus E_m(*) \qquad \text{and} \qquad E^m(X) \cong \tilde{E}^m(X) \oplus E^m(*).$$

Proof Let $X^* = X \cup C*$. Then $E_m(X, *) = \tilde{E}_m(X^*)$ and $E^m(X, *) = \tilde{E}^m(X^*)$. It is sufficient to show that if $X \in \mathcal{N}$, $(X, *) \simeq (X^*, 1)$. There is an obvious map $\alpha\colon X^* \to X$ in $\mathcal{CG}^*$, and since $X \in \mathcal{N}$ there is a retraction $\beta\colon X \times I \to X^*$. Let $\gamma(x) = \beta(x, 1)$. Then $\gamma\colon X \to K^*$ is in $\mathcal{CG}^*$. Now $\alpha\beta\colon 1 \sim \alpha\gamma$. Consider $H\colon X^* \times I \to X \times I$ defined by $H(x, s) = (x, s)$, $H(t, s) = (*, s + t(1 - s))$. Then $\beta H\colon 1 \sim \gamma\alpha$ and hence $(X^*, 1) \simeq (X, *)$. The second statement follows by applying 19.6. ∎

Corollary 19.8 $E_m(*) \cong \tilde{E}_m(S^0) \cong E^{-m}(*)$. ∎

Corollary 19.9 If $X \in \mathcal{N}$,

$$H_m(X, \pi) \cong \begin{cases} \tilde{H}_m(X; \pi) & m \neq 0 \\ \tilde{H}_0(X; \pi) \oplus \pi, & m = 0, \end{cases}$$

$$H^m(X; \pi) \cong \begin{cases} \tilde{H}^m(X; \pi), & m \neq 0 \\ \tilde{H}^0(X; \pi) \oplus \pi, & m = 0. \end{cases}$$ ∎

Proposition 19.10 If X is a CW complex, $\tilde{H}^m(X; \pi) \cong \tilde{H}_m(X; \pi) = 0$ for $m < 0$.

Proof $X \wedge K(\pi, n)$ is a CW complex with all cells in dimensions n and larger, except for 0-cells. Hence $\pi_{m+n}(X \wedge K(\pi, n)) = 0$ for $m < 0$ and $n > 1$.

Thus $\tilde{H}_m(X; \pi) = 0$ for $m < 0$. Since $S^{m-n}X$ has all cells in dimensions $n - m$ and larger, except for 0-cells, $[S^{n-m}X, K(\pi, n)] = 0$ if $m < 0$ by 16.3 applied to the diagram

$$\begin{array}{ccc} C^*S^{n-m}X & & \\ \uparrow & \searrow & \\ S^{n-m}X & \longrightarrow & K(\pi, n) \end{array} \quad \blacksquare$$

The Eilenberg–Steenrod axioms on a positively graded functor are the laws: (A), (B), and (C) of 19.2 together with the dimension axiom:

(D)

$$H_i(*; \pi) = \begin{cases} 0, & i \neq 0 \\ \pi, & i = 0; \end{cases} \qquad H^i(*; \pi) = \begin{cases} 0, & i \neq 0 \\ \pi, & i = 0. \end{cases}$$

A functor satisfying these properties is called an ordinary homology or cohomology theory. They are characteristic properties and are often used as a starting point for making calculations with ordinary homology and cohomology. By 19.5, 19.9, and 19.10 ordinary spectral homology and cohomology satisfy the Eilenberg–Steenrod axioms on the category of CW pairs.

Exercises

1.* Prove that if $X \supset A \supset B$, there are long exact sequences

$$\cdots \to E_m(A, B) \to E_m(X, B) \to E_m(X, A) \xrightarrow{\partial} E_{m-1}(A, B) \to \cdots$$

$$\cdots \leftarrow E^m(A, B) \leftarrow E^m(X, B) \leftarrow E^m(X, A) \xleftarrow{\delta} E^{m-1}(A, B) \leftarrow \cdots.$$

(Compare with Exercise 5, Section 10.)

2. Let E_n (E^n) be an unreduced homology (cohomology) theory. Define $\tilde{E}_n(X) = E_n(X, *)$ ($\tilde{E}^n(X) = E^n(X, *)$). Show that $\tilde{E}_n(\tilde{E}^n)$ is a reduced homology (cohomology) theory with the modification that SX is replaced by ΣX in Axiom (B) and $X \cup C^*A$ is replaced by $X \cup CA$ in Axiom (C).

3. Show that $E_n(\coprod X_\alpha) \cong \bigoplus E_n(X_\alpha)$, and if X_α are CW complexes and E is an Ω-spectrum, $E^n(\coprod X_\alpha) \cong \prod E^n(X_\alpha)$. The assumption that the X_α are CW complexes may be dropped if we assume that E_n is a CW complex for each n (see footnote 15).

4. Let $\{U_\alpha\}$ be an open cover of X such that for all α, α' there exists α'' with $U_\alpha \cup U_{\alpha'} \subset U_{\alpha''}$. Show that $E_n(X) \cong \varinjlim E_n(U_\alpha)$. (26.22, 26.28)

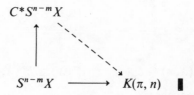

20

Ordinary Homology of CW Complexes

We begin this section with a description of chains, cycles, and homologies, and show how the ordinary homology groups of a complex provide us with a classification of cycles up to homology.

Suppose X is a CW complex. A chain is simply a collection of oriented cells with multiplicities. A chain is called a cycle if there is cancellation at the boundary of each cell. Thus consider the cells in the plane shown in Fig. 20.1.

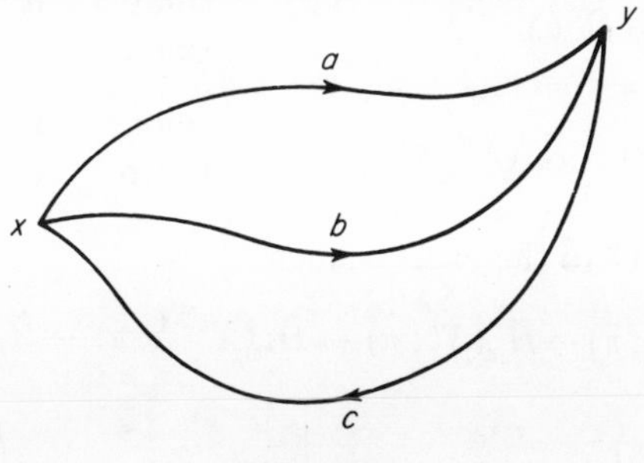

Figure 20.1

The chain consisting of a and b with multiplicity 1 and c with multiplicity 2 is a cycle since its boundary at y gets a contribution of $+1$ for a and b and a contribution of -2 for c. Similarly, we have cancellation at x.

To make this more precise, we define the n-dimensional chain group of X, $C_n(X)$ to be the free abelian group with one generator for each n-cell. For each oriented n-cell e_n, let ∂e_n be the $n-1$ chain that is its boundary. Then ∂ extends to a homomorphism $\partial\colon C_n(X) \to C_{n-1}(X)$ and the kernel of ∂

is the set of cycles. In the above example $\partial a = y - x$ $\partial b = y - x$ and $\partial c = x - y$. Thus $\partial(a + b + 2c) = 0$.

Two cycles will be called homologous if there is a chain whose boundary is their difference. Thus in Fig. 20.2 a and b are homologous since $\partial d = a - b$.

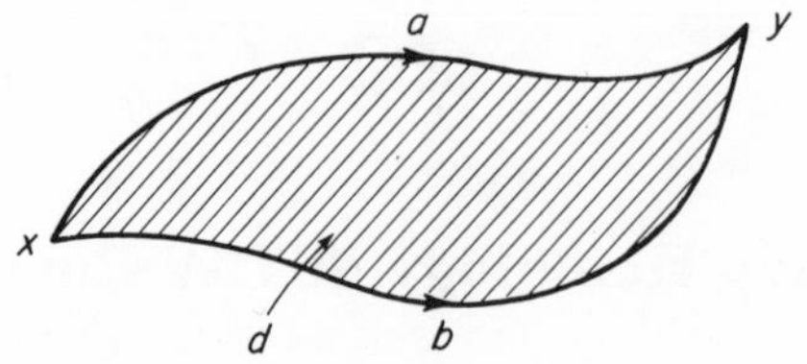

Figure 20.2

A homology is the obvious analogy of a homotopy among cycles. We write $a \sim b$ to indicate that a and b are homologous. If $Z_n(X)$ is the set of cycles, the n-dimensional homology group of X is just $Z_n(X)/\sim = Z_n(X)/\mathrm{Im}(\partial\colon C_{n+1}(X) \to C_n(X))$.

Our first task will be to prove that this quotient group is in fact $H_n(X)$. We proceed with a sequence of lemmas. That this description of homology is correct will follow from the calculations we do in the last part of this section.

Lemma 20.1 Let X be a CW complex. Then

$$i_*\colon \tilde{H}_m(X^n; \pi) \to \tilde{H}_m(X^{n+1}; \pi) \quad \text{is} \quad \begin{cases} \text{onto} & \text{if } m \neq n+1 \\ \text{mono} & \text{if } m \neq n; \end{cases}$$

$$i^*\colon \tilde{H}^m(X^{n+1}; \pi) \to \tilde{H}^m(X^n; \pi) \quad \text{is} \quad \begin{cases} \text{mono} & \text{if } m \neq n+1 \\ \text{onto} & \text{if } m \neq n. \end{cases}$$

Proof Apply 18.17 to the exact sequences

$$\begin{array}{ccccccc} \tilde{H}_{m+1}(X^{n+1}/X^n; \pi) & \to & \tilde{H}_m(X^n; \pi) & \to & \tilde{H}_m(X^{n+1}; \pi) & \to & \tilde{H}_m(X^{n+1}/X^n; \pi) \\ \wr\Vert & & & & & & \wr\Vert \\ \tilde{H}_{m+1}(\vee S_\alpha^{n+1}; \pi) & & & & & & \tilde{H}_m(\vee S_\alpha^{n+1}; \pi) \end{array}$$

$$\begin{array}{ccccccc} \tilde{H}^{m+1}(X^{n+1}/X^n; \pi) & \leftarrow & \tilde{H}^m(X^n; \pi) & \leftarrow & \tilde{H}^m(X^{n+1}; \pi) & \leftarrow & \tilde{H}^m(X^{n+1}/X^n; \pi) \\ \wr\Vert & & & & & & \wr\Vert \\ \tilde{H}^{m+1}(\vee S_\alpha^{n+1}; \pi) & & & & & & \tilde{H}^m(\vee S_\alpha^{n+1}; \pi). \end{array}$$ ∎

Lemma 20.2 $\tilde{H}_m(X^n; \pi) = 0$ if $m > n$.

$$i_*\colon \tilde{H}_m(X^n; \pi) \to \tilde{H}_m(X; \pi) \quad \text{is} \quad \begin{cases} \text{isomorphic} & \text{if } m < n \\ \text{onto} & \text{if } m = n. \end{cases}$$

$\tilde{H}^m(X^n; \pi) = 0$ if $m > n$.

$$i^*: \tilde{H}^m(X; \pi) \to \tilde{H}^m(X^n; \pi) \qquad \text{is} \qquad \begin{cases} \text{isomorphic} & \text{if} \quad m < n \\ \text{mono} & \text{if} \quad m = n. \end{cases}$$

Proof Consider the sequences (where we abbreviate by not writing π)

$$\xrightarrow{\approx} \tilde{H}_m(X^{m-2}) \xrightarrow{\approx} \tilde{H}_m(X^{m-1}) \xrightarrow{\text{mono}} \tilde{H}_m(X^m) \xrightarrow{\text{onto}} \tilde{H}_m(X^{m+1}) \xrightarrow{\approx} \tilde{H}_m(X^{m+2}) \xrightarrow{\approx} \cdots$$

$$\cdots \xleftarrow{\approx} \tilde{H}^m(X^{m-2}) \xleftarrow{\approx} \tilde{H}^m(X^{m-1}) \xleftarrow{\text{epi}} \tilde{H}^m(X^m) \xleftarrow{\text{mono}} \tilde{H}^m(X^{m+1}) \xleftarrow{\approx} \tilde{H}^m(X^{m+2}) \xleftarrow{\approx} \cdots.$$

The result for homology now follows from 15.6 and 18.15 since $\tilde{H}_m(X^0) = 0$ for $m > 0$. The result for cohomology follows similarly except here we argue that $i^*: \tilde{H}^m(X) \to \tilde{H}^m(X^{m+1})$ is an isomorphism from 18.13 and 16.3 applied to the diagrams

$$\begin{array}{ccc} X & & \\ \uparrow & \searrow & \\ X^{m+1} & \longrightarrow & K(\pi, m) \end{array} \qquad \begin{array}{ccc} X \times I & & \\ \uparrow & \searrow & \\ X \times 0 \cup X^{m+1} \times I \cup X \times 1 & \longrightarrow & K(\pi, m) \end{array} \quad \blacksquare$$

Define $C_n(X, A; \pi) = H_n(\bar{X}^n, \bar{X}^{n-1}; \pi)$ where $\bar{X}^n = A \cup X^n$. Define $\partial_n: C_n(X, A; \pi) \to C_{n-1}(X, A; \pi)$ as follows:

$$C_n(X, A; \pi) = H_n(\bar{X}^n, \bar{X}^{n-1}; \pi) \xrightarrow{\partial} H_{n-1}(\bar{X}^{n-1}, A; \pi) \xrightarrow{i_*} H_{n-1}(\bar{X}^{n-1}, \bar{X}^{n-2}; \pi) = C_{n-1}(X, A; \pi).$$

Observe that since

$$H_n(\bar{X}^n, A; \pi) \to H_n(\bar{X}^n, \bar{X}^{n-1}; \pi) \xrightarrow{\partial} H_{n-1}(\bar{X}^{n-1}, A; \pi)$$

is exact, $\partial_n \partial_{n+1} = 0$.

$C_n(X, A; \pi)$ is called the n-dimensional chain group of (X, A) with coefficients in π and its elements are called n-dimensional chains.

Theorem 20.3 $H_n(X, A; \pi) \cong \ker \partial_n / \operatorname{Im} \partial_{n+1}$.

Proof By Exercise 1, Section 19,

$$H_{n-1}(\bar{X}^{n-2}, A; \pi) \to \tilde{H}_{n-1}(\bar{X}^{n-1}, A; \pi) \xrightarrow{i_*} H_{n-1}(\bar{X}^{n-1}, \bar{X}^{n-2}; \pi)$$

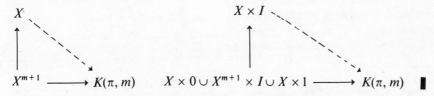

is exact. $H_{n-1}(\overline{X}^{n-2}, A; \pi) \cong H_{n-1}(\overline{X}^{n-2^+}/A^+; \pi) = 0$ so i_* is a monomorphism. Hence $\ker \partial_n = \ker \partial = \operatorname{Im} j_*$ in the diagram

$$\begin{array}{l} H_n(\overline{X}^{n-1}, A; \pi) \to H_n(\overline{X}^n, A; \pi) \xrightarrow{j_*} H_n(\overline{X}^n, \overline{X}^{n-1}; \pi) \xrightarrow{\partial} H_{n-1}(\overline{X}^{n-1}, A; \pi) \\ \qquad \wr\| \\ \tilde{H}_n(\overline{X}^{n-1^+}/A^+; \pi). \\ \qquad \| \\ \qquad 0 \end{array}$$

Reapplying 20.2 we see that j_* is a monomorphism so $\ker \partial_n \cong H_n(\overline{X}^n, A; \pi)$. Define ϕ: $\ker \partial_n \to H_n(X, A; \pi)$ by $\phi = k_* j_*^{-1}$ where k: $(\overline{X}^n, A) \to (X, A)$ is the inclusion. ϕ is onto by 20.2 and the diagram

$$\begin{array}{ccc} H_n(\overline{X}^n, A; \pi) & \to & H_n(X, A; \pi) \\ \wr\| & & \wr\| \\ \tilde{H}_n(\overline{X}^{n^+}/A^+; \pi) & \to & \tilde{H}_n(X^+/A^+; \pi). \end{array}$$

$\ker \phi = \operatorname{Im} j_* \partial$, where ∂: $H_{n+1}(X, \overline{X}^n; \pi) \to H_n(\overline{X}^n, A; \pi)$. Now consider the diagram

$$\begin{array}{ccc} H_{n+1}(X, \overline{X}^{n+1}; \pi) = 0 & & \\ \uparrow & & \\ H_{n+1}(X, \overline{X}^n; \pi) & \xrightarrow{\partial} & H_n(\overline{X}^n, A; \pi) \\ \uparrow \text{epi} & \nearrow \partial & \downarrow j_* \\ H_{n+1}(\overline{X}^{n+1}, \overline{X}^n; \pi) & \xrightarrow{\partial_{n+1}} & H_n(\overline{X}, \overline{X}^{n-1}; \pi) \end{array}$$

where it is proved that $H_n(X, \overline{X}^{n+1}; \pi) = 0$ by applying the long exact sequence and 20.2. It follows that $\ker \phi = \operatorname{Im} j_* \partial = \operatorname{Im} \partial_{n+1}$. ∎

Define $C^n(X, A; \pi) = H^n(\overline{X}^n, \overline{X}^{n-1}; \pi)$ and δ_n: $C^n(X, A; \pi) \to C^{n+1}(X, A; \pi)$ as the composition

$$\begin{array}{r} C^n(X, A; \pi) = H^n(\overline{X}^n, \overline{X}^{n-1}; \pi) \to H^n(\overline{X}^n, A; \pi) \xrightarrow{\delta} H^{n+1}(\overline{X}^{n+1}, \overline{X}^n; \pi) \\ \| \qquad\qquad \\ C^{n+1}(X, A; \pi). \qquad \end{array}$$

$C^n(X, A; \pi)$ is called the n-dimensional cochain group of (X, A) with coefficients in π and its elements elements are called n-dimensional cochains.

Theorem 20.4 $\delta_n \delta_{n-1} = 0$, $H^n(X, A; \pi) = \ker \delta_n / \operatorname{Im} \delta_{n-1}$.

The proof of 20.4 follows from 20.2 just as the case for homology. ∎

Definition 20.5 A collection of abelian groups $\{C_n\}$ ($\{C^n\}$) and homomorphisms ∂_n: $C_n \to C_{n-1}$ (δ_n: $C^n \to C^{n+1}$) such that $\partial_n \partial_{n+1} = 0$ ($\delta_n \delta_{n-1} = 0$) is called a chain complex (cochain complex). The groups $H_n(C) = \ker \partial_n/\operatorname{Im} \partial_{n+1}$ ($H^n(C) = \ker \delta_n/\operatorname{Im} \delta_{n-1}$) are called the homology (cohomology) groups of the chain (cochain) complex. A chain map (cochain map) $f: \{C_n\} \to \{D_n\}$ ($f: \{C^n\} \to \{D^n\}$) between two chain (cochain) complexes is a sequence of homomorphisms $f_n: C_n \to D_n$ ($f^n: C^n \to D^n$) such that $\partial_n f_n = f_{n-1}\partial_n$ ($f^{n+1}\delta_n = \delta_n f^n$). $Z_n = \ker \partial_n$ is called the group of cycles and $B_n = \operatorname{Im} \partial_{n+1}$ is called the group of boundaries. Two cycles (cocycles) are called homologous if their difference is a boundary (coboundary). Similarly, $Z^n = \ker \delta_n$ and $B^n = \operatorname{Im} \delta_{n-1}$ are called the groups of cocycles and coboundaries. A chain complex is called acyclic if $H_i(C) = 0$ for $i \neq 0$ and $H_0(C) = Z$.

Let us write $\mathcal{C}_\#$ for the category of chain complexes and chain maps.

Proposition 20.6 Taking homology defines a functor $H: \mathcal{C}_\# \to (\mathcal{M}_Z)_*$.

Proof If $f_\#: C_\# \to D_\#$ is a chain map, $f_\#(Z_n(C)) \subset Z_n(D)$ and $f_\#(B_n(C)) \subset B_n(D)$. Hence $f_\#$ induces a map $f_*: H_n(C) \to H_n(D)$ and this is clearly functorial. ∎

Let us write $\mathcal{K}_R$ for the category of relative CW complexes and cellular maps.

Theorem 20.7 $H_*: \mathcal{K}_R \to (\mathcal{M}_Z)_*$ factors into $H \cdot C_\#$ where $C_\#: \mathcal{K}_R \to \mathcal{C}_\#$ is a functor satisfying the axioms:

(a) If X is contractible, $C_\#(X)$ is acyclic.

(b) Let $i: (X, A) \to (X, \bar{X}^{n-1})$. Then $i_\#: C_n(X, A) \to C_n(X, \bar{X}^{n-1})$ is an isomorphism. In particular each characteristic map $\chi_\sigma: (B^n, S^{n-1}) \to (X, \bar{X}^{n-1})$ defines a homomorphism

$$(\chi_\sigma)_\#: C_n(B^n) \cong C_n(B^n, S^{n-1}) \xrightarrow{(\chi_\sigma)_*} C_n(X, \bar{X}^{n-1}) \cong C_n(X, A).$$

(c) $C_n(X, A)$ is a free abelian group with one generator for each n-cell of (X, A). Write B^n as a CW complex with only one n-cell and choose a generator $e^n \in C_n(B^n) \cong Z$. Then $\{(\chi_\sigma)_\#(e^n)\}$ is a free basis for $C_n(X, A)$ as σ varies over the n-cells of (X, A).

Proof If $f: (X, A) \to (Y, B)$ is cellular, $f(\bar{X}^n) \subset \bar{Y}^n$; hence f induces a map $f_*: H_n(\bar{X}^n, \bar{X}^{n-1}) \to H_n(\bar{Y}^n, \bar{Y}^{n-1})$ which we take to be $f_\#$. This is clearly a chain map and induces $f_*: Z_n(X, A) \to Z_n(Y, B)$ since $Z_n(X, A) \cong H_n(\bar{X}^n, A)$. Since the diagram

$$\begin{array}{ccc} H_n(\bar{X}^n, A) & \longrightarrow & H_n(X, A) \\ \big\downarrow f_* & & \big\downarrow f_* \\ H_n(\bar{Y}^n, B) & \longrightarrow & H_n(Y, B) \end{array}$$

commutes, $f_\#$ induces f_* in homology. (a) and (b) are immediate. (c) follows from

Lemma 20.8 Let A_n be the set of n-cells of $X - A$. The maps $(\chi_\sigma)_*: H_n(B^n, S^{n-1}; \pi) \to H_n(X^n, X^{n-1}; \pi)$ for $\sigma \in A_n$ determine an isomorphism

$$\chi: \bigoplus_{\sigma \in A_n} H_n(B_\sigma{}^n, S_\sigma^{n-1}; \pi) \xrightarrow{\cong} H_n(X^n, X^{n-1}; \pi).$$

Proof By the definition of unreduced homology, we have

$$\chi: \bigoplus_{\sigma \in A_n} \tilde{H}_n(S_\sigma{}^n; \pi) \to \tilde{H}_n(\bar{X}^n/\bar{X}^{n-1}; \pi)$$

which is an isomorphism by 18.17. ▮▮

If (X, A) has a finite number of cells in each dimension, $H_n(X, A)$ will be a finitely generated abelian group, and hence a direct sum of cyclic groups. The rank of the free part of $H_n(X, A)$ is called the nth Betti number. The orders of the finite cyclic summands are called the torsion coefficients. These invariants occurred historically before the notion of homology groups was formalized.

Observe that we used the existence of homology to define $C_\#(X, A)$. One might try to define $C_\#(X, A)$ by 20.7. The groups are well defined but the existence of ∂ and chain maps requires a lot of attention. In the end, it would be difficult to prove that homology is a topological invariant (i.e., it does not depend on the choice and number of cells). $C_\#$, for example, is not a topological invariant. $C_\#$ counts cells, and H_* makes $C_\#$ into a topological invariant. It is common philosophy to think of generators of ordinary homology as representatives of "natural cells."

We now study the chain complexes and homology of various cell complexes, based on treating 20.7 as axioms for $C_\#$.

1. CP^n has one cell in dimension $2k$ for $k \le n$ and hence

$$C_i(CP^n) = \begin{cases} Z, & i = 2k \le 2n \\ 0, & \text{otherwise} \end{cases}$$

Since all odd groups are 0, we must have $\partial = 0$. Thus

$$H_i(CP^n) = \begin{cases} Z, & i = 2k \le 2n \\ 0, & \text{otherwise;} \end{cases} \quad \text{and} \quad H_i(CP^\infty) = \begin{cases} Z, & i = 2k \\ 0 & \text{otherwise} \end{cases}$$

by 20.2.

2. HP^n is similar:

$$H_i(HP^n) = \begin{cases} Z, & i = 4k \leq 4n \\ 0, & \text{otherwise}; \end{cases} \qquad H_i(HP^\infty) = \begin{cases} Z, & i = 4k \\ 0, & \text{otherwise} \end{cases}$$

The case of RP^n is much harder since $\partial \neq 0$ in general. We will save this case until we deal a little more with complexes in which $\partial \neq 0$.

3. Let P be a one-point space. Then

$$C_i(P) = \begin{cases} Z, & i = 0 \\ 0, & \text{otherwise.} \end{cases}$$

Choose a generator $e^0 \in C_0(P)$. Clearly

$$H_i(P) = \begin{cases} Z, & i = 0 \\ 0, & \text{otherwise.} \end{cases}$$

4. Consider I as a complex with two 0-cells $\sigma_0{}^0 = 0$ and $\sigma_1{}^0 = 1$, and one 1-cell σ^1. Let $i_\varepsilon \colon P \to I$ be the cellular map onto $\sigma_\varepsilon{}^0$ for $\varepsilon = 0$ or 1. Define $e_\varepsilon{}^0 = (i_\varepsilon)_\#(e^0)$. These are generators of $C_0(I)$ by (c). Let $p_I \colon I \to p$. Then $(p_I)_\#(e_\varepsilon{}^0) = (p_I i_\varepsilon)_\#(e^0) = e^0$. Since $(p_I)_\#(\partial e^1) = 0$, $\partial e^1 = k(e_1{}^0 - e_0{}^0)$. It follows that $k = \pm 1$ since $C_\#(I)$ is acyclic. A choice of k corresponds to choosing the generator e^1. We think of this as "orienting" the simplex and express this by associating a direction to the 1-cell. Thus the choice $\partial e^1 = e_1{}^0 - e_0{}^0$ corresponds to the picture

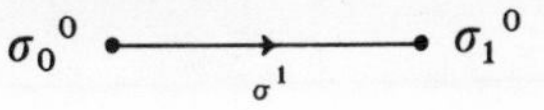

Similarly $\partial e^1 = e_0{}^0 - e_1{}^0$ corresponds to the picture

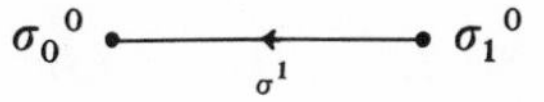

5. Consider I^2 as a CW complex with 0-cells $\sigma_i{}^0$ for $1 \leq i \leq 4$, 1-cells $\sigma_i{}^1$ for $1 \leq i \leq 4$, and a 2-cell σ^2

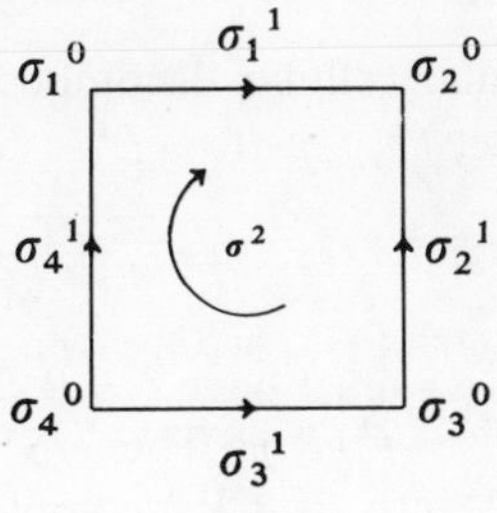

Choose cellular maps $g_i \colon I \to I^2$ such that g_i is a linear characteristic map for σ_i^1, and the sense of increasing is indicated by the arrow in the diagram.

Define $e_i^1 = (g_i)_\#(e^1)$ and $e_i^0 = (h_i)_\#(e^0)$ where $h_i\colon P \to \sigma_i^0$. Then $\{e_i^1\}$ and $\{e_i^0\}$ are generators for $C_1(I^2)$ and $C_0(I^2)$ respectively. One calculates

$$\partial e_1^1 = \partial((g_1)_\#(e^1)) = (g_1)_\#(\partial e^1) = (g_1)_\#(e_1^0 - e_0^0) = e_2^0 - e_1^0,$$

and similarly

$$\partial e_2^1 = e_2^0 - e_3^0, \qquad \partial e_3^1 = e_3^0 - e_4^0, \qquad \partial e_4^1 = e_1^0 - e_4^0.$$

Now $Z_1(I^2)$ is infinite cyclic generated by $e_1^1 - e_2^1 - e_3^1 + e_4^1$. Hence we may find $e^2 \in C_2$ with $\partial e^2 = e_1^1 - e_2^1 - e_3^1 + e_4^1$. e^2 is clearly a generator. e^2 corresponds to the clockwise orientation indicated by the curved arrow. This direction is consistent with σ_1^1 and σ_4^1 and opposite to that of σ_2^1 and σ_3^1. The other choice corresponds to the counterclockwise orientation.

6. We now calculate $H_*(T)$ where $T = S^1 \times S^1$ is the torus. We make use of a cellular map $f\colon I^2 \to T$. T in fact is a quotient space of I^2 under the identification $(x, y) \sim (x + 1, y) \sim (x, y + 1)$. T has a 0-cell $\sigma^0 = f(\sigma_i^0)$, two 1-cells, $\sigma_1^1 = f(\sigma_1^1) = f(\sigma_3^1)$ and $\sigma_2^1 = f(\sigma_2^1) = f(\sigma_4^1)$, and a 2-cell $\sigma^2 = f(\sigma^2)$. As characteristic maps for these cells we may choose $fg_i, f \cdot h_i$; this determines generators $e^0 = f_\#(e_i^0)$, $e_1^1 = f_\#(e_1^1) = f_\#(e_3^1)$, $e_2^1 = f_\#(e_2^1) = f_\#(e_4^1)$, and $e^2 = f_\#(e^2)$. Hence we calculate

$$\begin{aligned}\partial e^2 &= \partial f_\#(e^2) = f_\#(\partial e^2) = f_\#(e_1^1 - e_2^1 - e_3^1 + e_4^1)\\ &= e_1^1 - e_2^1 - e_1^1 + e_2^1 = 0\\ \partial e_1^1 &= f_\#(\partial e_1^1) = f_\#(e_2^0 - e_1^0) = 0\\ \partial e_2^1 &= 0.\end{aligned}$$

Hence:

Proposition 20.9

$$H_i(T) = \begin{cases} Z, & i = 0\\ Z \oplus Z, & i = 1\\ Z, & i = 2\\ 0, & \text{otherwise.} \end{cases} \quad \blacksquare$$

7. We choose a particular cellular decomposition of S^n in order to calculate $H_*(RP^n)$ as in (6). Let $f_+^k, f_-^k\colon B^k \to E^{k+1}$ be given by

$$\begin{aligned} f_+^k(x_1, \ldots, x_k) &= (x_1, \ldots, x_k, \sqrt{1 - \Sigma x_i^2})\\ f_-^k(x_1, \ldots, x_k) &= (-x_1, \ldots, -x_k, -\sqrt{1 - \Sigma x_i^2}). \end{aligned}$$

Let $B_+^k = f_+^k(B^k)$ and $B_-^k = f_-^k(B^k)$. Then $S^n = B_+^n \cup B_-^n$ and $S^{n-1} = B_+^n \cap B_-^n$. This makes S^n into a CW complex with cells B_+^k, B_-^k for $0 \leq k \leq n$. The antipodal map is a cellular homeomorphism $a\colon S^n \to S^n$ and hence induces a chain automorphism $a_\#$.

Lemma 20.10 There are generators e_+^k, e_-^k for $C_k(S^n)$ for $k \le n$ such that

$$\partial e_+^{2k} = \partial e_-^{2k} = e_+^{2k-1} + e_-^{2k-1}$$
$$\partial e_+^{2k+1} = -\partial e_-^{2k+1} = e_+^{2k} - e_-^{2k}.$$

Proof e_+^0 must be homologous to e_-^0 by axiom (a) applied to the subcomplex B_+^1. Hence $e_+^0 - e_-^0 \in \partial C_1(B_+^1)$. Choose a generator e_+^1 so that $\partial e_+^1 = e_+^0 - e_-^0$. Choose a generator e^1 of $C_1(B^1)$ so that $(f_+^1)_\#(e^1) = e_+^1$. Define $e_-^1 = (f_-^1)_\#(e^1)$. Then $a_\# e_+^1 = e_-^1$. Consequently

$$\partial e_-^1 = \partial a_\#(e_+^1) = a_\#(\partial e_+^1) = a_\#(e_+^0 - e_-^0) = e_-^0 - e_+^0.$$

Suppose by induction the generators e_+^k and e_-^k are picked for $k \le 2m < n$ satisfying the formulas in the conclusion and $a_\#(e_+^k) = e_-^k$. Now $\partial(e_+^{2m} - e_-^{2m}) = 0$. Since $H_{2m}(B_+^{2m+1}) = 0$, there is a generator e_+^{2m+1} of $C_{2m+1}(B_+^{2m+1})$ with $\partial e_+^{2m+1} = e_+^{2m} - e_-^{2m}$. Choose a generator e^{2m+1} of $C_{2m+1}(B^{2m+1})$ such that $(f_+^{2m+1})_\#(e^{2m+1}) = e_+^{2m+1}$. Let $e_-^{2m+1} = (f_-^{2m+1})_\#(e^{2m+1})$. Then $e_-^{2m+1} = a_\#(e_+^{2m+1})$. Hence

$$\partial e_-^{2m+1} = \partial a_\#(e_+^{2m+1}) = a_\#(\partial e_+^{2m+1}) = a_\#(e_+^{2m} - e_-^{2m}) = e_-^{2m} - e_+^{2m}$$

(by induction). Now $\partial(e_+^{2m+1} + e_-^{2m+1}) = 0$. Since $H_{2m+1}(B_+^{2m+2}) = 0$, we may choose a generator e_+^{2m+2} with $\partial e_+^{2m+2} = e_+^{2m+1} + e_-^{2m+1}$. As above, we choose e_-^{2m+2} and find

$$\partial e_-^{2m+2} = \partial a_\#(e_+^{2m+2}) = a_\#(\partial e_+^{2m+2}) = a_\#(e_+^{2m+1} + e_-^{2m+1}) = e_-^{2m+1} + e_+^{2m+1}.$$

This completes the proof of 20.10.

Let $\Pi_n\colon S^n \to RP^n$. Then Π_n is cellular. In fact (see Section 14) $\Pi_n f_+^k = \Pi_n f_-^k =$ the characteristic for the k cell of RP^n. Let $\bar{e}^k = (\Pi_n)_\#(e_+^k) = (\Pi_n)_\#(e_-^k)$, which is consequently a generator of $C_k(RP^n)$.

We have

$$\partial \bar{e}^k = \partial(\Pi_n)_\#(e_+^k) = (\Pi_n)_\#(\partial e_+^k) = \begin{cases} 0, & k \text{ odd} \\ 2\bar{e}^{k-1}, & k \text{ even.} \end{cases}$$

We have proven:

Proposition 20.11

$$H_i(RP^n) = \begin{cases} Z, & i = 0 \\ Z_2, & i \text{ odd}, \quad i < n \\ Z, & i \text{ odd}, \quad i = n \\ 0, & \text{otherwise} \end{cases} \qquad H_i(RP^\infty) = \begin{cases} Z, & i = 0 \\ Z_2, & i \text{ odd} \\ 0, & \text{otherwise.} \end{cases}$$ ∎

8. Let $X = \Delta^n$ be the standard n-simplex. Define

$$\eta_k\colon \Delta^{n-1} \to \Delta^n \qquad \text{by} \qquad \eta_k(x_0, \ldots, x_{n-1}) = (x_0, \ldots, x_{k-1}, 0, x_k, \ldots, x_{n-1})$$

for $0 \le k \le n$. X is a simplicial complex and hence a CW complex with $\binom{n+1}{k+1}$ k-cells for $0 \le k \le n$. $\eta_k(\partial\Delta^{n-1}) \subset X^{n-2}$, so η_k: $(\Delta^{n-1}, \partial\Delta^{n-1}) \to (X^{n-1}, X^{n-2})$. Let $e_{n-1} \in C_{n-1}(\Delta^{n-1})$ be a generator. Then $H_{n-1}(X^{n-1}, X^{n-2})$ is freely generated by $(\eta_k)_\#(e_{n-1})$ for $0 \le k \le n$.

Theorem 20.12 There are generators $e_n \in C_n(\Delta^n)$ such that

$$\partial e_n = \sum_{k=0}^{n} (-1)^k (\eta_k)_\#(e_{n-1})$$

for $n \ge 1$.

Proof The case $n = 1$ is Example 4 above. Suppose by induction that the formula is valid for $1, \ldots, n$. Observe that $\eta_k \eta_j = \eta_{j+1}\eta_k$ if $j \ge k$ and consequently $\eta_k \eta_j = \eta_j \eta_{k-1}$ if $j < k$. We now prove

$$\partial\left(\sum_{k=0}^{n+1} (-1)^k (\eta_k)_\#(e_n)\right) = 0.$$

Since η_k is a cellular map, $\partial(\eta_k)_\#(e_n) = (\eta_k)_\#(\partial e_n)$. Hence

$$\begin{aligned}
\partial\left(\sum_{k=0}^{n+1} (-1)^k (\eta_k)_\#(e_n)\right) &= \sum_{k=0}^{n+1} (-1)^k (\eta_k)_\#(\partial e_n) \\
&= \sum_{k=0}^{n+1} \sum_{j=0}^{n} (-1)^{k+j} (\eta_k \eta_j)_\#(e_{n-1}) \\
&= \sum_{n+1 \ge k > j \ge 0} (-1)^{k+j} (\eta_k \eta_j)_\#(e_{n-1}) \\
&\quad + \sum_{n \ge j \ge k \ge 0} (-1)^{k+j} (\eta_k \eta_j)_\#(e_{n-1}) \\
&= \sum_{n+1 \ge k \ge j \ge 0} (-1)^{k+j} (\eta_j \eta_{k-1})_\#(e_{n-1}) \\
&\quad + \sum_{n \ge l-1 \ge k \ge 0} (-1)^{k+l-1} (\eta_k \eta_{l-1})_\#(e_{n-1}) \\
&= 0.
\end{aligned}$$

Since Δ^{n+1} is acyclic, one may find $e_{n+1} \in C_{n+1}(\Delta^{n+1})$ with

$$\partial e_{n+1} = \sum_{k=0}^{n+1} (-1)^k (\eta_k)_\#(e_n).$$

e_{n+1} is a generator, since $\sum_{k=0}^{n+1}(-1)^k(\eta_k)_\#(e_n)$ may be chosen to be a generator of C_n. ∎

Let K be a simplicial complex. Order the vertices of K, $v_0, \ldots, v_n$. If σ is an n-simplex, there is a unique order preserving linear homeomorphism $\chi_\sigma\colon \Delta^n \to \sigma$ given by $\chi_\sigma(a_0, \ldots, a_n) = \sum_{k=0}^n a_k v_{i_k}$ where $\sigma = (v_{i_0} v_{i_1} \cdots v_{i_n})$ and $i_0 < i_1 < \cdots < i_n$. $\chi_\sigma((\Delta^n)^{n-1}) \subset K^{n-1}$ so χ_σ determines a generator $e_\sigma = (\chi_\sigma)_*(e_n) \in C_n(K)$. Since χ_σ is cellular, we have

$$\partial e_\sigma = \partial(\chi_\sigma)_\#(e_n) = (\chi_\sigma)_\#(\partial e_n) = (\chi_\sigma)_\#\left(\sum_{k=0}^n (-1)^k(\eta_k)_\#(e_{n-1})\right)$$

$$= \sum_{k=0}^n (-1)^k(\chi_\sigma \eta_k)_\#(e_{n-1}).$$

If $\sigma = (v_{i_0} \cdots v_{i_n})$ write

$$\sigma_k = (v_{i_0} v_{i_1} \cdots v_{i_{k-1}} v_{i_{k+1}} \cdots v_{i_n}).$$

Now $\chi_\sigma \eta_k = \chi_{\sigma_k}$, so $(\chi_\sigma \eta_k)_*(e_{n-1}) = e_{\sigma_k}$. Hence

$$\partial e_\sigma = \sum_{k=0}^n (-1)^k e_{\sigma_k}$$

Thus given a simplicial complex K, the chain complex is completely determined, and we have proven

Theorem 20.13 Let K be a simplicial complex with vertices ordered $v_0, \ldots, v_n$. For each n-simplex $\sigma = (v_{i_0}, \cdots v_{i_n})$ with $i_0 < i_1 < \cdots < i_n$, let $\sigma_k = (v_{i_0} v_{i_1} \cdots v_{i_{k-1}} v_{i_{k+1}} \cdots v_{i_n})$ be the kth face for $0 \leq k \leq n$. Then $C_n(K)$ has free generators $\{e_\sigma\}$, one for each n-simplex σ, and ∂ is determined by the formula

$$\partial e_\sigma = \sum_{k=0}^n (-1)^k e_{\sigma_k}. \quad \blacksquare$$

Note that by 20.13, one can define $H_*(K)$ for any simplicial complex, but it is very hard to prove directly that what is defined is a topological invariant.

As an example we will calculate $H_*(S^2)$ by this method. Write $S^2 \equiv \partial\Delta^3$. We have vertices $v_0 = (1, 0, 0, 0)$, $v_1 = (0, 1, 0, 0)$, $v_2 = (0, 0, 1, 0)$ and $v_3 = (0, 0, 0, 1)$. Write $e_{v_{i_0} \cdots v_{i_n}} = e_{i_0 \cdots i_n}$. Then

C_0 is freely generated by e_0, e_1, e_2, e_3.
C_1 is freely generated by $e_{01}, e_{02}, e_{03}, e_{12}, e_{13}, e_{23}$.
C_2 is freely generated by $e_{012}, e_{013}, e_{023}, e_{123}$.
$C_n = 0$ if $n > 2$.

We have

$$\partial e_{012} = e_{12} - e_{02} + e_{01} \qquad \partial e_{023} = e_{23} - e_{03} + e_{02}$$
$$\partial e_{013} = e_{13} - e_{03} + e_{01} \qquad \partial e_{123} = e_{23} - e_{13} + e_{12}.$$

$Z_2(\partial\Delta^3)$ is freely generated by

$$e_{012} - e_{013} + e_{023} - e_{123}.$$

Hence $H_2 \cong Z$.

$$\partial e_{01} = e_1 - e_0 \qquad \partial e_{12} = e_2 - e_1$$
$$\partial e_{02} = e_2 - e_0 \qquad \partial e_{23} = e_3 - e_2$$
$$\partial e_{03} = e_3 - e_0$$

It is easy to see that $C_2 \xrightarrow{\partial} C_1 \xrightarrow{\partial} C_0$ is exact at C_1. B_0 is generated by $e_1 - e_0$, $e_2 - e_0$, and $e_3 - e_0$. Hence $H_0 \cong Z$ is generated by e_0.

Exercises

1. Give details for a proof of 20.4.

2. Prove $H_n(X, X^m; \pi) = 0 = H^n(X, X^m; \pi)$ if $n \leq m$.

3. Show that

$$\tilde{H}_k(M(\pi, n)) \cong \begin{cases} \pi/[\pi, \pi], & n = k \\ 0, & n \neq k \end{cases}$$

where $[\pi, \pi]$ is the commutator subgroup ($=1$ if $n \geq 2$). (Exercise 6, Section 22)

4. Let $(X, \{\chi_\alpha\})$ be a semisimplicial CW complex (16.37). Show that one may choose generators $e_\alpha \in C_*(X)$, one for each cell such that

$$\partial e_\alpha = \sum_{s=0}^{n} (-1)^s e_{\partial_s \alpha}.$$

(Exercise 9, Section 21)

5. The Klein bottle K is defined to be a quotient space of $S^1 \times I$ under the identification $(z, 0) \sim (z^{-1}, 1)$. Calculate $H_*(K)$. (Hint: K is a cellular quotient space of I^2.) (See Exercise 14, Section 7.)

6. Calculate $(\Pi_n)_*\colon H_n(S^n) \to H_n(RP^n)$ if n is odd.

7. Show that if (X, A) is a CW pair, there is a short exact sequence of chain complexes:

$$0 \to C_\#(A) \to C_\#(X) \to C_\#(X, A) \to 0$$

(i.e., the maps are chain maps, and it is exact in each dimension).

8. Calculate $C_\#(I^3)$ where I^3 has the standard cellular decomposition with 8 0-cells, 12 1-cells, 6 2-cells, and 1 3-cell.

9. Let $0 \to C \xrightarrow{\alpha} D \xrightarrow{\beta} E \to 0$ be an exact sequence of chain complexes and chain maps as in Exercise 7 above. Prove that there is a long exact sequence

$$\cdots \xrightarrow{\partial} H(C) \xrightarrow{H(\alpha)} H(D) \xrightarrow{H(\beta)} H(E) \xrightarrow{\partial} H(C) \to \cdots$$

where $\partial\{e\}$ is defined as follows. Let $\beta(d) = e$. Then $\beta(\partial d) = 0$ so there exists $c \in C$ with $\alpha(c) = \partial d$. Define $\partial\{e\} = \{c\}$. Show that c is a cycle and $\{c\}$ depends only on $\{e\}$. (Section 25; Exercise 4, Section 30)

10.* Calculate $H_i(S^n \times S^m)$.

11. Given a graded vector space $\{V_n\}$ with $V_n \neq 0$ for only finitely many values of n, define the Euler characteristic of $\{V_n\}$ by

$$\chi(\{V_n\}) = \sum (-1)^n \dim V_n$$

Show that if $\{V_n, \partial_n\}$ is a chain complex with $\{V_n\}$ as above, $\chi(\{V_n\}) = \chi(H(\{V_n, \partial_n\}))$.

Define the Euler characteristic of a finite CW complex X as $\chi(H_*(X; k))$ where k is any field. Thus

$$\sum (-1)^n \operatorname{rank} C_n(X) = \sum (-1)^n \dim C_n(X; k) = \chi(X)$$

is a homotopy type invariant and does not depend on k. Note that $\chi(S^2) = 2$, hence for any CW decomposition of S^2, $\dim C_2 - \dim C_1 + \dim C_0 = 2$. (26.25)

12. Let $\sigma = (v_0, \ldots, v_n)$ be a simplex with ordered vertices and let $\sigma_T = (v_{T(0)}, \ldots, v_{T(n)})$ where T is a permutation of n letters. Show that $\{e_\sigma\} = \operatorname{sgn} T \cdot \{e_\sigma T\}$ in $H_n(\sigma, \partial\sigma)$. (Section 26)

13. Let X be a CW complex. Show that $H_0(X)$ is a free abelian group whose rank is the number of arc components. (26.29)

14. Consider the simplicial complex K with vertices v_0, v_1, v_2, v_3, v_4. As simplices, take all proper subsets of $(v_0 v_1 v_2)$ and all proper subsets of $(v_0 v_3 v_4)$. Calculate the homology of K with integer coefficients. Check your answer by verifying that $K \equiv S^1 \vee S^1$.

21

Homology and Cohomology Groups of More General Spaces

In this section we discuss the existence and properties of homology and cohomology theories when applied to more general spaces than CW complexes. The easiest method to obtain such theories is the singular extension. This is described in general. Its historical predecessor, the ordinary singular homology and cohomology functors are described in Exercise 9.

Assuming that for n sufficiently large $(E_n, *) \in \mathcal{N}$, we then prove that singular homology agrees with spectral homology under mild assumptions. Finally, assuming that for n sufficiently large $(E_n, *)$ is the homotopy type of a CW complex with 0-cell as base point, we derive the properties of spectral cohomology theories when applied to paracompact compactly generated spaces.

Definition 21.1 If $(X, A) \in \mathcal{C}^2$ define $SE_n(X, A) = E_n(S(X, A))$ and $SE^n(X, A) = E^n(S(X, A))$ where $S(X, A)$ is the singular complex (16.21). Similarly $S\tilde{E}_n(X) = \tilde{E}_n(S(X))$ and $S\tilde{E}^n(X) = \tilde{E}^n(S(X))$. These are called the singular homology and cohomology theories associated with the spectrum E.

Ordinary singular homology and cohomology $E = H\pi$ are classically defined in a different way. (See Exercise 11.)

Proposition 21.2 SE_n and SE^n are unreduced homology and cohomology theories on $\mathcal{C}^2$ with type 1 excision. That is, $SH_n(X, A; \pi)$ and $SH^n(X, A; \pi)$ satisfy the Eilenberg–Steenrod axioms (A), (B), and (C) of 19.2 on $\mathcal{C}^2$.

Proof Axioms (A) and (C) are clear. To prove (B) we prove that if (X_1, X_2) is excisive, $SE^n(X_1 \cup X_2, X_1) \cong SE^n(X_2, X_1 \cap X_2)$ and $SE_n(X_1 \cup X_2, X_1) \cong SE_n(X_2, X_1 \cap X_2)$. We construct a resolution $f_{12}: K_{12} \to X_1 \cap X_2$ and extend this to resolution $f_1: K_1 \to X_1$ and $f_2: K_2 \to X_2$ where $K_1 \cap K_2 = K_{12}$. As in the proof of 16.27, we define $f: K_1 \cup K_2 \to X_1 \cup X_2$ and by 16.24, f is a resolution. The isomorphisms follow. ∎

Singular homology and cohomology theories have the following characteristic property.

Proposition 21.3 Let $f: (X, A) \to (Y, B)$ be a weak homotopy equivalence. Then $f_*: SE_n(X, A) \to SE_n(Y, B)$ and $f^*: SE^n(Y, B) \to SE^n(X, A)$ are isomorphisms.

Proof Since f is a weak homotopy equivalence, $S(f)$ is a homotopy equivalence. Hence f_* and f^* are isomorphisms. ∎

Proposition 21.4 Let $S\tilde{E}_m$ and $S\tilde{E}^m$ be the reduced singular homology and cohomology theories associated with E (21.1). Then

$$SE_m(X) \cong S\tilde{E}_m(X) \oplus E_m(*), \qquad SE^m(X) \cong S\tilde{E}^m(X) \oplus E^m(*).$$

Proof By 19.6 it is sufficient to show that $E_m(S(X), *) \cong \tilde{E}_m(S(X))$ and $E^m(S(X), *) \cong \tilde{E}^m(S(X))$. This follows from 19.7. ∎

An important and useful property of singular theory is given by the Mayer–Vietoris sequences.

Proposition 21.5 (*Mayer–Vietoris*) If (X_1, X_2) is excisive in X, there are long exact sequences

$$\cdots \leftarrow SE^n(X_1 \cap X_2) \xleftarrow{i^*} SE^n(X_1) \oplus SE^n(X_2) \xleftarrow{\phi^*} SE^n(X_1 \cup X_2) \xleftarrow{\delta} SE^{n-1}(X_1 \cap X_2) \leftarrow \cdots$$

$$\cdots \to SE_n(X_1 \cap X_2) \xrightarrow{i^*} SE_n(X_1) \oplus SE_n(X_2) \xrightarrow{\phi^*} SE_n(X_1 \cup X_2) \xrightarrow{\partial} SE_{n-1}(X_1 \cap X_2) \to \cdots$$

∂ and δ are natural. The other homomorphisms are given by

$$\phi^*(\alpha) = (j_1{}^*(\alpha), j_2{}^*(\alpha)), \qquad i^*(\alpha, \beta) = i_1{}^*(\alpha) - i_2{}^*(\beta),$$
$$i_*(\alpha) = (i_{1*}(\alpha), i_{2*}(\alpha)), \qquad \phi_*(\alpha, \beta) = j_1{}^*(\alpha) - j_2{}^*(\beta)$$

where $j_\varepsilon: X_\varepsilon \to X_1 \cup X_2$ and $i_\varepsilon: X_1 \cap X_2 \to X_\varepsilon$.

Proof Since as the proofs are similar, we will do the homology case only: $SE_n(X_1 \cup X_2, X_1) \cong SE_n(X_2, X_1 \cap X_2)$. Consider the diagram

$$\begin{array}{ccccccccc}
\cdots \longrightarrow & SE_n(X_1) & \xrightarrow[j_{1*}]{} & SE_n(X_1 \cup X_2) & \longrightarrow & SE_n(X_1 \cup X_2, X_1) & \longrightarrow \cdots \\
& \uparrow{\scriptstyle i_{1*}} & & \uparrow{\scriptstyle j_{2*}} & & \uparrow{\scriptstyle \cong} & \\
\cdots \longrightarrow & SE_n(X_1 \cap X_2) & \xrightarrow{i_{2*}} & SE_n(X_2) & \longrightarrow & SE_n(X_2, X_1 \cap X_2) & \longrightarrow \cdots
\end{array}$$

The proof follows from:

Lemma 21.6 (*Barratt–Whitehead*) Given a commutative diagram

$$\begin{array}{ccccccccccccc}
\cdots \longrightarrow & A_n & \xrightarrow{i_n} & B_n & \xrightarrow{j_n} & C_n & \xrightarrow{k_n} & A_{n-1} & \longrightarrow & B_{n-1} & \longrightarrow & C_{n-1} & \longrightarrow \cdots \\
& \downarrow{\scriptstyle \alpha_n} & & \downarrow{\scriptstyle \beta_n} & & \downarrow{\scriptstyle \gamma_n} & & \downarrow{\scriptstyle \alpha_{n-1}} & & \downarrow{\scriptstyle \beta_{n-1}} & & \downarrow{\scriptstyle \gamma_{n-1}} & \\
\cdots \longrightarrow & A_n' & \xrightarrow{i_n'} & B_n' & \xrightarrow{j_n'} & C_n' & \xrightarrow{k_n'} & A_{n-1}' & \longrightarrow & B_{n-1}' & \longrightarrow & C_{n-1}' & \longrightarrow \cdots
\end{array}$$

in which γ_n is an isomorphism, and the rows are exact, there is an exact sequence

$$\cdots \to A_n \xrightarrow[f_n]{} A_n' \oplus B_n \xrightarrow[g_n]{} B_n' \xrightarrow[h_n]{} A_{n-1} \to \cdots$$

where $f_n(x) = (\alpha_n(x), i_n(x))$, $g_n(x, y) = i_n'(x) - \beta_n(y)$, and $h_n = k_n \gamma_n^{-1} j_n'$.

Proof This is an elementary diagram chase and is omitted. ▮▮

We now assume that for n sufficiently large $(E_n, *) \in \mathcal{N}$. We will first discuss the consequences for the spectral homology theories.

Theorem 21.7 For any $(X, *) \in \mathcal{N}$, $S\tilde{E}_m(X) \cong \tilde{E}_m(X)$. For any pair $(X, A) \in \mathcal{CG}^2$, $SE_m(X, A) \cong E_m(X, A)$. In particular, E_m is a homology theory on $\mathcal{CG}^2$. If X is Hausdorff, $SE_m(X, A) = E_m(k(X), k(A))$ where $k(X)$ is the associated compactly generated space (8.7).

The proof will depend on:

Proposition 21.8 Let $(X, *)$, $(Y, *)$, and $(E, *) \in \mathcal{N}$. Suppose $f: (X, *) \to (Y, *)$ is a weak homotopy equivalence. Then $f \wedge 1: X \wedge E \to Y \wedge E$ is a weak homotopy equivalence.

Proof of 21.7 Suppose $(X, *), (Y, *) \in \mathcal{N}$ and $f: (X, *) \to (Y, *)$ is a weak homotopy equivalence. Then

$$(f \wedge 1)_*: \pi_{n+m}(X \wedge E_n) \to \pi_{n+m}(Y \wedge E_n)$$

is an isomorphism for n sufficiently large. Considering the ladder that defines $f_*\colon \tilde{E}_m(X) \to \tilde{E}_m(Y)$ (18.5), it follows that f_* is an isomorphism. Hence $S\tilde{E}_m(X) = \tilde{E}_m(S(X)) \cong \tilde{E}_m(X)$. If $f\colon (X, A) \to (Y, B)$ is a weak homotopy equivalence, $\tilde{f}\colon X \cup CA \to Y \cup CB$ is a weak homotopy equivalence by Exercise 29, Section 16. Since $X \cup CA$ and $Y \cup CB$ belong to $\mathcal{N}$, $f_*\colon E_m(X, A) \to E_m(Y, B)$ is an isomorphism. Thus $SE_m(X, A) = E_m(X, A)$. Finally by 8.8, $(k(X), k(A)) \to (X, A)$ is a weak homotopy equivalence. Hence

$$SE_m(X, A) \cong SE_m(k(X), k(A)) \cong E_m(k(X), k(A)). \quad \blacksquare$$

Proof of 21.8 By Exercise 18, Section 14, $(X \times E, X \vee E)$ and $(Y \times E, Y \vee E)$ have the AHEP. Hence by 16.31,

$$X \wedge E \simeq (X \times E) \cup C(X \vee E) \qquad \text{and} \qquad Y \wedge E \simeq (Y \times E) \cup C(Y \vee E).$$

Consequently it is sufficient to show that

$$(X \times E) \cup C(X \vee E) \to (Y \times E) \cup C(Y \vee E)$$

is a weak homotopy equivalence. By Exercise 29, Section 16 it is sufficient to show that $f \times 1\colon X \times E \to Y \times E$ and $f \vee 1\colon X \vee E \to Y \vee E$ are weak homotopy equivalences. It is obvious that $f \times 1$ is a weak homotopy equivalence. 21.8 thus follows from:

Lemma 21.9 Suppose $(X, *)$, $(Y, *)$, and $(E, *) \in \mathcal{N}$ and $f\colon (X, *) \to (Y, *)$ is a weak homotopy equivalence. Then $(f \vee 1)\colon X \vee E \to (Y \vee E)$ is a weak homotopy equivalence.

Proof For any two spaces A and B with nondegenerate base points a and b,

$$A \vee B \simeq A \cup I \cup B \Big/ \begin{matrix} a \sim 0, \\ b \sim 1 \end{matrix}$$

where the homotopy equivalence is obtained by pinching I to a point. This follows from 16.31 since

$$A \cup I \cup B \Big/ \begin{matrix} a \sim 0 \\ b \sim 1 \end{matrix} \equiv (A \amalg B) \cup C(\{a, b\}).$$

Thus it is sufficient to show that the map $X \cup I \cup E/\!\sim\; \to Y \cup I \cup E/\!\sim$ is a weak homotopy equivalence. This follows directly from 16.24 with

$$X \cup I \cup E/\!\sim\; = (X \cup I/\!\sim) \cup (I \cup E/\!\sim)$$

and

$$Y \cup I \cup E/\!\sim\; = (Y \cup I/\!\sim) \cup (I \cup E/\!\sim). \quad \blacksquare\blacksquare$$

Spectral cohomology behaves somewhat differently than spectral homology. There is an arcwise connected compact subset X of R^2, with $\pi_i(X, *) = 0$ for all $i \geq 0$ but $H^1(X) \cong Z$ (see 21.21). Since $SH^1(X) = 0$, we do not have $SE^n(X) \cong E^n(X)$ in general. *In order to proceed we assume that for n sufficiently large* $(E_n, *)$ *is a CW complex with a* 0*-cell as base point.* We prove only that E^n is a cohomology theory on paracompact compactly generated spaces. We must exploit a special property of $(E_n, *)$ in order to achieve this. This property is contained in the following.

Definition 21.10 $(Y, *)$ is called a weak absolute neighborhood extensor (WANE) if for each paracompact space X, each closed subspace $A \subset X$, and each continuous map $f: A \to Y$, there is a neighborhood U of A in X and a map $g: U \to Y$ such that $g|_A \sim f\,(\text{rel}\, f^{-1}(*))$.

To utilize this concept we make some observations about paracompact spaces.

Lemma 21.11 Let X be paracompact, $A \subset X$ be a closed subset, and K be a compact Hausdorff space. Then A, X/A, and $X \times K$ are paracompact.

Proof (1) Given an open cover $\{\mathcal{U}_\alpha\}$ of A choose open sets $\mathcal{V}_\alpha$ with $\mathcal{V}_\alpha \cap A = \mathcal{U}_\alpha$. Then $\{\mathcal{V}_\alpha, X - A\}$ is an open cover of X. A locally finite refinement of this, when restricted to A, is a locally finite refinement of $\{\mathcal{U}_\alpha\}$.

(2) Let $\{\mathcal{U}_\alpha\}$ be an open cover of X/A. Suppose $\{A\} \in \mathcal{U}_{\alpha_0}$. Let $\mathcal{V}_\alpha = p_A^{-1}(\mathcal{U}_\alpha)$. Since X is paracompact it is normal. Choose $f: X \to I$ with $f(A) = 1$, $f(X - \mathcal{V}_{\alpha_0}) = 0$. Let $W = f^{-1}([0, \frac{1}{2}))$ and $D = f^{-1}([0, \frac{3}{4}])$. Then $X - \mathcal{V}_{\alpha_0} \subset W \subset D \subset X - A$. Since D is closed, it a paracompact. Choose a locally finite refinement $\{T_\gamma\}$ of $\{D \cap \mathcal{V}_\alpha\}$. $p_A^{-1} p_A(T_\gamma \cap W) = T_\gamma \cap W$. Since W is open, $p_A(T_\gamma \cap W)$ is open. Suppose $T_\gamma \subset \mathcal{V}_\alpha$. Then $p_A(T_\gamma \cap W) \subset p_A(\mathcal{V}_\alpha) = \mathcal{U}_\alpha$. Hence $\{p_A(T_\gamma \cap W), \mathcal{U}_{\alpha_0}\}$ is a refinement of $\{\mathcal{U}_\alpha\}$ by open sets. It covers X/A, since if $x \notin \mathcal{U}_{\alpha_0}$, $x \in p_A(W)$ and hence $x \in p_A(W \cap T_\gamma)$ for some γ. Finally, we claim that $\{p_A(T_\gamma \cap W), \mathcal{U}_{\alpha_0}\}$ is locally finite. If $x \notin p_A(D)$, $p_A(X - D)$ is a neighborhood of x which intersects only $\mathcal{U}_{\alpha_0}$. Suppose $x \in p_A(D)$. Let $y = p_A^{-1}(x) \in D$. There exists a neighborhood $\mathcal{U}_y$ of y in D such that $\mathcal{U}_y$ intersects only finitely many of the T_γ. Let $\mathcal{V}_y$ be an open set in X such that $\mathcal{V}_y \cap D = \mathcal{U}_y$. Then $p_A(\mathcal{V}_y - A)$ is a neighborhood of x and $p_A(\mathcal{V}_y - A) \cap p_A(T_\gamma \cap W) = p_A(\mathcal{U}_y \cap W \cap T_\gamma)$ and this is only nonempty for finitely many γ.

3. Clearly $X \times K$ is Hausdorff. Let $\{\mathcal{U}_\alpha\}$ be an open cover of $X \times K$. Choose a refinement $\{\mathcal{A}_\beta \times \mathcal{B}_\beta\}$ with $\mathcal{A}_\beta$ open in X and $\mathcal{B}_\beta$ open in K. For each $x \in X$ there exist $\beta_1(x), \ldots, \beta_{n(x)}(x)$ such that

$$x \times I \subset \mathcal{A}_{\beta_1(x)} \times \mathcal{B}_{\beta_1(x)} \cup \cdots \cup \mathcal{A}_{\beta_{n(x)}(x)} \times \mathcal{B}_{\beta_{n(x)}(x)}.$$

Let $V(x) = \bigcap \mathcal{A}_{\beta_i(x)}$. Choose a locally finite refinement $\{C_\gamma\}$ of the open cover $\{V(x)\}$ of X. For each γ, choose $x_\gamma \in X$ such that $C_\gamma \subset V(x_\gamma)$. Then

$$C_\gamma \times \mathcal{B}_{\beta_i(x_\gamma)} \subset \mathcal{A}_{\beta_i(x_\gamma)} \times \mathcal{B}_{\beta_i(x_\gamma)} \subset \mathcal{U}_\alpha$$

for some α, so $\{C_\gamma \times \mathcal{B}_{\beta_i(x_\gamma)}\}$ refines $\{\mathcal{U}_\alpha\}$. Given $(x, t) \in X \times I$. Choose γ with $x \in C_\gamma$. Now $t \in \mathcal{B}_{\beta_i(x_\gamma)}$ for some i since $x_\gamma \times I$ is covered by the sets $\mathcal{A}_{\beta_i(x_\gamma)} \times \mathcal{B}_{\beta_i(x_\gamma)}$. Thus $(x, t) \in C_\gamma \times \mathcal{B}_{\beta_i(x_\gamma)}$ and $\{C_\gamma \times \mathcal{B}_{\beta_i(x_\gamma)}\}$ is an open cover. To show that it is locally finite, let $(x, t) \in X \times I$. Let $\mathcal{U}_x$ be a neighborhood of x in X which meets only finitely many C_γ. Then $\mathcal{U}_x \times I$ can meet only finitely many $C_\gamma \times \mathcal{B}_{\beta_i(x_\gamma)}$. ∎

In particular if X is paracompact and A is closed, $X \cup CA$, $X \cup C^*A$, ΣX, and SX are all paracompact.

Proposition 21.12 (a) If $(Y, *) \simeq (Z, *)$ and $(Y, *)$ is a WANE, $(Z, *)$ is a WANE.

(b) If $(Y, *)$ is a WANE, $(\Omega(Y, *), *)$ with the compact open topology is a WANE.

Proof (a) Let $f\colon A \to Z$ and let $\phi\colon Z \to Y$ and $\phi'\colon Y \to Z$ be homotopy inverses in $\mathcal{C}^*$. Then there is a neighborhood U of A and a map $g\colon U \to Y$ such that $g|_A \sim \phi f$ $(\mathrm{rel}(\phi f)^{-1}(*))$. $(\phi f)^{-1}(*) \supset f^{-1}(*)$ so $\phi' g|_A \sim \phi'\phi f$ $(\mathrm{rel}\, f^{-1}(*))$. But $\phi\phi' \sim 1$ (rel $*$), so $\phi' g|_A \sim f\,(\mathrm{rel}\, f^{-1}(*))$.

(b) Let $f\colon A \to \Omega(Y, *)$. Then

$$f^*\colon (A \times I, f^{-1}(*) \times I \cup A \times 0 \cup A \times 1) \to (Y, *)$$

is continuous. f^* extends to a map

$$\bar{f}\colon (X \times 0 \cup A \times I \cup X \times 1, f^{-1}(*) \times I \cup X \times 0 \cup X \times 1) \to (Y, *)$$

Since $X \times 0 \cup A \times I \cup X \times 1$ is closed in $X \times I$ and $X \times I$ is paracompact, there is a neighborhood U of $X \times 0 \cup A \times I \cup X \times 1$ in $X \times I$ and a map $g\colon U \to (Y, *)$ such that $g|_{X \times 0 \cup A \times I \cup X \times 1} \sim \bar{f}\,(\mathrm{rel}\, f^{-1}(*) \times I \cup X \times 0 \cup X \times 1)$. Since I is compact, there is a neighborhood V of A in X with $V \times I \subset U$. Let $h = (g|_{V \times I})^*\colon V \to \Omega(Y, *)$. Then $h|_A \sim f\,(\mathrm{rel}\, f^{-1}(*))$. ∎

The reason for introducing WANE's is:

Theorem 21.13 Every CW complex is a WANE.

This result will allow us to derive many properties of $E^n(X, A)$ when the spaces $\{E_n\}$ are CW complexes for n sufficiently large. We will prove 21.13 by showing that every simplicial complex is a WANE. 21.13 then follows from 16.44 and 21.12.

Recall (Exercise 26, Section 16) the definition of an abstract simplicial complex.

Proposition 21.14 Every abstract simplicial complex $K = (V, S)$ determines a simplicial CW complex.

Proof Let V be partially ordered in such a way that each $\sigma \in S$ is linearly ordered. Given $f: V \to I$ write $\sup f = \{v \in V \mid f(v) \neq 0\}$. We define the realization of K by

$$|K| = \{f: V \to I \mid \sup f \in S \text{ and } \sum_{v \in V} f(v) = 1\}.$$

The sum is finite since every set in S is finite. For each $\sigma \in S$ write $|\sigma| = \{f \in |K| \mid \sup f \subset \sigma\}$. Suppose $\sigma = (v_0, \ldots, v_n)$. Then σ is called an n-simplex. There is a 1–1 correspondence $\chi_\sigma: \Delta^n \to |\sigma|$ given by $\chi_\sigma(x_1, \ldots, x_{n+1})(v_i) = x_i$. We topologize $|\sigma|$ by making χ_σ a homeomorphism. This does not depend on the ordering of the vertices. $|K| = \bigcup_{\sigma \in S} |\sigma|$. Topologize $|K|$ with the quotient topology:

$$\chi: \coprod_{\sigma \in S} |\sigma| \to |K|.$$

The inclusion $|K| \subset I^V$ is continuous, where I^V has the product topology, although $|K|$ does not have the induced topology in general. In any case $|K|$ is Hausdorff. As n-cells we take $\chi_\sigma(\Delta^n - \partial\Delta^n)$ for all n-simplexes σ. For each σ, χ_σ is a homeomorphism, hence $|K|$ is a cell complex. $|K|$ is obviously closure finite and by 14.5 it has the weak topology. By construction it is a simplicial CW complex. ∎

Corollary 21.15 Every simplicial CW complex is homeomorphic to $|K|$ for some abstract simplicial complex K.

Proof This follows immediately from Exercise 26, Section 16. ∎

Theorem 21.16 Let v_0 be a vertex of K. Then $(|K|, |v_0|)$ is a WANE.

Proof Let X be a paracompact, $A \subset X$ closed, and $f: A \to |K|$. Given $v \in V$ define the star of v by

$$\text{st } v = \{f \in |K| \mid f(v) \neq 0\}.$$

Clearly, $\{\text{st } v \mid v \in V\}$ is an open cover of $|K|$. $\{f^{-1}(\text{st } v)\}$ is thus an open cover of A. Choose, for each v, an open subset $\mathfrak{U}_v \subset X$ such that $\mathfrak{U}_v \cap A = f^{-1}(\text{st } v)$. $\{\mathfrak{U}_v, X - A\}$ is an open cover of X. Choose a subordinate partition of unity

$$p: X \to I, \qquad p_v: X \to I,$$

with $\{x \mid p(x) = 0\} \subset X - A$, $\{x \mid p_v(x) \neq 0\} \subset \mathfrak{U}_v$. Let $\mathfrak{U}' = \{x \mid p(x) \neq 1\}$. Define functions

$$q_v: \mathfrak{U}' \to I$$

by $q_v(x) = p_v(x)/(1 - p(x))$. Then $\{q_v\}$ is a partition of unity of $\mathcal{U}'$ subordinate to $\mathcal{U}_v \cap \mathcal{U}'$. Define a simplicial complex K' with vertex set V and simplex set

$$S' = \{\sigma \subset V \mid \sigma \in S \quad \text{or} \quad q_v(x) \neq 0 \quad \text{for some} \quad x \in \mathcal{U} \quad \text{and all} \quad v \in \sigma\}.$$

Then $K' = (V, S')$ is a simplicial complex. $|K| \subset |K'|$ as a subcomplex. We define $g'\colon \mathcal{U}' \to |K'|$ by $g'(x)(v) = q_v(x)$. This is well defined for $\{v \mid g'(x)(v) \neq 0\} \in S'$. To see that g' is continuous at $x \in U'$ choose a neighborhood $\mathcal{U}_x$ of x such that only finitely many $q_v \neq 0$ on $\mathcal{U}_x$. Then $g'(\overline{\mathcal{U}_x})$ is compact. Thus the topology on $g'(\overline{\mathcal{U}_x})$ is the induced topology as a subset of I^V. Since g' is clearly continuous with this topology, $g'|_{\overline{\mathcal{U}_x}}$ is continuous. Thus g' is continuous.

Now $g'(A) \subset |K|$, for if $g(a)(v) \neq 0$, $a \in \mathcal{U}_v$ and thus $f(a)(v) \neq 0$. Define $g\colon A \to |K|$ by $g(a) = g'(a)$. We show that $g \sim f\,(\mathrm{rel}\, f^{-1}(|v|)$. Define H by

$$H(x, t)(v) = tf(x) + (1 - t)g(x).$$

$H(x, t) \in |K|$ for if $H(x, t)(v) \neq 0, f(x)(v) \neq 0$. A proof of continuity for H is similar to that of g'. H is a homotopy rel $f^{-1}(|v|)$, for if $f(x) = |v|$, $p_v(x) = 1$ and hence $g_v(x) = 1$, so $g(x) = |v|$. It is only necessary to show that g extends to a map $\bar{g}\colon \mathcal{U} \to |K|$. By Exercise 20, Section 14 there is a neighborhood V of $|K|$ in $|K'|$ and a retraction $r\colon V \to |K|$. Let $\mathcal{U} = \mathcal{U}' \cap g'^{-1}(V)$. Define $\bar{g} = rg'$. ▮ ▮

Proposition 21.17 Suppose X is paracompact and $A \subset X$ is closed. If $(Y, *)$ is a WANE and

$$f\colon X \times 0 \cup A \times I \to Y$$

$f \sim g\,(\mathrm{rel}\, f^{-1}(*))$ and g extends to a map $G\colon X \times I \to Y$.

Proof By 21.10 and 21.11 we can find g, a homotopy $f \sim g\,(\mathrm{rel}\, f^{-1}(*))$ and an extension $\tilde{g}$ of g over a neighborhood $\mathcal{U}$ of $X \times 0 \cup A \times I$ in $X \times I$. Choose a neighborhood V of A in X such that $V \times I \subset \mathcal{U}$. Let $\alpha\colon X \to I$ satisfy $\alpha(X - V) = 0$ and $\alpha(A) = 1$. See Fig. 21.1. This is possible since every paracompact space is normal. Then define $G\colon X \times I \to Y$ by

$$G(x, t) = \begin{cases} \tilde{g}(x, t), & \alpha(x) \geq t \\ \tilde{g}(x, \alpha(x)), & \alpha(x) \leq t. \end{cases} \quad ▮$$

Proposition 21.18 Let $(Y, *)$ be a WANE, X paracompact, and C a closed subset of X that is contractible in $\mathcal{C}^*$, Then

$$(p_C)^*\colon [(X/C, \{C\}), (Y, *)] \to [(X, *), (Y, *)]$$

is a 1–1 correspondence.

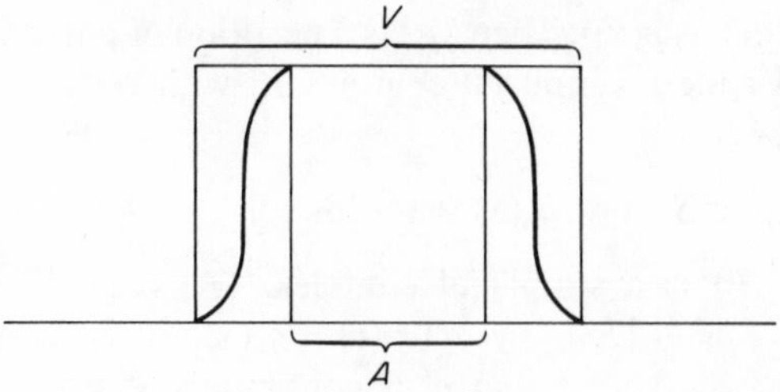

Figure 21.1

Proof Let $K\colon C \times I \to C$ be a contraction of C in $\mathcal{C}^*$. $K(x, 0) = x$, $K(x, 1) = *$, and $K(*, t) = *$.

To prove that $(p_C)^*$ is onto, let $f\colon (X, *) \to (Y, *)$. Define $F\colon X \times 0 \cup C \times I \to Y$ by $F(x, 0) = f(x)$ and $F(c, t) = f(K(c, t))$. By 21.17 there exists $G\colon X \times I \to Y$ and a homotopy

$$F \sim G|_{X \times 0 \cup C \times I} \text{ (rel } * \times I \cup C \times 1)$$

Let $g(x) = G(x, 0)$; then $f \sim g$ (rel $*$). If $g'(x) = G(x, 1)$, $g'(C) = *$, so $\{g'\} \in \operatorname{Im}(p_C)^*$. However $G\colon g \sim g'$ (rel $*$), so $(p_C)^*$ is onto.

Suppose now that $H\colon fp_C \sim gp_C$ (rel $*$). Define

$$\bar{H}\colon X \times I \times 0 \cup C \times I \times I \to Y$$

by

$$\bar{H}(x, s, 0) = H(x, s), \qquad \bar{H}(c, s, t) = H(K(c, t), s).$$

Then $\bar{H}(C \times 0 \times I \cup C \times 1 \times I \cup * \times I \times I \cup C \times I \times 1) = *$. Thus there exists

$$G\colon X \times I \times I \to Y$$

and a homotopy

$$L\colon \bar{H} \sim G|_{X \times I \times 0 \cup C \times I \times I}$$
$$\text{(rel } C \times 0 \times I \cup C \times 1 \times I \cup * \times I \times I \cup C \times I \times 1).$$

$L(x, 0, 0, 0) = fp_C(x)$, and

$$L|_{X \times 0 \times 0 \times I}\colon fp_C \sim L|_{X \times 0 \times 0 \times 1} \text{ (rel } C).$$

Similarly

$$gp_C \sim L|_{X \times 1 \times 0 \times 1} \text{ (rel } C).$$

Let $f', g'\colon X/C \to Y$ be given by $f'(x) = L(x, 0, 0, 1)$ and $g'(x) = L(x, 1, 0, 1)$. Then $\{f\} = \{f'\}$ and $\{g\} = \{g'\}$. Now define $M\colon X \times I \to Y$ by

$$M(x, t) = \begin{cases} G(x, 0, 3t), & 0 \le t \le \frac{1}{3} \\ G(x, 3t - 1, 1), & \frac{1}{3} \le t \le \frac{2}{3} \\ G(x, 1, 3 - 3t), & \frac{2}{3} \le t \le 1. \end{cases}$$

$M(C \times I) \subset G(C \times 0 \times I \cup C \times I \times 1 \cup C \times 1 \times I) = *$. Hence M defines a homotopy

$$\overline{M}\colon (X/C) \times I \to Y \ (\text{rel}\ \{C\})$$

and $\overline{M}\colon f' \sim g'$ (rel $\{C\}$). Thus $\{f\} = \{f'\} = \{g'\} = \{g\}$. ∎

Let $\mathscr{P}$ be the category of paracompact Hausdorff spaces.

Corollary 21.19 Let $X \in \mathscr{P} \cap \mathcal{CG}$. Suppose $* \in C \subset X$ and C is contractible in $\mathcal{CG}^*$ and closed. Then

$$(p_C)^*\colon \tilde{E}^m(X/C) \to \tilde{E}^m(X)$$

is an isomorphism.

Proof $S^{n-m}C$ is a closed contractible subset of $S^{n-m}X$, and $S^{n-m}X/S^{n-m}C \equiv S^{n-m}(X/C)$. Hence

$$[S^{n-m}(X/C), E_n] \xrightarrow{(p_C)^*} [S^{n-m}X, E_n]$$

is an isomorphism by 21.18. ∎

Theorem 21.20 The functors $E^n\colon \mathscr{P} \cap \mathcal{CG}^2 \to \mathcal{M}_Z$ satisfy the following properties:

1. *Relative Homeomorphism* Let $f\colon (X, A) \to (Y, B)$ and assume that f is a closed map, A and B are closed subsets, and $f|_{X-A}\colon X - A \to Y - B$ is a 1–1 correspondence. Then $f^*\colon E^m(Y, B) \to E^m(X, A)$ is an isomorphism.

2. *Neighborhood Extension Property* Let $\xi \in E^m(X, A)$ and suppose A is closed. Then there is a neighborhood U_ξ of A in X and a class $\xi' \in E^m(X, \overline{U}_\xi)$ such that $i^*(\xi') = \xi$ where $i\colon (X, A) \to (X, \overline{U}_\xi)$ is the inclusion.

3. *Excision* Suppose U is an open subset of X and $\overline{U} \subset A$. Then the inclusion $i\colon (X - U, A - U) \to (X, A)$ induces isomorphisms. In particular, $\{E^n\}$ is a cohomology theory. (This is type 2 excision.)

4. *Continuity* Let $\{X_\alpha\}$ be an inverse system of compact Hausdorff spaces. Then there is a natural isomorphism

$$\varinjlim E^m(X_\alpha) \xrightarrow{\cong} E^m(\varprojlim X_\alpha).$$

5. If C is closed and contractible, $(p_C)^*\colon E^m(X/C) \to E^m(X)$ is an isomorphism.

Proof 1. By 21.11 and 21.19, $E^m(X, A) = \tilde{E}^m(X \cup CA) \cong \tilde{E}^m(X/A)$. The hypothesis implies that f induces a homeomorphism

$$\tilde{f}\colon X/A \to Y/B.$$

2. Let $\xi \in E^m(X, A)$ be represented by a map

$$f\colon (X \cup CA) \wedge S^{n-m} \to E_n$$

Define $\alpha\colon X \times 0 \cup A \times I \cup X \times 1 \to \Omega^{n-m}E_n$ by

$$\alpha(x, 0) = f^*(x), \qquad \alpha(a, t) = f^*(a, t), \qquad \alpha(x, 1) = *,$$

where $f^*\colon X \cup CA \to \Omega^{n-m}E_n$ is the adjoint of f. This is continuous in the compact open topology. Hence there is a neighborhood $\mathcal{U}$ of $X \times 0 \cup A \times I \cup X \times 1$ in $X \times I$, a map $g \sim \alpha$ (rel $X \times 1$), and an extension $\tilde{g}$ of g over $\mathcal{U}$. There is a neighborhood $\bar{U}_\xi$ of A in X such that $\bar{U}_\xi \times I \subset \mathcal{U}$. Hence $\tilde{g}\,|\,\bar{U}_\xi \times I \cup X \times 0$ defines a map $g'\colon X \cup C\bar{U}_\xi \to \Omega^{n-m}E_n$ such that $g'\,|_{X \cup CA} \sim f^*$ (rel $*$). Since $X \cup C\bar{U}_\xi$ is compactly generated, g' is continuous in the compactly generated topology on $\Omega^{n-m}E_n$. Hence the adjoint of g' represents a cohomology class in $E^m(X, \bar{U}_\xi)$ which restricts to ξ.

3. Suppose $B \subset C \subset D$ are inclusions of spaces. Then the map $C/B \to D/B$ induced by the inclusion of C into D is 1–1 and continuous. We show that C/B has the induced topology. Let $p_1\colon C \to C/B$ and $p_2\colon D \to D/B$ be the projections, and suppose $\mathcal{U}$ is open in C/B. Then $p_1^{-1}(\mathcal{U})$ is open in C and we can thus find an open set V in D with $V \cap C = p_1^{-1}(\mathcal{U})$. Now $p_2^{-1}p_2(V) = V$ so $p_2(V)$ is open in D/B. Since $p_2(V) \cap C/B = \mathcal{U}$, we have accomplished this task.

It follows that $A/\bar{U} \subset X/\bar{U}$ and $A - U/\bar{U} - U \subset X - U/\bar{U} - U$. All of these spaces belong to $\mathscr{P} \cap \mathcal{CG}$ by 21.11, so we consider the diagram

$$\begin{array}{ccc}
E^m(X, A) & \longleftarrow & E^m(X/\bar{U}, A/\bar{U}) \\
\downarrow & & \downarrow \\
E^m(X - U, A - U) & \longleftarrow & E^m(X - U/\bar{U} - U, A - U/\bar{U} - U)
\end{array}$$

Now we claim that $(X - U/\bar{U} - U,\ A - U/\bar{U} - U) \equiv (X/\bar{U},\ A/\bar{U})$. There is clearly a 1–1 continuous map from $(X - U/\bar{U} - U,\ A - U/\bar{U} - U)$ onto $(X/\bar{U}, A/\bar{U})$ induced by the inclusion of $X - U$ into X. To see that this map is closed, observe that the composite $X - U \to X \to X/\bar{U}$ is closed. Thus the right-hand vertical map is an isomorphism. It is thus sufficient to show that the horizontal maps are isomorphisms. The lower one is a special case of the upper one by replacing $\bar{U} \subset A \subset X$ by $\bar{U} - U \subset A - U \subset X - U$. Consider the diagram

$$\begin{array}{ccccccccccc}
\cdots \to & E^{m-1}(X, \bar{U}) & \to & E^{m-1}(A, \bar{U}) & \to & E^m(X, A) & \to & E^m(X, \bar{U}) & \to & E^m(A, \bar{U}) & \to \cdots \\
& \uparrow & & \uparrow & & \uparrow & & \uparrow & & \uparrow & \\
\cdots \to & E^{m-1}(X/\bar{U}, *) & \to & E^{m-1}(A/\bar{U}, *) & \to & E^m(X/\bar{U}, A/\bar{U}) & \to & E^m(X/\bar{U}, *) & \to & E^m(A/\bar{U}, *) & \to \cdots
\end{array}$$

The horizontal sequences are exact by the techniques of 19.5, and the result follows from the 5-lemma and part 1.

As in 19.5 the other properties of a cohomology theory are easily proven.

4. By Exercise 9, Section 15

$$\varprojlim X_\alpha = \{\{x_\alpha\} \in \prod X_\alpha \mid f_{\alpha\alpha'}(x_{\alpha'}) = x_\alpha \quad \text{for} \quad \alpha' \geq \alpha\}.$$

There is a continuous map $\rho_\alpha\colon \varprojlim X_\alpha \to X_\alpha$ given by $\rho_\sigma(\{x_\alpha\}) = x_\alpha$. Since $f_{\alpha\alpha'} \rho_{\alpha'} = \rho_\alpha$, $\rho_\alpha{}^* = \rho_{\alpha'}^* f_{\alpha\alpha'}^*$. Hence $\{\rho_\alpha{}^*\colon E^n(X_\alpha) \to E^n(\varprojlim X_\alpha)\}$ defines a homomorphism

$$\rho\colon \varinjlim E^n(X_\alpha) \to E^n(\varprojlim X_\alpha).$$

Define $B_\beta = \{\{x_\alpha\} \in \prod_{\alpha \leq \beta} X_\alpha \mid f_{\alpha\alpha'}(x_{\alpha'}) = x_\alpha \text{ for } \alpha' \geq \alpha\}$. Clearly $B_\beta \equiv X_\beta$. Let $C_\beta = B_\beta \times \prod_{\alpha > \beta} CX_\alpha$. Then $g_\beta\colon C_\beta \to X_\beta$ given by projection is a homotopy equivalence. Furthermore if $\beta > \beta'$, $C_\beta \subset C_{\beta'}$, and $\bigcap C_\beta = \varprojlim X_\alpha$.

Now it is sufficient to show that

$$\theta\colon \varinjlim E^n(C_\beta) \to E^n(\bigcap C_\beta)$$

is an isomorphism where θ is defined similarly to ρ, for we have a commutative diagram

$$\begin{array}{ccc}
\varinjlim E^n(C_\beta) & \xrightarrow{\theta} & E^n(\bigcap C_\beta) \\
{\scriptstyle \{g_\beta{}^*\}} \uparrow \cong & & \uparrow \cong \\
\varinjlim E^n(X_\alpha) & \xrightarrow{\rho} & E^n(\varprojlim X_\alpha)
\end{array}$$

Let $C = \bigcap C_\beta$ and note that C is compact and Hausdorff. Let $\xi \in E^n(C)$. By part 2 there is a neighborhood $\bar{U}_\xi$ of C in C_β and a class $\xi' \in E^n(C_\beta, \bar{U}_\xi)$ with $i^*(\xi') = \delta\xi$. By contemplating the diagram

$$\begin{array}{ccccc}
E^{n+1}(C_\beta, C) & \xleftarrow{\delta} & E^n(C) & \longleftarrow & E^n(C_\beta) \\
\uparrow i^* & & \uparrow i^* & & \uparrow \cong \\
E^{n+1}(C_\beta, \bar{U}) & \xleftarrow{\delta} & E^n(\bar{U}) & \longleftarrow & E^n(C_\beta)
\end{array}$$

one sees that there exists $\xi'' \in E^n(\overline{U})$ with $i^*(\xi'') = \xi$. The open sets U and $X - C_\beta$ for all β with $C_\beta \not\supset C_\alpha$ an open cover of C. Since C is compact, there is a finite subcover and hence for some β, $C_\beta \subset U$. Thus $\xi = \theta_\beta{}^*(\xi'')$ for some $\xi'' \in E^n(C_\beta)$ and hence θ is onto.

Suppose $\theta(\xi) = 0$, and ξ is represented by $e \in E^n(C_\alpha)$. Then $i^*(e) = 0$ where $i\colon C \to C_\alpha$ is the inclusion. Choose $d \in E^n(C_\alpha, C)$ such that $j^*(d) = e$. By part 2 we can choose $\overline{U}_d \supset C$ and a class $d' \in E^n(C_\alpha, \overline{U}_d)$ such that $i^*(d') = d$. Again by compactness, one sees that there exists β with $C \subset C_\beta \subset U_d$. Thus we can find $d'' \subset E^n(C_\alpha, C_\beta)$ with $i^*(d'') = d$. Hence $a^*(e) = 0$ where $a\colon C_\beta \to C_\alpha$ is the inclusion. Thus $\xi = 0$.

5. The map $p_C\colon (X, C) \to (X/C, \{C\})$ is a relative homeomorphism. Since $E^n(C) \cong E^n(\{C\})$, the result follows by the 5-lemma and part 1. ∎

The properties expressed in 21.20 are characteristic for what is often called a continuous cohomology theory. Two continuous versions of ordinary cohomology are common in the literature. They are the Čech cohomology groups—written $\check{H}^n(X, A; \pi)$—and the Alexander cohomology groups—written $\overline{H}^n(X, A; \pi)$ (see [64]). These agree with each other and with ordinary spectral cohomology on paracompact compactly generated spaces. (All three are initial objects in the category of ordinary cohomology theories with coefficients in π.) [32]

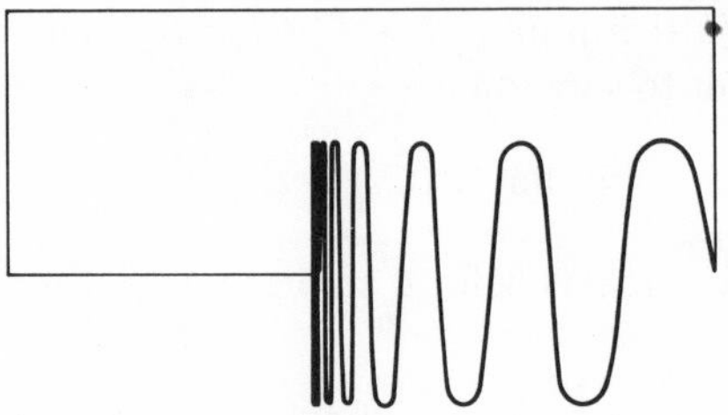

Figure 21.2

Example 21.21 Let X be the union of the closure of the graph of $y = \sin \pi/x$ for $0 < x \leq 1$ and the sets $[-1, 0] \times 0$, $[-1, 1] \times 2$, $1 \times [0, 2]$, and $-1 \times [0, 2]$; see Fig. 21.2. There is a continuous map from X to the rectangle determined by the lines $y = 0$, $y = 2$, $x = \pm 1$. This is defined by pinching the closure of the curve $\sin \pi/x$ down to the line $y = 0$. Let $Y = 0 \times [-1, 1]$. Then $Y \subset X$ and the above rectangle is homeomorphic to X/Y, since X is compact. Since Y is contractible, $E^n(X) \cong E^n(X/Y) \cong E^n(S^1)$; in particular $H^1(X) \cong Z$. $SH^1(X) = 0$ since every homotopy group of X is 0.

Corollary 21.22 If $X \in \mathscr{P}$, $\tilde{E}^n(X) \cong E^n(X, *)$.

Proof By 21.20 $\tilde{E}^n(X \cup C(*)) \cong \tilde{E}^n(X)$. ∎

Exercises

1. If $\{E_n\}$ is an Ω-spectrum, show that $SE^n(\coprod X_\alpha) \cong \prod SE^n(X_\alpha)$.

2. Prove that if $X \in \mathscr{P}$ and A, B are closed subsets, there is a Mayer–Vietoris sequence

$$\cdots \leftarrow E^n(X \cap B) \leftarrow E^n(A) \oplus E^n(B) \leftarrow E^n(A \cup B) \leftarrow E^{n-1}(A \cap B) \leftarrow \cdots$$

3. Let $X = S^1 \times [0, \infty) \subset R^3$ and A_α be a plane in R^3 through the line $(1, y, 0)$ and making an angle α with the x, y plane. Choose an increasing sequence of numbers $\alpha_i \geq 0$ with $\lim \alpha_i = \pi/2$. Define X_i as the points of X above the plane A_{α_i}. Show that $\varprojlim X_i = \bigcap X_i$ is the line $(1, 0, z)$ and hence $H^1(\varprojlim X_i) = 0$. Show that $H^1(X_i) \cong Z$ and $\varinjlim H^1(X_i) \cong Z$.

4. Let $A \subset B \subset X$ be closed subsets and suppose $\{(U_\alpha, V_\alpha)\}$ is the inverse system of all neighborhoods of (B, A) directed by inclusion. Show that the natural map

$$\varinjlim E^n(U_\alpha, V_\alpha) \to E^n(B, A)$$

is an isomorphism. (Hint: Consider first the case $B = X$.)

5. Suppose (U, V) is excisive in Z and $W \subset U \cap V$. Using the commutative diagram containing three exact sequences

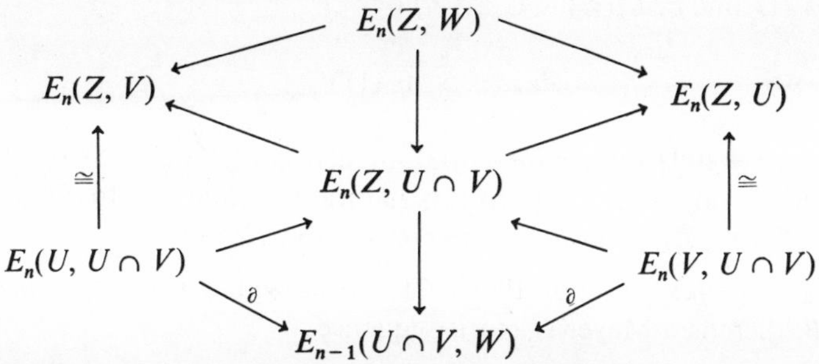

show that the sum of the two exterior homomorphisms:

$$E_n(Z, W) \to E_{n-1}(U \cap V, W)$$

is 0. (Exercise 13, Section 23)

6. Let $F \xrightarrow{i} E \xrightarrow{\pi} B$ be a Serre fibering. Assume that F is $(n-1)$-connected and B is $(m-1)$-connected with $m \geq 2$. Use Exercises 7 and 23 of Section 16 to construct exact sequences

$$S\tilde{H}_{m+n-1}(F) \to S\tilde{H}_{m+n-1}(E) \to \cdots$$

$$\to S\tilde{H}_i(F) \xrightarrow{i_*} S\tilde{H}_i(E) \xrightarrow{\pi_*} S\tilde{H}_i(B) \xrightarrow{\tau} S\tilde{H}_{i-1}(F) \to \cdots$$

and

$$\cdots \to S\tilde{H}^{i-1}(F) \xrightarrow{\tau} S\tilde{H}^i(B) \xrightarrow{\pi^*} S\tilde{H}^i(E) \xrightarrow{i^*} S\tilde{H}^i(F) \to \cdots$$
$$\to S\tilde{H}^{m+n-1}(E) \to S\tilde{H}^{m+n-1}(F)$$

with arbitrary coefficients. These are called the Serre exact sequences and τ is called the transgression. (30.7)

7. Generalize Exercise 8, Section 16 to the case that X and Y are arbitrary well-pointed spaces.

8. Show that there are isomorphisms, natural in all the variables:

(a) if X is well pointed:

$$\tilde{E}_i(X;\ \oplus A_\alpha) \cong \oplus \tilde{E}_i(X; A_\alpha);$$

(b) if the indexing set is finite:

$$\tilde{E}^i(X;\ \oplus A_\alpha) \cong \oplus \tilde{E}^i(X; A_\alpha).$$

(Hint: It is sufficient to prove (a) for CW complexes X. Apply Exercise 10, Section 18. For (b), apply Exercise 13, Section 18.)

9. Define the singular chain complex of a topological space as follows. Let $C_n(X)$ be a free abelian group with one generator e_f for each continuous map $f\colon \Delta^n \to X$. Define $\partial\colon C_n(X) \to C_{n-1}(X)$ by

$$\partial e_f = \sum_{s=0}^{n} (-1)^s e_{f i_s}.$$

Show that $\partial^2 = 0$. Using the functorial singular complex (16.35) and Exercise 4, Section 20 show that $H_*(C_*(X))$ is naturally isomorphic to $SH_*(X)$, and $H^*(\mathrm{Hom}(C_*(X), Z))$ is naturally isomorphic to $SH^*(X)$.

10. Use Exercise 1, Section 19 and 21.5 to show that if A and B are excisive in $A \cup B$, there is a Mayer–Vietoris sequence

$$\cdots \to SE_i(X, A \cap B) \to SE_i(X, A) \oplus SE_i(X, B) \to SE_i(X, A \cup B) \to \cdots.$$

(26.7; 26.8; 26.13)

11. Show that the hypothesis that $* \in X$ is nondegenerate may be dropped from Exercise 28, Section 16 if Y is a WANE and X is paracompact. In particular if E is an Ω-spectrum and E_n is connected and the homotopy type of a CW complex for each n, $\tilde{E}^n(X) \simeq [X, E_n]$. (29.13)

22

The Relation between Homotopy and Ordinary Homology

In this section we shall make some general observations about ordinary homology and cohomology. In particular they vanish in negative dimensions when applied to reasonable spaces, and are related to the components in dimension 0. In higher dimensions we prove the Hurewicz theorem which illuminates the close relation between homology and homotopy.

Proposition 22.1 $\tilde{H}^m(X; \pi) = 0$ for $m < 0$. If $(X, *) \in \mathcal{N}$, $\tilde{H}_m(X; \pi) = 0$ for $m < 0$.

Proof[19] By 18.13, $\tilde{H}^{-m}(X; \pi) \cong [(X, *), (\Omega^m K(\pi, 0), *)] = 0$ for $m > 0$ since in this case $\Omega^m K(\pi, 0) \simeq *$. Now for any CW complex L, $\tilde{H}_m(L; \pi) = 0$ for $m < 0$. Thus $S\tilde{H}_m(X; \pi) = \tilde{H}_m(S(X); \pi) = 0$ for $m < 0$ by 21.7. ▌

Corollary 22.2 For any pair $(X, A) \in \mathcal{CG}^2$, $H_m(X, A; \pi) = 0$ and[19] $H^m(X, A; \pi) = 0$ for $m < 0$.

Proposition 22.3 $H_0(X)$ is a free abelian group whose rank is the number of arc components of X. $H^0(X) \cong \prod_\alpha Z_\alpha$ where $Z_\alpha \cong Z$ and there is one copy for each component.

Proof[19] Let $K \xrightarrow{f} X$ be a resolution of X. Then X and K have the same number of arc components, and if $\{K_\alpha\}$ are the arc components of K, $K \equiv$

[19] The proofs we offer for the cohomology statements depend on the fact that $\Omega K(\pi, n) \simeq K(\pi, n-1)$ (see the remark after 18.13). Without reference to this fact the proofs are valid only when the spaces under consideration are of the homotopy type of a CW complex.

$\coprod K_\alpha$. By Exercise 3, Section 19 it is sufficient to show that $H_0(K_\alpha) \cong Z$. To do this we show that $\tilde{H}_0(K_\alpha) = 0$. In fact $K_\alpha \wedge K(Z, n)$ is n-connected so this is immediate. If $\{X_\alpha\}$ are the components of X, $X = \coprod_\alpha X$. By Exercise 3, Section 19, $H^0(X) \cong \prod H^0(X_\alpha)$. Thus it is sufficient to show that $H^0(X_\alpha) \cong Z$. But

$$H^0(X_\alpha) \cong [(X_\alpha{}^+, +), (Z, 0)] = [X_\alpha, Z] \cong Z$$

since $Z \simeq K(Z, 0)$. ∎

If $m > 0$, define $h\colon \pi_m(X) \to \tilde{H}_m(X) \cong H_m(X)$ as follows. Note that $S^1 \simeq K(Z, 1)$ and let h be the composition

$$\pi_m(X) \xrightarrow{E} \pi_{m+1}(X \wedge S^1) \cong \pi_{m+1}(X \wedge K(Z, 1)) \to \tilde{H}_m(X);$$

h is a natural homomorphism, and is called the Hurewicz homomorphism.

Theorem 22.4 (*Hurewicz Theorem*) If X is simply connected and well pointed, the following are equivalent:

(a) $\pi_i(X) = 0$ for $i < n$;
(b) $\tilde{H}_i(X) = 0$ for $i < n$.

Furthermore they imply that $h\colon \pi_r(X) \to \tilde{H}_r(X)$ is an $(n + 1)$-isomorphism.

Proof Since X is well pointed and h is natural, we may assume that X is a CW complex by 21.7. Suppose $\pi_i(X) = 0$ for $i < n$. Then $E\colon \pi_r(X) \to \pi_{r+1}(SX)$ is an $(n + 1)$-isomorphism since $n > 1$. (E is a $(2n - 1)$-isomorphism by 16.34.) We consider the composition γ_m

$$\pi_{r+m}(X \wedge K(Z, m)) \xrightarrow{E} \pi_{r+m+1}(X \wedge K(Z, m) \wedge S^1) \xrightarrow{(1 \wedge h_m)_*} \pi_{r+m+1}(X \wedge K(Z, m + 1)).$$

Since $X \wedge K(Z, m)$ is $(m + n - 1)$-connected, E is an isomorphism if $r < m + 2n - 1$ and is onto if $r = m + 2n - 1$. We now appeal to

Lemma 22.5 Let $f\colon X \to \Omega Y$ and suppose $f^*\colon SX \to Y$ is adjoint to f. Then the diagram

$$\begin{array}{ccc} \pi_{r+1}(SX) & \xrightarrow{(f^*)_*} & \pi_{r+1}(Y) \\ \uparrow E & & \uparrow \cong \\ \pi_r(X) & \xrightarrow{f_*} & \pi_r(\Omega Y) \end{array}$$

commutes. ∎

$E\colon \pi_r(K(Z, m)) \to \pi_{r+1}(SK(Z, m))$ is a $(2m - 1)$-isomorphism by 16.34. Hence by 22.5, $(h_m)_*\colon \pi_r(SK(Z, m)) \to \pi_r(K(Z, m + 1))$ is a $(2m + 1)$-isomorphism. By Exercise 7, Section 16 we may assume that $(K(Z, m + 1), SK(Z, m))$ is a relative CW complex with cells in dimensions greater than $2m + 1$.

Hence $(X \wedge K(Z, m+1), X \wedge K(Z, m) \wedge S^1)$ is a relative CW complex with cells in dimensions greater than $n + 2m + 1$. It follows that $(1 \wedge h_m)_*$ is an isomorphism if $r + m + 1 < n + 2m + 1$ and is onto if $r + m + 1 = n + 2m + 1$. Thus γ_m is an isomorphism if $r < m + n$ and is onto if $r = m + n$; i.e., h is an $(n + 1)$-isomorphism.

We have proven that under condition (a), h is an $(n + 1)$-isomorphism. It follows immediately that (a) is equivalent to (b). ∎

Corollary 22.6 If $X \in \mathcal{CG}$ and is simply connected, the conclusion of 22.4 remains valid with $\tilde{H}_i(X)$ replaced by $S\tilde{H}_i(X)$. ∎

If $A \neq \varnothing$, define $h\colon \pi_i(X, A) \to H_i(X, A)$ to be the composition

$$\pi_i(X, A) \xrightarrow{(p_A)_*} \pi_i(X/A, *) \to \tilde{H}_i(X/A) = H_i(X, A).$$

Proposition 22.7 (*Relative Hurewicz Theorem*) Suppose A is simply connected, and $\pi_1(X, A) = 0$. Then the following are equivalent:

(a) $\pi_i(X, A) = 0$ for $i < n$;
(b) $H_i(X, A) = 0$ for $i < n$.

Either implies that $h\colon \pi_i(X, A) \to H_i(X, A)$ is an isomorphism for $i \leq n$ and onto if $i = n + 1$.

Proof By 21.7 we may assume that (X, A) is a CW pair. As in the case of 22.4, we show that (a) implies the final conclusion. But by 16.30, $\pi_i(X, A) \xrightarrow{(p_A)_*} \pi_i(X/A)$ is an $(n + 1)$-isomorphism. ∎

Corollary 22.8 (*Whitehead Theorem*) Let $f\colon X \to Y$ and suppose both X and Y are simply connected CW complexes and $f_*\colon H_m(X) \to H_m(Y)$ is an isomorphism. Then f is a homotopy equivalence.

Proof Let Z be the mapping cylinder of f. Then $X \subset Z$ and $H_m(X) \to H_m(Z)$ is an isomorphism. Consequently $H_m(Z, X) = 0$ for all m. X is simply connected and $\pi_1(Z, X) = 0$ since we have an exact sequence in $\mathcal{S}^*$

$$\pi_1(Z) \to \pi_1(Z, X) \to \pi_0(X)$$

in which both ends are 0. By 22.7, $\pi_m(Z, X) = 0$ so $i_*\colon \pi_m(X) \to \pi_m(Z)$ induces isomorphisms in homotopy. Consequently i is a homotopy equivalence. It follows that $f = \pi i$ is a homotopy equivalence

$$\begin{array}{ccc} X & \xrightarrow{\ i\ } & Z \\ & \searrow^{f} & \downarrow{\scriptstyle \pi}\,\cong \\ & & Y \end{array}$$ ∎

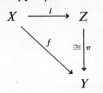

Proposition 22.9 There is a diagram

$$\begin{array}{ccccccccccc}
\pi_{n+1}(X, A) & \xrightarrow{\partial} & \pi_n(A) & \xrightarrow{i_*} & \pi_n(X) & \xrightarrow{j_*} & \pi_n(X, A) & \xrightarrow{\partial} & \pi_{n-1}(A) & \longrightarrow \\
\downarrow h & \pm 1 & \downarrow h & & \downarrow h & & \downarrow h & \pm 1 & \downarrow h & \\
H_{n+1}(X, A) & \xrightarrow{\partial} & H_n(A) & \xrightarrow{i_*} & H_n(X) & \xrightarrow{j_*} & H_n(X, A) & \xrightarrow{\partial} & H_{n-1}(A) & \longrightarrow
\end{array}$$

where all squares commute except the one involving ∂, which commutes up to sign.

Proof The square involving i_* commutes by naturality, and the one involving j_* commutes because of the commutative diagram

$$\begin{array}{ccc}
\pi_n(X) & \xrightarrow{j_*} & \pi_n(X, A) \\
\downarrow h & \searrow{(p_A)_*} & \downarrow (p_A)_* \\
 & & \pi_n(X/A, *) \\
 & & \downarrow h \\
\tilde{H}_n(X) & \xrightarrow{(p_A)_*} & \tilde{H}_n(X/A) \\
\| & & \| \\
H_n(X) & \xrightarrow{j_*} & H_n(X, A)
\end{array}$$

To prove the other square commutes up to sign, we use the following:

Lemma 22.10 There are generators $\alpha \in H_n(S^n)$ and $\beta \in H_n(B^n, S^{n-1})$ such that if $f\colon S^n \to X$, $h(\{f\}) = f_*(\alpha)$ and if $g\colon (B^n, S^{n-1}) \to (X, A)$, $h(\{g\}) = g_*(\beta)$.

Proof Let $\alpha = h(1_{S^n})$ and $\beta = h(1_{(B^n, S^{n-1})})$ where

$$1_{S^n}\colon S^n \to S^n \qquad \text{and} \qquad 1_{(B^n, S^{n-1})}\colon (B^n, S^{n-1}) \to (B^n, S^{n-1})$$

are the identity maps. The result follows from naturality by considering the diagrams

$$\begin{array}{ccc}
\pi_n(S^n) & \xrightarrow{f_*} & \pi_n(X) \\
\downarrow h & & \downarrow h \\
H_n(S^n) & \xrightarrow{f_*} & H_n(X),
\end{array}
\qquad
\begin{array}{ccc}
\pi_n(B^n, S^{n-1}) & \xrightarrow{g_*} & \pi_n(X, A) \\
\downarrow h & & \downarrow h \\
H_n(B^n, S^{n-1}) & \xrightarrow{g_*} & H_n(X, A)
\end{array}$$

α and β are generators by 22.4 and 22.7. ∎

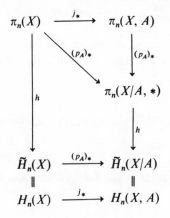

To finish the proof of 22.9, observe that $\partial\beta = \pm\alpha$. Hence $\partial h(\{g\}) = \partial g_*(\beta) = g_*(\partial\beta) = \pm f_*(\alpha) = \pm h(\{f\}) = \pm h(\partial\{g\})$ where $f = \partial g = g|_{S^{n-1}}: S^{n-1} \to A$. ▮

Exercises

1. A generator $\alpha \in \pi_n(K(Z, n))$ provides a map α_n: $S^n \to K(Z, n)$. Show that if the generators are chosen properly, this defines a map of spectra $\underline{S} \to H$ and hence a homomorphism H: $\pi_n^S(X) \to \tilde{H}_n(X)$ (see Exercise 4, Section 18). (Exercise 2)

2. Prove that there is a commutative diagram (up to sign)

$$\begin{array}{ccc} \pi_n(X) & \xrightarrow{h} & \pi_n^S(X) \\ & {\scriptstyle h}\searrow \quad \swarrow{\scriptstyle H} & \\ & \tilde{H}_n(X) & \end{array}$$

(see Exercise 1).

3. Show that the diagram

$$\begin{array}{ccc} \pi_n(X) & \xrightarrow{E} & \pi_{n+1}(SX) \\ \downarrow{\scriptstyle h} & & \downarrow{\scriptstyle h} \\ \tilde{H}_n(X) & \underset{\cong}{\xrightarrow{\sigma}} & \tilde{H}_{n+1}(SX) \end{array}$$

commutes up to sign.

4. Show that given $* \in A \subset B \subset X$ there is a commutative ladder of exact sequences

$$\begin{array}{ccccccccccc} \cdots \longrightarrow & \pi_n(X, A) & \longrightarrow & \pi_n(X, B) & \xrightarrow{\partial} & \pi_{n-1}(B, A) & \longrightarrow & \pi_{n-1}(X, A) & \longrightarrow \cdots \\ & \downarrow{\scriptstyle h} & & \downarrow{\scriptstyle h} & \pm 1 & \downarrow{\scriptstyle h} & & \downarrow{\scriptstyle h} & \\ \cdots \longrightarrow & H_n(X, A) & \longrightarrow & H_n(X, B) & \longrightarrow & H_{n-1}(B, A) & \longrightarrow & H_{n-1}(X, A) & \longrightarrow \cdots \end{array}$$

in which all squares commute except the one involving ∂, and that commutes up to sign.

5.* Show that if X is a simply connected CW complex and

$$\tilde{H}_n(X) = \begin{cases} Z, & n = k \\ 0, & n \neq k, \end{cases}$$

then $X \simeq S^k$.

6. Use Exercise 3, Section 20 to show that the kernel of $h\colon \pi_1(X, *) \to H_1(X)$ is the commutator subgroup for any space X.

7.* Show that if X and Y are simply connected and $f_*\colon H_i(X) \to H_i(Y)$ is a k-isomorphism, f is a k-equivalence.

8. Let $f\colon S^n \to S^n$ and define the degree of f to be n if $f_*(x) = nx$ for $x \in H_n(S^n)$. Prove the Brouwer degree theorem: $f \sim g$ iff $\deg f = \deg g$.

9. Consider the construction 16.14 of a resolution (K, f) of a space Y. Suppose Y is simply connected and $0 \to B_m \xrightarrow{\alpha_m} Z_m \xrightarrow{\beta_n} H_m(Y) \to 0$ is an arbitrary resolution of $H_m(Y)$ as an abelian group in which $Z_1 = B_0 = 0$ (i.e., B_m and Z_m are free and the sequence is exact). Show that the construction (K, f) can be done so that there is a commutative diagram of exact sequences

$$\begin{array}{ccccccccc}
0 & \longrightarrow & Z_m & \longrightarrow & H_m(K^m, K^{m-1}) & \searrow & & & \\
 & & \downarrow{\scriptstyle \beta_m} & & \downarrow{\scriptstyle f_*} & & H_{m-1}(K^{m-1}) & \longrightarrow & H_m(K^m) \\
0 & \longrightarrow & H_m(Z) & \longrightarrow & H_m(Z, K^{m-1}) & \nearrow & & &
\end{array}$$

and hence $C_m \cong Z_m \oplus B_{m-1}$, and $\partial_m(x, y) = (\alpha_{m-1}(y), 0)$ giving $\ker \partial_m \cong Z_m$ and $\operatorname{Im} \partial_{m+1} \cong B_m$.

10. Show that if $M(G, n)$ and $Y(G, n)$ are two Moore spaces for the group G and integer n, they have the same homotopy type.

11. Prove a generalized Hurewicz theorem. Consider $h_r\colon \pi_r{}^S(X; G) \to \tilde{H}_r(X; G)$ (see Exercise 12, Section 18). Show that if X is $(r-1)$-connected, h_r is an isomorphism and h_{r+1} is an epimorphism. Let $f\colon X \to Y$, and conclude that $f_*\colon \tilde{H}_r(X; G) \to \tilde{H}_r(Y; G)$ is a k-isomorphism iff $f_*\colon \pi_r{}^S(X; G) \to \pi_r{}^S(Y; G)$ is a k-isomorphism. (30.13)

12. Suppose E is a properly convergent spectrum and X is an $(n-1)$-connected well-pointed space. Show that $\pi_{n+1}(X \wedge E_1) \to \tilde{E}_r(X)$ is an isomorphism for $r \leq n$ and is onto if $r = n + 1$. (See Exercise 14, Section 18.) (Exercise 20, Section 23; 27.5)

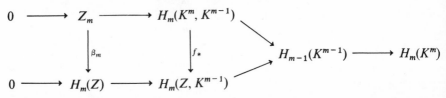

23

Multiplicative Structure

In this section we shall show how pairings of spectra (maps $\mu_{m,n}: E_m \wedge F_n \to G_{m+n}$ satisfying compatibility conditions), lead to pairings of the various homology and cohomology theories defined by the spectra. In particular, we establish that under certain conditions on the spectra, the cohomology forms a "graded ring," and both the homology and cohomology are modules over the "coefficient ring," $\tilde{E}^*(S^0)$. These conditions are satisfied for many spectra, including S and HR where R is a ring.

Proposition 23.1 Let c: $[SY, Z] \times [SX, Y] \to [S^2X, Z]$ be given by $c(\alpha, \beta) = \alpha \circ E\beta$. c is bilinear (and hence defines a homomorphism

$$c: [SY, Z] \otimes [SX, Y] \to [S^2X, Z]$$

if the groups involved are abelian).

Proof $c(g, f)(x, s, t) = g(f(x, s), t)$. Hence

$$\begin{aligned}
c(g_1 + g_2, f)(x, s, t) &= \begin{cases} g_1(f(x, s, 2t), & 0 \le t \le \frac{1}{2} \\ g_2(f(x, s), 2t - 1), & \frac{1}{2} \le t \le 1 \end{cases} \\
&= \begin{cases} c(g_1, f)(x, s, 2t), & 0 \le t \le \frac{1}{2} \\ c(g_2, f)(x, s, 2t - 1), & \frac{1}{2} \le t \le 1 \end{cases} \\
&= (c(g_1, f) + c(g_2, f))(x, s, t), \\
c(g, f_1 + f_2)(x, s, t) &= \begin{cases} g(f_1(x, 2s), t), & 0 \le s \le \frac{1}{2} \\ g(f_2(x, 2s - 1), t), & \frac{1}{2} \le s \le 1 \end{cases} \\
&= \begin{cases} c(g, f_1)(x, 2s, t), & 0 \le s \le \frac{1}{2} \\ c(g, f_2)(x, 2s - 1, t), & \frac{1}{2} \le s \le 1 \end{cases} \\
&= (c(g, f_1) + c(g, f_2))(x, s, t),
\end{aligned}$$

since addition may be defined using any suspension coordinate by 9.14. ∎

Proposition 23.2 $c(E\alpha \otimes E\beta) = Ec(\alpha \otimes \beta)$.

Proof $c(E\alpha \otimes E\beta) = E\alpha \circ E^2\beta = E(\alpha \circ E\beta) = Ec(\alpha \otimes \beta)$. ∎

We shall often deal with spaces of the form $S^nX \wedge S^mY$, and will choose a fixed standard homeomorphism of this with $S^{n+m}(X \wedge Y)$. This is defined by considering $S^n \equiv I^n/\partial I^n$. Consequently there is a homeomorphism $\varphi_{n,m}: S^n \wedge S^m \to S^{n+m}$ which is defined by

$$\varphi_{n,m}((s_1, \ldots, s_n), (t_1, \ldots, t_m)) = (s_1, \ldots, s_n, t_1, \ldots, t_m).$$

This is associative in the sense that the diagram

$$\begin{array}{ccc} S^n \wedge S^m \wedge S^k & \xrightarrow{\varphi_{n,m} \wedge 1} & S^{n+m} \wedge S^k \\ \Big\downarrow {\scriptstyle 1 \wedge \varphi_{m,k}} & & \Big\downarrow {\scriptstyle \varphi_{n+m,k}} \\ S^n \wedge S^{m+k} & \xrightarrow{\varphi_{n,m+k}} & S^{n+m+k} \end{array}$$

commutes. We define

$$\bar{\varphi}_{n,m}: (S^nX) \wedge (S^mY) \to S^{n+m}(X \wedge Y)$$

by

$$\bar{\varphi}_{n,m}(x, u, y, v) = (x, y, \varphi_{n,m}(u, v)).$$

Now define

$$\Sigma_Z^n: [S^mX, S^kY] \to [S^{n+m}(Z \wedge X), S^{n+k}(Z \wedge Y)]$$

and

$$E_Z^n: [S^mX, S^kY] \to [S^{n+m}(X \wedge Z), S^{n+k}(Y \wedge Z)]$$

by smashing a map on the left (right) by the identity map of S^nZ and resorting using the maps $\bar{\varphi}_{n,m}$. Explicitly, Σ_Z^n is the composition

$$[S^mX, S^kY] \to [S^nZ \wedge S^mX, S^nZ \wedge S^kY] \xrightarrow{(\bar{\varphi}_{n,m}{}^{-1})^* \circ (\bar{\varphi}_{n,k})_*} [S^{n+m}(Z \wedge Y), S^{n+k}(Z \wedge Y)]$$

and E_Z^n is the composition

$$[S^mX, S^kY] \to [S^mX \wedge S^nZ, S^kY \wedge S^nZ] \xrightarrow{(\bar{\varphi}_{m,n}{}^{-1})^* \circ (\bar{\varphi}_{k,n})_*} [S^{n+m}(X \wedge Z), S^{n+k}(Y \wedge Z)]$$

Proposition 23.3 ${\Sigma_Z}^n$ and ${E_Z}^n$ are homomorphisms. They satisfy the identities:

(a) $\Sigma_Z^n \circ E_{Z'}^{n'} = E_{Z'}^{n'} \circ {\Sigma_Z}^n$.
(b) $E_Z^n \circ E_{Z'}^{n'} = E_{Z \wedge Z'}^{n+n'}$.
(c) $\Sigma_Z^n \circ \Sigma_{Z'}^{n'} = \Sigma_{Z' \wedge Z}^{n+n'}$.
(d) $E_{S^0}^1 = E$.
(e) If $\Sigma = \Sigma_{S^0}^1$, $\Sigma = (-1)^{m-k}E$.

Proof $\Sigma_Z{}^0(f)(z, x, s_1, \ldots, s_m) = (z, f(x, s_1, \ldots, s_m))$, hence Σ_Z^0 is a homomorphism. Similarly E_Z^0 is a homomorphism. By (a)–(e), $\Sigma_Z^n = \Sigma^n \circ \Sigma_Z^0 = (-1)^{n(m-k)}E^n\Sigma_Z^0$ and $E_Z^n = E^n \circ E_Z^0$; it follows that Σ_Z^n and E_Z^n are homomorphisms, once we establish (a)–(e). (a), (b), and (c) follow immediately from the associativity of $\bar{\varphi}_{n,m}$ and (d) is the definition of E. To prove (e) we need a lemma.

If σ is a permutation of $(1, \ldots, n)$, σ induces homeomorphisms $T_\sigma: (I^n, \partial I^n) \to (I^n, \partial I^n)$ by $T_\sigma(x_1, \ldots, x_n) = (x_{\sigma(1)}, \ldots, x_{\sigma(n)})$ and $\bar{T}_\sigma: S^n \equiv I^n/\partial I^n \to I^n/\partial I^n \equiv S^n$.

Lemma 23.4 $(\bar{T}_\sigma)_*(x) = \operatorname{sgn} \sigma \cdot x$ for $x \in \pi_n(S^n)$.

Proof The transformation $\sigma \to (\bar{T}_\sigma)_*$ is a homomorphism from the symmetric group on n letters to $\operatorname{Aut}(Z) = \{\pm 1\}$. There are two such homomorphisms, sgn and 1, since every permutation is a product of transpositions. Now $\bar{T}^{(n)}_{(1,2)} = E^{n-2}\bar{T}^{(2)}_{(1,2)}$, so it is sufficient to show that $(T^{(2)}_{(1,2)})_* = -1$. We define $\psi: I^2 \to I^2$ by letting $\psi(u, v)$ be the point whose distance from the

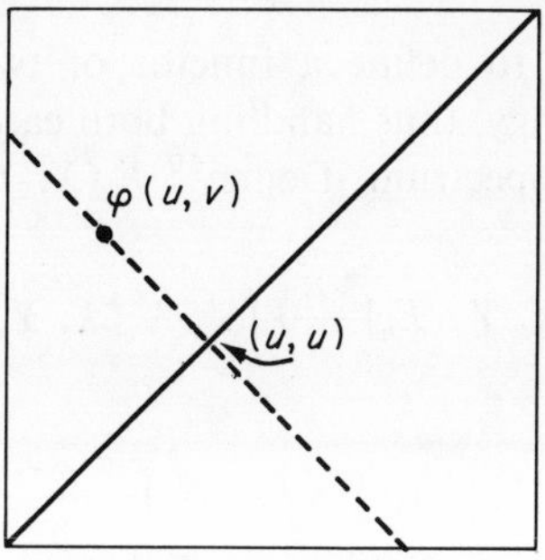

Figure 23.1

diagonal varies from the extremes linearly, as v varies from 0 to 1 and whose projection onto the diagonal is (u, u); see Fig. 23.1. Explicitly,

$$\psi(u, v) = \begin{cases} (2uv, 2u(1-v)), & 0 \le u \le \tfrac{1}{2} \\ (1 - 2(1-u)(1-v), 1 - 2(1-u)v), & \tfrac{1}{2} \le u \le 1. \end{cases}$$

ψ is 1–1, onto, and continuous and $\psi(\partial I^2) \subset \partial I^2$. Now $\psi^{-1}T_{(1,2)}\psi(u, v) = (u, 1-v)$, so $\{\psi^{-1}T_{(1,2)}\psi\} = \{-1\} \in \pi_2(I^2, \partial I^2)$. Hence $(T_{(1,2)})_* = -1$ and this implies $(\bar{T}_{(1,2)})_* = -1$. ∎

To finish the proof of 23.3, we observe that

$$\begin{aligned} \Sigma(f) &= (1_Y \wedge T_{(1,\ldots,k+1)}) \circ (f \wedge \iota_1) \circ (1_X \wedge T_{(1,\ldots,m+1)}) \\ &\sim (1_{S^kY} \wedge (-1)^k \iota_1) \circ (f \wedge \iota_1) \circ (1_{S^mX} \wedge (-1)^m \iota_1) \\ &\sim (f \wedge \iota_1) \circ (1_{S^mX} \wedge (-1)^{m-k} \iota_1) \\ &= (-1)^{m-k} E(f). \quad ∎ \end{aligned}$$

We now define a general composition pairing

$$C_{m,k}\colon [S^m(V\wedge W), X]\otimes[S^nY, S^k(W\wedge Z)]\to[S^{n+m}(V\wedge Y), S^k(X\wedge Z)]$$

by $C_{m,k}=c\circ(E_Z^k\otimes\Sigma_V^{m-1})$.

Proposition 23.5 $E\circ C_{m,k}=C_{m,k+1}\circ(1\otimes E)=(-1)^{n-k}C_{m+1,k}\circ(E\otimes 1)$.

Proof

$$\begin{aligned}E\circ C_{m,k}(\alpha\otimes\beta)&=E\circ c(E_Z^k\alpha\otimes\Sigma_V^{m-1}\beta)=c(E\otimes E)(E_Z^k\alpha\otimes\Sigma_V^{m-1}\beta)\\&=c(E_Z^{k+1}\alpha\otimes E\Sigma_V^{m-1}\beta)=c(E_Z^{k+1}\alpha\otimes\Sigma_V^{m-1}E\beta)\\&=C_{m,k+1}(\alpha\otimes E\beta)=C_{m,k+1}\circ(1\otimes E)(\alpha\otimes\beta)\\E\circ C_{m,k}(\alpha\otimes\beta)&=E\circ c(E_Z^k\alpha\otimes\Sigma_V^{m-1}\beta)=c(E\otimes E)(E_Z^k\alpha\otimes\Sigma_V^{m-1}\beta)\\&=c(E_Z^{k+1}\alpha\otimes(-1)^{n-k}\Sigma_V^m\beta)\\&=(-1)^{n-k}c(E_Z^k\otimes\Sigma_V^m)(E\alpha\otimes\beta)\\&=(-1)^{n-k}C_{m+1,k}\circ(E\otimes 1)(\alpha\otimes\beta).\quad\blacksquare\end{aligned}$$

It will be convenient to define a functor of two variables generalizing homology and cohomology, thus handling both cases at once.

Let $E=\{E_n, e_n\}$ be a spectrum. Define[20] $\tilde{E}_k(X, Y)$ as the direct limit of the system

$$\cdots\to[S^{k+n}X, Y\wedge E_n]\xrightarrow{\lambda_n}[S^{k+n+1}X, Y\wedge E_{n+1}]\to\cdots$$

where λ_n is the composite

$$[S^{k+n}X, Y\wedge E_n]\xrightarrow{E}[S^{k+n+1}X, Y\wedge E_n\wedge S^1]\xrightarrow{(1\wedge e_n)_*}[S^{k+n+1}X, Y\wedge E_{n+1}]$$

Then $\tilde{E}_k(S^0, Y)=\tilde{E}_k(Y)$ and $\tilde{E}_k(X, S^0)=\tilde{E}^{-k}(X)$.

Definition 23.6 Given spectra $E=\{E_n, e_n\}$, $F=\{F_n, f_n\}$ and $G=\{G_n, g_n\}$, a pairing from E and F to G is a collection of maps

$$\mu_{m,n}\colon E_m\wedge F_n\to G_{n+m}$$

such that the diagrams

$$\begin{array}{ccc}E_m\wedge F_n\wedge S^1 & \xrightarrow{\mu_{m,n}\wedge 1} & G_{m+n}\wedge S^1\\ \downarrow{\scriptstyle 1\wedge f_n} & & \downarrow{\scriptstyle g_{m+n}}\\ E_m\wedge F_{n+1} & \xrightarrow{\mu_{m,n+1}} & G_{m+n+1}\end{array}\qquad\begin{array}{ccc}E_m\wedge F_n\wedge S^1 & \xrightarrow{\mu_{m,n}\wedge 1} & G_{m+n}\wedge S^1\\ \downarrow{\scriptstyle \equiv} & & \\ E_m\wedge S^1\wedge F_n & (-1)^n & \downarrow{\scriptstyle g_{m+n}}\\ \downarrow{\scriptstyle e_m\wedge 1} & & \\ E_{m+1}\wedge F_n & \xrightarrow{\mu_{m+1,n}} & F_{m+n+1}\end{array}$$

[20] One is tempted to call the value of this bifunctor the biology groups of the pair of spaces.

commute up to homotopy (with the indicated sign). A spectrum E is called a ring spectrum if there is a pairing from E and E to E and a mapping $u\colon \underset{\sim}{S} \to E$ such that the diagrams

$$\begin{array}{ccc} E_m \wedge E_n \wedge E_k & \xrightarrow{1 \wedge \mu_{n,k}} & E_m \wedge E_{n+k} \\ \downarrow{\scriptstyle \mu_{m,n} \wedge 1} & & \downarrow{\scriptstyle \mu_{m,n+k}} \\ E_{m+n} \wedge E_k & \xrightarrow{\mu_{m+n,k}} & E_{m+n+k} \end{array}$$

$$\begin{array}{ccc} E_m \wedge E_n & \searrow{\scriptstyle \mu_{m,n}} & \\ \downarrow{\scriptstyle T} \quad (-1)^{mn} & & E_{m+n} \\ E_n \wedge E_m & \nearrow{\scriptstyle \mu_{n,m}} & \end{array} \qquad \begin{array}{ccc} E_m \wedge S^n & \xrightarrow{1 \wedge u_n} & E_m \wedge E_n \\ & {\scriptstyle e_m^{\,n}} \searrow \quad \swarrow {\scriptstyle \mu_{m,n}} & \\ & E_{m+n} & \end{array}$$

homotopy commute, with the indicated sign (after suspension), where $e_m{}^n = e_{m+n-1} \cdots S^{n-1} e_m$. If E is a ring spectrum, a spectrum F is called a module spectrum over E if there is a pairing from F and E to F such that the diagrams

$$\begin{array}{ccc} F_m \wedge E_n \wedge E_k & \xrightarrow{1 \wedge \mu_{n,k}} & F_m \wedge E_{n+k} \\ \downarrow{\scriptstyle \bar{\mu}_{m,n} \wedge 1} & & \downarrow{\scriptstyle \bar{\mu}_{m,n+k}} \\ F_{m+n} \wedge E_k & \xrightarrow{\bar{\mu}_{m+n,k}} & F_{m+n+k} \end{array} \qquad \begin{array}{ccc} F_m \wedge S^n & \xrightarrow{1 \wedge u_n} & F_m \wedge E_n \\ & {\scriptstyle f_m^{\,n}} \searrow \quad \swarrow {\scriptstyle \bar{\mu}_{m,n}} & \\ & F_{m+n} & \end{array}$$

homotopy commute.

Proposition 23.7 $\underset{\sim}{S}$ is a ring spectrum and every spectrum E is a module over $\underset{\sim}{S}$.

Proof Let $\mu_{m,n} = \varphi_{m,n}\colon S^m \wedge S^n \to S^{m+n}$, and $u_n = 1$; since $e_m = \varphi_{m,1}$, $e_m{}^n = \varphi_{m,n}$. All diagrams not involving signs commute pointwise. Those involving signs follow from 23.4. For any spectrum E, define $\bar{\mu}_{m,n} = e_m{}^n\colon E_m \wedge S^n \to E_{m+n}$. Then the diagrams required to homotopy commute do so pointwise. ∎

We now establish a pairing from $H\pi$ and $H\rho$ to $H\pi \otimes \rho$. We consider

$$C\colon \pi_m(X) \otimes \pi_n(Z) \to \pi_{n+m}(X \wedge Z)$$

given by setting $V = W = Y = S^0$ and $k = 0$. One easily checks that $C(\{f\} \otimes \{g\}) = \{f \wedge g\}$.

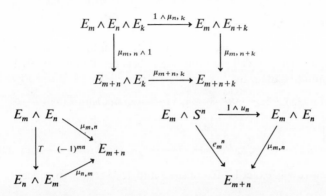

Proposition 23.8 If X is an $(n-1)$-connected well-pointed compactly generated space and $m, n > 1$,

$$C\colon \pi_m(M(\pi, m)) \otimes \pi_n(X) \to \pi_{n+m}(M(\pi, m) \wedge X)$$

is an isomorphism.

Proof Since X is well pointed and C is natural, we may assume X is a CW complex with no cells in dimensions between 0 and n by 21.8. Consider the commutative diagram

$$\begin{array}{ccccccc}
\pi_m(M(R, m)) \otimes \pi_n(X) & \xrightarrow{f_*\otimes 1} & \pi_m(M(F, m)) \otimes \pi_n(X) & \longrightarrow & \pi_m(M(\pi, m)) \otimes \pi_n(X) & \longrightarrow & 0 \\
\downarrow{\scriptstyle C_1} & & \downarrow{\scriptstyle C_2} & & \downarrow{\scriptstyle C} & & \\
\pi_{n+m}(M(R, m) \wedge X) & \xrightarrow{(f\wedge 1)_*} & \pi_{n+m}(M(F, m) \wedge X) & \longrightarrow & \pi_{n+m}(M(\pi, m) \wedge X) & \longrightarrow & \pi_{n+m-1}(M(R, m) \wedge X)
\end{array}$$

the bottom row is exact by Exercises 3 and 15, Section 16, since

$$M(\pi, m) \wedge X \simeq M(F, m) \wedge X \cup_{f\wedge 1} C^*\{M(R, m) \wedge X\}.$$

By 16.34 and Exercise 8, Section 16, C_1 and C_2 are isomorphisms. Furthermore $\pi_{n+m-1}(M(R, m) \wedge X) = 0$. We wish to conclude that C is an isomorphism by the 5-lemma. Now

$$0 \to \pi_m(M(R, m)) \to \pi_m(M(F, m)) \to \pi_m(M(\pi, m)) \to 0$$

is exact so the result follows from:

Lemma 23.9 If $A \xrightarrow{\alpha} B \xrightarrow{\beta} C \to 0$ is exact in $\mathcal{M}_R$ so is

$$A \otimes_R D \xrightarrow{\alpha\otimes 1} B \otimes_R D \xrightarrow{\beta\otimes 1} C \otimes_R D \to 0.$$

This fact is often stated by saying that the functor $\otimes_R D$ is right exact. It is a standard result of homological algebra. A proof will be found in the Appendix. ▌

Corollary 23.10 If $m, n > 1$,

$$C\colon \pi_m(K(\pi, m)) \otimes \pi_n(K(\rho, n)) \to \pi_{n+m}(K(\pi, m) \wedge K(\rho, n))$$

is an isomorphism.

Proof Consider the commutative diagram in which all maps are isomorphisms

$$\begin{array}{ccc}
\pi_m(M(\pi, m)) \otimes \pi_n((K(\rho, n)) & \longrightarrow & \pi_m(K(\pi, m)) \otimes \pi_n(K(\rho, n)) \\
\downarrow{\scriptstyle C} & & \downarrow{\scriptstyle C} \\
\pi_{n+m}(M(\pi, m) \wedge K(\rho, n)) & \longrightarrow & \pi_{n+m}(K(\pi, m) \wedge K(\rho, n))
\end{array}$$ ▌

We will now choose fixed isomorphisms

$$\alpha_n\colon \pi \xrightarrow{\cong} \pi_n(K(\pi, n))$$

inductively for $n \geq 1$ in such a way that the diagrams

$$\begin{array}{ccc} \pi_n(K(\pi, n)) & \xrightarrow{E} & \pi_{n+1}(K(\pi, n) \wedge S^1) \\ \alpha_n \uparrow \cong & & \downarrow (f_n)_* \\ \pi & \xrightarrow[\alpha_{n+1}]{\cong} & \pi_{n+1}(K(\pi, n+1)) \end{array}$$

commute.

Corollary 23.11 There are maps $\mu_{m,n}\colon K(\pi, m) \wedge K(\rho, n) \to K(\pi \otimes \rho, m + n)$ such that the composite

$$\pi \otimes \rho \xrightarrow{\alpha_m \otimes \alpha_n} \pi_m(K(\pi, m)) \otimes \pi_n(K(\rho, n)) \xrightarrow{C} \pi_{n+m}(K(\pi, m) \wedge K(\rho, n)) \xrightarrow{(\mu_{m,n})_*}$$
$$\pi_{n+m}(K(\pi \otimes \rho, m + n)) \xrightarrow{\alpha_{m+n}{}^{-1}} \pi \otimes \rho$$

is the identity (i.e., $\alpha_{m+n}(a \otimes b) = (\mu_{m,n})_*(C(\alpha_m(a) \otimes a_n(b)))$.

Proof $K(\pi, m) \wedge K(\rho, n)$ is $(n + m - 1)$-connected by construction and hence $\{K(\pi, m) \wedge K(\rho, n)\}^{[m+n]} = K(\pi \otimes \rho, m + n)$ by 23.10. Following $i_{n+m}\colon K(\pi, m) \wedge K(\rho, n) \to K(\pi \otimes \rho, m + n)$ by $f_\varphi\colon K(\pi \otimes \rho, m + n) \to K(\pi \otimes \rho, m + n)$ where φ is an appropriate automorphism of $\pi \otimes \rho$, we construct $\mu_{m,n}$. ∎

Lemma 23.12 If X is an $(n - 1)$-connected CW complex, a class $\{f\} \in [X, K(\pi, n)]$ is completely determined by $f_*\colon \pi_n(X) \to \pi$.

Proof We assume without loss of generality that X has no cells in dimension $< n$ except a 0-cell. Given $f_1, f_2\colon X \to K(\pi, n)$ with $(f_1)_* = (f_2)_*$, we certainly have $f_1|_{X^n} \sim f_2|_{X^n}$ since $X^n = \bigvee S_\alpha{}^n$. 16.3 now implies that $f_1 \sim f_2$ when applied to the diagram

$$\begin{array}{ccc} X \times I & & \\ \uparrow & \searrow & \\ X \times 0 \cup X^n \times I \cup X \times 1 & \longrightarrow & K(\pi, n) \end{array}$$ ∎

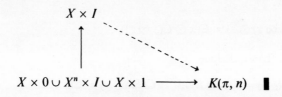

We use this lemma to prove:

Proposition 23.13 The following diagrams homotopy commute

$$\begin{array}{ccccc}
K(\pi, m) \wedge K(\rho, n) \wedge S^1 & \xrightarrow{\mu_{m,n} \wedge 1} & K(\pi \otimes \rho, m+n) \wedge S^1 & \xleftarrow{\mu_{m,n} \wedge 1} & K(\pi, m) \wedge K(\rho, n) \wedge S^1 \\
 & & & (-1)^n & \downarrow{\scriptstyle 1 \wedge T} \\
\downarrow{\scriptstyle 1 \wedge f_n} & & \downarrow{\scriptstyle f_{m+n}} & & K(\pi, m) \wedge S^1 \wedge K(\rho, n) \\
 & & & & \downarrow{\scriptstyle f_m \wedge 1} \\
K(\pi, m) \wedge K(\rho, n+1) & \xrightarrow{\mu_{m,n+1}} & K(\pi \otimes \rho, m+n+1) & \xleftarrow{\mu_{m+1,n}} & K(\pi, m+1) \wedge K(\rho, n)
\end{array}$$

$$\begin{array}{ccc}
K(\pi, m) \wedge K(\rho, n) \wedge K(\sigma, p) & \xrightarrow{\mu_{m,n} \wedge 1} & K(\pi \otimes \rho, m+n) \wedge K(\sigma, p) \\
\downarrow{\scriptstyle 1 \wedge \mu_{n,p}} & & \downarrow{\scriptstyle \mu_{m+n,p}} \\
K(\pi, m) \wedge K(\rho \otimes \sigma, n+p) & \xrightarrow{\mu_{m,n+p}} & K(\pi \otimes \rho \otimes \sigma, m+n+p)
\end{array}$$

$$\begin{array}{ccc}
K(\pi, m) \wedge K(\rho, n) & \xrightarrow{\mu_{m,n}} & K(\pi \otimes \rho, m+n) \\
\downarrow{\scriptstyle T} & (-1)^{mn} & \downarrow{\scriptstyle f_T} \\
K(\rho, n) \wedge K(\pi, m) & \xrightarrow{\mu_{n,m}} & K(\rho \otimes \pi, m+n)
\end{array}
\qquad
\begin{array}{ccc}
K(\pi, m) \wedge K(\rho, n) & \xrightarrow{\mu_{m,n}} & K(\pi \otimes \rho, m+n) \\
\downarrow{\scriptstyle f_\varphi \wedge f_{\varphi'}} & & \downarrow{\scriptstyle f_{\varphi \otimes \varphi'}} \\
K(\pi', m) \wedge K(\rho', n) & \xrightarrow{\mu_{m,n}} & K(\pi' \otimes \rho', m+n)
\end{array}$$

for homomorphisms $\varphi\colon \pi \to \pi'$ and $\varphi'\colon \rho \to \rho'$.

$$\begin{array}{ccc}
K(\pi, m) \wedge S^n & \xrightarrow{1 \wedge e_n} & K(\pi, m) \wedge K(Z, n) \\
 & \searrow{\scriptstyle f_m^n} & \downarrow{\scriptstyle \mu_{m,n}} \\
 & & K(\pi, m+n)
\end{array}$$

where $e_n\colon S^n \to K(Z, n)$ is such that $\{e_n\} = \alpha_n(1)$.

Proof In each case we will apply 23.12. To evaluate the various homomorphisms we will use some formulas, the proof of which is easy.

(a) If $\iota_1 \in \pi_1(S^1)$ is the class of the identity map, $C(\iota \otimes x) = \Sigma x$ and $C(x \otimes \iota) = Ex$.

(b) $C(C(x \otimes y) \otimes z) = C(x \otimes C(y \otimes z))$.

From (a) and (b) we deduce

(c) $EC(x \otimes y) = C(x \otimes Ey), \qquad C(Ex \otimes y) = C(x \otimes \Sigma y)$.

(d) Let $T\colon X \wedge Y \to Y \wedge X$ be the switching homeomorphism. Then

$$T_* C(x \otimes y) = (-1)^{mn} C(y \otimes x)$$

for $x \in \pi_m(X)$ and $y \in \pi_n(Y)$.

From (a) and (d) we conclude that

(e) $T_* Ex = (-1)^n \Sigma x$ for $x \in \pi_n(X)$.
(f) $(f \wedge g)_* C(x \otimes y) = C(f_*(x) \otimes g_*(y))$.
(g) If $\varphi\colon \pi \to \rho$, the maps $f_\varphi\colon K(\pi, n) \to K(\rho, n)$ can be chosen so that

$$(f_\varphi)_*(\alpha_n(x)) = \alpha_n(\varphi(x)).$$

The commutativity of the various diagrams then follows quite easily. We carry out the details in the top diagram only.

Elements of the form $x = EC(\alpha_m(a) \otimes \alpha_n(b))$ generate $\pi_{m+n+1}(K(\pi, m) \wedge K(\rho, n) \wedge S^1)$. We evaluate $(f_{m+n})_*(\mu_{m,n} \wedge 1)_*$, $(\mu_{m,n+1})_*(1 \wedge f_n)_*$, and $(\mu_{m+1,n})_*(f_m \wedge 1)_*(1 \wedge T)_*$ on these elements.

$$\begin{aligned}(f_{m+n})_*(\mu_{m,n} \wedge 1)_*(EC(\alpha_m(a) \otimes \alpha_n(b))) &= (f_{m+n})_*(E((\mu_{m,n})_* C(\alpha_m(a) \otimes \alpha_n(b))))\\ &= (f_{m+n})_*(E\alpha_{m+n}(a \otimes b)) = \alpha_{m+n+1}(a \otimes b)\end{aligned}$$

by 23.11, and the choice of α_{m+n+1}.

$$\begin{aligned}(\mu_{m,n+1})_*(1 \wedge f_n)_*(EC(\alpha_m(a) \otimes \alpha_n(b))) &= (\mu_{m,n+1})_*(1 \wedge f_n)_*(C(\alpha_m(a) \otimes E\alpha_n(b)))\\ &= (\mu_{m+n+1})_*(C(\alpha_m(a) \otimes (f_n)_* E(\alpha_n(b))))\\ &= (\mu_{m+n+1})_*(C(\alpha_m(a) \otimes \alpha_{n+1}(b)))\\ &= \alpha_{m+n+1}(a \otimes b)\end{aligned}$$

by (c), the choice of α_{n+1}, and 23.11.

$$\begin{aligned}&(\mu_{m+1,n})_*(f_m \wedge 1)_*(1 \wedge T)_*(EC(\alpha_m(a) \otimes \alpha_n(b)))\\ &\quad= (\mu_{m+n,n})_*(f_m \wedge 1)_*(C(\alpha_m(a) \otimes T_* E(\alpha_n(b))))\\ &\quad= (-1)^n(\mu_{m+1,n})_*(f_m \wedge 1)_*(C(\alpha_m(a) \otimes \Sigma\alpha_n(b)))\\ &\quad= (-1)^n(\mu_{m+1,n})_*(f_m \wedge 1)_* C(E\alpha_m(a) \otimes \alpha_n(b))\\ &\quad= (-1)^n(\mu_{m+1,n})_* C((f_m)_* E\alpha_m(a) \otimes \alpha_n(b))\\ &\quad= (-1)^n(\mu_{m+1,n})_* C(\alpha_{m+1}(a) \otimes \alpha_n(b)) = (-1)^n\alpha_{m+n+1}(a \otimes b),\end{aligned}$$

by (a), (f), (e), (c), the choice of α_{m+1}, and 23.11. ∎

Corollary 23.14 If R is a commutative ring, HR is a ring spectrum. If M is a right R-module, HM is an HR-module.

Proof Let $c\colon R \otimes R \to R$ be the multiplication. Define

$$\mu_{m,n}\colon K(R, m) \wedge K(R, n) \to K(R, m+n)$$

to be the composite

$$K(R, m) \wedge K(R, n) \xrightarrow{\mu_{m,n}} K(R \otimes R, m+n) \xrightarrow{f_c} K(R, m+n).$$

Then $\mu_{m,n}$ is a pairing. Let $e\colon Z \to R$ be defined by $e(1) = 1$. Define $u_n\colon S^n \to K(R, n)$ as the composite

$$S^n \xrightarrow{e_n} K(Z, n) \xrightarrow{f_e} K(R, n).$$

All diagrams of 23.6 follow immediately from 23.13. For example

$$\begin{array}{ccccc}
K(R,m)\wedge K(R,n)\wedge K(R,k) & \xrightarrow{\mu_{m,n}\wedge 1} & K(R\otimes R, m+n)\wedge K(R,k) & \xrightarrow{f_c\wedge 1} & K(R, m+n)\wedge K(R,k) \\
\downarrow{\scriptstyle 1\wedge\mu_{n,k}} & & \downarrow{\scriptstyle \mu_{m+n,k}} & & \downarrow{\scriptstyle \mu_{m+n,k}} \\
K(R,m)\wedge K(R\otimes R, m+k) & \xrightarrow{\mu_{m,n+k}} & K(R\otimes R\otimes R, m+n+k) & \xrightarrow{f_c\otimes 1} & K(R\otimes R, m+n+k) \\
\downarrow{\scriptstyle 1\wedge f_c} & & \downarrow{\scriptstyle f_{1\otimes c}} & & \downarrow{\scriptstyle f_c} \\
K(R,m)\wedge K(R, n+k) & \xrightarrow{\mu_{m,n+k}} & K(R\otimes R, m+n+k) & \xrightarrow{f_c} & K(R, m+n+k)
\end{array}$$

homotopy commutes. The case of a right R-module is similar, using the homomorphism $M \otimes R \to M$. ▌

Theorem 23.15 A pairing from E and F to G induces a homomorphism

$$M\colon \tilde{E}_s(V \wedge W, X) \otimes \tilde{F}_t(Y, W \wedge Z) \to \tilde{G}_{s+t}(V \wedge Y, X \wedge Z)$$

which is natural[21] in V, Y, X, W, and Z.

Proof Let $\alpha_{m,n}$ be the composite

$$\begin{aligned}
&[S^{s+m}(V \wedge W), X \wedge E_m] \otimes [S^{t+n}Y, W \wedge Z \wedge F_n] \\
&\qquad \xrightarrow{C} [S^{s+t+m+n}(V \wedge Y), X \wedge E_m \wedge Z \wedge F_n] \\
&\qquad \xrightarrow{f_*} [S^{s+t+m+n}(V \wedge Y), X \wedge Z \wedge G_{m+n}]
\end{aligned}$$

where $f = (1 \wedge \mu_{m,n}) \circ (1 \wedge T \wedge 1)\colon X \wedge E_m \wedge Z \wedge F_n \to X \wedge Z \wedge G_{m+n}$. We have a diagram which commutes with the indicated signs (see page 231).

[21] Naturality in W means that if $f\colon W \to W'$, the diagram

$$\begin{array}{ccc}
\tilde{E}_s(V \wedge W', X) \otimes \tilde{F}_t(Y, W \wedge Z) & \xrightarrow{1\otimes(f\wedge 1)_*} & \tilde{E}_s(V \wedge W', X) \otimes \tilde{F}_t(Y, W' \wedge Z) \\
\downarrow{\scriptstyle (1\wedge f)^*\otimes 1} & & \downarrow{\scriptstyle M} \\
\tilde{E}_s(V \wedge W, X) \otimes F_t(Y, W \wedge Z) & \xrightarrow{M} & G_{s+t}(V \wedge Y, X \wedge Z)
\end{array}$$

commutes.

$$
\begin{array}{ccccc}
[S^{s+m+1}(V \wedge W), X \wedge E_{m+1}] \otimes [S^{t+n}Y, W \wedge Z \wedge F_n] & \xrightarrow{C} & [S^{s+t+m+n+1}(V \wedge W), X \wedge E_{m+1} \wedge Z \wedge F_n] & \xrightarrow{f_*} & [S^{s+t+m+n+1}(V \wedge Y), X \wedge Z \wedge G_{m+n+1}] \\
\Big\uparrow {\scriptstyle (1 \wedge e_m)_* \otimes 1} & & \Big\uparrow {\scriptstyle (1 \wedge e_m \wedge 1)_*} & (-1)^n & \Big\uparrow {\scriptstyle (1 \wedge g_{m+n})_*} \\
[S^{s+m+1}(V \wedge W), S(X \wedge E_m)] \otimes [S^{t+n}Y, W \wedge Z \wedge F_n] & \xrightarrow{C} & [S^{s+t+m+n+1}(V \wedge W), S(X \wedge E_m \wedge Z \wedge F_n)] & \xrightarrow{(Ef)_*} & [S^{s+t+m+n+1}(V \wedge Y), S(X \wedge Z \wedge G_{m+n})] \\
\Big\uparrow {\scriptstyle E \otimes 1} & (-1)^{t+n} & \Big\uparrow {\scriptstyle E} & & \Big\uparrow {\scriptstyle E} \\
[S^{s+m}(V \wedge W), X \wedge E_m] \otimes [S^{t+n}Y, W \wedge Z \wedge F_n)] & \xrightarrow{C} & [S^{s+t+m+n}(V \wedge Y), X \wedge E_m \wedge Z \wedge F_n] & \xrightarrow{f_*} & [S^{s+t+m+n}(V \wedge Y), X \wedge Z \wedge G_{m+n}] \\
\Big\downarrow {\scriptstyle 1 \otimes E} & & \Big\downarrow {\scriptstyle E} & & \Big\downarrow {\scriptstyle E} \\
[S^{s+m}(V \wedge W), X \wedge E_m] \otimes [S^{t+n+1}Y, S(W \wedge Z \wedge F_n)] & \xrightarrow{C} & [S^{s+t+m+n+1}(V \wedge Y), S(X \wedge E_m \wedge Z \wedge F_n)] & \xrightarrow{(Ef)_*} & [S^{s+t+m+n+1}(V \wedge Y), S(X \wedge Z \wedge G_{m+n})] \\
\Big\downarrow {\scriptstyle 1 \otimes (1 \wedge f_n)_*} & & \Big\downarrow {\scriptstyle (1 \wedge f_n)_*} & & \Big\downarrow {\scriptstyle (1 \wedge g_{m+n})_*} \\
[S^{s+m}(V \wedge W), X \wedge E_m] \otimes [S^{t+n+1}Y, W \wedge Z \wedge F_{n+1}] & \xrightarrow{C} & [S^{s+t+m+n+1}(V \wedge Y), X \wedge E_m \wedge Z \wedge E_{n+1}] & \xrightarrow{f_*} & [S^{s+t+m+n+1}(V \wedge Y), X \wedge Z \wedge G_{m+n+1}]
\end{array}
$$

Let $M_m = [S^{s+m}(V \wedge W), X \wedge E_m]$, $N_n = [S^{t+n}Y,\ W \wedge Z \wedge F_n]$ and $P_p = [S^{s+t+p}(V \wedge Y), X \wedge Z \wedge G_p]$. The diagram on page 231 then reduces, to

$$\begin{array}{ccc}
M_m \otimes N_{n+1} & \xrightarrow{\alpha_{m,n+1}} & P_{m+n+1} \\
\uparrow{\scriptstyle 1 \otimes \lambda_n} & & \uparrow{\scriptstyle \lambda_{m+n}} \\
M_m \otimes N_n & \xrightarrow{\alpha_{m,n}} & P_{m+n} \\
\downarrow{\scriptstyle \lambda_m \otimes 1} & (-1)^t & \downarrow{\scriptstyle \lambda_{m+n}} \\
M_{m+1} \otimes N_n & \xrightarrow{\alpha_{m+1,n}} & P_{m+n+1}
\end{array}$$

Let $M_{m,n} = (-1)^{tm}\alpha_{m,n}$. This induces a homomorphism

$$M\colon \varinjlim M_m \otimes \varinjlim N_n \to \varinjlim P_{m+n}$$

by

Lemma 23.16 Let M_u, N_v, and P_w be direct systems in $\mathcal{M}_R$ directed over Z^+ and suppose we have homomorphisms

$$M_{u,v}\colon M_u \otimes_R N_v \to P_{u+v}$$

such that the diagrams

$$\begin{array}{ccccc}
M_{u+1} \otimes N_v & \xleftarrow{\lambda_u \otimes 1} & M_u \otimes_R N_v & \xrightarrow{1 \otimes \lambda_v} & M_u \otimes_R N_{v+1} \\
\downarrow{\scriptstyle M_{u+1,v}} & & \downarrow{\scriptstyle M_{u,v}} & & \downarrow{\scriptstyle M_{u,v+1}} \\
P_{u+v+1} & \longleftarrow & P_{u+v} & \longrightarrow & P_{u+v+1}
\end{array}$$

commute. Then there is an induced map

$$M\colon (\varinjlim M_u) \otimes (\varinjlim N_v) \to \varinjlim P_w.$$

Proof Define $M(x \otimes y)$ as follows. If x is represented by $\alpha \in M_u$ and y by $\beta \in N_v$, let $M(x \otimes y) = \{M_{u,v}(\alpha \otimes \beta)\}$. To see that this is well defined, suppose x and y are also represented by α' and β' with $\alpha' \in M_{u'}$ and $\beta' \in N_{v'}$. Then $\lambda^r(\alpha) = \lambda^{r'}(\alpha')$ and $\lambda^s(\beta) = \lambda^{s'}(\beta')$ where $\lambda^n = \overbrace{\lambda \cdots \lambda}^{n}$. Now x and y are also represented by $\lambda^r(\alpha)$ and $\lambda^s(\beta)$. Furthermore

$$M_{u+r,\,v+s}(\lambda^r(\alpha) \otimes \lambda^s(\beta)) = \lambda^{r+s} M_{u,v}(\alpha \otimes \beta).$$

Similarly,

$$M_{u'+r',\,v'+s'}(\lambda^{r'}(\alpha) \otimes \lambda^{s'}(\beta')) = \lambda^{r'+s'} M_{u',v'}(\alpha' \otimes \beta').$$

Hence $\{M_{u,v}(\alpha \otimes \beta)\} = \{M_{u',v'}(\alpha' \otimes \beta')\}$. Since M is bilinear, it defines a homomorphism on the tensor product. ∎∎

Definition 23.17 A pairing from E and F to G defines four natural homomorphisms

(a) $\triangle : \tilde{E}_s(A) \otimes \tilde{F}_t(B) \to \tilde{G}_{s+t}(A \wedge B)$
(b) $\overline{\wedge} : \tilde{E}^s(A) \otimes \tilde{F}^t(B) \to \tilde{G}^{s+t}(A \wedge B)$
(c) $/ : \tilde{E}^{-s}(A \wedge B) \otimes \tilde{F}_t(B) \to \tilde{G}^{-s-t}(A)$
(d) $\backslash : \tilde{E}^{-s}(A) \otimes \tilde{F}_t(A \wedge B) \to \tilde{G}_{s+t}(B)$

by applying 23.15 with:

(a) $V = W = Y = S^0$, $X = A$, $Z = B$
(b) $X = W = Z = S^0$, $V = A$, $Y = B$
(c) $X = Y = Z = S^0$, $V = A$, $W = B$
(d) $V = X = Y = S^0$, $W = A$, $Z = B$.

The images of an element $x \otimes y$ are denoted $x \triangle y$, $x \overline{\wedge} y$, x/y and $x\backslash y$ respectively. The first two are called external products in homology and cohomology and the last two are called slant products.

As in the case of homology and cohomology, a map of spectra

$$\{\alpha_n\} : \{E_n\} \to \{E_n'\}$$

induces a homomorphism between the bifunctors

$$\alpha : \tilde{E}_s(A, B) \to \tilde{E}_s'(A, B)$$

which is natural in A and B (see Exercise 4, Section 18).

Theorem 23.18 Suppose there are pairings and maps of spectra which make the diagram

$$\begin{array}{ccc} E_m \wedge F_n & \longrightarrow & G_{n+k} \\ \downarrow{\scriptstyle \alpha_m \wedge \beta_n} & & \downarrow{\scriptstyle \gamma_{n+k}} \\ E_m' \wedge F_n' & \longrightarrow & G'_{n+k} \end{array}$$

homotopy commute. Then the diagram

$$\begin{array}{ccc} \tilde{E}_s(V \wedge W, X) \otimes \tilde{E}_t(Y, W \wedge Z) & \xrightarrow{M} & \tilde{G}_{s+t}(V \wedge Y, X \wedge Z) \\ \downarrow{\scriptstyle \alpha \otimes \beta} & & \downarrow{\scriptstyle \gamma} \\ \tilde{E}_s(V \wedge W, X) \otimes \tilde{F}_t'(Y, W \wedge Z) & \xrightarrow{M'} & \tilde{G}_{s+t}(V \wedge Y, X \wedge Z) \end{array}$$

commutes.

Proof This follows immediately by substituting in the various definitions. ▮

Theorem 23.19 Suppose there is a pairing from E and E to F such that the diagram

$$\begin{array}{ccc} E_m \wedge E_n & & \\ & \searrow^{\mu_{m,n}} & \\ \Big| T \quad (-1)^{mn} & & F_{m+n} \\ & \nearrow_{\mu_{n,m}} & \\ E_n \wedge E_m & & \end{array}$$

homotopy commutes with the indicated sign (after suspension). This happens, for example, if $E = F$ is a ring spectrum. Then

$$\begin{array}{ccc} \tilde{E}_s(A, B) \otimes \tilde{E}_t(C, D) & \xrightarrow{M} & \tilde{F}_{s+t}(A \wedge C, B \wedge D) \\ \Big\downarrow T & (-1)^{st} & \Big\downarrow T^* \circ T_* \\ \tilde{E}_t(C, D) \otimes E_s(A, B) & \xrightarrow[M]{} & \tilde{F}_{s+t}(C \wedge A, D \wedge B) \end{array}$$

commutes with the indicated signs. In particular

$$x \,\overline{\wedge}\, y = (-1)^{st} T^*(y \,\overline{\wedge}\, x) \qquad \text{and} \qquad x \,\underline{\wedge}\, y = (-1)^{st} T_*(y \,\underline{\wedge}\, x).$$

Proof If $h\colon S^{u+s}A \to B \wedge E_u$ and $g\colon S^{v+t}C \to D \wedge E_v$, then $(-1)^{ut}$ $M(\{h\} \otimes \{g\})$ is represented by

$$S^{u+s+v+t}(A \wedge C) \equiv (S^{u+s}A) \wedge (S^{v+t}C) \xrightarrow{h \wedge g} (B \wedge E_u) \wedge (D \wedge E_v) \xrightarrow{f} B \wedge D \wedge F_{u+v}$$

while $(-1)^{vs} M(\{g\} \otimes \{h\})$ is represented by

$$S^{u+s+v+t}(A \wedge C) \equiv (S^{v+t}C) \wedge (S^{u+s}A) \xrightarrow{g \wedge h} (D \wedge E_v) \wedge (B \wedge E_u) \xrightarrow{f} D \wedge B \wedge F_{u+v}$$

These differ by the sign $(-1)^{uv}$ and $(-1)^{(u+s)(v+t)}$ (which comes from the homeomorphism $S^{u+s} \wedge S^{t+v} \equiv S^{u+s+t+v} \equiv S^{t+v} \wedge S^{u+s}$). Together with the signs $(-1)^{us}$ and $(-1)^{ut}$ these combine to give $(-1)^{st}$. ∎

Theorem 23.20 Suppose there are pairings so that the diagram

$$\begin{array}{ccc} E_m \wedge F_n \wedge G_k & \xrightarrow{\mu_{m,n} \wedge 1} & H_{m+n} \wedge G_k \\ \Big\downarrow 1 \wedge \mu_{n,k}' & & \Big\downarrow \mu_{m+n,k}''' \\ E_m \wedge I_{n+k} & \xrightarrow[\mu_{m,n+k}'']{} & J_{m+n+k} \end{array}$$

homotopy commutes. This happens, for example, if $E = F = G$ is a ring spectrum.

Then the diagram

$$\begin{array}{ccc}
\tilde{E}_r(S\wedge T, U)\otimes\tilde{F}_s(V\wedge W, T\wedge X)\otimes\tilde{G}_t(Y, W\wedge Z) & \xrightarrow{M\otimes 1} & \tilde{H}_{r+s}(S\wedge V\wedge W, U\wedge X)\otimes\tilde{G}_t(Y, W\wedge Z)\\
\Big\downarrow{\scriptstyle 1\otimes M'} & & \Big\downarrow{\scriptstyle M'''}\\
\tilde{E}_r(S\wedge T, U)\otimes\tilde{I}_{s+t}(V\wedge Y, T\wedge X\wedge Z) & \xrightarrow{M''} & \tilde{J}_{r+s+t}(S\wedge V\wedge Y, U\wedge X\wedge Z)
\end{array}$$

commutes.

Proof This follows immediately by substituting the definitions of the various homomorphisms, as in 23.19. The combined sign contribution of either composite is $(-1)^{us+ut+vt}$. ∎

Diagrams of the above form occur in 23.7 and 23.13.

Corollary 23.21 Suppose there are pairings as in 23.20; then the following formulas hold:[22]

(a) $(x\,\underline{\wedge}\,y)\,\underline{\wedge}\,z = x\,\underline{\wedge}\,(y\,\underline{\wedge}\,z)\in J_*(X\wedge Y\wedge Z)$,
(b) $(x\,\overline{\wedge}\,y)\,\overline{\wedge}\,z = x\,\overline{\wedge}\,(y\,\overline{\wedge}\,z)\in J^*(X\wedge Y\wedge Z)$,
(c) $x\backslash(u\,\underline{\wedge}\,z) = (x\backslash u)\,\underline{\wedge}\,z\in J_*(Y\wedge Z)$,
(d) $x\,\overline{\wedge}\,(v/z) = (x\,\overline{\wedge}\,v)/z\in J^*(X\wedge Y)$,

with $u\in F_*(X\wedge Y)$ and $v\in F^*(Y\wedge Z)$, and the other variables belong to evident groups.

Proof Apply 23.20 with

(a) $S=T=V=W=Y=S^0$
(b) $U=T=X=W=Z=S^0$
(c) $S=U=V=W=Y=S^0$
(d) $T=U=X=Y=Z=S^0$. ∎

Definition 23.22 A graded ring is a graded abelian group $R=\{R_n\}$ together with an associative multiplication

$$R_n\otimes R_k\to R_{n+k}.$$

It will be called graded commutative if $x\cdot y=(-1)^{nk}y\cdot x$. R need not have a unit. $M=\{M_n\}$ is a module over R if there is an associative action $M_n\otimes R_k\to M_{n+k}$. If R has a unit, it is required to act as a unit on M.

Theorem 23.23 If E is a ring spectrum, $\tilde{E}^*(X)$ is a graded commutative ring with unit. If $f\colon X\to Y$, $f^*\colon \tilde{E}^*(Y)\to\tilde{E}^*(X)$ is a ring homomorphism. If

[22] A comprehensive list of formulas of the type given here, in 23.35, and in Exercises 6, 7, 10, 13, and 14 is given in Chapter 9 of [4], to which we are indebted.

F is a module spectrum over E, $\tilde{F}^*(X)$ is a module over $\tilde{E}^*(X)$ such that the diagram

$$\begin{array}{ccc} \tilde{F}^*(X) \otimes \tilde{E}^*(X) & \longrightarrow & \tilde{F}^*(X) \\ \uparrow {\scriptstyle f^* \otimes f^*} & & \uparrow {\scriptstyle f^*} \\ \tilde{E}^*(Y) \otimes \tilde{E}^*(Y) & \longrightarrow & \tilde{F}^*(Y) \end{array}$$

commutes.

Proof Let $\Delta\colon X \to X \wedge X$ be the diagonal map. We then have compositions

$$\tilde{E}^n(X) \otimes \tilde{E}^k(X) \xrightarrow{\bar{\wedge}} \tilde{E}^{n+k}(X \wedge X) \xrightarrow{\Delta^*} \tilde{E}^{n+k}(X)$$
$$\tilde{F}^n(X) \otimes \tilde{E}^k(X) \xrightarrow{\bar{\wedge}} \tilde{F}^{n+k}(X \wedge X) \xrightarrow{\Delta^*} \tilde{F}^{n+k}(X);$$

i.e., $x \cdot y = \Delta^*(x \bar{\wedge} y)$.

The proof of associativity and graded commutativity follows immediately from 23.19 and 23.21. ∎

This multiplication is called the cup product and is written either $x \cup y$ or simply xy.

Theorem 23.24 If E is a ring spectrum, $\tilde{E}^*(S^0)$ and $\tilde{E}_*(S^0)$ are graded commutative rings with unit and $\tilde{E}^*(X)$ and $\tilde{E}_*(X)$ are modules over $\tilde{E}^*(S^0)$ and $\tilde{E}_*(S^0)$ in a natural way.

Proof $\{u_n\} \in [S^n, E_n]$ determines an element $u \in \tilde{E}^0(S^0)$. If $f\colon S^{k-m} \to E_k$ represents $x \in \tilde{E}^m(S^0)$, $x \cdot u$ is represented by

$$\mu_{k,n}(f \wedge u_n)\colon S^{k-m} \wedge S^n \to E_k \wedge E_n \to E_{k+n}.$$

However, we have a homotopy commutative diagram

$$\begin{array}{ccc} S^{k-m} \wedge S^n & \xrightarrow{f \wedge u_n} & E_k \wedge E_n \\ \downarrow {\scriptstyle f \wedge 1} & \nearrow {\scriptstyle 1 \wedge u_n} & \downarrow {\scriptstyle \mu_{k,n}} \\ E_k \wedge S^n & \xrightarrow{e_k^n} & E_{n+k} \end{array}$$

and $e_k^n \circ f \wedge 1 = \lambda_{n+k+1} \circ \cdots \circ \lambda_k(\{f\}) \in [S^{k-m+n}, E_{n+k}]$.

Since $\tilde{E}^k(S^0) = \tilde{E}_{-k}(S^0)$, $\tilde{E}_*(S^0)$ is a ring with unit. Alternatively, the multiplications can be defined by

$$\bar{\wedge}\colon \tilde{E}^k(S^0) \otimes \tilde{E}^n(S^0) \to \tilde{E}^{k+n}(S^0 \wedge S^0) = \tilde{E}^{k+n}(S^0)$$
$$\underline{\wedge}\colon \tilde{E}_k(S^0) \otimes \tilde{E}_n(S^0) \to \tilde{E}_{k+n}(S^0 \wedge S^0) = \tilde{E}_{k+n}(S^0),$$

since $\Delta\colon S^0 \to S^0 \wedge S^0$ is a homeomorphism.

The module structure is given by the maps

$$\tilde{E}^m(X) \otimes \tilde{E}^n(S^0) \to \tilde{E}^{m+n}(X \wedge S^0) = \tilde{E}^{m+n}(X)$$
$$\tilde{E}_m(X) \otimes \tilde{E}_m(S^0) \to \tilde{E}_{m+n}(X \wedge S^0) = \tilde{E}_{m+n}(X).$$

Verification of the various diagrams is an easy exercise. ∎

Corollary 23.25 If R is a commutative ring, $\tilde{H}^k(X; R)$ and $\tilde{H}_k(X; R)$ are R-modules for each k. ∎

Proposition 23.26 Suppose there is a pairing from E and F to G. Let $|x|$ be the dimension of x. Then:

(a) $(1 \wedge T)_*((\sigma x) \underline{\wedge} y) = \sigma(x \underline{\wedge} y) = (-1)^{|x|} x \underline{\wedge} (\sigma y) \in \tilde{G}_*(S(X \wedge Y))$.
(b) $(1 \wedge T)^*((\sigma x) \overline{\wedge} y) = \sigma(x \overline{\wedge} y) = (-1)^{|x|} x \overline{\wedge} (\sigma y) \in \tilde{G}^*(S(X \wedge Y))$.
(c) $(-1)^{|x|} \sigma x / \sigma y = x/y \in \tilde{G}^*(X)$.
(d) $(-1)^{|x|} \sigma x \backslash (1 \wedge T)_* \sigma y = x \backslash y \in \tilde{G}_*(Y)$.

Proof This is most easily seen by direct substitution. If x is represented by $\alpha\colon S^{m+s} \to X \wedge E_m$ and y by $\beta\colon S^{n+t} \to Y \wedge F_n$, then $(-1)^{mt}\sigma(x \underline{\wedge} y)$ is represented by

$$S^{1+s+m+t+n} \xrightarrow{1 \wedge \alpha \wedge \beta} S^1 \wedge X \wedge E_m \wedge Y \wedge F_n \xrightarrow{f} S^1 \wedge X \wedge Y \wedge G_{m+n}.$$

This also represents $(-1)^{mt}(\sigma x) \underline{\wedge} y$, since $\sigma\colon \tilde{E}_*(X) \to \tilde{E}_*(S^1 \wedge X)$ and $\sigma\colon \tilde{G}_*(X \wedge Y) \to \tilde{G}_*(S^1 \wedge X \wedge Y)$ are given by smashing on the left with S^1. However, $(-1)^{m(t+1)} x \underline{\wedge} (\sigma y)$ is represented by

$$S^{s+m+1+t+n} \xrightarrow{\alpha \wedge 1 \wedge \beta} X \wedge E_m \wedge S^1 \wedge Y \wedge F_n \xrightarrow{f} X \wedge S^1 \wedge Y \wedge G_{m+n}.$$

Thus $(-1)^{m(t+1)} x \underline{\wedge} (\sigma y)$ differs from $(-1)^{mt}\sigma(x \underline{\wedge} y)$ by a permutation of the coordinates of $S^{s+m+1+t+n}$ of degree $(-1)^{s+m}$. Similar computations prove the other formulas. ∎

The above compositions in reduced theory define corresponding compositions in unreduced theory.

Definition 23.27 A pairing from E and F to G defines four natural transformations

$$\underline{\times}: E_s(X, A) \otimes F_t(Y, B) \to G_{s+t}(X \times Y, X \times B \cup A \times Y)$$
$$\overline{\times}: E^s(X, A) \otimes F^t(Y, B) \to G^{s+t}(X \times Y, X \times B \cup A \times Y)$$
$$/: E^{-s}(X \times Y, X \times B \cup A \times Y) \otimes F_t(Y, B) \to G^{-s-t}(X, A)$$
$$\backslash: E^{-s}(X, A) \otimes F_t(X \times Y, X \times B \cup A \times Y) \to G_{s+t}(Y, B)$$

since there is a natural homeomorphism in $\mathcal{CG}$:

$$(X \cup CA) \wedge (Y \cup CB) \equiv X \times Y \cup C(X \times B \cup A \times Y).$$

Proposition 23.28 If E is a ring spectrum,

$$\begin{array}{rr} x \,\underline{\times}\, y = (-1)^{st} T_* y \,\underline{\times}\, x, & x \,\overline{\times}\, y = (-1)^{st} T^* y \,\overline{\times}\, x. \\ (x \,\overline{\times}\, y) \,\overline{\times}\, z = x \,\overline{\times}\, (y \,\overline{\times}\, z), & (x \,\underline{\times}\, y) \,\underline{\times}\, z = x \,\underline{\times}\, (y \,\underline{\times}\, z). \\ x \backslash (u \,\underline{\times}\, z) = (x \backslash u) \,\underline{\times}\, z, & x \,\overline{\times}\, (v/z) = (x \,\overline{\times}\, v)/z. \end{array}$$

Proof Apply 23.19 and 23.21, substituting $\tilde{E}_s(X \cup CA)$ for $E_s(X, A)$, etc. ▮

Proposition 23.29 If E is a ring spectrum, there is a natural multiplication

$$E^s(X, A) \otimes E^t(X, B) \to E^{s+t}(X, A \cup B)$$

(called the cup product) which is associative and graded commutative. Hence if $A = B$, $E^*(X, A)$ is a graded commutative ring. If $A = B = \emptyset$, this ring has a unit.

Proof As before we use $\Delta\colon (X, A \cup B) \to (X \times X, X \times B \cup A \times X)$ and define $x \cup y = \Delta^*(x \,\overline{\times}\, y)$. Since $E^0(P) = \tilde{E}^0(S^0)$, where P is a one-point space, $E^*(P)$ is a ring with unit. Define $1 \in E^0(X)$ by $1 = (p_X)^*(1)$ where $p_X\colon X \to P$. Consideration of the commutative diagram

$$\begin{array}{ccccc} E^k(X) \otimes E^0(X) & \longrightarrow & E^k(X \times X) & \xrightarrow{\Delta^*} & E^k(X) \\ \uparrow{\scriptstyle 1 \otimes (p_X)^*} & & \uparrow{\scriptstyle (1 \times p_X)^*} & \nearrow{\scriptstyle 1} & \\ E^k(X) \otimes E^0(P) & \longrightarrow & E^k(X \times P) & & \end{array}$$

proves that 1 is a unit. ▮

Corollary 23.30 If F is a module spectrum over E, $F^*(X, A)$ is a module over $E^*(X, A)$ in a natural way. ▮

Corollary 23.31 If E is a ring spectrum, $E^*(X, A)$ is a module over $E^*(P)$ and $E_*(X, A)$ is a module over $E_*(P)$. ▮

Corollary 23.32 If R is a commutative ring $H^k(X, A; R)$ and $H_k(X, A; R)$ are R-modules. ▮

One can also define $\overline{\times}$ in terms of $\cup$.

Proposition 23.33 Let $\pi_1: X \times Y \to X$ and $\pi_2: X \times Y \to Y$ be the projections. Let $\alpha \in E^*(X)$ and $\beta \in E^*(Y)$. Then

$$\alpha \,\bar{\times}\, \beta = \pi_1^*(\alpha) \cup \pi_2^*(\beta).$$

Proof

$$\pi_1^*(\alpha) \cup \pi_2^*(\beta) = \Delta^*(\pi_1^*(\alpha) \,\bar{\times}\, \pi_2^*(\beta)) = \Delta^*(\pi_1 \times \pi_2)^*(\alpha \,\bar{\times}\, \beta) = \alpha \,\bar{\times}\, \beta$$

Since $(\pi_1 \times \pi_2) \cdot \Delta = 1$. ∎

Corollary 23.34 $H^*(S^n \times S^m)$ is freely generated by elements 1, e_n, e_m, and $e_n \cup e_m$.

Proof By Exercise 10, Section 20, $H^*(S^n \times S^m)$ is free and has generators 1, e_n, e_m, and e_{n+m} of dimensions 0, n, m, and $n + m$ respectively. We now claim $e_n \cup e_m \pm e_{n+m}$. Let $\alpha \in H^n(S^n)$ and $\beta \in H^m(S^m)$ be generators. Then $e_n = \pi_1^*(\alpha)$ and $e_m = \pi_2^*(\beta)$. Thus $e_n \cup e_m = \alpha \,\bar{\times}\, \beta$. Let μ: $S^n \times S^m \to S^{n+m}$ be the projection. Clearly $e_{n+m} = \pm\mu^*(\gamma) = \pm\alpha \,\bar{\times}\, \beta$ where $\gamma \in H^{n+m}(S^{n+m})$ is a generator. ∎

This example illustrates the geometric meaning of the cup product. The classes e_n and e_m represent the n- and m-cells in $S^n \times S^m$. Their cup product represents the cartesian product of the cells—the $(n + m)$-cell in $S^n \times S^m$.

Theorem 23.35 Let $(X, A), (Y, B) \in \mathcal{CG}^2$ and $C \subset A$. Let

$$e: (A \times Y, A \times B \cup C \times Y) \to (A \times Y \cup X \times B, C \times Y \cup X \times B)$$
$$e_1: (X \times B, A \times B) \to (A \times Y \cup X \times B, A \times Y)$$
$$e_2: (A \times Y, A \times B) \to (A \times Y \cup X \times B, X \times B)$$

be the excision maps. Then if $x \in E_p(X, A)$ and $y \in F_q(Y, B)$,

(a) $e_{2*}(\partial x \,\underline{\times}\, y) = \partial(x \,\underline{\times}\, y)$.
(b) $(-1)^p e_{1*}(x \,\underline{\times}\, \partial y) = \partial(x \,\underline{\times}\, y)$.

If $\alpha \in E^p(A)$, $\beta \in F^q(Y, B)$, $\gamma \in E^p(X, A)$, and $\varepsilon \in F^q(B)$,

(c) $(\delta\alpha) \,\bar{\times}\, \beta = \delta(e_2^*)^{-1}(\alpha \,\bar{\times}\, \beta)$ if A is closed or $B = \varnothing$.
(d) $(-1)^p \gamma \,\bar{\times}\, (\delta\varepsilon) = \delta(e_1^*)^{-1}(\gamma \,\bar{\times}\, \varepsilon)$ if B is closed or $A = \varnothing$.

If $u \in E^p(A \times Y \cup X \times B,\ X \times B)$, $v \in F_q(Y, B)$, $w \in F_q(X \times Y,\ A \times Y \cup X \times B)$, and $z \in E^p(A, C)$,

(e) $\delta u/v = \delta(e_2^* u/v)$.
(f) $\qquad = e_1^* u/\partial v$.
(g) $(-1)^p z\backslash(e_*)^{-1}(\partial w) = (\delta z/w)$ if A and B are open or $B = \varnothing$.
(h) $(-1)^p \gamma\backslash(e_{1*})^{-1}(\partial w) = \partial(\gamma\backslash w)$ if A and B are open or $A = \varnothing$.

Each of these formulas may be rewritten as a commutative diagram. For example, (a) may be expressed by

$$\begin{array}{ccc}
E_p(X, A) \otimes F_q(Y, B) & \xrightarrow{\partial \otimes 1} & E_{p-1}(A) \otimes F_q(Y, B) \\
& & \downarrow \underline{\times} \\
\big\downarrow \underline{\times} & & G_{p+q-1}(A \times Y, A \times B) \\
& & \cong \big\downarrow (e_2)_* \\
G_{p+q}(X \times B, A \times Y \cup X \times B) & \xrightarrow{\partial} & G_{p+q-1}(A \times Y \cup X \times B, X \times B)
\end{array}$$

Proof The formulas follow directly from the definitions. For example, to prove (a) we expand the above diagram as follows. Let $j\colon X \cup CA \to S(A^+)$ be the natural map identifying X with the vertex of the cone. Then $\partial\colon E_p(X, A) \to E_{p-1}(A)$ is given by the composition

$$E_p(X, A) = \tilde{E}_p(X \cup CA) \xrightarrow{j_*} \tilde{E}_p(S(A^+)) \underset{\cong}{\xleftarrow{\sigma}} \tilde{E}_{p-1}(A^+) = E_{p-1}(A).$$

Observe that $A^+ \wedge (Y \cup CB) \equiv A \times Y \cup C(A \times B)$. Consider the diagram

$$\begin{array}{ccccc}
\tilde{E}_p(X \cup CA) \otimes \tilde{F}_q(Y \cup CB) & \xrightarrow{j_* \otimes 1} & E_p(S(A^+)) \otimes F_q(Y \cup CB) & \xleftarrow{\sigma \otimes 1} & E_{p-1}(A^+) \otimes F_q(Y \cup CB) \\
\big\downarrow \underline{\wedge} & & \big\downarrow \underline{\wedge} & & \big\downarrow \underline{\wedge} \\
\tilde{G}_{p+q}((X \cup CA) \wedge (Y \cup CB)) & \xrightarrow{(j \wedge 1)_*} & \tilde{G}_{p+q}(S(A \times Y \cup C(A \times B))) & \xleftarrow{\sigma} & \tilde{G}_{p+q-1}(A \times Y \cup C(A \times B)) \\
\big\downarrow & & \big\downarrow & & \big\downarrow \\
\tilde{G}_{p+q}(X \times Y \cup C(A \times Y \cup X \times B)) & \longrightarrow & \tilde{G}_{p+q}(S(A \times Y \cup X \times B \cup C(X \times B))) & \xleftarrow{\sigma} & \tilde{G}_{p+q-1}(A \times Y \cup X \times B \cup C(X \times B))
\end{array}$$

which commutes by the naturality of $\triangle$ and σ, and 23.26(a).

Observe that the extra conditions in (c), (d), (g), and (h) imply that the inverted homomorphisms are isomorphisms. In (g) and (h) one need only assume that A and B are deformation retracts of neighborhoods in X and Y to draw the same conclusion. ▌

Appendix

Proof of 23.9 Clearly $\beta \otimes 1$ is onto. Let $L = B \otimes_R D/\text{Im}(\alpha \otimes 1)$. Then $A \otimes_R D \xrightarrow{\alpha \otimes 1} B \otimes_R D \xrightarrow{\gamma} L$ is exact. We must prove that $C \otimes_R D \cong L$. One can construct $f\colon L \to C \otimes_R D$ such that $f\gamma = \beta \otimes 1$ since $(\beta \otimes 1)(\alpha \otimes 1) = 0$. To construct a map $g\colon C \otimes_R D \to L$ it suffices to find a bilinear map $\bar{g}\colon C \times D \to L$ such that $\bar{g}(c, rd) = \bar{g}(cr, d)$. For each $c \in C$, choose b with $\beta(b) = c$. b is well defined mod $\alpha(A)$ and hence $b \otimes d$ is well defined mod $(\alpha \otimes 1)(A \otimes_R D)$. Hence $\gamma(b \otimes d)$ is well defined. Let $\bar{g}(c,d) = \gamma(b \otimes d)$. Clearly $\bar{g}$ is bilinear and $\bar{g}(c, rd) = \bar{g}(cr, d)$. Thus $\bar{g}$ defines g. Now by definition, $g \circ (\beta \otimes 1) = \gamma$. Consequently $gf\gamma = \gamma$ and $fg(\beta \otimes 1) = \beta \otimes 1$. Since both γ and $\beta \otimes 1$ are onto $gf = 1$ and $fg = 1$. Thus $L \cong C \otimes_R D$ and the sequence is exact. ▌

Even if we have a short exact sequence $0 \to A \xrightarrow{\alpha} B \xrightarrow{\beta} C \to 0$ it does not follow that $0 \to A \otimes_R D \to B \otimes_R D$ is exact, i.e., $\alpha \otimes 1$ is not a monomorphism. This is seen by an example with $R = Z$; the short exact sequence is $0 \to Z \xrightarrow{\alpha} Z \xrightarrow{\beta} Z_2 \to 0$ where $\alpha(n) = 2n$ and β is reduction modulo 2. If $D = Z_2$, we have that $Z \otimes Z_2 \cong Z_2$ is generated by $1 \otimes 1$ and $(\alpha \otimes 1)(1 \otimes 1) = 2 \otimes 1 = 1 \otimes 2 = 0$. Thus $\alpha \otimes 1 = 0$.

Exercises

1. Suppose X and Y are well pointed. Show that if X is $(m-1)$-connected and Y is $(n-1)$-connected with $n, m > 1$,

$$C\colon \pi_m(X) \otimes \pi_n(Y) \to \pi_{m+n}(X \wedge Y)$$

is an isomorphism. (Exercise 20)

2. Give the omitted details for the proof of 23.13.

3. Give the omitted details for the proof of 23.14.

4. Prove the naturality assertions in 23.15.

5. Show that there are natural homomorphisms

$$\Sigma_X\colon \tilde{E}_m(Y, Z) \to \tilde{E}_m(X \wedge Y, X \wedge Z), \qquad E_X\colon \tilde{E}_n(Y, Z) \to \tilde{E}_n(Y \wedge X, Z \wedge X)$$

such that $E_{ST} = E_T E_S$, $\Sigma_{ST} = \Sigma_S \Sigma_T$ and such that the diagrams

$$\begin{array}{ccc}
\tilde{E}_m(V \wedge W, X) \otimes \tilde{F}_n(Y, W \wedge Z) & \xrightarrow{M} & \\
\Big\downarrow {\scriptstyle E_Z \otimes \Sigma_V} & & \tilde{G}_{m+n}(V \wedge Y, X \wedge Y) \\
\tilde{E}_m(V \wedge W \wedge Z, X \wedge Z) \otimes \tilde{F}_n(V \wedge Y, V \wedge W \wedge Z) & \xrightarrow{M} &
\end{array}$$

$$\begin{array}{ccc}
 & \tilde{E}_m(Y, Z) & \\
{\scriptstyle \Sigma_X} \swarrow & & \searrow {\scriptstyle E_X} \\
\tilde{E}_m(X \wedge Y, X \wedge Z) & \xrightarrow{T^* \circ T_*} & \tilde{E}_m(Y \wedge X, Z \wedge X)
\end{array}$$

commute.

6. Using Exercise 5 above and 23.20 or by direct substitution, prove the formulas:

(a) $v/(y\backslash u) = [(1 \wedge T)^*(v \wedge y)]/u \in \tilde{J}^*(X)$

$$v \in \tilde{E}^*(X \wedge Z),\ u \in \tilde{G}_*(Y \wedge Z),\ y \in \tilde{F}^*(Y).$$

(b) $(t/z)/y = t/(T_*(z \,\triangle\, y)) \in \tilde{J}^*(X)$

$$t \in \tilde{E}^*(X \wedge Y \wedge Z),\ z \in \tilde{F}_*(Z),\ y \in \tilde{G}_*(Y).$$

(c) $y\backslash(x\backslash t) = (T^*(y \mathbin{\overline{\wedge}} x))\backslash t \in \tilde{J}_*(Z)$

$$t \in \tilde{G}_*(X \wedge Y \wedge Z),\ x \in \tilde{F}^*(X),\ y \in \tilde{E}^*(Y). \quad (26.21)$$

(d) $(w/y)\backslash v = w\backslash[(T \wedge 1)_*(y \,\triangle\, v)] \in \tilde{J}_*(Z)$

$$w \in \tilde{E}^*(X \wedge Y),\ v \in \tilde{G}_*(X \wedge Z),\ y \in \tilde{F}_*(Y).$$

7. Given a pairing from E and F to G, one can define a homomorphism

$$\cap\colon E^s(X, A) \otimes F_t(X, A \cup B) \to G_{t-s}(X, B)$$

by $x \cap y = x\backslash\Delta_*(y)$. $x \cap y$ is called the cap product of x and y. Show that the cap product is natural in the sense that if $f\colon X \to Y$ is a map such that $f(A) \subset A'$ and $f(B) \subset B'$

$$f_*(f^*(y) \cap x) = y \cap f_*(x)$$

for $x \in F_t(X, A \cup B)$ and $y \in E^s(Y, A')$. One can also define a natural homomorphism

$$\langle\ ,\ \rangle\colon E^s(X, A) \otimes E_s(X, A) \to E_0(P)$$

by $\langle x, y\rangle = x\backslash y$ (23.27 with $Y = P$ and $B = \varnothing$). $\langle x, y\rangle$ is called the Kronecker product. Show that if $E = F = G$ is a ring spectrum

$$x \cap (y \cap z) = (x \cap y) \cap z, \qquad \langle x, y \cap z\rangle = \langle x \cup y, z\rangle.$$

(24.9 Section 26)

8. Write the details down for the last four formulas of 23.26.

9. Give complete proofs for the last seven formulas in 23.35.

10. Keeping the notation of 23.35, prove the formula

$$i^*(e_1{}^*)^{-1}((\delta\alpha) \mathbin{\overline{\times}} \varepsilon) = -j^*(e_2{}^*)^{-1}(\alpha \mathbin{\overline{\times}} (\delta\varepsilon)),$$

where

$$i\colon (A \times Y \cup X \times B, \varnothing) \to (A \times Y \cup X \times B, A \times Y)$$

and

$$j\colon (A \times Y \cup X \times B, \varnothing) \to (A \times Y \cup X \times B, X \times B)$$

are inclusions, and A and B are closed.

If $u \in E^p(X \times Y, A \times B)$, prove that

$$\delta((f^*u)/y) = (-1)^{p+1}(g^*u)/\partial y$$

where

$$f\colon (A \times Y, A \times B) \to (X \times Y, A \times B)$$

and

$$g\colon (X \times B, A \times B) \to (X \times Y, A \times B)$$

are inclusions.

11. Use 23.4 to determine the sign in 18.6.

12. Let X be a homotopy associative, homotopy commutative H-space. Show that $\underline{X}^+$ is a ring spectrum.

13. Suppose $X \supset A \supset C$, $Y \supset B \supset D$, with A, B, C, and D open. Using (g), (h) and Exercise 5, Section 21 with

$$Z = A \times Y \cup X \times B, \qquad U = C \times Y \cup X \times B, \qquad V = A \times Y \cup X \times D$$

and

$$W = C \times Y \cup X \times D$$

prove that the diagram

$$\begin{array}{ccc}
 & R^i(A, C) \otimes R_n(A \times Y \cup X \times B, C \times Y \cup X \times D) & \\
 \swarrow^{i_*} & & \searrow^{\delta \otimes i_*} \\
R^i(A, C) \otimes R_n(A \times Y \cup X \times B, C \times Y \cup X \times B) & & R^{i+1}(X, A) \otimes R_n(A \times Y \cup X \times B, A \times Y \cup X \times D) \\
\uparrow \cong & & \uparrow \cong \\
R^i(A, C) \otimes R_n(A \times Y, C \times Y \cup A \times B) & (-1) & R^{i+1}(X, A) \otimes R_n(X \times B, A \times B \cup X \times D) \\
\downarrow & & \downarrow \backslash \\
R_{n-i}(Y, B) & \xrightarrow{\quad \partial \quad} & R_{n-i-1}(B, D)
\end{array}$$

commutes with a sign -1. (Exercise 14)

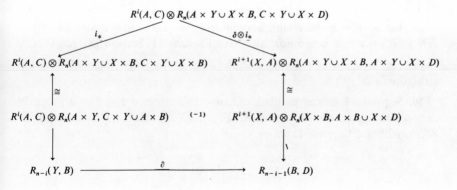

14. Suppose $X = A \cup B$, $C \subset A$, $D \subset B$, and A, B, C, and D are open. Using Exercises 7, 13, and the naturality of $\backslash$, show that there is a commutative diagram

$$\begin{array}{ccccc}
R^i(A, C) \otimes R_n(X, D) & \xrightarrow{\delta \otimes i_*} & & & R^{i+1}(X, A) \otimes R_n(X, A \cup D) \\
\downarrow & & & & \uparrow \cong \\
R^i(A, C) \otimes R_n(X, B \cup C) & & & & R^{i+1}(B, A \cap B) \otimes R_n(B, (A \cap B) \cup D) \\
\uparrow \cong & & (-1) & & \\
R^i(A, C) \otimes R_n(A, (A \cap B) \cup C) & & & & \\
\downarrow \cap & & & & \downarrow \\
R_{n-i}(A, A \cap B) & \longrightarrow & R_{n-i}(X, B) & \xrightarrow{\partial} & R_{n-i-1}(B, D)
\end{array} \qquad (26.20)$$

15. Use

$$\begin{array}{ccc}
E^n(SX, C_1 X) \otimes E^m(SX, C_2 X) & \longrightarrow & E^{m+n}(SX, SX) \\
\downarrow & & \downarrow \\
E^n(SX, *) \otimes E^m(SX, *) & \longrightarrow & E^{m+n}(SX, *)
\end{array}$$

to show that all cup products in $\tilde{E}^*(SX)$ are 0.

16. Let $x \in E^*(A)$ and $y \in F_*(A \wedge B \wedge S^1)$. Show that $x \backslash y = \sigma(\sigma x \backslash (1 \wedge T)_* y)$. Let $u \in E^*(A \wedge B)$ and $v \in F_*(B)$. Show that $(1 \wedge T)^* \sigma(u)/v = \sigma(u/v)$.

17. Show that $(x \,\bar{\times}\, y) \cup (u \,\bar{\times}\, v) = (-1)^{|y||u|}(x \cup u) \,\bar{\times}\, (y \cup v)$. (27.15; 27.16)

18. Let $\rho\colon R \to R'$ be a ring homomorphism. Show that $c_\rho\colon H^*(X; R) \to H^*(X; R')$ is a ring homomorphism (see Exercise 11, Section 18). (28.18)

19. Let X consist of n points with the discrete topology. Calculate the ring structure in $H^0(X)$.

20. Suppose X is well pointed and $(n-1)$-connected and E is a properly convergent spectrum. Show that $\tilde{E}_n(X) \cong \pi_n(X) \otimes \tilde{E}_n(S^n)$. (Use Exercise 12, Section 22 and Exercise 1, above.)

24

Relations between Chain Complexes

In this section we shall develop some relationships between chain and cochain complexes for spaces with various coefficients, and between the chain and cochain complexes of a product of relative CW complexes with the original chain and cochain complexes. We will use this to develop the Künneth theorem and universal coefficient theorems in the next section.

Exercise 10, Section 20 suggests that $H_n(X \times Y)$ bears some relation to $\bigoplus_{i+j=n} H_i(X) \otimes H_j(Y)$. The pairing of $H\pi$ and $H\rho$ to $H\pi \otimes \rho$ provides us with transformations

$$\underline{\times} : \bigoplus_{i+j=n} H_i(X, A; \pi) \otimes H_j(Y, B; \rho) \to H_n(X \times Y, X \times B \cup A \times Y; \pi \otimes \rho)$$

$$\overline{\times} : \bigoplus_{i+j=n} H^i(X, A; \pi) \otimes H^j(Y, B; \rho) \to H^n(X \times Y, X \times B \cup A \times Y; \pi \otimes \rho).$$

For simplicity, we define the tensor product of two graded groups G_* and G_*' by:

$$(G \otimes G')_n = \bigoplus_{i+j=n} G_i \otimes G_j'.$$

With this notation we have

$$\underline{\times} : H_*(X, A; \pi) \otimes H_*(Y, B; \rho) \to H_*(X \times Y, X \times B \cup A \times Y; \pi \otimes \rho)$$
$$\overline{\times} : H^*(X, A; \pi) \otimes H^*(Y, B; \rho) \to H^*(X \times Y, X \times B \cup A \times Y; \pi \otimes \rho)$$

as homomorphisms of graded groups. Similarly, one has such maps in reduced theory:

$$\underline{\wedge} : \tilde{H}_*(X; \pi) \otimes \tilde{H}_*(Y; \rho) \to \tilde{H}_*(X \wedge Y; \pi \otimes \rho)$$
$$\overline{\wedge} : \tilde{H}^*(X; \pi) \otimes \tilde{H}^*(Y; \rho) \to \tilde{H}^*(X \wedge Y; \pi \otimes \rho).$$

Lemma 24.1 If X and Y are one-point unions of spheres, $\triangle$ is an isomorphism. If X and Y are finite one-point unions of spheres, $\overline{\wedge}$ is an isomorphism as well.

Proof If both X and Y are spheres, both of these results reduce to 23.10 since the maps

$$\pi_n(K(\pi, n)) \xrightarrow{\Sigma^m} \pi_{n+m}(S^m \wedge K(\pi, n)) \to \tilde{H}_m(S^m; \pi)$$
$$\pi_n(K(\pi, n)) \to \tilde{H}^m(S^m; \pi)$$

are isomorphisms. Suppose now that $X = \bigvee_{\alpha \in A} S_\alpha^n$. Then we have a commutative diagram

$$\begin{array}{ccc} \tilde{H}_*(X; \pi) \otimes \tilde{H}_*(Y; \rho) & \xrightarrow{\triangle} & \tilde{H}_*(X \wedge Y; \pi \otimes \rho) \\ \uparrow \cong & & \uparrow \cong \\ \bigoplus_{\alpha \in A} \tilde{H}_*(S_\alpha^n; \pi) \otimes \tilde{H}_*(Y; \rho) & \xrightarrow{\triangle} & \bigoplus_{\alpha \in A} \tilde{H}_*(S_\alpha^n \wedge Y; \pi \otimes \rho) \end{array}$$

with the vertical maps isomorphisms by 18.17 and Exercise 12, Section 15. If Y is a sphere, the bottom map is an isomorphism by the above argument and Exercise 13, Section 15; hence the top map is also an isomorphism. Similarly, if X is a sphere and Y is a wedge of spheres, $\triangle$ is an isomorphism. Thus applying the diagram again we see that if X and Y are wedges of spheres, $\triangle$ is an isomorphism. In the case of cohomology we consider the diagram

$$\begin{array}{ccc} \tilde{H}^*(X; \pi) \otimes \tilde{H}^*(Y; \rho) & \xrightarrow{\overline{\wedge}} & \tilde{H}^*(X \wedge Y; \pi \otimes \rho) \\ \uparrow \cong & & \uparrow \cong \\ \left(\prod_{\alpha \in A} \tilde{H}^*(S_\alpha^n; \pi)\right) \otimes \tilde{H}^*(Y; \rho) & \longrightarrow & \prod_{\alpha \in A} \tilde{H}^*(S_\alpha^n \wedge Y; \pi \otimes \rho) \end{array}$$

where the vertical maps isomorphisms by 18.16.

This is sufficient to prove the second part of 24.1 similarly to the first part. The loss in generality is due to the fact that in general we do not have $(\prod_{\alpha \in A} A_\alpha) \otimes B = \prod_{\alpha \in A}(A_\alpha \otimes B)$. (See Exercise 1.) ∎

We apply this result to the chain and cochain complexes of relative CW complexes.

Lemma 24.2 Let (X, A) and (Y, B) be relative CW complexes. Then $(X \times Y, X \times B \cup A \times Y)$ is a relative CW complex with $\overline{X \times Y}^k = \bigcup_{m+n=k} \overline{X}^m \times \overline{Y}^n \cup X \times B \cup A \times Y$, and

$$\bigvee_{m+n=k} (\overline{X}^m/\overline{X}^{m-1} \wedge \overline{Y}^n/\overline{Y}^{n-1}) \equiv \overline{X \times Y}^k/\overline{X \times Y}^{k-1}.$$

Proof $\bar{X}^m/\bar{X}^{m-1} \wedge \bar{Y}^n/\bar{Y}^{n-1} \equiv \bar{X}^m \times \bar{Y}^n/\bar{X}^m \times \bar{Y}^{n-1} \cup \bar{X}^{m-1} \times \bar{Y}^n \subset \overline{X \times Y}^k/\overline{X \times Y}^{k-1}$. This defines a map

$$f: \bigvee_{m+n=k} (\bar{X}^m/\bar{X}^{m-1} \wedge \bar{Y}^n/\bar{Y}^{n-1}) \to \overline{X \times Y}^k/\overline{X \times Y}^{k-1}.$$

It is easy to see that it is 1–1 and onto. In fact both spaces have the quotient topology on the same identifications applied to the disjoint union $\bigcup_{k=m+n} \bar{X}^m \times \bar{Y}^n$. Hence f is a homeomorphism. ▮

We define an isomorphism

$$\wedge_{\#}: C_{\#}(X, A; \pi) \otimes C_{\#}(Y, B; \rho) \to C_{\#}(X \times Y, X \times B \cup A \times Y; \pi \otimes \rho)$$

by

$$\begin{aligned} &\bigoplus_{m+n=k} \tilde{H}_*(\bar{X}^m/\bar{X}^{m-1}; \pi) \otimes \tilde{H}_*(\bar{Y}^n/\bar{Y}^{n-1}; \rho) \\ &\xrightarrow[\cong]{} \bigoplus_{m+n=k} \tilde{H}_*(\bar{X}^m/\bar{X}^{m-1} \wedge \bar{Y}^n/\bar{Y}^{n-1}; \pi \otimes \rho) \\ &\cong \tilde{H}_*\left(\bigvee_{m+n=k} (\bar{X}^m/\bar{X}^{m-1} \wedge \bar{Y}^n/\bar{Y}^{n-1}); \pi \otimes \rho\right) \\ &\cong \tilde{H}_*(\overline{X \times Y}^k/\overline{X \times Y}^{k-1}; \pi \otimes \rho). \end{aligned}$$

Similarly, if (X, A) and (Y, B) have a finite number of cells in each dimension, we define an isomorphism

$$\wedge^{\#}: C^{\#}(X, A; \pi) \otimes C^{\#}(Y, B; \rho) \to C^{\#}(X \times Y, X \times B \cup A \times Y; \pi \otimes \rho).$$

Now $C_{\#}(X, A; \pi) \otimes C_{\#}(Y, B; \rho)$ is given here only as a graded abelian group. One would like to make it into a chain complex such that $\wedge_{\#}$ is an isomorphism of chain complexes.

Definition 24.3 If $\{C_n, \partial_n\}$ and $\{C_n', \partial_n\}$ are chain complexes, one makes $C \otimes C'$ into a chain complex by

$$\partial(x \otimes y) = \partial x \otimes y + (-1)^{|x|} x \otimes \partial y.$$

(Observe that $\partial^2 = 0$.)

Theorem 24.4 $\wedge_{\#}$ is a chain isomorphism (i.e., a chain map which is an isomorphism).

The only point to be checked is that $\wedge_{\#}$ is a chain map, and this follows from:

Lemma 24.5 Let $X \supset A \supset A_0$ and $Y \supset B \supset B_0$ be relative CW complexes, and suppose there is a pairing from E and F to G. Then the diagram

$$\begin{array}{ccc}
E_*(X, A) \otimes F_*(Y, B) & \xrightarrow{\times} & G_*(X \times Y, X \times B \cup A \times Y) \\
\downarrow{\scriptstyle \partial \otimes 1 + 1 \otimes \partial} & & \downarrow{\scriptstyle \partial} \\
\{E_*(A, A_0) \otimes F_*(Y, B)\} \oplus \{E_*(X, A) \otimes F_*(B, B_0)\} & \xrightarrow{\times} & G_*(X \times B \cup A \times Y, A_0 \times Y \cup A \times B \cup X \times B_0)
\end{array}$$

commutes.[23]

Proof We expand the diagram into a larger one, where all unlabeled homomorphisms are induced by inclusion mappings. Let, $W = A \times Y \cup X \times B$ $U = A_0 \times Y \cup X \times B$, $V = X \times B_0 \cup A \times Y$, $U_0 = A_0 \times Y \cup A \times B$ and $V_0 = X \times B_0 \cup A \times B$. We consider the diagram shown on page 249.

We apply 23.35 and naturality to prove that the part of the diagram not involving α and β commutes. α is induced from a homeomorphism

$$A \times Y/U_0 \vee X \times B/V_0 \equiv W/U_0 \cup V_0$$

(see 24.2). Thus α is an isomorphism and is given by adding the values of the homomorphisms induced by the inclusions. β is given on each coordinate by the induced homomorphisms of the inclusions. Since

$$(A \times Y, U_0) \subset (W, U_0 \cup V_0) \subset (W, V)$$

factors through the pair $(A \times Y, A \times Y)$, and $G_*(A \times Y, A \times Y) = 0$, it induces the zero homomorphism. Similarly

$$(X \times B, V_0) \subset (W, U)$$

induces the zero homomorphism. Hence the triangle commutes, and β is an isomorphism. The other isomorphism is an excision. Clearly $\beta\partial = \partial \oplus \partial$. Since β is a monomorphism, the diagram in 24.5 commutes. ∎

An analogous result is true for cohomology with essentially the same proof.

[23] In order to makes this commutative we define a degree k homomorphism φ between graded groups as a sequence of homomorphisms: $\varphi_k\colon G_n \to G'_{n+k}$. Write $|\varphi| = k$. Then define the tensor product of homomorphisms $\varphi \otimes \phi$ by

$$(\varphi \otimes \phi)(x \otimes y) = (-1)^{|x||\phi|}\, \varphi(x) \otimes \phi(y).$$

Thus

$$(\partial \otimes 1 + 1 \otimes \partial)(x \otimes y) = \partial x \otimes y + (-1)^{|x|}\, x \otimes \partial y.$$

$$
\begin{array}{ccccc}
E_*(X, A) \otimes F_*(Y, B) & & \xrightarrow{\quad\underline{\times}\quad} & & G_*(X \times Y, W) \\
\downarrow{\scriptstyle \partial \otimes 1 + 1 \otimes \partial} & & & \swarrow{\scriptstyle \partial + \partial} & \\
\{E_*(A) \otimes F_*(Y, B)\} \oplus \{E_*(X, A) \otimes F_*(B)\} & \xrightarrow{\underline{\times}} & G_*(A \times Y, A \times B) \oplus G_*(X \times B, A \times B) & \xrightarrow{\cong} & G_*(W, X \times B) \oplus G_*(W, A \times Y) \quad \Big\downarrow{\scriptstyle \partial} \\
\downarrow & & \downarrow & & \downarrow \\
\{E_*(A, A_0) \otimes F_*(Y, B)\} \oplus \{E_*(X, A) \otimes F_*(B, B_0)\} & \xrightarrow{\underline{\times}} & G_*(A \times Y, U_0) \oplus G_*(X \times B, V_0) & \xrightarrow{\cong} & G_*(W, U) \oplus G_*(W, V) \\
 & & & \searrow{\scriptstyle \alpha} & \downarrow{\scriptstyle \beta} \\
 & & & & G_*(W, U \cup V)
\end{array}
$$

Lemma 24.6 Let $X \supset A \supset A_0$ and $Y \supset B \supset B_0$ be relative CW complexes, and suppose there is a pairing from E and F to G. Then the diagram

$$\begin{array}{ccc} E^*(X,A) \otimes F^*(Y,B) & \xrightarrow{\bar{\times}} & G^*(X \times Y, X \times B \cup A \times Y) \\ \downarrow{\scriptstyle \delta\otimes 1 + 1 \otimes \delta} & & \downarrow{\scriptstyle \delta} \\ \{E^*(A,A_0) \otimes F^*(Y,B)\} \oplus \{E^*(X,A) \otimes F^*(B,B_0)\} & \xrightarrow{\bar{\times}} & G^*(X \times B \cup A \times Y, A_0 \times Y \cup A \times B \cup X \times B_0) \end{array}$$

commutes.[24] ∎

From this we conclude:

Corollary 24.7 Let (X, A) and (Y, B) be relative CW complexes with a finite number of cells in each dimension. Then

$$\Lambda^{\#}: C^{\#}(X, A; \pi) \otimes C^{\#}(Y, B; \rho) \to C^{\#}(X \times Y, X \times B \cup A \times Y; \pi \otimes \rho)$$

is a chain isomorphism. ∎

An important special case of 24.4 is when $Y = P$, $B = \varnothing$, and $\pi = Z$.

Corollary 24.8 There is a chain isomorphism

$$C_{\#}(X, A) \otimes \rho \to C_{\#}(X, A; \rho). \quad ∎$$

This determines $C_{\#}(X, A; \rho)$ given $C_{\#}(X, A)$.

One can also determine $C^{\#}(X, A; \pi)$ by studying the Kronecker product (Exercise 7, Section 23):

$$\langle\ ,\ \rangle: E^s(X, A) \otimes F_s(X, A) \to G_0(P).$$

By 23.35(g) we have

$$(-1)^p\langle z, \partial\omega\rangle = \langle \delta z, \omega\rangle$$

for $z \in E^p(A)$ and $\omega \in F_p(X, A)$. The adjoint of $\langle\ ,\ \rangle$

$$d: E^s(X, A) \to \hom(F_s(X, A), G_0(P))$$

is given by

$$d(x)(y) = \langle x, y\rangle,$$

and consequently

$$d(\delta\alpha)(\omega) = \langle \delta\alpha, \omega\rangle = (-1)^p\langle \alpha, \partial\omega\rangle = (-1)^p d(\alpha)(\partial\omega).$$

If C_n is a chain complex and D is a group, we make $C^n = \hom(C_n, D)$ into a cochain complex by defining

$$\delta^n: C^n \to C^{n+1}$$

[24] See footnote 23.

via $\delta^n(f)(c) = (-1)^n f(\partial c)$. Applying this with $E = G = H\pi$ and $F = HZ$, we have proved:

Lemma 24.9

$$d\colon C^n(X, A; \pi) \to \hom(C_n(X, A), \pi)$$

is a chain map. ▌

Proposition 24.10

$$d\colon C^n(X, A; \pi) \to \hom(C_n(X, A), \pi)$$

is a chain isomorphism.

Proof We show that in each degree, d is an isomorphism. This follows from

Lemma 24.11 If X is a wedge of spheres,

$$d\colon H^n(X; \pi) \to \hom(H_n(X), \pi)$$

is an isomorphism.

Proof In case $X = S^n$, this is an isomorphism since

$$\langle\ ,\ \rangle\colon \tilde{H}^n(S^n; \pi) \otimes \tilde{H}_n(S^n) \to \pi$$

is an isomorphism by 23.11. Now consider the commutative diagram

$$\begin{array}{ccc}
\tilde{H}^n\Big(\bigvee_{\alpha \in A} S_\alpha{}^n; \pi\Big) & \xrightarrow{\ d\ } & \hom\Big(\tilde{H}_n\Big(\bigvee_{\alpha \in A} S_\alpha{}^n\Big), \pi\Big) \\
\Big\downarrow \cong & & \Big\downarrow \cong \\
 & & \hom\Big(\bigoplus_{\alpha \in A} \tilde{H}_n(S_\alpha{}^n), \pi\Big) \\
 & & \Big\downarrow \cong \\
\prod_{\alpha \in A} \tilde{H}^n(S_\alpha{}^n; \pi) & \xrightarrow{\ \Pi d\ } & \prod_{\alpha \in A} \hom(\tilde{H}_n(S_\alpha{}^n), \pi)
\end{array}$$

in which we apply Exercises 13, Section 15 and 18.17 to see that d is an isomorphism. Note, however, that

$$\langle\ ,\ \rangle\colon \tilde{H}^n\Big(\bigvee_{\alpha \in A} S_\alpha{}^n; \pi\Big) \otimes \tilde{H}_n\Big(\bigvee_{\alpha \in A} S_\alpha{}^n\Big) \longrightarrow \pi$$

is not in general an isomorphism. ▌▌

24.4, 24.7, 24.8, and 24.10 are useful in calculating various homology and cohomology groups as we will see with an example.

Proposition 24.12 Let p be a prime.

(a)

$$H^i(RP^n) = \begin{cases} Z & \text{if } i=0 \text{ or } i=n \text{ and } n \text{ is odd} \\ Z_2 & \text{if } i \text{ is even and } 0<i<n \\ 0 & \text{otherwise.} \end{cases}$$

(b)

$$H_i(RP^n; Z_p) = \begin{cases} Z_2 & \text{if } p=2 \text{ and } 0 \le i \le n \\ Z_p & \text{if } i=0 \text{ or } i=n \text{ and } n \text{ is odd} \\ 0 & \text{otherwise.} \end{cases}$$

(c)

$$H^i(RP^n; Z_p) = \begin{cases} Z_2 & \text{if } p=2 \text{ and } 0 \le i \le n \\ Z_p & \text{if } i=0 \text{ or } i=n \text{ and } n \text{ is odd} \\ 0 & \text{otherwise.} \end{cases}$$

(d)

$$H_i(RP^2 \times RP^3) = \begin{cases} Z & \text{if } i=0 \\ Z_2 \oplus Z_2 & \text{if } i=1 \\ Z_2 & \text{if } i=2 \\ Z \oplus Z_2 & \text{if } i=3 \\ Z_2 & \text{if } i=4. \end{cases}$$

Proof (a) $C^i(RP^n) \cong Z$ with a generator e^i dual to e_i for $0 \le i \le n$ and $\delta e^{i-1} = -(1+(-1)^i)e^i$ by 24.9; for

$$\begin{aligned}(\delta e^{i-1})(e_i) &= (-1)^{i-1}e^{i-1}(\partial e_i) \\ &= (-1)^{i-1}e^{i-1}((1+(-1)^i)e_{i-1}) \\ &= -(1+(-1)^i)e^{i-1}(e_{i-1}) \\ &= -(1+(-1))^i.\end{aligned}$$

Hence $Z^i = C^i$ if i is even or $i=n$ and $Z^i = 0$ otherwise, $B^i = 2C^i$ if i is even and $i>0$, and $B^i = 0$ otherwise. Thus (a) follows. To prove (b), we note that if $p=2$, $\partial = 0$. Hence $H_i(RP^n; Z_2) = C_i(RP^n; Z_2)$. If $p \ne 2$, $\partial\colon C_{2i} \to C_{2i-1}$ is an isomorphism. Hence the only cycles are in dimension 0 and n if n is odd. To prove (c), observe that the cochain complex has the same form as (a) except all calculations take place in Z_p instead of Z. Thus if $p=2$, $\delta=0$, and if $p \ne 2$, $\delta\colon C^{2i-1} \to C^{2i}$ is an isomorphism. Thus (c) follows.

(d) is proved by writing down explicit generators for

$$C_\#(RP^2) \otimes C_\#(RP^3).$$

Write $e_i \otimes e_j = e_{i,j}$. Then we have a table of generators:

i			
0	$e_{0,0}$		
1	$e_{0,1}$	$e_{1,0}$	
2	$e_{0,2}$	$e_{1,1}$	$e_{2,0}$
3	$e_{0,3}$	$e_{1,2}$	$e_{2,1}$
4	$e_{1,3}$	$e_{2,2}$	
5	$e_{2,3}$		

The boundary is determined by the formulas:

$$\begin{aligned}
\partial e_{2,3} &= \partial(e_2 \otimes e_3) = \partial e_2 \otimes e_3 + e_2 \otimes \partial e_3 = 2e_1 \otimes e_3 = 2e_{1,3}\\
\partial e_{1,3} &= \partial(e_1 \otimes e_3) = \partial e_1 \otimes e_3 - e_1 \otimes \partial e_3 = 0\\
\partial e_{2,2} &= \partial(e_2 \otimes e_2) = \partial e_2 \otimes e_2 + e_2 \otimes \partial e_2 = 2e_1 \otimes e_2 + 2e_2 \otimes e_1\\
&= 2(e_{1,2} + e_{2,1})\\
\partial e_{0,3} &= \partial(e_0 \otimes e_3) = \partial e_0 \otimes e_3 + e_0 \otimes \partial e_3 = 0.
\end{aligned}$$

Similarly we have

$$\begin{aligned}
\partial e_{1,2} &= -2e_{1,1} & \partial e_{2,0} &= 2e_{1,0}\\
\partial e_{2,1} &= 2e_{1,1} & \partial e_{0,1} &= 0\\
\partial e_{0,2} &= 2e_{0,1} & \partial e_{1,0} &= 0\\
\partial e_{1,1} &= 0 & \partial e_{0,0} &= 0.
\end{aligned}$$

As a free basis for ker ∂ we have

i	$Z_i = \ker \partial_i$	
0	$e_{0,0}$	
1	$e_{0,1}$	$e_{1,0}$
2	$e_{1,1}$	
3	$e_{0,3}$	$e_{1,2} + e_{2,1}$
4	$e_{1,3}$	
5	0	

As a basis for $B_i = \operatorname{Im} \partial_{i+1}$ we have

i	$B_i = \operatorname{Im} \partial_{i+1}$	
0	0	
1	$2e_{0,1}$	$2e_{1,0}$
2	$2e_{1,1}$	
3	$2(e_{1,2} + e_{2,1})$	
4	$2e_{1,3}$	
5	0	

The quotient, $H_i = Z_i/B_i$ is thus generated by the following classes:

i	H_i	order of generators
0	$\{e_{0,0}\}$	∞
1	$\{e_{0,1}\}$, $\{e_{1,0}\}$	2, 2
2	$\{e_{1,1}\}$	2
3	$\{e_{0,3}\}$, $\{e_{1,2}+e_{2,1}\}$	∞, 2
4	$\{e_{1,3}\}$	2
5	0	

and this completes the calculation. ▮

These calculations demonstrate how knowledge of $C_\#(X, A)$ determines $H_*(X, A; \pi)$ and $H^*(X, A; \pi)$ and knowledge of both $C_\#(X, A)$ and $C_\#(Y, B)$ determines $H_*(X \times Y, X \times B \cup A \times Y; \pi)$ and $H^*(X \times Y, X \times B \cup A \times Y; \pi)$. They do not hint at the remarkable fact from homological algebra that these homology and cohomology groups depend only on $H_*(X, A)$ and that $H_*(Y, B)$. This is the subject of the next section.

Exercises

1. Let Q be the rational numbers. Show that if $A \subset B$, $Q \otimes A \subset Q \otimes B$. (Use Exercise 6, Section 15.) Conclude that $Q \otimes (\prod_{n=2}^{\infty} Z_n) \neq 0$. Hence the natural transformation $H \otimes (\prod Z_n) \to \prod (H \otimes Z_n)$ is not 1–1 in general since $Q \otimes Z_n = 0$.

2. Calculate $H^i(RP^2 \times RP^3)$.

3. Let Q be the rational numbers. Calculate $H_i(RP^n; Q)$ and $H^i(RP^n; Q)$.

4. Let $f\colon C \to C'$ and $g\colon D \to D'$ be chain maps. Show that $f \otimes g\colon C \otimes C' \to D \otimes D'$ is a chain map. (25.5)

5. Show that if C is a free chain complex, $Z(C)$ is a direct summand. Conclude that there is a chain map $f\colon C \to H(C)$ (where $H(C)$ is considered as a chain complex with 0 differential) which induces an isomorphism in homology.

6. Show that there is a chain isomorphism

$$C^\#(X, A) \otimes \pi \to C^\#(X, A; \pi)$$

if (X, A) is a relative CW complex with a finite number of cells in each dimension. (Hint: Find a natural isomorphism $\hom(A, B) \otimes C \to \hom(A, B \otimes C)$ when A is free and finitely generated.)

7. Let R be a commutative ring. Show that there is a natural isomorphism

$$C_\#(X, A; R) \otimes_R C_\#(Y, B; R) \to C_\#(X \times Y, X \times B \cup A \times Y; R)$$

where (X, A) and (Y, B) are relative CW complexes. Furthermore, if (X, A) and (Y, B) have a finite number of cells in each dimension, there is a natural isomorphism

$$C^{\#}(X, A; R) \otimes_R C^{\#}(Y, B; R) \to C^{\#}(X \times Y, X \times B \cup A \times Y; R).$$

25

Homological Algebra over a Principal Ideal Domain: Künneth and Universal Coefficient Theorems

In this section we develop homological algebra over a principal ideal domain R. This will be applied to the chain isomorphisms of Section 24 to prove the Künneth formulas and universal coefficient theorems. A more general treatment of homological algebra can be found in any of the standard texts on homological algebra (see [18, 44, 55]). Since a principal ideal domain by definition is commutative, we make no distinction between left and right R-modules. A principal ideal domain has the following characteristic property.

Proposition 25.1 Let R be a principal ideal domain, M a free R-module, and $N \subset M$ a submodule. Then N is free.

Proof Let $\{x_\alpha\}$, $\alpha \in A$ be a basis for M and suppose A is well ordered. Let M_β be the submodule generated by $\{x_\alpha \mid \alpha \leq \beta\}$. Let $f_\alpha\colon M \to R$ be given by $f_\alpha(x_\alpha) = 1$, $f_\alpha(x_\beta) = 0$ for $\beta \neq \alpha$. f_α extends to an R-module homomorphism since M is free. Thus $f_\alpha(N \cap M_\alpha)$ is an ideal in R and we have $f_\alpha(N \cap M_\alpha) = (r_\alpha)$. Let $T = \{\alpha \in A \mid r_\alpha \neq 0\}$ and choose for each $\alpha \in T$, $n_\alpha \in N \cap M_\alpha$ with $f_\alpha(n_\alpha) = r_\alpha$. Let $\overline{N}$ be a free R-module with one generator c_α for each $\alpha \in T$. Define $f\colon \overline{N} \to N$ by

$$f\left(\sum_{i=1}^{p} \tilde{r}_{\alpha_i} c_{\alpha_i}\right) = \sum_{i=1}^{p} \tilde{r}_{\alpha_i} n_{\alpha_i}.$$

We claim that f is an isomorphism. f is clearly a homomorphism of R-modules since $\bar{N}$ is free. Suppose

$$f\left(\sum_{i=1}^{k} \tilde{r}_{\alpha_i} c_{\alpha_i}\right) = 0 \qquad \text{and} \qquad \alpha_1 < \alpha_2 < \cdots < \alpha_k.$$

Then $\sum_{i=1}^{k-1} \tilde{r}_{\alpha_i} n_{\alpha_i} = -\tilde{r}_{\alpha_k} n_{\alpha_k}$. Since $f_{\alpha_k}(M_\alpha) = 0$ for $\alpha < \alpha_k$, $f_{\alpha_k}(n_\alpha) = 0$ for $\alpha < \alpha_k$. Hence

$$\tilde{r}_{\alpha_k} r_{\alpha_k} = f_{\alpha_k}(\tilde{r}_{\alpha_k} n_{\alpha_k}) = -f_{\alpha_k}\left(\sum_{i=1}^{k-1} \tilde{r}_{\alpha_i} n_{\alpha_i}\right) = 0,$$

and thus $\tilde{r}_{\alpha_k} = 0$. Continuing in this way we see that $\tilde{r}_{\alpha_i} = 0$ for all i. Hence $\sum_{i=1}^{k} \tilde{r}_{\alpha_i} c_{\alpha_i} = 0$. f is consequently 1–1. If f is not onto, choose $\bar{\alpha}$ to be the smallest element of A for which $f(\bar{N}) \not\supset M_{\bar{\alpha}} \cap N$. Choose $x \in M_{\bar{\alpha}} \cap N - f(\bar{N})$. Write $x = \sum_{i=1}^{k} \tilde{r}_{\alpha_i} x_{\alpha_i}$. If $\alpha_1 \le \cdots \le \alpha_k$ it follows that $\alpha_k = \bar{\alpha}$. Let $f_{\bar{\alpha}}(x) = r r_{\bar{\alpha}}$. Then $f_{\bar{\alpha}}(x - r n_{\bar{\alpha}}) = 0$. Hence $x - r n_{\bar{\alpha}} \in M_{\bar{\alpha}} \cap N$ for some $\alpha < \bar{\alpha}$. By the choice of $\bar{\alpha}$, $x - r n_{\bar{\alpha}} \in f(\bar{N})$, so $x \in f(\bar{N})$ and we have a contradiction. Thus f is an isomorphism. ∎

This is the only property of a principal ideal domain that we require. In most applications we have $R = Z$ or R will be a field.

Given modules M and N we define R-modules $\operatorname{Tor}_R(M, N)$ and $\operatorname{Ext}_R(M, N)$ as follows. According to 25.1 we can find a short exact sequence

$$0 \to F_1 \xrightarrow{\alpha} F_2 \xrightarrow{\beta} M \to 0$$

of R-modules with F_1 and F_2 free. Such an exact sequence will be called a resolution of M. Define $\operatorname{Tor}_R(M, N)$ to be the kernel of

$$F_1 \otimes_R N \xrightarrow{\alpha \otimes 1} F_2 \otimes_R N$$

and $\operatorname{Ext}_R(M, N)$ to be the cokernel of

$$\hom_R(F_2, N) \xrightarrow{\alpha^*} \hom_R(F_1, N).$$

Proposition 25.2 $\operatorname{Tor}_R(M, N)$ and $\operatorname{Ext}_R(M, N)$ do not depend on the resolution of M. $\operatorname{Tor}_R(M, N)$ is a covariant functor of M and N. $\operatorname{Ext}_R(M, N)$ is contravariant in M and covariant in N.

Proof Let $f\colon M \to M'$ and suppose we are given resolutions

$$(E) \qquad 0 \to F_1 \xrightarrow{\alpha} F_2 \xrightarrow{\beta} M \to 0$$

$$(E') \qquad 0 \to F_1' \xrightarrow{\alpha'} F_2' \xrightarrow{\beta'} M' \to 0$$

with which we calculate Tor and Ext. We will construct homomorphisms

$$f_*: \mathrm{Tor}_R(M, N) \to \mathrm{Tor}_R(M', N)$$

$$f^*: \mathrm{Ext}_R(M', N) \to \mathrm{Ext}_R(M, N)$$

by using the following:

Lemma 25.3 Let $\phi: N \to N'$ be an epimorphism of R-modules, let F be a free R-module, and let $\psi: F \to N'$: then there is an R-module morphism $\lambda: F \to N$ such that $\phi\lambda = \psi$.

Proof Let $\{x_\alpha\}$ be a basis for F. A homomorphism $\lambda: F \to N$ is determined by the images $\lambda(x_\alpha)$. Define $\lambda(x_\alpha)$ to be any element in $\phi^{-1}(\psi(x_\alpha))$. This defines λ and $\phi\lambda(x_\alpha) = \psi(x_\alpha)$. Hence $\phi\lambda = \psi$. ∎

We now construct homomorphisms f_1 and f_2 forming a commutative diagram

$$(D) \qquad \begin{array}{ccccccccc} 0 & \longrightarrow & F_1 & \xrightarrow{\alpha} & F_2 & \xrightarrow{\beta} & M & \longrightarrow & 0 \\ & & \downarrow{\scriptstyle f_1} & & \downarrow{\scriptstyle f_2} & & \downarrow{\scriptstyle f} & & \\ 0 & \longrightarrow & F_1' & \xrightarrow{\alpha'} & F_2' & \xrightarrow{\beta'} & M' & \longrightarrow & 0 \end{array}$$

Since β' is onto and F_2 is free 25.3 implies that f_2 exists such that $\beta' f_2 = f\beta$. Since $0 = f\beta\alpha = \beta' f_2 \alpha, f_2(\alpha(F_1)) \subset \alpha'(F_1')$. Thus $f_2 \alpha: F_1 \to \mathrm{Im}\, \alpha'$ and applying 25.3 again we can find f_1 such that $\alpha' f_1 = f_2 \alpha$.

Now for any such commutative diagram D we can define homomorphisms

$$f_D = f_1 \otimes 1: \ker \alpha \otimes 1 \to \ker \alpha' \otimes 1$$

$$f^D = f_1^*: \mathrm{ckr}\, (\alpha')^* \to \mathrm{ckr}\, \alpha^*.$$

Thus choosing E and E' to calculate Tor and Ext, and choosing D, we get transformations

$$\mathrm{Tor}_R(M, N) \to \mathrm{Tor}_R(M', N), \qquad \mathrm{Ext}_R(M', N) \to \mathrm{Ext}_R(M, N).$$

We claim that these homomorphisms do not depend on the choice of D. Suppose f_1' and f_2' are chosen instead of f_1 and f_2 (keeping the same resolutions E and E' as before). Then $\beta'(f_2 - f_2') = 0$, so applying 25.3 we can construct $\phi: F_2 \to F_1'$ such that $\alpha'\phi = f_2 - f_2'$. Now $\alpha'\phi\alpha = f_2\alpha - f_2'\alpha = \alpha'(f_1 - f_1')$, hence $\phi\alpha = f_1 - f_1'$ since α' is a monomorphism. Consequently $f_1 \otimes 1 - f_1' \otimes 1 = 0$ on $\ker \alpha \otimes 1$ and $f_1^* - f_1'^* = 0$ on $\mathrm{ckr}\, \alpha^*$. Thus f_D and f^D do not depend on the choice of f_1 and f_2.

We have shown that for any choice of resolutions E and E', there are well-defined induced homomorphisms

$$\mathrm{Tor}_R(M, N) \xrightarrow{f_*} \mathrm{Tor}_R(M', N), \qquad \mathrm{Ext}_R(M', N) \xrightarrow{f_*} \mathrm{Ext}(M, N).$$

Clearly by definition $(fg)_* = f_* g_*$, $1_* = 1$, $(fg)^* = g^* f^*$, and $1^* = 1$. We can now compare the values of Tor or Ext using two different resolutions of M by taking $f = 1$. It follows that there are homomorphisms going both ways such that the composites are the identity. Thus Ext and Tor do not depend on the resolution and are functors of M. That they are covariant functors of N follows easily from the definition. Hence 25.2 is proved. ∎

Clearly if R is a field $\mathrm{Tor}_R(M, N) = \mathrm{Ext}_R(M, N) = 0$ for we may choose $F_2 = M$ and $F_1 = 0$. If $R = Z$, we abbreviate $\mathrm{Tor}_Z(M, N)$ and $\mathrm{Ext}_Z(M, N)$ by $\mathrm{Tor}(M, N)$ and $\mathrm{Ext}(M, N)$.

Proposition 25.4

(a) $\mathrm{Tor}_R(M, N) \cong \mathrm{Tor}_R(N, M) \cong \mathrm{Ext}_R(M, N) = 0$ if M is free.
(b) $\mathrm{Ext}(Z_n, Z) \cong Z_n$.
(c) $\mathrm{Tor}(Z_m, Z_n) \cong \mathrm{Ext}(Z_m, Z_n) \cong Z_k$ where k is the greatest common divisor of m and n.
(d) $\mathrm{Tor}_R(M, N \oplus N') \cong \mathrm{Tor}_R(M, N) \oplus \mathrm{Tor}_R(M, N')$.
$\mathrm{Ext}_R(M, N \oplus N') \cong \mathrm{Ext}_R(M, N) \oplus \mathrm{Ext}_R(M, N')$.
(e) $\mathrm{Tor}_R(M \oplus M', N) \cong \mathrm{Tor}_R(M, N) \oplus \mathrm{Tor}_R(M', N)$.
$\mathrm{Ext}_R(M \oplus M', N) \cong \mathrm{Ext}_R(M, N) \oplus \mathrm{Ext}_R(M', N)$.

Proof (a) To calculate $\mathrm{Tor}_R(M, N)$ and $\mathrm{Ext}_R(M, N)$ take $F_1 = 0$ and $F_2 = M$. To calculate $\mathrm{Tor}_R(N, M)$ choose a basis $\{m_\alpha\}$ for M and note that if N is another module, every element in $N \otimes_R M$ can be written uniquely in the form $\sum n_\alpha \otimes m_\alpha$. Thus if $\sum \alpha(x_\alpha) \otimes m_\alpha = 0$, $\alpha(x_\alpha) = 0$ and hence $x_\alpha = 0$ so $\sum x_\alpha \otimes m_\alpha = 0$. Consequently $\alpha \otimes 1$ is a monomorphism.

(b) Take $0 \to Z \to Z \to Z_n \to 0$ as a resolution of Z_n and observe that $\hom(Z, Z) = Z$, so $\mathrm{Ext}(Z, Z_n)$ is the cokernel of $Z \xrightarrow{\times n} Z$.

(c) By considering the resolution $0 \to Z \to Z \to Z_m \to 0$ one sees that $\mathrm{Tor}(Z_m, Z_n)$ is the kernel of $Z_n \xrightarrow{\times m} Z_n$ and $\mathrm{Ext}(Z_m, Z_n)$ is the cokernel of this map. Since any subgroup or quotient group of a cyclic group is cyclic, one sees that the kernel and cokernel are isomorphic, for they have the same order. Let $k = \gcd(m, n)$. There is a homomorphism $Z_k \to Z_n$ mapping 1 to n/k. The sequence

$$Z_k \longrightarrow Z_n \xrightarrow{\times m} Z_n$$

is exact, for $mn/k = n(m/k)$ and hence the composite is 0; on the other hand if $x \in Z_n$ is such that $mx \equiv 0$, we have $kx \equiv amx + bnx \equiv 0$ since $k = am + bn$ for some a and b. Hence n divides kx and $x = c(n/k)$.

To prove (d), consider the distributive laws

$$F \otimes_R (N \oplus N') = F \otimes_R N \oplus F \otimes_R N'$$
$$\hom_R(F, N \oplus N') = \hom_R(F, N) \oplus \hom_R(F, N')$$

applied to $F = F_1$ or F_2 to see that ker $\alpha \otimes 1$ and coker α^* split into a direct sum. To prove (e), consider resolutions $0 \to F_1 \to F_2 \to M \to 0$ and $0 \to F_1' \to F_2' \to M' \to 0$ and observe that

$$0 \to F_1 \oplus F_1' \to F_2 \oplus F_2' \to M \oplus M' \to 0$$

is a resolution. Applying the distributive laws above proves (e). ∎

Let C be a chain complex of R-modules and consider $Z(C)$ as a subcomplex with 0 differential. Define a chain complex B by $B_n = B_{n-1}(C)$ with 0 differential. We then have a short exact sequence of chain complexes:

$$0 \to Z(C) \xrightarrow{\iota} C \xrightarrow{\delta} B \to 0.$$

Suppose now D is a chain complex and consider the sequence:

$$C \otimes_R Z(D) \xrightarrow{\iota \otimes 1} C \otimes_R D \xrightarrow{\partial \otimes 1} C \otimes_R D \to 0.$$

$\iota \otimes 1$ and $\partial \otimes 1$ are chain maps by Exercise 4, Section 24. This is exact by 23.9. Thus by 25.4(a) we have:

Lemma 25.5 If C or D is R-free,

$$0 \to C \otimes_R Z(D) \xrightarrow{\iota \otimes 1} C \otimes_R D \xrightarrow{\partial \otimes 1} C \otimes_R D \to 0$$

is an exact sequence of chain complexes. ∎

We now apply Exercise 9, Section 20 to produce a long exact sequence

$$\cdots \to H(B \otimes_R D) \xrightarrow{\partial} H(Z(C) \otimes_R D) \xrightarrow{H(\iota \otimes 1)} H(C \otimes_R D)$$
$$\xrightarrow{H(\partial \otimes 1)} H(B \otimes_R D) \xrightarrow{\bar{\partial}} H(Z(C) \otimes_R D) \to \cdots$$

We will utilize this to calculate $H(C \otimes_R D)$.

Lemma 25.6 Let C be an R-free chain complex with 0 differential, and D be an arbitrary chain complex. Then

$$H(C \otimes_R D) = C \otimes_R H(D).$$

Proof The differential in $C \otimes_R D$ is $\pm 1 \otimes \partial$. We have short exact sequences

$$0 \to C \otimes_R Z(D) \xrightarrow{1 \otimes \iota} C \otimes_R D \xrightarrow{1 \otimes \partial} C \otimes_R B(D) \longrightarrow 0$$

$$0 \to C \otimes_R B(D) \longrightarrow C \otimes_R D \longrightarrow C \otimes_R (D/B(D)) \to 0$$

by 23.9 and 25.4(a).

Hence Im $1 \otimes \partial = C \otimes_R B(D)$ and ker $1 \otimes \partial = C \otimes_R Z(D)$. Applying 23.9 and 25.4(a) again, we have a short exact sequence

$$0 \to C \otimes_R B(D) \to C \otimes_R Z(D) \to C \otimes_R H(D) \to 0. \quad \blacksquare$$

By 25.6, $H(Z(C) \otimes_R D) = Z(C) \otimes_R H(D)$ and $H(B \otimes_R D) = B \otimes_R H(D)$. Now consider the composite

$$\{B(C) \otimes_R H(D)\}_k = \{B \otimes_R H(D)\}_{k+1} \xrightarrow{\bar{\partial}} \{Z(C) \otimes_R H(D)\}_k$$

where $\bar{\partial}$ is the homomorphism defined in Exercise 9, Section 20. $\bar{\partial}$ is calculated as follows. Given $x \otimes \{y\} \in B(C) \otimes_R H(D)$, choose $u \in C$ so that $x = \partial u$. Then

$$(\partial \otimes 1)(u \otimes \{y\}) = x \otimes \{y\}.$$

Now calculate $\partial(u \otimes y)$ in $C \otimes_R D$. $\partial(u \otimes y) = \partial u \otimes y \pm u \otimes \partial y = x \otimes y$ since $y \in Z(D)$. $x \otimes y$ is in the image of $Z(C) \otimes_R D \to C \otimes_R D$ and $\bar{\partial}(x \otimes \{y\})$ is its homology class in $H(Z(C) \otimes_R D)$. Thus $\bar{\partial} = j \otimes 1$ where j: $B(C) \to Z(C)$ is the inclusion. To calculate ker $j \otimes 1$ and coker $j \otimes 1$ we observe that $0 \to B(C) \xrightarrow{j} Z(C) \to H(C) \to 0$ is a resolution. Hence we have an exact sequence

$$0 \to \operatorname{Tor}_R(H(C), H(D)) \to B(C) \otimes_R H(D) \xrightarrow{j \otimes 1} Z(C) \otimes_R H(D)$$
$$\to H(C) \otimes_R H(D) \to 0.$$

This proves the first part of:

Theorem 25.7 If C is R-free, there is a natural exact sequence

$$0 \to H(C) \otimes_R H(D) \xrightarrow{\nu} H(C \otimes_R D) \xrightarrow{\Delta} \operatorname{Tor}_R(H(C), H(D)) \to 0$$

where $\nu(\{x\} \otimes \{y\}) = \{x \otimes y\}$ and Δ has degree -1.

Proof The statement about ν follows since

$$\nu(\{x\} \otimes \{y\}) = H(\iota \otimes 1)(x \otimes \{y\}) = \{\iota x \otimes y\} = \{x \otimes y\}.$$

Δ has degree -1 since $\partial \otimes 1$ has degree -1. Naturality in chain maps is clear since all the constructions are natural. $\blacksquare$

To be more explicit, we write

$$0 \to \bigoplus_{n=i+j} H_i(C) \otimes_R H_j(D) \to H_n(C \otimes_R D) \to \bigoplus_{i+j=n-1} \mathrm{Tor}_R(H_i(C), H_j(D)) \to 0.$$

In fact, this sequence splits. That is:

Theorem 25.8 If both C and D are R-free chain complexes,

$$H_n(C \otimes_R D) \cong \left\{ \bigoplus_{n=i+j} H_i(C) \otimes_R H_j(D) \right\} \oplus \left\{ \bigoplus_{i+j=n-1} \mathrm{Tor}_R(H_i(C), H_j(D)) \right\}.$$

The isomorphism, however, is *not* natural as we shall see by an example. This follows from:

Proposition 25.9 If C and D are each either R-free or have 0 differential,

$$\nu: \bigoplus_{i+j=n} \{H_i(C) \otimes_R H_j(D)\} \to H_n(C \otimes_R D)$$

is the inclusion of a direct summand.

Proof We will find an R-module homomorphism

$$\gamma: H(C \otimes_R D) \to H(C) \otimes_R H(D)$$

such that $\gamma\nu = 1$. This is enough by Exercise 11, Section 11. We claim that if a chain complex C is either R-free or has 0 differential, $Z(C)$ is a direct summand. Under the second hypothesis this is trivial. Suppose C is R-free. Since R is a principal ideal domain, $B(C)$ is R-free and hence by 25.3 and Exercise 11, Section 11, $C_n = Z_n(C) \oplus B_{n-1}(C)$.

Suppose now that $Z(C)$ and $Z(D)$ are direct summands in C and D respectively. Let $\alpha: C \to Z(C)$ and $\beta: D \to Z(D)$ be the projections. They define $\phi: C \to H(C)$ and $\psi: D \to H(D)$ by $\phi(x) = \{\alpha(x)\}$ and $\psi(y) = \{\beta(y)\}$. Then $\phi \otimes \psi: C \otimes_R D \to H(C) \otimes_R H(D)$. $\phi \otimes \psi|_{B(C \otimes_R D)} = 0$ since $\phi(B(C)) = 0$ and $\psi(B(D)) = 0$. Hence $\phi \otimes \psi$ determines γ and clearly $\gamma\nu = 1$. ▮ ▮

Theorem 25.10 (*Universal Coefficient Theorem*) If C is R-free and π is an R-module, there is a natural exact sequence

$$0 \to H_n(C) \otimes_R \pi \to H_n(C \otimes_R \pi) \xrightarrow{\Delta} \mathrm{Tor}_R(H_{n-1}(C), \pi) \to 0$$

which splits (nonnaturally).

Proof We apply 25.7 with

$$D_n = \begin{cases} \pi, & n = 0 \\ 0, & n \neq 0. \end{cases}$$

D has 0 differential, so we may apply 25.9. ▮

By 24.4, 24.7, and 24.8, we have the corollaries:

Corollary 25.11 (*Künneth Formula*) If (X, A) and (Y, B) are relative CW complexes, there is a natural short exact sequence

$$0 \to \bigoplus_{i+j=n} H_i(X, A) \otimes H_j(Y, B) \xrightarrow{\times} H_n(X \times Y, X \times B \cup A \times Y)$$
$$\to \bigoplus_{i+j=n-1} \operatorname{Tor}(H_i(X, A), H_j(Y, B)) \to 0$$

which splits (nonnaturally). ▮

Corollary 25.12 (*Künneth Formula*) If (X, A) and (Y, B) are relative CW complexes with a finite number of cells in each dimension, there is a natural short exact sequence

$$0 \to \bigoplus_{i=j=n} H^i(X, A) \otimes H^j(Y, B) \xrightarrow{\overline{\times}} H^n(X \times Y, X \times B \cup A \times Y)$$
$$\to \bigoplus_{i+j=n-1} \operatorname{Tor}(H^i(X, A), H^i(Y, B)) \to 0$$

which splits (nonnaturally). ▮

Corollary 25.13 (*Universal Coefficient Theorem I*) There is a natural short exact sequence

$$0 \to H_n(X, A) \otimes \pi \xrightarrow{d} H_n(X, A; \pi) \xrightarrow{\Delta} \operatorname{Tor}(H_{n-1}(X, A), \pi) \to 0$$

which splits (nonnaturally). ▮

Corollary 25.14 (*Universal Coefficient Theorem II*) Let (X, A) be a relative CW complex and assume that π is finitely generated. Then there is a natural short exact sequence

$$0 \to H^n(X, A) \otimes \pi \to H^n(X, A; \pi) \to \operatorname{Tor}(H^{n+1}(X, A), \pi) \to 0$$

which splits (nonnaturally).

Proof There is an isomorphism of chain complexes

$$C^n(X, A) \otimes \pi \to C^n(X, A; \pi)$$

given by

$$C^n(X, A) \otimes \pi \cong \hom(C_n(X, A), Z) \otimes \pi \xrightarrow{\phi} \hom(C_n(X, A), \pi) \cong C^n(X, A; \pi)$$

where $\phi(f \otimes x)(c) = f(c) \cdot x$. ϕ is an isomorphism since π is finitely generated. 25.14 now follows from 25.10. ▮

Let k be a field. Then there are chain isomorphisms

$$C_{\#}(X, A; k) \otimes_k C_{\#}(Y, B; k) \xrightarrow{\cong} C_{\#}(X \times Y, X \times B \cup A \times Y; k)$$
$$C^{\#}(X, A; k) \otimes_k C^{\#}(Y, B; k) \xrightarrow{\cong} C^{\#}(X \times Y, X \times B \cup A \times Y; k)$$

induced by 24.4 and 24.7. These induce Künneth formulas:

Corollary 25.15 Let k be a field and suppose (X, A) and (Y, B) are relative CW complexes. Then

$$\bigoplus_{i+j=n} H_i(X, A; k) \otimes_k H_j(Y, B; k) \xrightarrow{\times} H_n(X \times Y, X \times B \cup A \times Y; k)$$

is an isomorphism. If (X, A) and (Y, B) have a finite number of cells in each dimension,

$$\bigoplus_{i+j=n} H^i(X, A; k) \otimes_k H^j(Y, B; k) \xrightarrow{\overline{\times}} H^n(X \times Y, X \times B \cup A \times Y; k)$$

is an isomorphism. ▮

Finally we will exploit 24.10 to prove:

Theorem 25.16 (*Universal Coefficient Theorem III*) Let (X, A) be a relative CW complex. Then there is a natural short exact sequence

$$0 \to \text{Ext}(H_{n-1}(X, A), \pi) \xrightarrow{\Delta} H^n(X, A; \pi) \xrightarrow{d} \text{hom}(H_n(X, A), \pi) \to 0$$

which splits (nonnaturally). d is the map adjoint to the Kronecker product (see the discussion before 24.9).

This follows from:

Lemma 25.17 Let C be an R-free chain complex and π be an R-module. Then there is a natural short exact sequence

$$0 \to \text{Ext}_R(H_{n-1}(C), \pi) \to H_n(\text{hom}_R(C, \pi)) \xrightarrow{d} \text{hom}_R(H_n(C), \pi) \to 0$$

which splits (nonnaturally). Furthermore

$$d(\{f\})(\{c\}) = \{f(c)\}.$$

The proof will be much the same as 25.7 and will depend on the following simple lemma which is the analogue of 23.9.

Lemma 25.18 Let $A \xrightarrow{\alpha} B \xrightarrow{\beta} C \to 0$ be exact. Then

$$0 \to \text{hom}_R(C, D) \xrightarrow{\beta^*} \text{hom}_R(B, D) \xrightarrow{\alpha^*} \text{hom}_R(A, D)$$

is exact for any D.

Proof Let $f: C \to D$ and suppose $\beta^*(f) = f\beta = 0$. Since β is onto $f(x) = f\beta(y) = 0$, for some $y \in B$. Hence $f = 0$. Clearly $\alpha^*\beta^* = (\beta\alpha)^* = 0$. If $\alpha^* f = 0$ for $f: B \to D$, $f|_{\mathrm{Im}\, A} = 0$ and hence f extends to a map $\bar{f}: B/\mathrm{Im}\ A \to D$. Since $C \cong B/\mathrm{Im}\ A$, $f \in \mathrm{Im}\ \beta^*$. ∎

Proof of 25.17 As in 25.7 we consider the exact sequence of chain complexes

$$0 \to Z(C) \xrightarrow{\iota} C \xrightarrow{\partial} B \to 0.$$

This yields an exact sequence

$$0 \to \hom_R(B, \pi) \xrightarrow{\partial^*} \hom_R(C, \pi) \xrightarrow{\iota^*} \hom_R(Z(C), \pi) \to 0$$

by 25.18 and 25.4(a) since B is R-free and

$$0 \to Z(C) \to C \to B \to 0$$

is a resolution of B. We now apply Exercise 9, Section 20 to produce a long exact sequence

$$\cdots \to H(\hom_R(Z(C), \pi)) \xrightarrow{\bar{\partial}} H(\hom_R(B, \pi)) \xrightarrow{(\partial^*)_*} H(\hom_R(C, \pi))$$
$$\xrightarrow{(\iota^*)_*} H(\hom_R(Z(C), \pi)) \to \cdots.$$

Now $H(\hom_R(Z(C), \pi)) \cong \hom_R(Z(C), \pi)$ and $H(\hom_R(B, \pi)) \cong \hom_R(B, \pi)$ since in each case the differential is 0.

Consider the composite

$$\hom_R(B_k(C), \pi) \cong \hom_R(B_{k+1}, \pi) \xleftarrow{\partial} \hom_R(Z_k(C), \pi).$$

This is $\pm j^*$ for let $f: Z_k(C) \to \pi$ and choose $\bar{f}: C \to \pi$ so that $\iota^*(\bar{f}) = f$. This is possible since ι^* is onto. Hence $\bar{\partial}\{f\} = \pm\{\bar{f}|B\} = \pm\{f|B\} = \pm\{j^*(f)\}$. We now have a short exact sequence

$$0 \to \mathrm{ckr}\, j^* \to H(\hom_R(C, \pi)) \to \ker j^* \to 0.$$

But since $0 \to B(C) \xrightarrow{j} Z(C) \to H(C)$ is a resolution, we have an exact sequence

$$0 \to \hom_R(H(C), \pi) \to \hom_R(Z(C), \pi) \to \hom_R(B(C), \pi) \to \mathrm{Ext}_R(H(C), \pi) \to 0$$

by 25.18 and the definition of Ext. The exact sequence of 25.17 follows. It is easy to check the formula for d, and that Δ increases degrees by 1 since ∂^* does. To find a splitting, as in 25.9 we see that $Z(C)$ is a direct summand in C. Let $\gamma: C \to Z(C)$ satisfy $\gamma\bar{\iota} = 1$. γ determines $\bar{\gamma}: C \to H(C)$ and thus

$$\bar{\gamma}^*: \hom_R(H(C), \pi) \to \hom_R(C, \pi).$$

Since $\bar{\gamma}(B(C)) = 0$, $\delta\bar{\gamma}^* = 0$ and $\bar{\gamma}^*$ determines a map $\tilde{\gamma}: \hom_R(H(C), \pi) \to H(\hom_R(C, \pi))$. Since $d\tilde{\gamma} = 1$, the sequence splits by 11.11. ∎ ∎

One simple consequence of 25.16 is the following:

Corollary 25.19 Suppose $H_n(X, A) \cong F_n \oplus T_n$ where F_n is a free abelian group of finite rank and T_n is a finite abelian group. Then $H^n(X, A) = F_n \oplus T_{n-1}$.

Proof $\hom(H_n(X, A), Z) = \hom(F_n, Z) = F_n$ and by Proposition 25.4, $\operatorname{Ext}(H_{n-1}(X, A), Z) = \operatorname{Ext}(T_{n-1}, Z) = T_{n-1}$ since T_{n-1} is a direct sum of cyclic groups. Hence the result follows from 25.16. ▮

Observe that the hypothesis of 25.19 is satisfied if (X, A) has a finite number of cells in each dimension.

Another universal coefficient theorem is given by the natural isomorphisms

$$C^{\#}(X, A; k) \cong \{C_{\#}(X, A; k)\}^*$$

for k a field, where $\{C_{\#}(X, A; k)\}^*$ is the dual space to $C_{\#}(X, A; k)$. This proves:

Corollary 25.20 Let k be a field and (X, A) a relative CW complex. Then there is a natural isomorphism

$$H^n(X, A; k) \cong \{H_n(X, A; k)\}^*. \quad ▮$$

We now give an example to illuminate the nonnaturality of the splitting.

Let $\xi: RP^2 \to CP^\infty = K(Z, 2)$ be a map whose homotopy class $\{\xi\} \in H^2(RP^2) = Z_2$ (by 24.12(a)) is nonzero. Thus $\xi^*: H^2(CP^\infty) \to H^2(RP^2)$ is nonzero for it maps $\{1\} \in [CP^\infty, CP^\infty]$ to $\{\xi\}$.

Consider the exact sequences (from 25.16)

$$\begin{array}{ccccccccc}
0 & \longrightarrow & \operatorname{Ext}(H_1(CP^\infty), Z) & \longrightarrow & H^2(CP^\infty) & \longrightarrow & \hom(H_2(CP^\infty), Z) & \longrightarrow & 0 \\
 & & \downarrow & & \downarrow & & \downarrow & & \\
0 & \longrightarrow & \operatorname{Ext}(H_1(RP^2), Z) & \longrightarrow & H^2(RP^2) & \longrightarrow & \hom(H_2(RP^2), Z) & \longrightarrow & 0
\end{array}$$

This has the form

$$\begin{array}{ccccccccc}
0 & \longrightarrow & 0 & \longrightarrow & Z & \longrightarrow & Z & \longrightarrow & 0 \\
 & & \downarrow & & \downarrow & & \downarrow & & \\
0 & \longrightarrow & Z_2 & \longrightarrow & Z_2 & \longrightarrow & 0 & \longrightarrow & 0
\end{array}$$

since $H_1(CP^\infty) = 0$, $H_2(CP^\infty) = Z$, $H_1(RP^2) = Z_2$, and $H_2(RP^2) = 0$ by 20.11, 24.12, Example 1, Section 20, and 25.4(b). Thus $\xi^*: H^2(CP^\infty) \to H^2(RP^2)$ is not equal to

$$\operatorname{Ext}(\xi_*, 1) \oplus \hom(\xi_*, 1): \operatorname{Ext}(H_1(CP^\infty), Z) \oplus \hom(H_2(CP^\infty), Z) \to \operatorname{Ext}(H_1(RP^2), Z) \oplus \hom(H_2(RP^2), Z)$$

for this map is 0. Thus the isomorphism

$$H^n(X, \pi) \cong \text{Ext}(H_{n-1}(X), \pi) \oplus \hom(H_n(X), \pi)$$

is not in general natural. In homology one can use the same example. Since ξ^* is nonzero, it is onto. Consider the exact sequences (from 25.14):

$$\begin{array}{ccccccccc}
0 & \longrightarrow & H^2(CP^\infty) \otimes Z_2 & \longrightarrow & H^2(CP^\infty; Z_2) & \longrightarrow & \text{Tor}(H^3(CP^\infty), Z_2) & \longrightarrow & 0 \\
 & & \downarrow & & \downarrow & & \downarrow & & \\
0 & \longrightarrow & H^2(RP^2) \otimes Z_2 & \longrightarrow & H^2(RP^2; Z_2) & \longrightarrow & \text{Tor}(H^3(RP^2), Z_2) & \longrightarrow & 0
\end{array}$$

This implies that $\xi^*\colon H^2(CP^\infty; Z_2) \to H^2(RP^2; Z_2)$ is an isomorphism. By 25.20, $\xi_*\colon H_2(RP^2; Z_2) \to H_2(CP^\infty; Z_2)$ is an isomorphism. Thus from 25.13 we have

$$\begin{array}{ccccccccc}
0 & \longrightarrow & H_2(RP^2) \otimes Z_2 & \longrightarrow & H_2(RP^2; Z_2) & \longrightarrow & \text{Tor}(H_1(RP^2), Z_2) & \longrightarrow & 0 \\
 & & \downarrow & & \downarrow & & \downarrow & & \\
0 & \longrightarrow & H_2(CP^\infty) \otimes Z_2 & \longrightarrow & H_2(CP^\infty; Z_2) & \longrightarrow & \text{Tor}(H_1(CP^\infty), Z_2) & \longrightarrow & 0
\end{array}$$

which reduces to

$$\begin{array}{ccccccccc}
0 & \longrightarrow & 0 & \longrightarrow & Z_2 & \longrightarrow & Z_2 & \longrightarrow & 0 \\
 & & \downarrow & & \downarrow \cong & & \downarrow & & \\
0 & \longrightarrow & Z_2 & \longrightarrow & Z_2 & \longrightarrow & 0 & \longrightarrow & 0
\end{array}$$

Consequently the splitting in 25.13 is not natural. One can also see that the splittings in 25.11 and 25.12 are not natural by considering the map

$$\xi \times 1\colon RP^2\, R \times P^2 \to CP^\infty \times RP^2,$$

and in 25.14 by considering the inclusion $S^1 \equiv RP^1 \subset RP^2$.

Exercises

1. Show that if R is a principal ideal domain, there is a natural exact sequence

$$0 \to \bigoplus_{i+j=n} H_i(X, A; R) \otimes_R H_j(Y, B; R) \to H_n(X \times Y, X \times B \cup A \times Y; R)$$
$$\to \bigoplus_{i+j=n-1} \text{Tor}_R(H_i(X, A; R), H_j(Y, B; R)) \to 0$$

which splits, where (X, A) and (Y, B) are relative CW complexes and that there is a similar split exact sequence for cohomology if (X, A) and (Y, B) have a finite number of cells in each dimension.

2. Let R be a principal ideal domain. Prove there is a natural exact sequence

$$0 \to \bigoplus_{i+j=n} \tilde{H}_i(X; R) \otimes_R \tilde{H}_j(Y; R) \to \tilde{H}_n(X \wedge Y; R)$$
$$\to \bigoplus_{i+j=n-1} \operatorname{Tor}_R(\tilde{H}_i(X; R), \tilde{H}_j(Y; R)) \to 0$$

which splits. Derive a similar exact sequence for cohomology.

3. Suppose $f: (K, L) \to (X, A)$ and $g: (K', L') \to (Y, B)$ are resolutions. Prove that $f \times g$ is a resolution. Prove that Exercises 1 and 2 hold for singular homology.

4. Reprove 24.12 using the results of this section.

5. Show that for any group G, $\operatorname{Tor}(Q, G) = 0$ where Q is the rational numbers.

6. Show by example that the splittings in 25.11 and 25.12 are not natural.

7. Prove analogues to 25.10–25.16 using singular homology and cohomology.

8. Suppose X is well pointed and $0 \to R \xrightarrow{\varphi} F \xrightarrow{\psi} G \to 0$ is a resolution. Use Exercise 8, Section 21 and Exercise 13, Section 18 to construct natural long exact sequences

$$\cdots \to \tilde{E}_i(X) \otimes R \xrightarrow{1 \otimes \varphi} \tilde{E}_i(X) \otimes F \to \tilde{E}_i(X; G) \to \tilde{E}_{i-1}(X) \otimes R \to \cdots$$
$$\cdots \to \tilde{E}^i(X) \otimes R \xrightarrow{1 \otimes \varphi} \tilde{E}^i(X) \otimes F \to \tilde{E}^i(X; G) \to \tilde{E}^{i+1}(X) \otimes R \to \cdots,$$

and hence construct universal coefficient exact sequences

$$0 \to \tilde{E}_n(X) \otimes G \to \tilde{E}_n(X; G) \to \operatorname{Tor}(\tilde{E}_{n-1}(X), G) \to 0$$
$$0 \to \tilde{E}^n(X) \otimes G \to \tilde{E}^n(X; G) \to \operatorname{Tor}(\tilde{E}^{n+1}(X), G) \to 0$$

generalizing 25.13 and 25.14. (Note: These sequences do not split in general.) (30.13)

9. Let $0 \to R \to F \to \pi \to 0$ be a resolution. Show that the Bockstein $\beta: H_n(X; \pi) \to H_{n-1}(X; R)$ has a factorization

$$H_n(X; \pi) \xrightarrow{\Delta} \operatorname{Tor}(H_{n-1}(X), \pi) \to H_{n-1}(X) \otimes R \cong H_{n-1}(X; R)$$

(see Exercise 13, Section 18). State and prove a similar result for cohomology if π is finitely generated.

10. Using Exercise 9 calculate the Bockstein homomorphism

$$H_r(RP^n; Z_2) \to H_{r-1}(RP^n)$$

corresponding to the sequence $0 \to Z \to Z \to Z_2 \to 0$.

26

Orientation and Duality

In this section we discuss orientation of manifolds and duality. Manifolds arise naturally in many analysis problems, and historically homology theory was first applied to manifolds.

A k-dimensional subspace V of R^n determines an $(n-k)$-dimensional subspace $V^\perp$ of $(R^n)^*$ by $V^\perp = \{f \mid f(V) = 0\}$. In an n-dimensional orientable manifold we will generalize this to determine for each k-dimensional cycle an $(n-k)$-dimensional cocycle. This will induce an isomorphism $H_k(M) \cong H^{n-k}(M)$. We prove a relative version of this duality theorem for an arbitrary ring spectrum E and manifolds that are orientable (in an appropriate sense) with respect to the ring spectrum. This has a number of applications to geometric problems and gives information about the ring structure in the cohomology of manifolds. The exposition we give here has been influenced by [20; 28; 48].

Definition 26.1 An n-manifold is a Hausdroff topological space M such that every point has a neighborhood homeomorphic to R^n.

All manifolds that we consider are assumed to be paracompact.[25]

The notion of orientation is quite familiar. A line has two orientations, corresponding to the two directions. Similarly, a plane has both a clockwise and counterclockwise sense. In making measurements along the line, or measuring angles, an arbitrary choice of orientation has to be made. Similarly, the "right-hand rule" for calculating the vector product in R^3 corresponds the choice of one of two orientations of R^3. In general, we can orient a simplex σ^n by ordering its vertices. Such an ordering $v_0, \ldots, v_n$ determines

[25] This is sufficient to guarantee that M is a separable metric space.

a homology class $\{e_\sigma\} \in H_n(\sigma, \partial\sigma) = Z$ which is a generator by 20.13. Two orderings determine the same generator iff they correspond to each other under the action of the alternating group (Exercise 12, Section 20). Thus the two generators of Z correspond to the two possible orientations of σ in intuitive sense.

Now we have isomorphisms

$$H_n(\sigma, \partial\sigma) \cong H_n(\sigma, \sigma - x) \cong H_n(R^n, R^n - x)$$

for $x \in \operatorname{Int} \sigma$; hence a choice of an orientation depends only on some point x of σ. However for any two points x and y of R^n, there is a simplex containing them. Using this simplex one can determine an orientation at y from one at x and vice versa.[26] We express this by saying that R^n is orientable. Such a choice will be called an orientation. In general, it is not true that a "local orientation" of a manifold extends to the whole manifold as above. The simplest examples of this phenomenon are the Möbius band $M = (0, 1) \times [0, 1]/(x, 0) \sim (1 - x, 1)$ (Fig. 26.1) and the Klein bottle (see Exercise 14, Section 7). In these

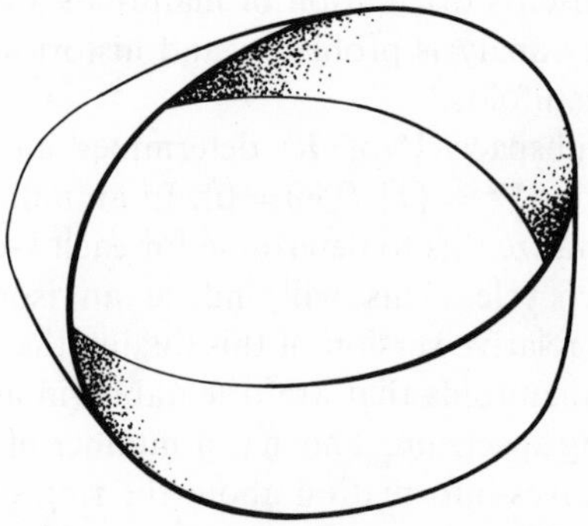

Figure 26.1

two-dimensional manifolds it is impossible to choose a clockwise direction "continuously" over the whole manifold.

Before making precise definitions, we will generalize to arbitrary theories defined by a ring spectrum. For the rest of this section E will denote an arbitrary ring spectrum. If M is a manifold and $x \in U \subset M$ where $U \equiv R^n$, there are isomorphisms

$$\begin{aligned} E_n(M, M - x) &\cong E_n(U, U - x) \cong E_n(R^n, R^n - x') \\ &\cong \tilde{E}_{n-1}(R^n - x') \cong \tilde{E}_{n-1}(S^{n-1}) \cong \tilde{E}_0(P) \end{aligned}$$

of $E_0(P)$ modules. Thus $E_n(M, M - x)$ is a free $E_0(P)$ module on one generator.

[26] Thus for example, to choose a clockwise direction at one point in the plane determines a clockwise direction at every point.

Definition 26.2 An orientation of M at x with respect to E is a choice of $E_0(P)$ module generators of $E_n(M, M - x)$. Given a collection $\{X\}$ of subsets of M, M is said to be consistently oriented along $\{X\}$ with respect to E if there is chosen a collection of classes $[X] \in E_n(M, M - X)$ such that

(a) $(\rho^X_{X \cap Y})_*[X] = (\rho^Y_{X \cap Y})_*[Y]$

where $\rho^A_B\colon (M, M - A) \to (M, M - B)$ is the inclusion, and
(b) $(\rho^X_x)_*[X]$ is an orientation at x.

A manifold is called E-orientable if it can be consistently oriented along all compact subsets. A collection of such classes is called an E-orientation.

If $E = HZ$, it is customary to delete reference to E in the above definitions. This is the intuitive notion discussed above. Notice that if M is compact, an E-orientation is determined by $[M] \in E_n(M)$. Such a homology class is called the fundamental class of M (with respect to E). In this case the only requirement put on $[M]$ is that $(\rho^M_x)_*[M]$ is an orientation at x for all x.

Proposition 26.3 R^n is E-orientable for all E. There is one orientation for each unit in $E_0(P)$.

Proof We first define $[b^n(r)]$ where $b^n(r) = \{x \in R^n \mid \|x\| < r\}$. We use the sequence of isomorphisms

$$E_n(R^n, R^n - b^n(r)) \cong E_n(B^n(r), S^{n-1}(r)) \cong \tilde{E}_{n-1}(S^{n-1}(r)) \cong \tilde{E}_0(S^0(r)) \cong E_0(P).$$

Thus a choice of a generator $g \in E_0(P)$ determines a class $[b^n(r)] \in E_n(R^n, R^n - b^n(r))$ for all r. If K is compact, $K \subset b^n(r)$ for some r so we can define $[K] = (\rho_K)_*[b^n(r)]$. This may conceivably depend on the choice of r. To show that it does not, it is only necessary to show that if $r < r'$. $\rho_*[b^n(r')] = [b^n(r)]$. Since there is a homotopy of pairs in R^n between the identity and the map φ which multiplies all vectors by the scalar r'/r, $\rho_*[b^n(r')] = \varphi_*[b^n(r')]$. Restriction of φ induces an obvious homeomorphism from $(B^n(r), S^{n-1}(r))$ to $(B^n(r'), S^{n-1}(r'))$ which induces the identity on $E_0(P)$ under the above isomorphisms. Thus $[K]$ is unambiguously defined. By definition, (a) of 26.2 is satisfied. To prove (b) one simply observes that if $x \in b^n(r)$, $(\rho_x)_*\colon E_n(R^n, R^n - b^n(r)) \to E_n(R^n, R^n - x)$ is an isomorphism. ∎

As an example of orientability we consider the spaces RP^n, CP^n, and HP^n. These are manifolds; in fact the sets V_i constructed in Example 4 of Section 11 are homeomorphic to R^n, R^{2n}, and R^{4n} respectively (see Exercise 25).

Proposition 26.4 RP^n is orientable iff n is odd. For each n, CP^n and HP^n are orientable.

Proof We do only the case of RP^n. The others are similar. If n is even, $H_n(RP^n) = 0$. It is thus impossible to choose $[RP^n]$ since $(\rho_x)_*([RP^n])$ must have infinite order. If n is odd, choose $[RP^n]$ to be a generator. Let $x_0 = [0|\cdots|0|1] \in RP^n$. Then $RP^n - x_0 \simeq RP^{n-1}$. Now the homomorphism $H_n(RP^n) \to H_n(RP^n, RP^{n-1})$ is an isomorphism by 20.11. Hence $(\rho_{x_0})_*$ is an isomorphism. To see that $(\rho_x)_*$ is isomorphism for all x, apply

Lemma 26.5 If M is arcwise connected, the homomorphisms

$$H_n(M) \xrightarrow{(\rho_x)_*} H_n(M, M - x) \cong Z$$

differ at most by a sign as x varies.

Proof If x_1 and x_2 belong to the same coordinate neighborhood, there is a line segment L with x_1 and x_2 as end points, lying inside the coordinate neighborhood. Hence there is a commutative diagram

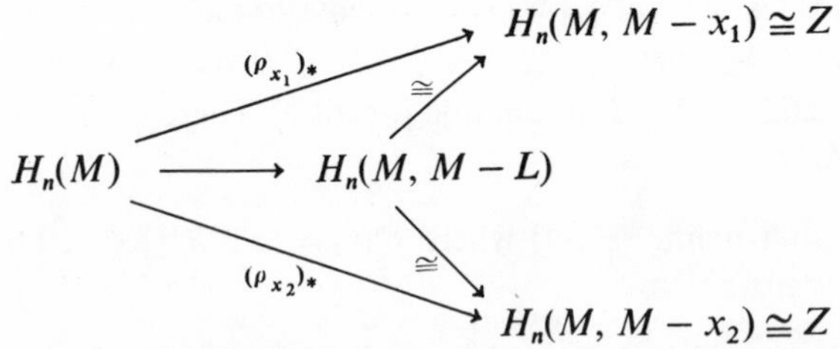

Thus the homomorphisms determined by x_1 and x_2 differ at most by a sign. Since any two points x and x' belong to a sequence $x = x_0, x_1, \ldots, x_k = x'$ with x_i and x_{i+1} belonging to some coordinate neighborhood, the lemma follows. ∎ ∎

Proposition 26.6 Every manifold is orientable with respect to HZ_2.

Proof Let $U \subset M$ be a coordinate neighborhood and suppose $D \subset U$ corresponds to B^n. Then $H_n(M, M - D; Z_2) \cong H_n(U, U - D; Z_2) \cong H_n(R^n, R^n - B^n; Z_2) \cong Z_2$. Thus there is a unique choice for $[D]$. The proposition follows from:

Lemma 26.7 Let A be a ring. If M is consistently oriented with respect to HA along a collection of sets whose interiors cover M, there is a unique extension to an orientation of M with respect to HA.

To prove this we need the following lemma.

Lemma 26.8 If $K \subset M$ is compact, and G is any abelian group, $H_i(M, M - K; G) = 0$ for $i > n$.

Proof We first prove this in the case $M = R^n$. Call an n-cube $[a_1, b_1] \times \cdots \times [a_n, b_n]$ a type k cube if $a_i = m_i/k$ for some integer m_i and $b_i = a_i + (1/k)$. R^n is a CW complex with the set of all type k cubes as n cells. Any open subset U of R^n can similarly be made into a CW complex as follows. Let K_1 be the union of all type 2 cubes contained in U. Having defined K_{l-1}, let K_l be the union of K_{l-1} and all type 2^l cubes contained in $U - K_{l-1}$. K_l is clearly a CW complex with K_{l-1} as a subcomplex. Hence $U = \bigcup K_l$ is a CW complex. Any element of $C_i(U; G)$ must lie in the image of $C_i(L; G)$ where L is some finite subcomplex of U. Hence any element of $H_i(R^n - K; G)$ is in the image of the homomorphism

$$H_i(L; G) \to H_i(R^n - K; G)$$

induced by the inclusion of L for some $L \subset R^n - K$. In the commutative diagram

$$\begin{array}{ccc} H_i(L; G) & \longrightarrow & H_i(R^n - K; G) \\ \cong \uparrow \partial & & \cong \uparrow \partial \\ H_{i+1}(R^n, L; G) & \longrightarrow & H_{i+1}(R^n, R^n - K; G) \end{array}$$

observe that (R^n, L) is a relative CW complex with cells in dimensions less than or equal to n, and hence $H_{i+1}(R^n, L; G) = 0$ for $i \geq n$. Consequently, $H_i(R^n, R^n - K; G) = 0$ for $i > n$.

Suppose now that $K \subset U \subset M$ where U is homeomorphic to R^n. Then $H_i(M, M - K; G) \cong H_i(U, U - K; G) \cong H_i(R^n, R^n - K; G) = 0$ for $i > n$, by excision.

Suppose now that K is an arbitrary compact set. $K = K_1 \cup \cdots \cup K_s$ where each K_i is contained in a set homeomorphic to R^n. We show by induction on s that $H_i(M, M - (K_1 \cup \cdots \cup K_i); G) = 0$ for $i > n$ and $1 \leq i \leq s$. Let $K' = K_1 \cup \cdots \cup K_{i-1}$ and $K'' = K_i$. Then

$$H_i(M, M - K'; G) = H_i(M, M - K''; G) = H_i(M, M - (K' \cap K''); G) = 0$$

for $i > n$. Applying the Mayer–Vietoris sequence (Exercise 10, Section 21)

$$\begin{aligned} H_{i+1}(M, M - (K' \cap K''); G) &\to H_i(M, M - (K' \cup K''); G) \\ &\to H_i(M, M - K'; G) \oplus H_i(M, M - K''; G) \end{aligned}$$

one concludes that $H_i(M, M - (K' \cup K''); G) = 0$ completing the inductive step. ∎

Proof of 26.7 Let $\{U_\alpha\}$ be the interiors of the sets covering M along which M is consistently oriented. As in the case of 26.8, an arbitrary compact set K can be written as $K = K_1 \cup \cdots \cup K_k$ where $K_i \subset U_{\alpha_i}$. We will show that given

$[K']$ and $[K'']$ with $(\rho_{K' \cap K''})_* [K'] = (\rho_{K' \cap K''})_*[K'']$, there is a unique class $[K' \cup K'']$ with $(\rho_{K'})_*[K' \cup K''] = [K']$ and $(\rho_{K''})_*[K' \cup K''] = [K'']$. This is sufficient to prove that there is a unique class $[K]$ with $(\rho_{K_i})_*[K] = [K_i]$. Then the sets $[K]$ are compatible in the sense of 26.2(a) by uniqueness, and 26.2(b) follows immediately. To construct $[K' \cup K'']$ we apply the Mayer–Vietoris sequence again

$$\begin{aligned} H_{n+1}(M, M - (K' \cap K''); A) &\to H_n(M, M - (K' \cup K''); A) \\ &\xrightarrow{\varphi} H_n(M, M - K'; A) \oplus H_n(M, M - K''; A) \\ &\xrightarrow{\phi} H_n(M, M - (K' \cap K''); A) \to \end{aligned}$$

where $\varphi = ((\rho_{K'})^*, (\rho_{K''})^*)$ and $\phi = (\rho^{K'}_{K' \cap K''})_* - (\rho^{K''}_{K' \cap K''})_*$. By 26.8, $H_{n+1}(M, M - (K' \cap K''); A) = 0$, so $[K' \cup K'']$ exists uniquely. ▮▮

An important tool in studying orientability is the following result.

Proposition 26.9 Let M be a manifold. Then there is a double covering space $\pi: \tilde{M} \to M$ such that $\tilde{M}$ is an orientable manifold.

Applying Exercise 13, Section 7 we immediately conclude:

Corollary 26.10 Every simply connected manifold is orientable. ▮

Remark 26.11 We could use 26.10 to prove that CP^n and HP^n are orientable instead of the proof in 26.4.

Call a coordinate neighborhood $U \subset M$ special if there is another coordinate neighborhood $V \supset U$ such that $(V, U) \equiv (R^n, B^n - S^{n-1})$. Clearly the special coordinate neighborhoods form a basis for the topology.

Lemma 26.12 Let U be a special coordinate neighborhood. Then for all $x \in \bar{U}$,

$$(\rho_x{}^U)_*: H_n(M, M - \bar{U}) \to H_n(M, M - x)$$

is an isomorphism.

Proof Choose V as above. Then $(V, V - \bar{U}) \subset (M, M - \bar{U})$ and $(V, V - x) \subset (M, M - x)$ are excisions. Thus it is sufficient to consider the restriction

$$\rho: (V, V - \bar{U}) \to (V, V - x);$$

using the homeomorphism of V with R^n, this corresponds to

$$\rho: (R^n, R^n - B^n) \to (R^n, R^n - x)$$

This clearly induces isomorphisms in homology since $R^n - B^n \simeq R^n - x$. ▮

Proof of 26.9 Let $\tilde{M} = \{(x, [x]) \mid x \in M, [x] \text{ is an orientation at } x\}$. Let $\pi\colon \tilde{M} \to M$ be defined by $\pi(x, [x]) = x$. Let $U \subset M$ be a special coordinate neighborhood and $[\overline{U}]$ a generator for $H_n(M, M - \overline{U})$. Define

$$\overline{W}([\overline{U}]) = \{(x, [x]) \in \tilde{M} \mid x \in \overline{U}, (\rho_x^{\overline{U}})_*([U]) = [x]\}$$
$$W([\overline{U}]) = \overline{W}([\overline{U}]) \cap \pi^{-1}(U).$$

Now the sets $W([\overline{U}])$ cover $\tilde{M}$ since each $z \in M$ belongs to some special U and $(\rho_x{}^{\overline{U}})_*$ is onto. In fact we show that the sets $W([\overline{U}])$ form the basis for a topology. Suppose $(x, [x]) \in W([\overline{U}]) \cap W([\overline{U}'])$. Then $x \in U \cap U'$ and there exists a special coordinate neighborhood U'' with $x \in U'' \subset U \cap U'$. Define $[\overline{U}''] = (\rho_{\overline{U}''}^{\overline{U}})_*([\overline{U}])$. Now $(x, [x]) \in W([\overline{U}''])$ since $(\rho_x^{\overline{U}''})_*([\overline{U}'']) = (\rho_x{}^{\overline{U}})_*([\overline{U}]) = [x]$. Since $(\rho_x^{\overline{U}''})_*$ is an isomorphism, $(\rho_{\overline{U}''}^{\overline{U}})_*([\overline{U}]) = (\rho_{\overline{U}''}^{\overline{U}'})_*([\overline{U}'])$; hence $[\overline{U}''] = (\rho_{\overline{U}''}^{\overline{U}'})_*([\overline{U}'])$. Consequently if $(y, [y]) \in W([\overline{U}''])$,

$$[y] = (\rho_y^{\overline{U}''})_*([\overline{U}'']) = (\rho_y^{\overline{U}})_*([\overline{U}]) = (\rho_y^{\overline{U}'})_*([\overline{U}']),$$

and thus $W([\overline{U}'']) \subset W([\overline{U}]) \cap W([\overline{U}'])$, and the sets $W([\overline{U}])$ for U special form a basis for a topology.

With this topology π is continuous, for $\pi^{-1}(U) = W([U]) \cup W(-[U])$ and $W([U]) \cap W(-[U]) = \varnothing$. Thus π is a double covering space. Clearly $\tilde{M}$ is an n-manifold since $W([\overline{U}]) \equiv U \equiv R^n$.

We claim that $W([\overline{U}])$ is a special coordinate neighborhood. Since U is special, we may choose V such that $(V, U) \equiv (R^n, B^n - S^{n-1})$. Let U' correspond to $\{x \in R^n \mid \|x\| < 2\}$. Then U' is special. By 26.12 $(\rho_{\overline{U}}^{\overline{U}'})_*$ is an isomorphism. Let $[\overline{U}'] = (\rho_{\overline{U}}^{\overline{U}'})_*^{-1}([\overline{U}])$. Then $W([\overline{U}]) \subset W([\overline{U}'])$. Furthermore π establishes a homeomorphism $(W([\overline{U}']), W([\overline{U}])) \equiv (U', U)$ so $W([\overline{U}])$ is indeed special.

In the diagram

$$\begin{array}{ccc}
H_n(\tilde{M}, \tilde{M} - \overline{W}([\overline{U}])) & \xrightarrow{\pi_*} & H_n(M, M - \overline{U}) \\
\uparrow \cong & & \uparrow \cong \\
H_n(W([\overline{U}']), W([\overline{U}']) - \overline{W}([\overline{U}])) & \xrightarrow{\cong} & H_n(U', U' - \overline{U})
\end{array}$$

the vertical arrows are excisions and the bottom is an isomorphism by the above construction. Hence π_* is an isomorphism and we define an orientation along the sets $\overline{W}([\overline{U}])$ by

$$[\overline{W}([\overline{U}])] = \pi_*^{-1}([\overline{U}]).$$

This will orient $\tilde{M}$ by 26.7 if we show that these classes are consistent. Let $p = (x, [x]) \in \overline{W}([\overline{U}])$. Then

$$\pi_*(\rho_p)_*([\overline{W}([\overline{U}])]) = [x] \in H_n(M, M - x).$$

Since π_* is an isomorphism, condition (b) is satisfied and furthermore, if $p \in \overline{W}([\overline{U}]) \cap \overline{W}([\overline{V}])$,

$$(\rho_p)_*([\overline{W}([\overline{U}])]) = (\rho_p)_*([\overline{W}([\overline{V}])]).$$

Thus condition (a) follows from:

Lemma 26.13 Let K be a compact subset of an n-manifold M, and $\xi \in H_n(M, M - K)$. Then $\xi = 0$ iff $(\rho_x^K)_*(\xi) = 0$ for all $x \in K$.

Proof The proof will be based on case analysis and the use of the following.

Basic inductive step If $K = K_1 \cup K_2$ and 26.13 is true for K_1 and K_2, it is true for K.

Proof By the Mayer–Vietoris sequence (Exercise 10, Section 21) and 26.8 we have the exact sequence

$$0 \to H_n(M, M - K) \xrightarrow{s} H_n(M, M - K_1) \oplus H_n(M, M - K_2) \to \cdots$$

where $s(\xi) = ((\rho_{K_1}^K)_*(\xi), (\rho_{K_2}^K)_*(\xi))$. Since $\rho_x^K = \rho_x^{K_i} \rho_{K_i}^K$ for $x \in K_i$, it follows that if $(\rho_x^K)_*(\xi) = 0$ for all $x \in K$, we must have $s(\xi) = 0$ and hence $\alpha = 0$. ∎

Proof of 26.13 We observe that it is true if $M = R^n$ and K is a ball B, for $R^n - B \simeq R^n - x$ if $x \in B$. Hence by the inductive step, 26.13 is true if $M = R^n$ and K is a finite union of balls. Suppose now that $K \subset R^n$ is an arbitrary compact set, and $\alpha \in H_n(R^n, R^n - K)$. As in the proof of 26.8 one can find a complex $L \subset R^n - K$ such that ξ is in the image of the restriction homomorphism

$$H_n(R^n, L) \to H_n(R^n, R^n - K);$$

Since K is compact, there exist balls $B_1, \ldots, B_s$ with $K \subset B_1 \cup \cdots \cup B_s \subset R^n - L$. We will suppose in addition that each ball intersects K. Now ξ is in the image of the restriction homomorphism

$$H_n(R^n, R^n - (B_1 \cup \cdots \cup B_s)) \xrightarrow{\rho_*} H_n(R^n, R^n - K).$$

Let $\rho_*(\xi') = \xi$. Clearly $(\rho_x)_*(\xi') = 0$ for all $x \in K$. To see that $(\rho_x)_*(\xi') = 0$ for $x \in B_i - K$, join x with a point x' of $B_i \cap K$ by a straight line segment L. Then since

$$\rho_*: H_n(R^n, R^n - L) \to H_n(R^n, R^n - x')$$

is an isomorphism, $(\rho_L)_*(\xi') = 0$, hence $(\rho_x)_*(\xi') = 0$. Since $(\rho_x)_*(\xi') = 0$ for all $x \in B_1 \cup \cdots \cup B_s$, $\xi' = 0$ and hence $\xi = 0$. This completes the proof in the case that $M = R^n$. If M is arbitrary but K is contained in a coordinate neighborhood U, the result still holds because of the excision isomorphism

$$H_n(M, M - K) \simeq H_n(U, U - K).$$

Now if M and K are arbitrary, $K = K_1 \cup \cdots \cup K_i$ with K_i contained in a coordinate neighborhood. It follows from the basic inductive step that the lemma holds. ▮ ▮

Corollary 26.14 If M is a compact connected n-manifold, $H_n(M; Z_2) \simeq Z_2$.

Proof By 26.6 $H_n(M; Z_2) \neq 0$. But by 26.5 and 26.13

$$(\rho_x)_*: H_n(M; Z_2) \to H_n(M, M - x; Z_2) \cong Z_2$$

is a monomorphism. ▮

Theorem 26.15 If a manifold is $\underline{S}$-orientable it is E-orientable for every E ($\underline{S}$ is the sphere spectrum).

Proof The mapping $u: S \to E$ defines a natural transformation

$$u: \pi_*{}^S(X, A) \to E_*(X, A);$$

a choice $[K]_S \in \pi_n{}^S(M, M - K)$ thus determines $[K]_E = u[K]_S \in E_n(M, M - K)$. Condition (a) is easily seen to be satisfied. To check (b), we observe that

$$\pi_0{}^S(S^0) \cong \pi_n{}^S(M, M - x) \xrightarrow{u} E_n(M, M - x) \cong \tilde{E}_0(S^0)$$

is a ring homomorphism and hence sends generators to generators. ▮

Definition 26.16 An element $\xi \in H_i(X)$ will be called spherical if it is in the image of the Hurewicz homomorphism

$$h: \pi_i(X) \to H_i(X).$$

ξ will be called stably spherical if it is in the image of the stable Hurewicz homomorphism

$$H: \pi_i{}^S(X) \to \tilde{H}_i(X)$$

(see Exercise 1, Section 22).

Theorem 26.17 If M is compact and orientable, $[M] \in H_n(M)$ is stably spherical iff M is $\underline{S}$-orientable.

Proof Note that if $A = *$, $u = H$. Hence if M is $\underline{S}$-orientable, $[M] \in H_n(M)$ is stably spherical. Conversely, if $[M] \in H_n(M)$ is stably spherical we choose $[M]_S \in \pi_n{}^S(M, \varnothing)$ with $u([M]_S) = [M]$. $(\rho_x)_*([M]_S)$ is a generator since

$$\pi_n{}^S(M, M - x) \xrightarrow{u} H_n(M, M - x)$$

is an isomorphism. ▮

Corollary 26.18 S^n is $\underline{S}$-orientable.

Proof This follows immediately from the Hurewicz theorem. ▮

In order to prove the duality theorem we shall construct a duality operator. This will be based on the cap product (see Exercise 7, Section 23). Let E be a ring spectrum such that for n sufficiently large, E_n is a CW complex. Let X be paracompact and suppose that $L \subset K$ are closed subsets of X. We define a pairing

$$C\colon E^i(K, L) \otimes E_n(X, X-K) \to E_{n-i}(X-L, X-K).$$

Let (V, W) be an open pair of subsets of X containing (K, L). We will use the symbols $i_1, i_2, \ldots$ to denote various natural inclusion maps. Thus we have a homomorphism

$$E^i(V, W) \otimes E_n(X, X-K) \xrightarrow{i_1^* \otimes i_2^*} E^i(V-L, W-L) \otimes E_n(X, (X-K) \cup W)$$
$$\xrightarrow{(1 \otimes i_3^*)^{-1}} E^i(V-L, W-L) \otimes E_n(V-L, (V-K) \cup (W-L))$$

since i_3 is an excision. Define $\xi_{V,W}$ to be the composite of this homomorphism with

$$E^i(V-L, W-L) \otimes E_n(V-L, (V-K) \cup (W-L))$$
$$\xrightarrow{\cap} E_{n-i}(V-L, V-K) \xrightarrow{i_{4*}} E_{n-i}(X-L, X-K)$$

where $\cap$ is the homomorphism from Exercise 7, Section 23. $\xi_{V,W}$ is defined for all open pairs (V, W) containing (K, L). If $(V', W') \subset (V, W)$, there is a commutative diagram

$$\begin{array}{ccc} E^i(V, W) \otimes E_n(X, X-K) & \searrow^{\xi_{V,W}} & \\ \uparrow{\scriptstyle i_5^* \otimes 1} & & E_{n-i}(X-L, X-K) \\ E^i(V', W') \otimes E_n(X, X-K) & \nearrow_{\xi_{V',W'}} & \end{array}$$

Hence $\{\xi_{V,W}\}$ defines a mapping

$$\varinjlim \{E^i(V, W) \otimes E_n(X, X-K)\} \to E_{n-i}(X-L, X-K).$$

By Exercise 13, Section 15 and Exercise 4, Section 21, the left-hand group is naturally isomorphic with $E^i(K, L) \otimes E_n(X, X-K)$. We define C via this natural isomorphism. (Note that if $K = X$ and $L = \varnothing$, this is the cap product.)

Lemma 26.19 C is natural in the following sense. Let $(K, L) \subset (K', L')$. Then there is a commutative diagram

$$\begin{array}{ccc} E^i(K, L) \otimes E_n(X, X-K) & \xrightarrow{C} & E_{n-i}(X-L, X-K) \\ \uparrow{\scriptstyle i_6^* \otimes i_{7*}} & & \uparrow{\scriptstyle i_{8*}} \\ E^i(K', L') \otimes E_n(X, X-K') & \xrightarrow{C} & E_{n-i}(X-L', X-K') \end{array}$$

Proof Let (V, W) be a neighborhood of (K, L) and (V', W') a neighborhood of (K', L') containing (V, W). We can then prove commutativity in the diagram

$$\begin{array}{ccc}
E^i(V, W) \otimes E_n(X, X-K) & \xrightarrow{\xi_{V\ W}} & E_{n-i}(X-L, X-K) \\
\uparrow{\scriptstyle i_9{}^* \otimes i_{7*}} & & \uparrow{\scriptstyle i_{8*}} \\
E^i(V', W') \otimes E_n(X, X-K') & \xrightarrow{\xi_{V'W'}} & E_{n-i}(X-L', X-K')
\end{array}$$

by naturality of $\cap$. Taking limits first over all $(V, W) \supset (K, L)$ and then all $(V', W') \supset (K', L')$ one establishes the lemma. ∎

Lemma 26.20 There is a commutative diagram:

$$\begin{array}{ccccc}
 & & E^i(L) \otimes E_n(X, X-L) & \xrightarrow{C} & E_{n-i}(X, X-L) \\
 & \nearrow{\scriptstyle 1 \otimes i_{10*}} & & & \\
E^i(L) \otimes E_n(X, X-K) & & & & \downarrow{\scriptstyle \partial} \\
 & \searrow{\scriptstyle \delta \otimes 1} & & & \\
 & & E^{i+1}(K, L) \otimes E_n(X, X-K) & \xrightarrow{C} & E_{n-i-1}(X-L, X-K)
\end{array}$$

Proof Let (V, W) be a neighborhood (K, L). We will prove commutativity of the diagram

$$\begin{array}{ccccc}
 & & E^i(W) \otimes E_n(X, X-L) & \xrightarrow{\xi_{W,\varnothing}} & E_{n-i}(X, X-L) \\
 & \nearrow & & & \\
E^i(W) \otimes E_n(X, X-K) & & & & \downarrow{\scriptstyle \partial} \\
 & \searrow{\scriptstyle \delta \otimes 1} & & & \\
 & & E^{i+1}(V, W) \otimes E_n(X, X-K) & \xrightarrow{\xi_{V,W}} & E_{n-i-1}(X-L, X-K)
\end{array}$$

from which 26.20 will follow. Using the definitions of $\xi_{V,W}$ and $\xi_{W,\varnothing}$, we expand this diagram to the larger one (see page 280). Commutativity of the center diagram on page 280 follows from Exercise 14, Section 23 with $A = W$, $B = V - L$, $X = V$, $C = \varnothing$, and $D = V - K$. ∎

Suppose now that M is a manifold and $\{[K]\}$ is an orientation of M with respect to E. Suppose $K \supset L$ are compact subsets. Define

$$D: E^i(K, L) \to E_{n-i}(M - L, M - K)$$

by $D(x) = C(x \otimes [K])$.

$$
\begin{array}{ccccccc}
E^i(W)\otimes E_n(W, W-L) & \xrightarrow{\cap} & E_{n-i}(W, W-L) & \longrightarrow & E_{n-i}(V, V-L) & \longrightarrow & E_{n-i}(X, X-L) \\
\uparrow & \searrow^{\cong} & & & \Big\downarrow{\scriptstyle \partial} & & \Big\downarrow{\scriptstyle \partial} \\
E^i(W)\otimes E_n(X, X-L) & \xleftarrow{\cong} & E^i(W)\otimes E_n(V, V-L) & & & & \\
\uparrow & & \uparrow & & & & \\
E^i(W)\otimes E_n(X, X-L) & \xleftarrow{\cong} & E^i(W)\otimes E_n(V, V-L) & & & & \\
\Big\downarrow{\scriptstyle \delta\otimes 1} & & \Big\downarrow{\scriptstyle \delta\otimes i_*} & & & & \\
 & & E^{i+1}(V, W)\otimes E_n(V, (V-K)\cup W) & & E_{n-i-1}(V-L, V-K) & \longrightarrow & E_{n-i-1}(X-L, X-K) \\
 & & \Big\downarrow & \searrow^{i^*\otimes (i_*)^{-1}} & \Big\uparrow{\scriptstyle \cap} & & \\
E^{i+1}(V, W)\otimes E_n(X, X-K) & \longrightarrow & E^{i+1}(V-L, W-L)\otimes E_n(X, (X-K)\cup W) & \xleftarrow{\cong} & E^{i+1}(V-L, W-L)\otimes E_n(V-L, (V-K)\cup(W-L)) & &
\end{array}
$$

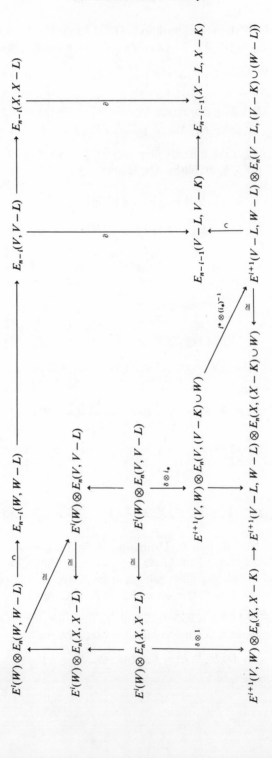

Duality Theorem 26.21 D is an isomorphism.

The proof will involve repeated use of the following:

Lemma 26.22 If D is an isomorphism for $(K_1, \varnothing)$, $(K_2, \varnothing)$, and $(K_1 \cap K_2, \varnothing)$, then it is an isomorphism for $(K_1 \cup K_2, \varnothing)$.

Proof Let $K \supset L$ and consider the diagram of exact sequences

$$\begin{array}{ccccccccc} \cdots \longrightarrow & E^i(K) & \longrightarrow & E^i(L) & \longrightarrow & E^{i+1}(K, L) & \longrightarrow & E^{i+1}(K) & \longrightarrow \cdots \\ & \downarrow D & & \downarrow D & & \downarrow D & & \downarrow D & \\ \cdots \longrightarrow & E_{n-i}(M, M-K) & \longrightarrow & E_{n-i}(M, M-L) & \longrightarrow & E_{n-i-1}(M-L, M-K) & \longrightarrow & E_{n-i-1}(M, M-K) & \longrightarrow \cdots \end{array}$$

By 26.19 and 26.20 this is commutative. Suppose D is an isomorphism for $(K_1, \varnothing)$ and $(K_1 \cap K_2, \varnothing)$. By the 5-lemma D is an isomorphism for $(K_1, K_1 \cap K_2)$. Now there are isomorphisms

$$E^i(K_1, K_1 \cap K_2) \xleftarrow{\cong} E^i(K_1 \cup K_2, K_2)$$

$$E_{n-i}(M - (K_1 \cap K_2), M - K_1) \xleftarrow{\cong} E_{n-i}(M - K_2, M - (K_1 \cup K_2))$$

by 21.7, 21.2, and 21.20(1). Hence by 26.19 D is an isomorphism for $(K_1 \cup K_2, K_2)$. Consequently by the above diagram and the 5-lemma D is an isomorphism for $(K_1 \cup K_2, \varnothing)$. ▮

Proof of 26.21 This will be divided into several cases.

Case 1 K consists of only one point: If $L \neq \varnothing$ all the groups are zero. Suppose $L = \varnothing$. By Exercise 6c, Section 23, D is an $E^*(P)$-module homomorphism. Since $1 \in E^0(P)$ is a generator and $D(1) = [P]$ generates $E_n(M, M - P)$ by 26.2, D is an isomorphism.

Case 2 $M = R^n$, K is compact and convex, $L = \varnothing$: For any point $P \in K$, we claim that the inclusions $P \to K$ and $R^n - K \to R^n - P$ are homotopy equivalences. The first statement follows from convexity. To prove the second one, note that K is contained in some ball B centered at P. Hence a linear homotopy away from P retracts both $R^n - K$ and $R^n - P$ onto $R^n - B$. Thus

$$E^i(K, \varnothing) \to E^i(P, \varnothing) \qquad \text{and} \qquad E_{n-i}(R^n, R^n - K) \to E_{n-i}(R^n, R^n - P)$$

are isomorphisms. By 26.19, D is an isomorphism in this case.

Case 3 $M = R^n$, K is a finite union of compact convex sets, and $L = \varnothing$: We prove this by induction. Suppose $K = C_1 \cup \cdots \cup C_r$ with each C_i convex.

The case $r = 1$ is case 2. Suppose the result proven for any union of less than r convex sets. Then it is true for

$$C_1 \cup \cdots \cup C_{r-1}$$

and

$$(C_1 \cup \cdots \cup C_{r-1}) \cap C_r = (C_1 \cap C_r) \cup \cdots \cup (C_{r-1} \cap C_r).$$

By 26.22, it is true for $C_1 \cup \cdots \cup C_r$.

Case 4 $M = R^n$, K is an arbitrary compact set, $L = \varnothing$: Let $\{V_\alpha\}$ be the collection of all compact neighborhoods of K that are the union of a finite number of compact convex sets. Since every neighborhood of K contains some V_α, $\bigcup_\alpha (M - V_\alpha) = M - K$. By Exercise 4, Section 19,

$$E_{n-i}(M, M - K) \cong \varinjlim E_{n-i}(M, M - V_\alpha).$$

By 21.20(4),

$$\varinjlim E^i(V_\alpha) \cong E^i(K).$$

Thus by 26.19, D is an isomorphism in this case.

Case 5 M is arbitrary, K is an arbitrary compact set, and $L = \varnothing$: Suppose K is contained in some coordinate neighborhood $U \equiv R^n$

$$E_{n-i}(U, U - K) \cong E_{n-i}(M, M - K)$$

by excision. If $i = 0$, this isomorphism determines an orientation on U from the orientation on M. Using this orientation and 26.19, it follows that D is an isomorphism in case K is contained in a coordinate neighborhood.

In general K is covered by a finite number of coordinate neighborhoods. Thus we can write $K = K_1 \cup \cdots \cup K_s$ where each K_i is compact and contained in some coordinate neighborhood. By 26.22, D is an isomorphism in this case.

Case 6 The general case: This follows from the 5-lemma applied to the diagram

$$\begin{array}{ccccccccccc}
\cdots \to & E^i(K) & \to & E^i(L) & \to & E^{i+1}(K, L) & \to & E^{i+1}(K) & \to & E^{i+1}(L) & \to \cdots \\
& \downarrow D & & \downarrow D & & \downarrow D & & \downarrow D & & \downarrow D & \\
\cdots \to & E_{n-i}(M, M-K) & \to & E_{n-i}(M, M-L) & \to & E_{n-i-1}(M-L, M-K) & \to & E_{n-i-1}(M, M-K) & \to & E_{n-i-1}(M, M-L) & \to \cdots
\end{array}$$

(see 26.22). ∎

Although $D\colon E^i(K, L) \to E_{n-i}(M - L, M - K)$ is only defined if K is compact, we can nevertheless prove

Corollary 26.23 If $\overline{K - L}$ is compact

$$E^i(K, L) \cong E_{n-i}(M - L, M - K).$$

Proof Let C be a compact set containing $K - L$. Then

$$\begin{aligned} E^i(K, L) \cong E^i(K \cap C, L \cap C) &\cong E_{n-i}(M - (L \cap C), M - (K \cap C)) \\ &\cong E_{n-i}(M - L, M - C) \end{aligned}$$

by 21.20(1) and excision (21.2 and 21.7). ▮

Corollary 26.24 (*Poincaré Duality Theorem*) If M is a compact manifold oriented with respect to E,

$$D\colon E^i(M) \to E_{n-i}(M)$$

is an isomorphism. ▮

There are a few simple observations one can make by applying the Poincaré duality theorem to ordinary homology. Let M be a compact manifold. Then $H_i(M;\, Z_2) \cong H^{n-i}(M;\, Z_2) \cong H_{n-i}(M;\, Z_2)^*$. Hence $H_i(M;\, Z_2) \cong H_i(M; Z_2)^{**}$. This implies that $H_i(M;\, Z_2)$ is a finite dimensional vector space.

If we suppose that M is compact and orientable it follows that it is orientable with respect to HG for any abelian group G. As in the case of Z_2 it follows that $H_i(M;\, Z_p)$ and $H_i(M;\, Q)$ are finite-dimensional vector spaces.

Suppose that M is compact, orientable, and $H_i(M) \cong F_i \oplus T_i$ where F_i is free and of finite rank, and T_i is a finite group. By 25.19, $H^i(M) \cong F_i \oplus T_{i-1}$. Hence we have $F_i \cong F_{n-i}$ and $T_i \cong T_{n-i-1}$.

Recall the Euler characteristic χ defined in Exercise 11, Section 20.

Corollary 26.25 Suppose M is a compact manifold of odd dimension. Then $\chi(M) = 0$.

Proof $\chi(M)$ is well defined since $H_i(M;\, Z_2)$ is finite dimensional. Since $\dim H_i(M; Z_2) = \dim H_{n-i}(M; Z_2)$,

$$\chi(M) = \sum (-1)^i \dim H_i(M; Z_2) = 0. \quad ▮$$

Corollary 26.26 (*Alexander Duality Theorem*) Let K be a compact subset of S^n. Then for each ring spectrum E,

$$D\colon E^i(K, L) \to E_{n-i}(S^n - L, S^n - K)$$

is an isomorphism.

Proof This follows from 26.21 and 26.17. ▮

Corollary 26.27 If K is a compact subset of S^n,

$$\tilde{E}^{i-1}(K) \cong \tilde{E}_{n-i}(S^n - K)$$

Proof

$$\tilde{E}^{i-1}(K) \cong E^{i-1}(K, *) \cong E_{n-i+1}(S^n - *, S^n - K) \cong E_{n-i}(S^n - K, *')$$

by 26.26 and the exact sequence of the triple $*' \subset S^n - K \subset S^n - *$. ▮

Another version of the duality theorem involves cohomology with compact supports. If X is locally compact and paracompact, and E is a spectrum, define

$$E^i_{\text{comp}}(X) = \tilde{E}^i(X^\infty)$$

where X^∞ is the one-point compactification.

Theorem 26.28 Let M be a manifold orientable with respect to E. Then

$$E^i_{\text{comp}}(M) \cong E_{n-i}(M).$$

Proof Consider all subsets $U \subset M$ with $\overline{U}$ compact. Since M is locally compact and Hausdorff

$$M^\infty \equiv \underrightarrow{\lim}\ M/M - U.$$

Thus $E^i_{\text{comp}}(M) \cong \underrightarrow{\lim}\ E^i(M, M - U)$ by 21.20(4, 1). Since $M/M - U$ is compact

$$\underrightarrow{\lim}\ E^i(M, M - U) \cong \underrightarrow{\lim}\ E_{n-i}(U) \cong E_{n-i}(M)$$

by 26.23 and Exercise 4, Section 19. ▮

Theorem 26.29 (*Jordan Separation Theorem*) Let $X \subset S^n$ be a subset homeomorphic to S^{n-1}. Then $S^n - X$ has exactly two components, and their boundary is X.

Proof $H_0(S^n - X) \cong H^n(S^n, X)$ by 26.26. The exact sequence

$$0 \leftarrow H^n(S^n) \leftarrow H^n(S^n, X) \leftarrow H^{n-1}(X) \leftarrow 0$$

thus yields $H_0(S^n - X) \cong Z \oplus Z$. By Exercise 13, Section 20 and the proof of 26.8, $S^n - X$ consists of two components U_1 and U_2. Clearly $\overline{U}_1 \subset S^n - U_2 = U_1 \cup X$. We now prove $X \subset \overline{U}_1 \cap \overline{U}_2$. Let $x \in X$ and suppose U is an open set containing x; we will show that both $U_1 \cap U$ and $U_2 \cap U$ are non-empty. If $U_1 \cap U = \varnothing$, then $U_2 \cup U$ and U_1 are disjoint open sets. Similarly if $U_2 \cap U = \varnothing$, U_2 and $U_1 \cup U$ are disjoint open sets. We dispose of these possibilities by showing that $U_1 \cup U_2 \cup U$ is connected. Indeed $U_1 \cup U_2 \cup U = S^n - (X - U)$ and $H_0(S^n - (X - U)) \cong H^n(S^n, X - U) \cong Z$. Thus $\overline{U}_1 = U_1 \cup X$ and $\overline{U}_2 = U_2 \cup X$. ∎

Theorem 26.30 (*Invariance of Domain*) Suppose $X \subset S^n$ and $X \equiv R^n$. Then X is open.

Proof Let $x \in X$ and $h: R^n \to X$ be a homeomorphism with $h(0) = x$. By Exercise 8, $S^n - h(B)$ is connected. Now $S^n - h(S^{n-1}) = \{S^n - h(B)\} \cup \{h(B) - h(S^{n-1})\}$. Since these are disjoint and connected they must be the components of $S^n - h(S^{n-1})$ by 26.29. In particular $h(B) - h(S^{n-1})$ is open and hence a neighborhood of x in X. Thus X is open. ∎

Corollary 26.31 If $f: R^m \to R^n$ is continuous and 1–1, $n \geq m$.

Proof If $n < m$, $R^m \xrightarrow{f} R^n \subset R^m$ is not open. ∎

Corollary 26.32 Let M and N be manifolds of dimension n. If $X \subset M$, $Y \subset N$, and $X \equiv Y$, then X is open iff Y is open.

Proof Suppose X is open. Let $h: X \to Y$ be a homeomorphism, let $y \in Y$ and choose an open neighborhood V of y and a homeomorphism $\gamma: E^n \to V$. $h^{-1}(V)$ is a neighborhood of $h^{-1}(y)$. Choose an open neighborhood U of $h^{-1}(y)$ with $U \subset h^{-1}(V)$ and a homeomorphism $\theta: R^n \to U$. By 26.30 applied to $R^n \xrightarrow{\theta} U \xrightarrow{h} V \to V^\infty \equiv S^n$, one sees that $h(U)$ is open in V. Since V is open, $h(U)$ is open. Since $y \in h(U) \subset Y$, Y is open. The converse is equivalent. ∎

Definition 26.33 A homomorphism $A \otimes B \to k$ where A and B are vector spaces over k will be called a dual pairing if its adjoint $A \to B^*$ is an isomorphism.

Theorem 26.34 Suppose k is a field and M is a compact n-manifold orientable with respect to Hk. Then there is a dual pairing

$$SH^i(M; k) \otimes H^{n-i}(M; k) \to k$$

given by $(x, y) = \langle x \cup y, [M] \rangle$ where $\langle\ ,\ \rangle$ is the Kronecker product (see Exercise 7, Section 23).

Proof $D: M^{n-i}(M; k) \to H_i(M; k)$ is given, in this case by $D(y) = y \cap [M]$. By Exercise 7, Section 23

$$(x, y) = \langle x \cup y, [M] \rangle = \langle x, D(y) \rangle.$$

However, D is an isomorphism and $\langle\ ,\ \rangle$ is a dual pairing by 25.20. ∎

This is useful in calculation. For example:

Corollary 26.35 (a) Let $x \in H^1(RP^n; Z_2)$ be the nonzero element. Then $x^k \neq 0$ for $k < n$ and hence generates $H^k(RP^n; Z_2)$.

(b) Choose a generator $y \in H^2(CP^n)$. Then y^k generates $H^{2k}(CP^n)$ for $k \leq n$.

(c) Choose a generator $z \in H^4(HP^n)$. Then z^k generates $H^{4k}(HP^n)$ for $k \leq n$.

Proof (a) We use induction on n. The inclusion $RP^{n-1} \subset RP^n$ induces an isomorphism in cohomology in dimensions less than n. Thus $x^k \neq 0$ for $k < n$. But $x \cdot x^{n-1} \neq 0$ by 26.34 since $x^{n-1} \neq 0$. (b) and (c) are proven similarly. In these cases we use integer coefficients. We cannot apply 26.34 but the proof of 26.34 applies since $H_*(CP^n)$ and $H_*(HP^n)$ are free and hence the Kronecker product is a dual pairing. ▮

Corollary 26.36 $H^*(RP^\infty; Z_2)$ is a polynomial ring over Z_2 generated by $x \in H^1(RP^\infty; Z_2)$. $H^*(CP^\infty)$ and $H^*(HP^\infty)$ are integral polynomial rings generated by $y \in H^2(CP^\infty)$ and $z \in H^4(HP^\infty)$ respectively. ▮

Theorem 26.37 (*Borsuk–Ulam*) If $n > m \geq 1$, there is no map $g: S^n \to S^m$ such that $g(-x) = -g(x)$.

Proof Such a map g would induce a map

$$f: RP^n \to RP^m.$$

Since a map $S^1 \xrightarrow{\alpha} RP^n$ is essential iff it is covered by a map $I \xrightarrow{\beta} S^n$ with $\beta(0) = -\beta(1)$, it follows that $f_*: \pi_1(RP^n) \to \pi_1(RP^m)$ is an isomorphism. Hence $f^*: H^1(RP^m; Z_2) \to H^1(RP^n; Z_2)$ is an isomorphism. Thus $f^*(x_1) = x_2$ where $x_1 \in H^1(RP^m; Z_2)$ and $x_2 \in H^1(RP^n; Z_2)$ are nonzero elements. But $x_1^{m+1} = 0$ and since $m + 1 \leq n$, $x_2^{m+1} \neq 0$ by 26.35. This is a contradiction since $f^*(x_1^{m+1}) = x_2^{m+1}$. ▮

Another version of this theorem is:

Theorem 26.38 (*Borsuk–Ulam*) If $f: S^n \to R^n$ is continuous and $n \geq 1$, there exists $x \in S^n$ with $f(x) = f(-x)$.

Proof If not,

$$g(x) = (f(x) - f(-x))/\|f(x) - f(-x)\|$$

defines a continuous map $g: S^n \to S^{n-1}$, and $g(-x) = -g(x)$ contradicting 26.37. ▮

The meaning of this result can be better understood if we assume that the surface of the earth is S^2 and that temperature and humidity are continuous functions of position. Then at any time there are two antipodal points on the earth where the temperature and humidity are the same.

Corollary 26.39 Let $A_1, \ldots, A_m$ be bounded measurable subsets in R^m. Then there is a hyperplane H that bisects each of the A_i.

Proof For each $x \in S^m$, let A_x be a hyperplane through $(0, \ldots, 0, 1) \in R^{m+1}$ and orthogonal to x; see Fig. 26.2. A_x intersects R^m in a hyperplane which

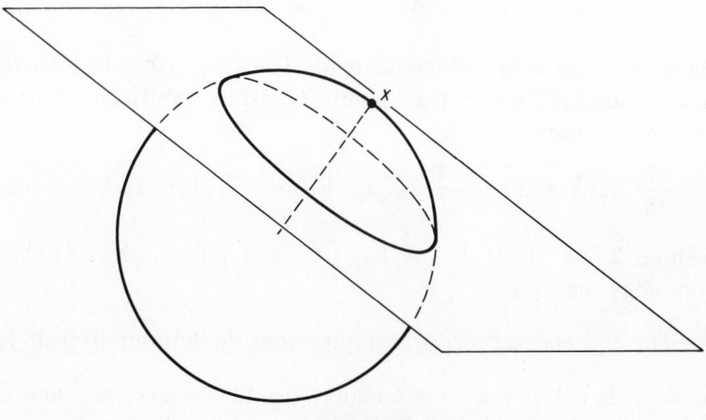

Figure 26.2

consequently breaks A_i into two pieces. Let $f_i(x)$ be the measure of the part of A_i on the same side of A_x as x. (Clearly $x \notin A_x$ unless $x = (0, \ldots, 0, 1.)$ In this case we define $f_i(x) = 0$.) Since A_i is bounded, f_i is continuous. $f_i(x)$ is the ith coordinate of a continuous function $f\colon S^m \to R^m$. Now $f_i(-x)$ is the measure of the other part of A_i. Hence a point x with $f(x) = f(-x)$ determines a cut of each A_i into two equal parts. ∎

In case $m = 3$, this theorem has been called the ham sandwich theorem since it indicates that there is a fair way to cut a three layer sandwich in half.

Proposition 26.40 If M is a compact orientable manifold of dimension $4k + 2$, $\chi(M)$ is even.

Proof $\chi(M) = \sum (-1)^i \dim(H_i(M; Q) \equiv \dim H_{2k+1}(M; Q) \pmod 2$ since $\dim H_k(M; Q) = \dim H_{4k+2-j}(M; Q)$. The cup product pairing is a skew symmetric dual pairing on $H_{2k+1}(M; Q)$, and hence is represented by a skew symmetric nonsingular matrix. This implies that $\dim H_{2k+1}(M; Q)$ is even since there are no $m \times m$ skew symmetric nonsingular matrices with m odd. Thus $\chi(M) \equiv 0 \pmod 2$. ∎

As a final application we will discuss fixed point theory. Let $f\colon X \to X$. We will consider conditions for the existence of an $x \in X$ with $f(x) = x$. Such an x exists iff the graph of f and the graph of the identity map intersect. This can be expressed diagrammatically as follows:

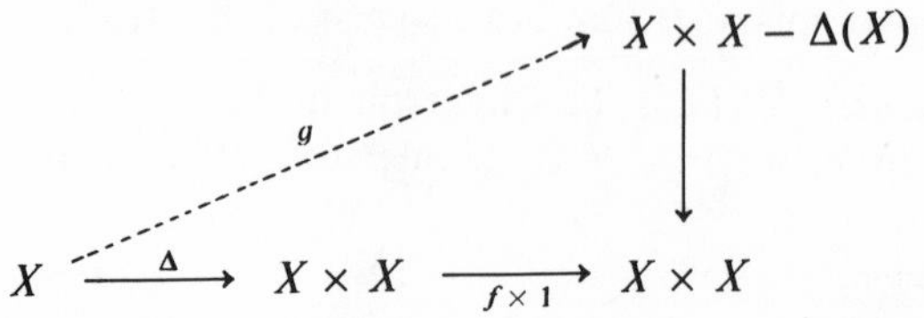

In the diagram Δ is the diagonal map. The map g exists such that the diagram commutes iff f has no fixed points. Consider now the homomorphism $L(f)$ given by the composition

$$E_*(X) \xrightarrow{\Delta_*} E_*(X \times X) \xrightarrow{(f \times 1)_*} E_*(X \times X) \to E_*(X \times X, X \times X - \Delta(x))$$

Proposition 26.41 If $f\colon X \to X$ has no fixed points, then $L(f) = 0$ for every homology theory E_*.

Proof The existence of g clearly implies that the homomorphism is 0. ▮

In the case that X is a compact manifold and $E = Hk$ for some field k, $L(f)([M])$ can easily be calculated. We begin with a definition.

Definition 26.42 Let V be a finite-dimensional vector space and $A\colon V \to V$ a linear transformation. Then

$$\theta\colon V^* \otimes V \to \operatorname{Hom}(V, V)$$

given by $\theta(x \otimes y)(z) = x(z) \cdot y$ is an isomorphism. Let $e\colon V^* \otimes V \to k$ be the evaluation. Define the trace of A by the formula

$$\operatorname{Tr}(A) = e(\theta^{-1}(A)).$$

One can easily check that if A is represented by a matrix, $\operatorname{Tr}(A)$ is the sum of the diagonal entries.

Theorem 26.43 (*Lefschetz Fixed Point Theorem*) If X is a compact n-manifold that is orientable with respect to Hk for some field k, and $f\colon X \to X$ is a map without fixed points, then

$$0 = L(f)[M] = \sum (-1)^{n+i} \operatorname{Tr}_i(f_*),$$

where $\operatorname{Tr}_i(f_*)$ is the trace of $f_*\colon H_i(X; k) \to H_i(X; k)$.

Proof of 26.43 Note that since X is compact, $D(x) = x \cap [X] = x \backslash \Delta_*([X])$. Let $\{x_\alpha\}$ be a basis for $H_*(X; k)$. For $x \in H_i(X; k)$, we will write x^* for the dual class in $H^i(X; k)$. Thus $x^* \backslash x = 1$. Now

$$\Delta_*([X]) = \sum_\alpha x_\alpha \underline{\times} y_\alpha$$

for some classes $y_\alpha \in H_*(X; k)$ by 25.15. Thus

$$y_\alpha = x_\alpha^* \backslash \left(\sum_\alpha x_\alpha \underline{\times} y_\alpha \right) = x_\alpha^* \backslash \Delta_*([X]) = D(x_\alpha^*),$$

so we have established the formula

$$\Delta_*([X]) = \sum x_\alpha \underline{\times} D(x_\alpha^*).$$

We will evaluate the composite

$$\begin{array}{ccccccccc}
H_n(X; k) & \xrightarrow{\Delta_*} & H_n(X \times X; k) & \xrightarrow{(f \times 1)_*} & H_n(X \times X; k) & \longrightarrow & H_n(X \times X, X \times X - \Delta(X); k) & & \\
 & & & & \cong \uparrow D_1 & & \cong \uparrow D_2 & & \\
 & & & & H^n(X \times X; k) & \xrightarrow{\Delta^*} & H^n(X; k) & \xrightarrow[\cong]{D} H_0(X; k) & \xrightarrow{P_*} H_0(P; k)
\end{array}$$

By Exercise 18, $D_1(x \overline{\times} y) = (-1)^{|x| \cdot n} D(x) \underline{\times} D(y)$. Hence

$$\begin{aligned}
D_1^{-1}((f \times 1)_*) \, \Delta_*([X]) &= \sum_\alpha D_1^{-1}(f_*(x_\alpha) \underline{\times} D(x_\alpha^*)) \\
&= \sum_\alpha (-1)^{|x_\alpha| \cdot n} D^{-1}(f_*(x_\alpha)) \overline{\times} x_\alpha^*.
\end{aligned}$$

Thus

$$\begin{aligned}
P_* D \, \Delta^* D_1^{-1}(f \times 1)_* \, \Delta_*([X]) &= P_* D \left(\sum_\alpha (-1)^{|x_\alpha| \cdot n} D^{-1}(f_*(x_\alpha)) \cdot x_\alpha^* \right) \\
&= P_* D \left(\sum_\alpha (-1)^{|x_\alpha|} x_\alpha^* \cdot D^{-1}(f_*(x_\alpha)) \right) \\
&= P_* \left(\sum_\alpha (-1)^{|x_\alpha|} x_\alpha^* \cap f_*(x_\alpha) \right)
\end{aligned}$$

(since $(x \cup y) \cap z = x \cap (y \cap z)$)

$$\begin{aligned}
&= P_* \left(\sum_i (-1)^i \sum_{|x_\alpha| = i} x_\alpha^* \cap f_*(x_\alpha) \right) \\
&= \sum_i (-1)^i \sum_{|x_\alpha| = i} x_\alpha^* \backslash f_*(x_\alpha) \\
&= \sum_i (-1)^i \operatorname{Tr}_i(f_*)
\end{aligned}$$

since $\theta(x_\alpha^* \otimes f_*(x_\alpha)) = f_*$. ∎

Exercises

1. If A is a closed subset of X, show that A^∞ is a closed subset of X^∞. If U is an open subset of X, construct a continuous mapping $f^\infty\colon X^\infty \to U^\infty$. Show that $X^\infty/A^\infty \equiv (X - A)^\infty$. Hence there is an exact sequence:

$$\cdots \leftarrow E^{m-1}_{\text{comp}}(X - A) \leftarrow E^m_{\text{comp}}(A) \leftarrow E^m_{\text{comp}}(X) \leftarrow E^m_{\text{comp}}(X - A) \leftarrow \cdots$$

(Section 30)

2. Let M be a manifold. Each open set $U \subset M$ homeomorphic to R^n determines a map $r_U\colon M^\infty \to S^n$. Show that if M is arcwise connected, the homotopy class of r_U does not depend on the choice of U.

3. Suppose M is a compact connected manifold. Show that M is orientable with respect to E iff the homomorphism

$$r_*\colon E_n(M) \to E_n(S^n)$$

is an isomorphism, where $r = r_U$ (see Exercise 2). Show that r_* is an isomorphism iff there is a class $\alpha \in E_n(M)$ such that $r^*([S^n])\backslash\alpha = 1$.

4. Suppose X is a compact subset of R^n. Prove that $H^i(X) = 0$ for $i \geq n$.

5. Let M be a connected n-manifold. Show that if M is compact and orientable, $H^n(M) \simeq Z$. Otherwise $H^n(M) = 0$.

6. Prove $(A \times B)^\infty \equiv A^\infty \wedge B^\infty$. Prove that if $M^n \subset R^{n+k}$ has a neighborhood $U \equiv M^n \times R^k$ such that M^n corresponds to $M^n \times 0$, then M is a π manifold. (30.4)

7. Show that if M and N are orientable with respect to E, $M \times N$ is orientable and there is a unique orientation such that $[K \times L] = [K] \,\underline{\times}\, [L]$.

8. Suppose $X \subset S^n$ and $X \equiv B^n$. Prove that $S^n - X$ is connected. (26.30)

9. Prove that a subset of the plane is simply connected iff its complement has no bounded components.

10. A manifold with boundary is a Hausdorff space M^n such that every point has a neighborhood homeomorphic to R^n or $H^n = \{(x_1, \ldots, x_n) \in R^n \mid x_n \geq 0\}$. The set of points with a neighborhood homeomorphic to R^n is called the interior of M (Int M). $\partial M = M -$ Int M is called the boundary of M. Show that Int M is open. Use 26.30 to show that ∂M is an $(n-1)$-manifold.

11. Let M be a manifold with boundary. Show that

$$L = M \cup \{\partial M \times [0, 1)\}/x \sim (x, 0)$$

is a manifold.

12. Using the notation of Exercise 11, let $K = M \cup \{\partial M \times [0, \frac{1}{2}]\}$ and $B = \partial M \times \frac{1}{2}$. Show that $M \simeq K - B$, $(K, B) \simeq (M, \partial M)$, $L \simeq M$, and $K - K \simeq \partial M$. Hence prove that if L is orientable with respect to E and M is compact there are isomorphisms:

$$E^i(M, \partial M) \cong E_{n-i}(M), \qquad E^i(M) \cong E_{n-i}(M, \partial M). \quad (30.26)$$

13. Use Exercise 12 to show that if M is a compact n-manifold with boundary, $\chi(\partial M) = (1 + (-1)^{n-1})\chi(M)$.

14. Suppose M^n is a compact connected n-manifold and $M^n \subset N^{n+1}$ where N^{n+1} is a simply connected $(n+1)$-manifold. Then M^n is orientable and $N - M$ has exactly two components. Conclude that $RP^2 \not\subset R^3$.

15. Prove via Exercise 13 that RP^{2n} is not the boundary of a compact manifold with boundary.

16. Calculate the multiplication structure in $H^*(RP^n)$ using the multiplicative homomorphisms $H^k(RP^n) \to H^k(RP^n; Z_2)$.

17. Suppose that if M is an orientable n-manifold with respect to Hk, where k is a field. Suppose $L \subset K \subset M$ are compact. Show that there is a dual pairing

$$H^i(K, L; k) \otimes SH^{n-i}(X - L, X - K; k) \to k.$$

18. Let M and N be compact manifolds of dimensions m and n that are orientable with respect to E. Using the orientation for $M \times N$ from Exercise 7, show that

$$D(x \mathbin{\bar{\times}} y) = (-1)^{|x| \cdot n} D(x) \mathbin{\underline{\times}} D(y)$$

by establishing the commutativity (with the sign $(-1)^{ml}$) of the diagram

$$\begin{array}{ccc}
E^{k+m}(A \times B) \otimes E_{l+n}(A \times B \times C \times D) & \xrightarrow{\ \backslash\ } & E_{l+n-k-m}(C \times D) \\
\uparrow{\scriptstyle 1 \otimes (1 \times T \times 1)_*} & & \\
E^{k+m}(A \times B) \otimes E_{l+n}(A \times C \times B \times D) & (-1)^{ml} & \\
\uparrow{\scriptstyle \bar{\times} \otimes \underline{\times}} & & \uparrow{\scriptstyle \underline{\times}} \\
E^k(A) \otimes E^m(B) \otimes E_l(A \times C) \otimes E_n(B \times D) & & \\
\uparrow{\scriptstyle 1 \otimes T \otimes 1} & & \\
E^k(A) \otimes E_l(A \times C) \otimes E^m(B) \otimes E_n(B \times D) & \xrightarrow{\ \backslash \otimes \backslash\ } & E_{l-k}(C) \otimes E_{n-m}(D)
\end{array} \quad (26.43)$$

19. Show that every map $f\colon X \to X$ has a fixed point where $X = CP^n$, HP^n, or RP^{2n}. Find a map $f\colon RP^{2n+1} \to RP^{2n+1}$ without a fixed point (Hint: Use a nonzero vector field on S^{2n+1}.)

20. Generalize 26.43 to an arbitrary homotopy commutative homotopy associative ring spectrum such that, for the manifold X in question:

(a) $E_*(X)$ is free and finitely generated over $E_*(P)$.

(b) $E^k(X) \otimes_{E_*(P)} E_k(X) \xrightarrow{\backslash} E_0(P)$ is a dual pairing.

(c) $E_n(X \times X) \cong \bigoplus_k E_k(X) \otimes_{E_*(P)} E_{n-k}(X)$.

21. Let $k\iota_m : S^m \to S^m$ be a map of degree k. Then $\{k\iota_2 \eta\} = n_k\{\eta\} = \{\eta \cdot n_k \iota_3\}$ since η generates $\pi_3(S^2)$. The homotopy commutative diagram

$$\begin{array}{ccc} S^3 & \xrightarrow{\eta} & S^2 \\ {\scriptstyle n_k \iota_3}\downarrow & & \downarrow{\scriptstyle k\iota_2} \\ S^3 & \xrightarrow{\eta} & S^2 \end{array}$$

defines a map $\theta : CP^2 \to CP^2$ by Exercise 5, Section 14. Using 26.35 show that $n_k = k^2$. Using Exercise 7, Section 13, prove that $2E\{\eta\} = 0$. (27.19)

22. Show that $L_{2n-1}(Z_p)$ is an orientable manifold (see Exercise 19, Section 7). Suppose $p > 2$. Use 26.34 to prove that $H^*(L_{2n-1}(Z_p); Z_p)$ has generators x and y of dimensions 1 and 2 and relations $x^2 = 0$ and $y^n = 0$. Show that $L(Z_p) = \bigcup_{n=1}^{\infty} L_{2n-1}(Z_p)$ is a space $K(Z_p, 1)$, and $H^*(K(Z_p, 1); Z_p) \cong Z_p[y] \otimes \Lambda(x)$ where $Z_p[y]$ is a polynomial algebra over Z_p and $\Lambda(x)$ is an exterior algebra ($\Lambda(x)$ has generators 1 and x and a single relation $x^2 = 0$). (Appendix, Section 27)

23. Let $f : M^n \to N^n$ be a 1–1 continuous mapping where M and N are n-manifolds. Prove that f is open. (Hint: Apply 26.32.)

24. Choose an orientation for R^n and hence for each open subset of R^n. Call a homeomorphism $h : U \to V$ orientation preserving if $h_*([K]) = [h(K)]$ for each compact $K \subset U$. Show that a manifold M is orientable iff there is a coordinate system $\{U_\alpha, h_\alpha\}$ such that $h_\alpha^{-1} h_\beta : h_\beta^{-1}(U_\alpha) \to h_\alpha^{-1}(U_\beta)$ is orientation preserving.

25. Consider the projective spaces RP^n, CP^n, and HP^n of Sections 7 and 11. Show that $[\xi_0 | \cdots | \xi_n] \to (\xi_i^{-1}\xi_0, \ldots, \xi_i^{-1}\xi_n)$ is a homeomorphism from V_i to R^n, R^{2n}, or R^{4n} respectively.

26. Let us consider D to be the basic duality operator in a manifold. Then there is a pairing dual to the cup product which is in fact its historical predecessor. This pairing, called the intersection pairing, is defined for any two open sets U and V in a manifold M^n that is orientable with respect to E

$$I : E_r(U) \otimes E_s(V) \to E_{r+s-n}(U \cap V)$$

is defined to be the composition

$$E_r(U)\otimes E_s(V)\xleftarrow[\cong]{D\otimes D} E^{n-r}(M, M-U)\otimes E^{n-s}(M, M-V)$$
$$\xrightarrow{\cup} E^{2n-r-s}(M, M-(U\cap V))$$
$$\xrightarrow{D} E_{r+s-n}(U\cap V).$$

Show that I is graded-commutative and associative. Show that if M is compact $I(\xi\otimes[M])=\xi$.

If ξ and η are ordinary homology classes represented by cycles $c(\xi)$ and $c(\eta)$ in "general position," their intersection will have dimension $r+s-n$. This intersection will represent $I(\xi\otimes\eta)$. As a simple example of this consider curves on a torus.

27. (*Leray–Hirsch Theorem*) Let $F\xrightarrow{i}E\xrightarrow{\pi}B$ be a locally trivial bundle with B compact. Let R be a principal ideal domain. Suppose there are classes $x_i\in SH^{n_i}(E; R)$ such that $\{i^*(x_i)\}$ is an R-free basis for $SH^*(F; R)$. Then $\{x_i\}$ is an $SH^*(B; R)$ free basis for $SH^*(E; R)$ (Hint: Construct a model functor $L^*(A)=$ free $H^*(A; R)$ module generated by $\{x_i\}$, for $A\subset B$, a reality functor $K^*(A)=SH^*(\pi^{-1}(A); R)$, and a natural transformation $\theta_A\colon L^*(A)\to K^*(A)$ which is the $SH^*(A; R)$ module homomorphism which sends the generator x_i of $L^*(A)$ into $(i_A)^*(x_i)$ where $i_A\colon\pi^{-1}(A)\to E$ is the inclusion. The object is to prove θ_B is an isomorphism. Construct Mayer–Vietous sequences for L^* and K^* and use induction over the open subsets of B, using the fact that θ_U is an isomorphism if U is a coordinate neighborhood by the Künneth theorem.) (30.7)

28. Prove that the only compact contractible manifold is a point.

27

Cohomology Operations

In this section we shall discuss natural transformations in homology and cohomology theories. Operations will be constructed in ordinary theory, and we will make applications to geometric problems.

In Exercise 4, Section 18 the notions of stable homology and cohomology operations were introduced. The simplest examples of such operations are coefficient transformations. Suppose that E is a ring spectrum and $\alpha \in \pi_n(E) = \tilde{E}_n(S^0) = \tilde{E}^{-n}(S^0)$. Then the transformations

$$\phi_\alpha(x) = \alpha \,\underline{\wedge}\, x, \qquad \phi^\alpha(x) = \alpha \,\overline{\wedge}\, x$$

define operations $\phi_\alpha \colon \tilde{E}_k(X) \to \tilde{E}_{n+k}(X)$ $\phi^\alpha \colon \tilde{E}^k(X) \to \tilde{E}^{k-n}(X)$. These are clearly natural and stable. In fact they are induced by a mapping of spectra:

$$E_k \equiv S^0 \wedge E_k \xrightarrow{\alpha \wedge 1} E_n \wedge E_k \to E_{n+k}.$$

These facts thus follow from Exercise 4, Section 18. In ordinary theory these operations correspond to the action of the coefficient ring R on the modules $\tilde{H}_*(X; R)$ and $\tilde{H}^*(X; R)$.

Proposition 27.1 In $\pi_*{}^S(X)$ and $\pi_S{}^*(X)$, all stable operations are coefficient operations.

Proof Let Θ_k be the set of stable homology operations of degree k in $\pi_*{}^S$. This set has a natural addition given by adding values. We have defined a homomorphism

$$\phi \colon \pi_k{}^S(S^0) \to \Theta_k.$$

A homomorphism $E \colon \Theta_k \to \pi_k{}^S(S^0)$ is defined by $E(\theta) = \theta(\iota)$ where $\iota \in \pi_0{}^S(S^0)$ is the class of the identity. It is easy to see that $E\phi = 1$. To prove that ϕ and

E are inverse isomorphisms, we prove that E is a monomorphism. Suppose $E(\theta) = 0$. Let $x \in \pi_k^S(X)$. Then there is a mapping $f\colon S^{n+k} \to S^n X$ such that $f_*(\sigma^{n+k}(\iota)) = \sigma^n x$. We then have

$$\sigma^n(\theta(x)) = \theta(\sigma^n(x)) = \theta(f_*(\sigma^{n+k}(\iota))) = f_*(\theta(\sigma^{n+k}(\iota))) = f_*(\sigma^{n+k}(\theta(\iota))) = 0.$$

Hence $\theta(x) = 0$ and thus $\theta = 0$. The proof in stable cohomotopy is similar. ∎

There are cohomology operations unrelated to stable operations. For example, the map $\mathrm{Sq}(x) = x^2$ defines a natural transformation

$$\tilde{H}^n(X) \to \tilde{H}^{2n}(X).$$

By 23.34, $\mathrm{Sq}(x + y) \neq \mathrm{Sq}(x) + \mathrm{Sq}(y)$, so this operation is not a homomorphism, and hence not a stable operation. It is, however, natural. $f^*(\mathrm{Sq}(x)) = f^*(x^2) = (f^*(x))^2 = \mathrm{Sq}(f^*(x))$.

Definition 27.2 A cohomology operation of type (E, m, F, n) is a natural transformation

$$\phi\colon \tilde{E}^m \to \tilde{F}^n.$$

Let $\{E, m, F, n\}$ be the abelian group of all cohomology operations of type (E, m, F, n). We define a homomorphism

$$R\colon \{E, m, F, n\} \to \tilde{F}^n(E_m)$$

by $R(\theta) = \theta(\iota)$ where $\iota \in \tilde{E}^m(E_m)$ is the class of the identity map.

Theorem 27.3 If E is an Ω-spectrum and each E_n is a CW complex, R is an isomorphism.

Proof We define a homomorphism $C\colon \tilde{F}^n(E_m) \to \{E, m, F, n\}$ as follows. Let

$$x \in \tilde{E}^m(X) = [X, E_m].$$

Let $f\colon X \to E_m$ represent x. Then define $C(\alpha)(x) = f^*(\alpha)$. C is clearly a homomorphism and is natural since

$$C(\alpha)(g^*(x)) = (fg)^*(\alpha) = g^*f^*(\alpha) = g^*(C(\alpha)(x)).$$

Furthermore, $R(C(\alpha)) = C(\alpha)(\iota) = \alpha$, and $C(R(\theta))(x) = f^*(R(\theta)) = f^*(\theta(\iota)) = \theta(f^*(\iota)) = \theta(x)$. ∎

Corollary 27.4 If E is an Ω-spectrum and each E_n is a CW complex, the graded abelian group of all stable cohomology operations of degree k from $\tilde{E}^*$ to $\tilde{F}^*$ is isomorphic to

$$\varprojlim \tilde{F}^{n+k}(E_n)$$

where the mappings in the limit are

$$\tilde{F}^{n+k}(E_n) \xrightarrow[\cong]{\sigma} \tilde{F}^{n+k+1}(SE_n) \xleftarrow{e_n^*} \tilde{F}^{n+k+1}(E_{n+1}).$$

Proof Since an element of $\varprojlim \tilde{F}^{n+k}(E_n)$ is a sequence of elements $x_n \in \tilde{F}^{n+k}(E_n)$ such that

$$\sigma(x_n) = e_n{}^*(x_{n+1})$$

(Exercise 9, Section 15), and a stable cohomology operation is a sequence ϕ_n of cohomology operations such that $\phi_{n+1}(\sigma(x)) = \sigma(\phi_n(x))$, we need only show that these two relations are equivalent, under 27.3. Suppose x_n corresponds to ϕ_n and x_{n+1} to ϕ_{n+1}. Let $x \in \tilde{E}^n(X)$ be represented by $f\colon X \to E_n$. Then $\sigma(x)$ is represented by $e_n(Sf)$. Thus $\phi_n(x) = f^*(x_n)$ and $\phi_{n+1}(\sigma(x)) = (Sf)^* e_n{}^*(x_{n+1}) = (Sf)^*\sigma(x_n) = \sigma(f^*(x_n)) = \sigma(\phi_n(x))$. ∎

Corollary 27.5 If E is an Ω-spectrum, each E_n is a connected CW complex, and F is properly convergent, the group of stable operations of degree k from E to F is isomorphic to $\tilde{F}^{2k+1}(E_{k+1})$ and

$$\sigma^{-1}e_k{}^*\colon \tilde{F}^{2k+1}(E_{k+1}) \to \tilde{F}^{2k}(E_k)$$

is a monomorphism.

Proof Apply 27.4, Exercise 14, Section 18, and Exercise 12, Section 22 to see that $\sigma^{-1}e_n{}^*\colon \tilde{F}^{n+k+1}(E_{n+1}) \to \tilde{F}^{n+k}(E_n)$ is an isomorphism for $n > k$ and is 1–1 if $n = k$. ∎

Corollary 27.6 If both E and F are Ω-spectra, and the spaces E_n and F_n are CW complexes, every stable cohomology operation is determined by a mapping of spectra.

Proof By 27.4 we have for each operation φ of degree k and each n a mapping $\varphi_n\colon E_n \to F_{n+k}$ such that $\sigma(\{\varphi_n\}) = e_n{}^*(\{\varphi_{n+1}\})$. This is precisely the diagram of Exercise 4, Section 18. ∎

We will investigate the stable cohomology operations in ordinary theory in the case that $\pi = \rho = Z_2$. However, the methods we use can be applied to other cases.

We construct stable operations Sq^i; $\tilde{H}^n(X; Z_2) \to \tilde{H}^{n+i}(X; Z_2)$ called the Steenrod squares. The method of construction in essence occurs in Steenrods paper [66], but we will follow a slick modification due to Milgram [47].

Definition 27.7 Let $X \rtimes Y = X \times Y/* \times Y$

$$\Gamma^n(X) = S^n \ltimes (X \wedge X)/(\theta, x, x') \sim (-\theta, x', x)$$

Proposition 27.8 (a) $\Gamma^n(X)$ is a covariant functor from $\mathcal{CG}^*$ to $\mathcal{CG}^*$.

(b) There is a natural inclusion

$$\Gamma^n(X) \to \Gamma^{n+1}(X)$$

write $\Gamma(X)$ for the union with the weak topology. $\Gamma(X)$ is called the quadratic construction on X.

(c) There is a natural map

$$H\colon \Gamma^n(X) \rtimes I \to \Gamma^n(X \rtimes I)$$

such that the diagrams

$$\begin{array}{ccc} \Gamma^n(X) \rtimes I & \xrightarrow{H} & \Gamma^n(X \rtimes I) \\ \downarrow & & \downarrow \\ \Gamma^{n+1}(X) \rtimes I & \xrightarrow{H} & \Gamma^{n+1}(X \rtimes I), \end{array} \qquad \begin{array}{ccc} \Gamma^n(X) \rtimes I & \xrightarrow{H} & \Gamma^n(X \rtimes I) \\ \uparrow{\scriptstyle i_\varepsilon} & \nearrow{\scriptstyle \Gamma^n(i_\varepsilon)} & \\ \Gamma^n(X) & & \end{array}$$

commute.

(d) If $f \sim g$, $\Gamma^n(f) \sim \Gamma^n(g)$ and the homotopies are compatible for various n.

(e) $\Gamma^0(X) \equiv X \wedge X$.

(f) There are compatible natural mappings

$$\Gamma^n(X \wedge Y) \xrightarrow{L^n} \Gamma^n(X) \wedge \Gamma^n(Y)$$

with L^0 the natural homeomorphism.

Proof (a) $\Gamma^n(X)$ clearly belongs to $\mathcal{CG}^*$ since it is a quotient space of $S^n \times (X \wedge X)$. If $f\colon X \to Y$,

$$1 \times (f \wedge f)\colon S^n \times (X \wedge X) \to S^n \times (Y \wedge Y)$$

induces a map $\Gamma^n(f)\colon \Gamma^n(X) \to \Gamma^n(Y)$ and the functorial identities are obvious.

(b) The equatorial inclusion $S^n \subset S^{n+1}$ induces a map $S^n \times (X \wedge X) \to S^{n+1} \times (X \wedge X)$ preserving the identifications. The induced map is clearly natural. It is 1–1 and closed, hence an inclusion map.

(c) Define H by

$$H(\theta, x, x', t) = (\theta, (x, t), (x', t));$$

this clearly preserves the identifications, and the commutativity of the diagrams is trivial.

(d) If $K\colon f \sim g$ is a homotopy, $\Gamma^n(K) \circ H\colon \Gamma^n(X) \rtimes I \to \Gamma^n(Y)$ is a homotopy between $\Gamma^n(f)$ and $\Gamma^n(g)$ by (c).

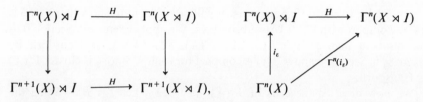

(e) This is obvious.
(f) Define L by

$$L(\theta, (x, y), (x', y')) = ((\theta, x, x'), (\theta, y, y'));$$

this is natural and preserves the identifications. ∎

Theorem 27.9 Suppose X is a CW complex with cells $\{e_\alpha{}^m, *\}$. Then $\Gamma(X)$ is a CW complex and $\Gamma^n(X)$ is a subcomplex. $\Gamma^0(X) \equiv X \wedge X$ has the cellular structure of $X \wedge X$. The cells of $\Gamma^n(X) - \Gamma^{n-1}(X)$ are of the form $e_{\alpha, \alpha'}$; one for each pair of cells $e_\alpha{}^m$, $e_{\alpha'}^{m'}$. $\dim e_{\alpha, \alpha'} = n + m + m'$.

Proof By Exercise 13, Section 14, it will be sufficient to show that $\Gamma^n(X)$ is a CW complex with $\Gamma^{n-1}(X)$ a subcomplex. We apply Exercise 4, Section 0 to the quotient map q: $S^n \times (X \wedge X) \to \Gamma^n(X)$. $S^n \times (X \wedge X)$ is Hausdorff by Exercise 6, Section 8 and $q^{-1}(y)$ is compact for each y. Since q is closed, $\Gamma^n(X)$ is Hausdorff.

We now describe the cells of $\Gamma^n(X) - \Gamma^{n-1}(X)$. Let f_+^n: $B^n \to S^n$ be the characteristic map for the upper hemisphere of S^n (see the proof of 20.10). For each pair of cells $e_\alpha{}^m$, $e_{\alpha'}^{m'}$ of X, not the base point, define a map $\chi_{\alpha, \alpha'}$: $B^{m+m'+n} \to \Gamma^n(X)$ by

$$B^{m+m'+n} \equiv B^n \times B^m \times B^{m'} \xrightarrow{f_+{}^n \times \chi_\alpha \times \chi_{\alpha'}} S^n \times (X \wedge X) \xrightarrow{q} \Gamma^n(X)$$

where q is the quotient map. Since every point of $\Gamma^n(X) - \Gamma^{n-1}(X)$ can be written uniquely in the form $q(\theta, x, x')$ for $\theta \in \operatorname{Int} B_+^n = f_+^n(\operatorname{Int} B^n)$, the sets $e_{\alpha, \alpha'} = \chi_{\alpha, \alpha'}(\operatorname{Int} B^{m+m'+n})$ cover $\Gamma^n(X) - \Gamma^{n-1}(X)$. In fact

$$q|_{\operatorname{Int} B_+{}^n \times (X - *) \times (X - *)}: \operatorname{Int} B_+^n \times (X - *) \times (X - *) \to \Gamma^n(X) - \Gamma^{n-1}(X)$$

is a homeomorphism. Hence $\chi_{\alpha, \alpha'}|_{\operatorname{Int} B^{m+m'+n}}$ is a homeomorphism onto $e_{\alpha, \alpha'}$. If we use the cellular structure of $X \wedge X$ for $\Gamma^0(X)$, we have a complete description of the cells of $\Gamma^n(X)$. $\chi_{\alpha, \alpha'}(S^{m+m'+n}) \subset \Gamma^n(X)^{m+m'+n-1}$ since $\chi_{\alpha, \alpha'}^{(n)}(S^n \times B^m \times B^{m'}) \subset \chi_{\alpha, \alpha'}^{(n-1)}(B^{m+m'+n-1})$. Thus $\Gamma^n(X)$ is a cell complex. It is closure finite, since if $\chi_\alpha(B^m) \subset K$ and $\chi_{\alpha'}(B^{m'}) \subset L$,

$$\chi_{\alpha, \alpha'}(B^{m+m'+n}) \subset \Gamma^n(K \cup L).$$

Since $S^n \times (X \wedge X)$ is a CW complex, $B(S^n \times (X \wedge X)) \xrightarrow{\chi} S^n \times (X \wedge X)$ is a quotient map. Moreover, there is a commutative diagram

$$\begin{array}{ccc} B(S^n \times (X \wedge X)) & \xrightarrow{\chi} & S^n \times (X \wedge X) \\ \downarrow & & \downarrow q \\ B(\Gamma^n(X)) & \xrightarrow{\chi} & \Gamma^n(X) \end{array}$$

Hence $\chi\colon B(\Gamma^n(X)) \to \Gamma^n(X)$ is a quotient map and thus $\Gamma^n(X)$ has the weak topology. ▌

Corollary 27.10 If X is an $(n-1)$-connected CW complex, $\Gamma^k(X)$ is $(2n-1)$-connected for all k.

Proof $X \simeq Y$ where Y has no cells in dimensions less than n, except for $*$. By 27.8(d), $\Gamma^n(X) \simeq \Gamma^n(Y)$. Now $\Gamma^n(Y)$ has no cells in dimensions less than $2n$, except for $*$. Thus $\Gamma^n(Y)$ is $(2n-1)$-connected. ▌

Lemma 27.11 There exists a unique homotopy class of maps

$$\gamma\colon \Gamma(K(Z_2, n)) \to K(Z_2, 2n)$$

such that

$$\mu_{n,n} = \gamma|_{\Gamma^0(K(Z_2, n))}\colon K(Z_2, n) \wedge K(Z_2, n) \to K(Z_2, 2n).$$

Proof

$$\Gamma^1(X) \equiv I \times (X \wedge X) \Big/ \begin{matrix} (0, x, y) \sim (1, y, x) \\ (t, *, *) \sim * \end{matrix}$$

Thus a map $\gamma^1\colon \Gamma^1(X) \to Y$ with $\gamma^1|_{\Gamma^0(X)} = \mu$ is determined by a diagram

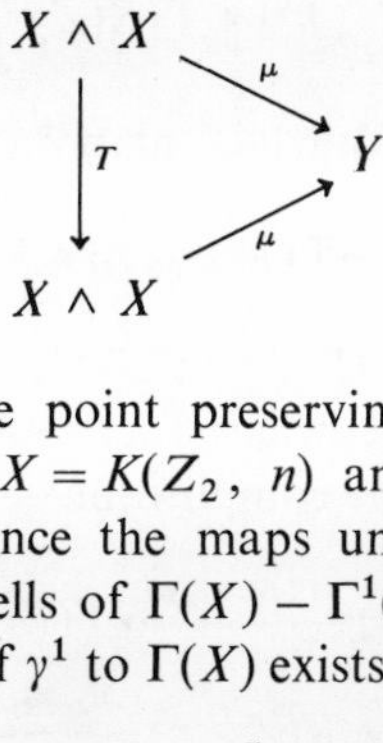

which commutes up to base point preserving homotopy; by 23.14 and 23.6, γ^1 exists in the case $X = K(Z_2, n)$ and $Y = K(Z_2, 2n)$ (the sign $(-1)^{n^2}$ is immaterial here, since the maps under consideration belong to $H^{2n}(X \wedge X; Z_2)$). Now the cells of $\Gamma(X) - \Gamma^1(X)$ have dimensions at least $2n + 2$. Hence an extension of γ^1 to $\Gamma(X)$ exists by 16.3. Now

$$[\Gamma(X), K(Z_2, 2n)] = H^{2n}(\Gamma(X); Z_2) = Z_2,$$

for there is only one cell of dimension $2n$ in $\Gamma(X)$. Consequently, there is only one nontrivial homotopy class of maps $\Gamma(X) \to K(Z_2, 2n)$. Uniqueness follows since $\mu_{n,n}\colon K(Z_2, n) \wedge K(Z_2, n) \to K(Z_2, n)$ is nontrivial. ▌

Theorem 27.12 There is a transformation

$$\gamma\colon \tilde{H}^n(X; Z_2) \to \tilde{H}^{2n}(\Gamma(X); Z_2)$$

such that

(a) $\iota^*(\gamma(x)) = x \barwedge x$ where $\iota: X \wedge X \to \Gamma(X)$ is the inclusion;
(b) if $f: X \to Y$, $\gamma(f^*(x)) = \Gamma(f)^*(\gamma(x))$;
(c) $\gamma(x \cdot y) = \gamma(x) \cdot \gamma(y)$.

Proof Let $\xi: X \to K(Z_2, n)$ represent x. Define $\gamma(x) = \{\gamma \circ \Gamma(\xi)\}$. By 27.8(d), this is well defined. Now $\iota^*(\gamma(x)) = \{\gamma \circ \Gamma(\xi) \circ \iota\} = \{\mu_{n,n} \circ (\xi \wedge \xi)\} = x \barwedge x$. If $f: X \to Y$,

$$\Gamma(f)^*(\gamma(x)) = \{\gamma \circ \Gamma(\xi) \circ \Gamma(f)\} = \{\gamma \circ \Gamma(\xi \circ f)\} = \gamma(\{\xi \circ f\}) = \gamma(f^*(x)).$$

Finally we observe that the diagram

$$\begin{array}{ccc} K(Z_2, 2n+2p) & \xleftarrow{\gamma} & \Gamma(K(Z_2, n+p)) \\ \uparrow{\scriptstyle \mu_{2n,2p}} & & \uparrow{\scriptstyle \Gamma(\mu_{n,p})} \\ K(Z_2, 2n) \wedge K(Z_2, 2p) & & \Gamma(K(Z_2, n) \wedge K(Z_2, p)) \\ {\scriptstyle \gamma \wedge \gamma}\nwarrow & & \swarrow{\scriptstyle L} \\ & \Gamma(K(Z_2, n)) \wedge \Gamma(K(Z_2, p)) & \end{array}$$

commutes by 23.12, for the nontrivial map

$$S^{2n+2p} \to \Gamma(K(Z_2, n) \wedge K(Z_2, p))$$

yields the nonzero element of $H^{2n+2p}(S^{2n+2p}; Z_2)$ under both compositions.

Now if $x = \{f\}$ and $y = \{g\}$, where $f: X \to K(Z_2, n)$ and $g: X \to K(Z_2, p)$, $\gamma(x) \cdot \gamma(y)$ is represented by the composition

$$\Gamma(X) \xrightarrow{\Delta} \Gamma(X) \wedge \Gamma(X) \xrightarrow{\Gamma(f) \wedge \Gamma(g)} \Gamma(K(Z_2, n)) \wedge \Gamma(K(Z_2, p))$$
$$\xrightarrow{\gamma \wedge \gamma} K(Z_2, n) \wedge K(Z_2, p) \xrightarrow{\mu_{2n,2p}} K(Z_2, 2n+2p).$$

One can easily see that the diagram

$$\begin{array}{ccc} \Gamma(X) & \xrightarrow{\Delta} & \Gamma(X) \wedge \Gamma(X) \\ & {\scriptstyle \Gamma(\Delta)}\searrow & \uparrow \\ & & \Gamma(X \wedge X) \end{array}$$

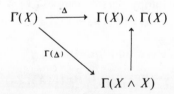

commutes. Hence $\gamma(x) \cdot \gamma(y)$ is also represented by

$$\Gamma(X) \xrightarrow{\Gamma(\Delta)} \Gamma(X \wedge X) \xrightarrow{\Gamma(f \wedge g)} \Gamma(K(Z_2, n) \wedge K(Z_2, p))$$
$$\xrightarrow{\Gamma(\mu_{n,p})} \Gamma(K(Z_2, 2n + 2p) \xrightarrow{\gamma} K(Z_2, 2n + 2p).$$

This, however, also represents $\gamma(x \cdot y)$. ∎

We now define $\tilde{\Delta}^n\colon RP^n \times X \to \Gamma^n(X)$ by $\tilde{\Delta}^n\{(\theta\}, x) = \{(\theta, x, x)\}$ for $\theta \in S^n$. This is well defined and continuous for $X \in \mathcal{CG}$. Furthermore, the maps $\tilde{\Delta}^n$ fit together to define a map $\tilde{\Delta}\colon RP^\infty \times X \to \Gamma(X)$.

Lemma 27.13 $\tilde{\Delta}$ is natural. The diagram

$$\begin{array}{ccc} X & \xrightarrow{\Delta} & X \wedge X \\ \downarrow{\scriptstyle \iota_2} & & \downarrow{\scriptstyle \iota} \\ RP^\infty \times X & \xrightarrow{\tilde{\Delta}} & \Gamma(X) \end{array}$$

commutes where $\iota_2(x) = (*, x)$.

Proof These facts are easy to verify. ∎

Definition 27.14 If $x \in \tilde{H}^n(X; Z_2)$, we define $\mathrm{Sq}^i(x) \in \tilde{H}^{n+i}(X; Z_2)$ (read: square-i) be the formula

$$\sum x^{n-i} \bar{\times} \mathrm{Sq}^i(x) = \tilde{\Delta}^*(\gamma(x)) \in H^{2n}(RP^\infty \times X; Z_2)$$

where $x^{n-i} \in H^{n-i}(RP^\infty; Z_2)$ is the nonzero element (see 26.35).

$\mathrm{Sq}^i(x)$ is well defined by 27.14 if X is a CW complex. This hypothesis is necessary to apply 25.15. It is easy to extend Sq^i to $\tilde{H}^n(X; Z_2)$ for any $X \in \mathcal{CG}^*$ by 27.3, and thus one may also define Sq^i on $S\tilde{H}^n(X; Z_2)$.

The operations Sq^i are called the Steenrod squaring operations or Steenrod squares; an operation that is a sum of products of Steenrod squares is called a Steenrod operation.

Theorem 27.15 $\mathrm{Sq}^i\colon \tilde{H}^n(X; Z_2) \to \tilde{H}^{n+i}(X; Z_2)$ is a cohomology operation satisfying:

(a) $\mathrm{Sq}^i(x) = 0$ if $i < 0$ or $i > n$.
(b) $\mathrm{Sq}^n(x) = x^2$.
(c) (Cartan Formula) $\mathrm{Sq}^k(x \cup y) = \Sigma\, \mathrm{Sq}^i(x) \cup \mathrm{Sq}^{k-i}(y)$.
(d) $\mathrm{Sq}^i(\sigma(x)) = \sigma(\mathrm{Sq}^i(x))$.
(e) $\mathrm{Sq}^0 = 1$, Sq^1 is the Bockstein homomorphism associated with the sequence $0 \to Z_2 \to Z_4 \to Z_2 \to 0$.

Proof Since $\tilde{\Delta}$ and γ are natural, Sq^i is natural. If $i > n$, $\mathrm{Sq}^i(x) = 0$ since $H^{n-i}(RP^\infty; Z_2) = 0$. Suppose $i < 0$. Let $X_{n-1} = X/X^{n-1}$. For $x \in \tilde{H}^n(X; Z_2)$ there exists $y \in \tilde{H}^n(X_{n-1}; Z_2)$ such that $(p_{X^{n-1}})^*(y) = x$. Thus $\mathrm{Sq}^i x = (p_{X^{n-1}})^*(\mathrm{Sq}^i(y)) = 0$, since $H^{n+1}(X_{n-1}; Z_2) = 0$. Now $\mathrm{Sq}^n x = \iota_2{}^*(x^{n-i} \,\bar{\times}\, \mathrm{Sq}^i(x)) = \iota_2{}^*\Delta^*(\gamma(x)) = \tilde{\Delta}^*\iota^*(\gamma(x)) = \tilde{\Delta}^*(x \,\bar{\wedge}\, x) = x^2$. Applying $\tilde{\Delta}^*$ to the equation $\gamma(x \cdot y) = \gamma(x) \cdot \gamma(y)$ yields

$$\sum_k x^{n+m-k} \,\bar{\times}\, \mathrm{Sq}^k(x \cdot y) = \Big(\sum_i x^{n-i} \,\bar{\times}\, \mathrm{Sq}^i(x)\Big)\Big(\sum_j x^{m-j} \,\bar{\times}\, \mathrm{Sq}^j(y)\Big).$$

(c) follows by calculating the coefficient of x^{m+n-k} on the right-hand side (using Exercise 17, Section 23). To prove (d) and (e) we need some lemmas.

Lemma 27.16 $\mathrm{Sq}^k(x \,\bar{\times}\, y) = \Sigma \mathrm{Sq}^i(x) \,\bar{\times}\, \mathrm{Sq}^{k-i}(y)$.

Proof This follows immediately from (c) and Exercise 17, Section 23, for $x \,\bar{\times}\, y = (x \,\bar{\times}\, 1) \cdot (1 \,\bar{\times}\, y) = p_1{}^*(x)\, p_2{}^*(y)$. ∎

Lemma 27.17 Let $u \in H^1(S^1; Z_2)$ be the nonzero element. Then $\mathrm{Sq}^0 u = u$.

Proof We will describe a cellular structure on $\Gamma^1(S^1)$ such that $\tilde{\Delta}\colon S^1 \times S^1 \to \Gamma^1(S^1)$ is cellular. We use this to evaluate $u \,\bar{\times}\, \mathrm{Sq}^0 u$. Now

$$\Gamma^0(S^1) \equiv S^1 \wedge S^1 \equiv I/\partial I \wedge I/\partial I \equiv I^2/\partial I^2.$$

We give this space a cellular structure so that the diagonal $\Delta\colon S^1 \to S^1 \wedge S^1$ is cellular; see Fig. 27.1.

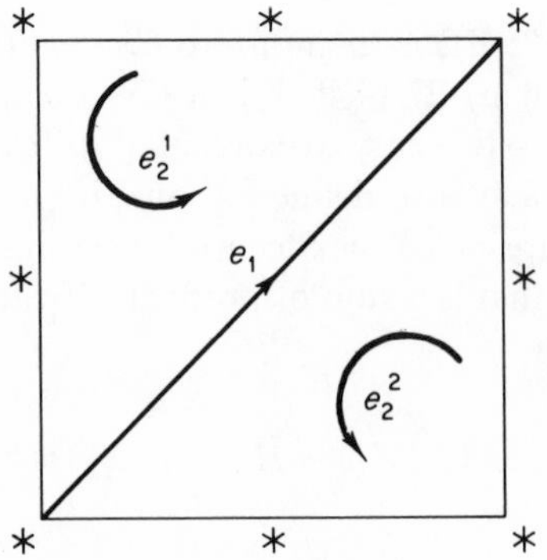

Figure 27.1

Let $*$ be a 0-cell. We define a 1-cell and two 2-cells with characteristic maps

$$\chi_1\colon I \to I^2/\partial I^2, \qquad \chi_2{}^1\colon \Delta^2 \to I^2/\partial I^2, \qquad \chi_2{}^2\colon \Delta^2 \to I^2/\partial I^2$$

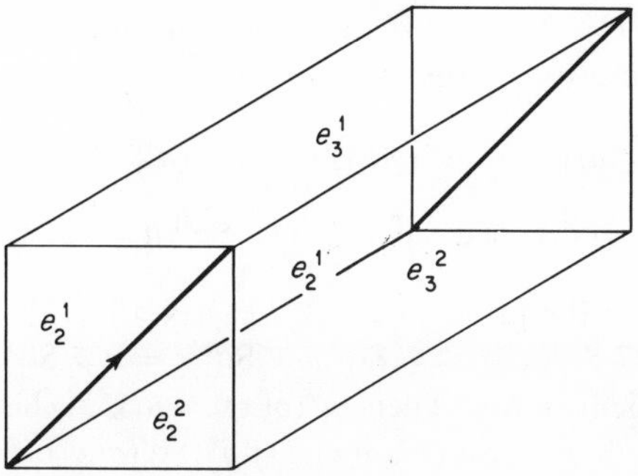

Figure 27.2

defined by $\chi_1(s) = (s, s)$, $\chi_2{}^1(t_0, t_1, t_2) = (t_2, t_0 + t_2)$, and $\chi_2{}^2(t_0, t_1, t_2) = (t_1 + t_2, t_1)$.

Now $\Gamma^1(S^1) = I \times (S^1 \wedge S^1)/\sim$ has an additional 2-cell and two 3-cells with characteristic maps

$$\chi_2{}^3 : I \times I \to \Gamma^1(S^1),$$
$$\chi_3{}^1 : I \times \Delta^2 \to \Gamma^1(S^1),$$
$$\chi_3{}^2 : I \times \Delta^2 \to \Gamma^1(S^1)$$

defined by $\chi_2{}^3(s, t) = (s, t, t)$, $\chi_3{}^1(s, t_0, t_1, t_2) = (s, t_2, t_0 + t_2)$, and $\chi_3{}^2(s, t_0, t_1, t_2) = (s, t_1 + t_2, t_1)$. We choose generators of $C_\#(\Gamma^1(S^1); Z_2)$ given by these characteristic maps:

$$e_0 \in C_0, \qquad e_1 \in C_1, \qquad e_2{}^1, e_2{}^2, e_2{}^3 \in C_2, \qquad \text{and} \qquad e_3{}^1, e_3{}^2 \in C_3.$$

One calculates

$$\partial e_3{}^1 = \partial e_3{}^2 = e_2{}^1 + e_2{}^2 + e_2{}^3,$$
$$\partial e_2{}^3 = 0, \qquad \partial e_2{}^1 = \partial e_2{}^2 = e_1,$$
$$\partial e_1 = 0 \qquad \partial e_0 = 0.$$

See Fig. 27.2.

Now $\tilde{\Delta}$: $S^1 \times S^1 \to \Gamma^1(S^1)$ is cellular. Choose generators x_0, $x_1{}^1$, $x_1{}^2$, and x_2 corresponding to the cells of $S^1 \times S^1$. Then $\tilde{\Delta}_\#(x_1{}^1) = 0$, $\tilde{\Delta}_\#(x_1{}^2) = e_1$ and $\tilde{\Delta}_\#(x_2) = e_2{}^3$.

We now evaluate $\tilde{\Delta}^*(\gamma(u))(\{x_2\})$:

$$\begin{aligned}\tilde{\Delta}^\#(\gamma(u))(x_2) &= \gamma(u)(\tilde{\Delta}_\#(x_2)) \\ &= \gamma(u)(e_2{}^3) \\ &= \gamma(u)(e_2{}^1 + e_2{}^2),\end{aligned}$$

since

$$0 = \delta(\gamma(u))(e_3{}^2) = \gamma(u)(\partial e_3{}^2) = \gamma(u)(e_2{}^1 + e_2{}^2 + e_2{}^3).$$

$e_2{}^1 + e_2{}^2 = \iota_\#(v \triangle v)$ where $v \in \tilde{H}_1(S^1; Z_2)$ is a nonzero class and $\iota: S^1 \wedge S^1 \to \Gamma^1(S^1)$ is the inclusion. Thus

$$\tilde{\Delta}^\#(\gamma(u))(x_2) = \iota^\#(\gamma(u))(v \triangle v) = (u \,\bar{\wedge}\, u)(v \triangle v) = 1.$$

Thus $\Delta^*(\gamma(u)) \neq 0$ and hence $\mathrm{Sq}^0\, u \neq 0$ so $\mathrm{Sq}^0\, u = u$. ∎

We continue with the proof of 27.15. Let $u \,\bar{\times}\, x \in \tilde{H}^{n+1}(S^1 \times X; Z_2)$. Then by 27.16 and 27.17 $\mathrm{Sq}^i(u \,\bar{\times}\, x) = \mathrm{Sq}^0\, u \,\bar{\times}\, \mathrm{Sq}^i\, x = u \,\bar{\times}\, \mathrm{Sq}^i\, x$. Let $p: S^1 \times X \to SX$ be the quotient map. Then $p^*(\sigma(x)) = u \,\bar{\times}\, x$, hence $p^*(\sigma(\mathrm{Sq}^i(x)) = u \,\bar{\times}\, \mathrm{Sq}^i\, x = \mathrm{Sq}^i(u \,\bar{\times}\, x) = \mathrm{Sq}^i(p^*(\sigma(x))) = p^*(\mathrm{Sq}^i(\sigma(x))$. Since p^* is a monomorphism, (d) follows. By 27.5 there is only one nonzero stable cohomology operation of degree 0. Since $\mathrm{Sq}^0\, u \neq 0$, $\mathrm{Sq}^0 = 1$.

Now let $x \in H^1(RP^\infty; Z_2)$ be the nonzero element. $\mathrm{Sq}^1 x = x^2 \neq 0$. To evaluate $\beta(x)$, observe that the monomorphism $Z_2 \to Z_4$ induces an isomorphism $H^1(RP^\infty; Z_2) \to H^1(RP^\infty; Z_4)$. Thus by Exercise 13, Section 18 $\beta(x) \neq 0$. Consequently $\beta(x) = \mathrm{Sq}^1\, x$. By 27.5, $\beta = \mathrm{Sq}^1$. ∎

Corollary 27.18 Sq^i is a homomorphism.

Proof This follows directly from 27.4 and Exercise 4, Section 18. ∎

Proposition 27.19 $\pi_{n+1}(S^n) = Z_2$ for $n > 2$. $S^{n-2}\eta: S^{n+1} \to S^n$ is essential.

Proof If $S^{n-2}\eta \sim *$, $C_{S^{n-2}\eta} \simeq S^n \vee S^{n+2}$ by Exercise 22, Section 14. Now by Exercise 5, Section 14, $S^2 \cup_\eta e^4 \equiv CP^2$. Hence

$$S^{n-2}CP^2 \equiv S^n \cup_{S^{n-2}\eta} e^{n+2} \simeq S^n \vee S^{n+2}.$$

Consequently there is a map $\alpha: S^{n-2}CP^2 \to S^n$ inducing an isomorphism in $\tilde{H}^n(\ ; Z_2)$. Choose $x \in \tilde{H}^n(CP; Z_2)$, and $y \in \tilde{H}^n(S^n; Z_2)$ with $\alpha^*(y) = x \neq 0$. $\mathrm{Sq}^2\, y = 0$, so $\mathrm{Sq}^2\, x = 0$. On the other hand $\mathrm{Sq}^2\, x = \mathrm{Sq}^2(\sigma^{n-2}(v)) = \sigma^{n-2}(\mathrm{Sq}^2\, v) = \sigma^{n-2}(v^2) \neq 0$, for $v \in \tilde{H}^2(CP^2; Z_2)$ a generator. This contradiction implies that $S^{n-2}\eta \not\sim *$. By 13.18 and 13.13, $\pi_{n+1}(S^n)$ is generated by $S^{n-2}\eta$, and by Exercise 21, Section 26, $S^{n-2}\eta$ has order 2. ∎

Proposition 27.20 Let $x \in H^1(RP^\infty; Z_2)$, $u \in H^2(CP^\infty; Z_2)$, and $v \in H^4(HP^\infty; Z_2)$ be nonzero elements. Then:

(1) $\mathrm{Sq}^i\, x = \binom{n}{i} x^{n+i}$.
(2) $\mathrm{Sq}^{2i}\, u^n = \binom{n}{i} u^{n+i}$.
(3) $\mathrm{Sq}^{4i}\, v^n = \binom{n}{i} v^{n+i}$.

Proof We use induction on n. If $n = 1$ $\mathrm{Sq}^1\, x = x^2$, $\mathrm{Sq}^2\, u = u^2$ and

$\mathrm{Sq}^4 v = v^2$, and these are the only possible nonzero values by 27.15(a). The inductive step applies the Cartan formula as follows:

$$\begin{aligned}\mathrm{Sq}^i x^{n+1} &= \mathrm{Sq}^i (x^n \cdot x) = \mathrm{Sq}^i x^n \cdot x + \mathrm{Sq}^{i-1} x^n \cdot x^2 \\ &= \left\{\binom{n}{i} + \binom{n}{i-1}\right\} x^{n+i+1} = \binom{n+1}{i} x^{n+i+1};\end{aligned}$$

the cases of u and v are similar. ∎

Thus for example, $x^{2^n} = \mathrm{Sq}^{2^{n-1}}\,\mathrm{Sq}^{2^{n-2}} \cdots \mathrm{Sq}^1 x$

One can define the operations Sq^i in $H^*(X, A; Z_2)$ and one easily proves:

Proposition 27.21 The operations $\mathrm{Sq}^i\colon H^n(X, A; Z_2) \to H^{n+i}(X, A; Z_2)$ satisfy 27.15 and furthermore $\mathrm{Sq}^i\delta = \delta\mathrm{Sq}^i$ where $\delta\colon H^n(A; Z_2) \to H^{n+1}(X, A; Z_2)$ is the coboundary. ∎

In particular, 27.16 also holds for $x \in H^n(X, A; Z_2)$ and $y \in H^m(Y, B; Z_2)$. By 27.6 one can also define $\mathrm{Sq}_i\colon H_n(X, A; Z_2) \to H_{n-i}(X, A; Z_2)$.

Theorem 27.22

(a) $\mathrm{Sq}_i\,(\sigma(x)) = \sigma(\mathrm{Sq}_i(x))$; $\mathrm{Sq}_i\,\partial x = \partial\,\mathrm{Sq}_i\,x$.

(b) $\mathrm{Sq}_k(x \underline{\times} y) = \sum_{i+j=k} \mathrm{Sq}_i\, x \underline{\times} \mathrm{Sq}_j\, y$.

(b′) $\mathrm{Sq}_k(x\backslash y) = \sum_{i+j=k} \mathrm{Sq}^i\, x\backslash \mathrm{Sq}_j\, y$.

(c) $\mathrm{Sq}_0 = 1$, Sq_1 is the Bockstein associated with the sequence $0 \to Z_2 \to Z_4 \to Z_2 \to 0$.

Proof The proof of (a) is easy. To prove (b) and (b′) observe that the Cartan formula can be written as a homotopy commutative diagram:

$$\begin{array}{ccc} K(Z_2, n) \wedge K(Z_2, m) & \xrightarrow{\Pi(\mathrm{Sq}^i \wedge \mathrm{Sq}^{k-i})} & \prod_{i=0}^{k} \{K(Z_2, n+i) \wedge K(Z_2, m+k-i)\} \\ & & \Big\downarrow {\scriptstyle \Pi\mu_{n+i,\, m+k-i}} \\ \Big\downarrow {\scriptstyle \mu_{n,m}} & & \prod K(Z_2, m+n+k) \\ & & \Big\downarrow \\ K(Z_2, m+n) & \xrightarrow{\mathrm{Sq}^k} & K(Z_2, m+n+k) \end{array}$$

Proof of this follows from evaluating $\mathrm{Sq}^k(\iota_1 \,\overline{\wedge}\, \iota_2)$ where $\iota_1 \in H^n(K(Z_2, n); Z_2)$ and $\iota_2 \in H^m(K(Z_2, m); Z_2)$.

Thus if $\alpha\colon S^{u+n} \to X \wedge K(Z_2, n)$ and $\beta\colon S^{v+m} \to Y \wedge K(Z_2, m)$ are representatives for $x \in H_u(X)$ and $y \in H_u(X)$, $\mathrm{Sq}_k(x \,\triangle\, y)$ is given by $(1 \wedge \mathrm{Sq}^k) \circ (1 \wedge \mu_{n,m}) \circ (1 \wedge T \wedge 1) \circ (\alpha \wedge \beta)$. Applying the diagram, this becomes $\sum \mathrm{Sq}_i x \,\triangle\, \mathrm{Sq}_{k-i} y$. Applying this to $H^n(X, A; Z_2) \cong \tilde{H}^n(X^+/A^+; Z_2)$ gives (b). A similar argument proves (b′). (c) follows for reasons very similar to those in 27.15. ∎

Corollary 27.23 Let $S^n \supset K \supset L$ with K and L compact. Then the diagram

$$\begin{array}{ccc} H^k(K, L; Z_2) & \xrightarrow[\simeq]{D} & H_{n-k}(S^n - L, S^n - K; Z_2) \\ \downarrow \mathrm{Sq}^i & & \downarrow \mathrm{Sq}_i \\ H^{k+i}(K, L; Z_2) & \xrightarrow[\simeq]{D} & H_{n-k-i}(S^n - L, S^n - K; Z_2) \end{array}$$

commutes.

Proof Let

$$C\colon H^i(K, L; Z_2) \otimes H_n(S^n, S^n - K; Z_2) \to H_{n-i}(S^n - L, S^n - K; Z_2)$$

be defined as in Section 26. Then clearly

$$\mathrm{Sq}_k\, C(x \otimes y) = \sum_{k=s+t} C(\mathrm{Sq}^s\, x \otimes \mathrm{Sq}_t\, y)$$

by 27.22(b′) and the definition of C. Now $[K] = i_*([S^n])$ and $\mathrm{Sq}_t([S^n]) = 0$ unless $t = 0$. Hence $\mathrm{Sq}_t([K]) = 0$ unless $t = 0$. Since $D(x) = C(x \otimes [K])$, $\mathrm{Sq}_i(D(x)) = C(\mathrm{Sq}^i\, x \otimes [K]) = D(\mathrm{Sq}^i\, x)$. ∎

In addition to Sq^i it is possible to define cohomology operations $(\mathrm{Sq}_i)^*$ by 25.20. These operations are different from Sq^i. More generally, for each stable cohomology operation ϕ^t of degree t, there is a corresponding homology operation ϕ_t of degree $-t$ by 27.6. Since $H^*(X; Z_2)$ and $H_*(X; Z_2)$ are dual vector spaces ϕ_t determines a stable cohomology operation $(\phi_t)^*$ of degree t. Let us write $\mathcal{A}(2)$ for the Z_2-algebra of stable cohomology operations with Z_2 coefficients. We can define $\chi\colon \mathcal{A}(2) \to \mathcal{A}(2)$ by $\chi(\phi^t) = (\phi_t)^*$.

Proposition 27.24

(1) $\chi(\alpha\beta) = \chi(\beta)\chi(\alpha)$, $\chi(1) = 1$, and χ is a homomorphism.
(2) $\sum_{t=0}^{n} \chi(\mathrm{Sq}^{n-t})\mathrm{Sq}^t = 0$ if $n > 0$.
(3) $\sum_{t=0}^{n} \mathrm{Sq}^{n-t}\, \chi(\mathrm{Sq}^t) = 0$ if $n > 0$.
(4) If α is a Steenrod operation,[27] $\chi^2(\alpha) = \alpha$.

Observe that Eq. (2) or (3) determines χ on Sq^n inductively, and hence on all Steenrod operations.

Proof (1) is immediate since $(\alpha\beta)_t = \alpha_t \beta_t$ and $(\alpha_t \beta_t)^* = \beta_t{}^* \alpha_t{}^*$. Let $x = \sum_{t=0}^{n} \chi\, (\mathrm{Sq}^{n-t})\, \mathrm{Sq}^t\, y$ and suppose $x \neq 0$. Let x^* be a dual homology class. Then

$$\begin{aligned} 1 = x\backslash x^* &= \sum (\mathrm{Sq}_{n-t})^*\, \mathrm{Sq}^t y \backslash x^* \\ &= \sum \mathrm{Sq}^t\, y \backslash \mathrm{Sq}_{n-t}\, x^* = \mathrm{Sq}_n(y \backslash x^*) \end{aligned}$$

by 27.22(b′).

[27] We will prove in Section 28 that every element of $\mathcal{A}(2)$ is a Steenrod operation.

Consideration of the diagram

$$\begin{array}{ccc} H_n(X; Z_2) & \xrightarrow{\mathrm{Sq}_n} & H_0(X; Z_2) \\ \downarrow p_* & & \downarrow p_* \\ H_n(*; Z_2) & \xrightarrow{\mathrm{Sq}_n} & H_0(*; Z_2) \end{array}$$

proves that there is no class $u \in H_n(X; Z_2)$ with $\mathrm{Sq}_n(u) = 1$. This contradiction implies $x = 0$ and establishes (2).

By using (2) and induction one proves that

$$\chi(\mathrm{Sq}^n) = \sum_{\substack{i_1 + \cdots + i_k = n \\ i_s > 0}} \mathrm{Sq}^{i_1} \ldots \mathrm{Sq}^{i_k}.$$

This function can easily be seen to satisfy Eq. (3) as well by symmetry.

Assume that $\chi^2(\mathrm{Sq}^i) = \mathrm{Sq}^i$ for $i < n$. Applying χ to (2) we get

$$\sum_{t>0} \chi(\mathrm{Sq}^t)\, \mathrm{Sq}^{n-t} + \chi^2(\mathrm{Sq}^n) = 0.$$

Comparing this with (2) we get $\chi^2(\mathrm{Sq}^n) = \mathrm{Sq}^n$. By iterated application of (1), we see that $\chi^2(\alpha) = \alpha$ if α is a Steenrod operation. ∎

Theorem 27.25 (*Thom*) If K is a compact subset of S^n, the homomorphism

$$\chi(\mathrm{Sq}^i)\colon H^{n-2i}(K; Z_2) \to H^{n-i}(K; Z_2)$$

is 0.

Proof We first show that for any $A \subset S^n$, $\mathrm{Sq}^i\colon H^i(S^n, A; Z_2) \to H^{2i}(S^n, A; Z_2)$ is 0. Now $\mathrm{Sq}^i x = x^2$. Let $f\colon (S^n, \varnothing) \to (S^n, K)$ be the inclusion. Then $x^2 = f^*(x) \cup x = 0$ unless $i = n$. In this case $\mathrm{Sq}^i x = 0$ since $H^{2n}(S^n, K; Z_2) \cong H_{-n}(S^n - K; Z_2) = 0$. Consequently $\chi(\mathrm{Sq}_i)\colon H_{2i}(S^n, A; Z_2) \to H_i(S^n, A; Z_2)$ is 0.

We now consider the commutative diagram

$$\begin{array}{ccc} H^k(K; Z_2) & \cong & H_{n-k}(S^n, S^n - K; Z_2) \\ \downarrow \chi(\mathrm{Sq}^i) & & \downarrow \chi(\mathrm{Sq}_i) \\ H^{k+i}(K; Z_2) & \cong & H_{n-k-i}(S^n, S^n - K; Z_2) \end{array}$$

with $A = S^n - K$, $k = n - 2i$. The result follows. ∎

Theorem 27.26 (*Peterson*) If $n = 2^s$, RP^n cannot be imbedded in S^{2n-1}, ([56])

Proof By 27.25, it is sufficient to evaluate

$$\chi(\mathrm{Sq}^i)\colon H^{2n-2i-1}(RP^n; Z_2) \to H^{2n-i-1}(RP^n; Z_2).$$

We show that $\chi(\mathrm{Sq}^{n-1})(x) = x^n$ iff $n = 2^s$ for $x \in H^1(RP^\infty; Z_2)$ a generator, by induction on n. The result is clear for $n = 0$. Suppose we have proven the result for all $r < n$. Then

$$\chi(\mathrm{Sq}^n)(x) = \sum_{t<n} \mathrm{Sq}^{n-t}\chi(\mathrm{Sq}^t)(x) = \sum_{t=2^\alpha-1<n} \mathrm{Sq}^{n-t}x^{t+1} = \sum_{t=2^\alpha-1<n} \binom{t+1}{n-t} x^{n+1}.$$

Now $\binom{2^\alpha}{\beta} \equiv 1 \pmod 2$ iff $\beta = 0$ or 2^α. Hence $\chi(\mathrm{Sq}^n)(x) = x^{n+1}$ iff there is a t satisfying $t = 2^\alpha - 1 < n$ and $t + 1 = n - t$. This is true iff $n = 2^s$. ∎

Appendix

In analogy with the squaring operations in Z_2 cohomology, one can define pth power operations in Z_p cohomology for p a prime.

Recall that Z_p acts without fixed points on S^{2n-1} (Exercise 19, Section 7). Define

$$\Gamma^n_{(p)}(X) = S^{2n-1} \times (\overbrace{X \wedge \cdots \wedge X}^{p})/(\theta, x_1, \ldots, x_p) \sim (\sigma,\theta\, x_2, \ldots, x_p, x_1)$$

where σ is the generator of Z_p. By analogy with 27.12 we can define, for each $x \in \tilde{H}^n(X; Z_p)$ a class $\gamma_{(p)}(x) \in H^{np}(\Gamma_{(p)}(X); Z_p)$. One then defines a natural transformation $\tilde{\Delta}\colon L_{2n-1}(Z_p) \times X \to \Gamma^n_{(p)}(X)$ by $\tilde{\Delta}(\{\theta\}, x) = \{(\theta, x, \ldots, x)\}$. This induces a cohomology homomorphism and we write

$$\tilde{\Delta}^*(\gamma_{(p)}(x)) = \sum D_k(x) \otimes w_k$$

where $w_{2k} = y^k$ and $w_{2k+1} = xy^k$ (see Exercise 22, Section 26). By a suitable choice of constants $a_{r,n} \in Z_p$ we define

$$P^r(x) = a_{r,n} D_{(n-2r)(p-1)}(x)$$

for $x \in \tilde{H}^n(X; Z_p)$. One can then prove

Theorem 27.27 $P^r\colon \tilde{H}^n(X; Z_p) \to \tilde{H}^{n+2r(p-1)}(X; Z_p)$ is a stable cohomology operation satisfying

(a) $P^0 = 1$;
(b) if $\dim x = 2r$, $P^r x = x^p$;
(c) if $\dim x < 2r$, $P^r x = 0$;
(d) $P^r(xy) = \sum P^i(x)P^{r-i}(y)$.

Exercises

1. Show that for any ring spectrum E, there are ring homomorphisms

$$E\colon \Theta_*(E) \to \tilde{E}_*(S^0), \qquad \phi\colon \tilde{E}_*(S^0) \to \Theta_*(E)$$

such that $E\phi = 1$. Prove a similar result for cohomology.

2. Show that the properties stated in 27.15 hold when Sq^i is applied to a class in $\tilde{H}^n(X; Z_2)$ for an arbitrary space $X \in \mathcal{CG}^*$.

3. Prove that $S^{n-4}\nu\colon S^{n+3} \to S^n$ is essential.

4. Fill in the details to 27.22(b′).

5. Show that $(\mathrm{Sq}_i)^*\colon \tilde{H}^{n-i}(X; Z_2) \to \tilde{H}^n(X, Z_2)$ given by the dual to Sq_i is a stable cohomology operation. Prove a Cartan formula:

$$(\mathrm{Sq}_k)^*(x \barwedge y) = \sum_{k=i+j} (\mathrm{Sq}_j)^*(x) \barwedge (\mathrm{Sq}_i)^*(y).$$

6. Let $x \in H^1(RP^\infty; Z_2)$. By computing $\mathrm{Sq}^2\mathrm{Sq}^1(x \barwedge x \barwedge x)$ and $\mathrm{Sq}^3(x \barwedge x \barwedge x)$, show that $H^{n+3}(K(Z_2, n); Z_2)$ has dimension at least 2 for $n > 3$.

7. Let $x \in H^1(RP^\infty; Z_2)$ be a generator. Show that

$$\mathrm{Sq}^{i_s}\,\mathrm{Sq}^{i_{s-1}} \cdots \mathrm{Sq}^{i_1} x = \begin{cases} x^{2^s}, & i_k = 2^{k-1} \quad \text{for each } k \\ 0, & \text{otherwise.} \end{cases}$$

(28.15)

8. Show that for a spectrum E,

$$\tilde{H}^k(E; Z_p) = \varprojlim \tilde{H}^{n+k}(E_n; Z_p)$$

is a module over $\mathcal{A}(p)$ and that maps of spectra induce $\mathcal{A}(p)$ module homomorphisms.

28

Adem Relations

It is the purpose of this section to determine the algebra $\mathcal{A}(2)$ of stable Z_2 cohomology operations. We do this by calculating $H^*(K(Z_2, n); Z_2)$. We will find that every operation is a Steenrod operation, i.e., can be written in the form $\mathrm{Sq}^{i_1} \circ \cdots \circ \mathrm{Sq}^{i_n}$. We also derive all relations among the Steenrod squares. There is, in fact, a family of nontrivial relations called the Adem relations. The existence of these relations makes the application of Steenrod operations very pungent (see 28.18 and 28.19).

We will base our calculation of $H^*(K(Z_2, n); Z_2)$ on a theorem of A. Borel [14] which we quote without proof. The proof is a straightforward application of the Serre spectral sequence [62; 64, 9.4, Corollary 9; 21; 41; 31] and the comparison theorem [3; 76]. See [3] for details.

Definition 28.1 A commutative algebra $\mathcal{A}$ over Z_2 is said to have a simple system of generators $\{x_\alpha\}$ if the monomials $x_{\alpha_1} \cdots x_{\alpha_n}$ form a vector space basis for $\mathcal{A}$ as $\{\alpha_1, \ldots, \alpha_n\}$ varies over all finite subsets of $\{\alpha\}$. (The empty subset corresponds to the monomial 1.)

Example 28.2 Let $Z_2[x_1, \ldots, x_n]$ be the polynomial ring over Z_2 generated by the indeterminates $x_1, \ldots, x_n$. Then $Z_2[x_1, \ldots, x_n]$ has a simple system of generators $\{x_1^{2k_1}, \ldots, x_n^{2k_n}\}$. For example,

$$x_1{}^3 + x_2{}^6 x_1{}^5 = x_1{}^1 \cdot x_1{}^2 + x_2{}^4 \cdot x_2{}^2 \cdot x_1{}^4 \cdot x_1{}^1.$$

To prove this in general observe that the monomials form a basis, and any monomial can be written in this form since for any n, $n = \sum_{i=0}^{\infty} \varepsilon_i 2^i$ where ε_i is 0 or 1.

As a second example we consider the exterior algebra $\Lambda(x_1, \ldots, x_n)$. This is generated as an algebra over Z_2 by $x_1, \ldots, x_k$ subject to the relations $x_i x_j = x_j x_i$ and $x_i^2 = 0$. It has dimension 2^n and $\{x_i\}$ forms a simple system of generators.

Definition 28.3 Let E be a spectrum. We define a homomorphism $\Sigma\colon \tilde{E}^i(X) \to \tilde{E}^{i-1}(\Omega X)$ called the suspension by

$$\tilde{E}^i(X) \xrightarrow{f^*} \tilde{E}^i(S\Omega X) \underset{\cong}{\overset{\sigma}{\longleftarrow}} \tilde{E}^{i-1}(\Omega X)$$

where $f\colon S\Omega X \to X$ is given by $f(s, \omega) = \omega(s)$.

Theorem 28.4 (*Borel*) Suppose that X is simply connected and there are elements $x_\alpha \in \tilde{H}^*(X; Z_2)$ such that $\Sigma(x_\alpha)$ form a simple system of generators for $H^*(\Omega X; Z_2)$. Assume that there are only finitely many x_α in each grading. Then $H^*(X, Z_2)$ is the polynomial ring $Z_2[\{x_\alpha\}]$—generated by the indeterminates x_α.

Example 28.5 Let $u \in H^2(CP^\infty; Z_2)$ be a generator. Then $\Sigma(u) \in H^1(S^1; Z_2)$ is a generator and $\{\Sigma(u)\}$ is a simple system of generators for $H^*(S^1; Z_2)$. Hence $H^*(CP^\infty; Z_2) \cong Z_2[u]$ which also follows from 26.35.

Let $I = (i_k, \ldots, i_1)$ be a finite sequence of integers. We define $\mathrm{Sq}^I = \mathrm{Sq}^{i_k} \cdots \mathrm{Sq}^{i_1}$. We define the dimension of I as $i_k + \cdots + i_1$. Furthermore we identify the sequence $(i_k, \ldots, i_1, 0)$ with the sequence $(i_k, \ldots, i_1)$. We call I (and Sq^I) admissible if $i_{s+1} \geq 2i_s$ for $k > s \geq 1$. We define the excess of I (and of Sq^I) by

$$\mathrm{ex}(I) = i_k - (i_{k-1} + \cdots + i_1) = (i_k - 2i_{k-1}) + \cdots + (i_2 - 2i_1) + i_1.$$

The notion of admissibility and excess are invariant under the above identification. Note that if I is admissible, $\mathrm{ex}(I) \geq 0$.

Theorem 28.6 (*Serre*) $H^*(K(Z_2; n); Z_2)$ is a polynomial algebra with one generator x_I for each admissible sequence of excess less that n. x_I has dimension $n + \dim I$ and $x_I = \mathrm{Sq}^I \iota$ where $\iota \in H^n(K(Z_2, n); Z_2)$ is a generator.

Before proving this result we will examine its contents for $n = 1$ and $n = 2$. If $\mathrm{ex}(I) = 0$, it follows that $i_1 = 0$ and $i_s = 2i_{s-1}$. Hence each $i_s = 0$. That is, the only admissible sequence I with $\mathrm{ex}(I) < 1$ is the sequence $(0, \ldots, 0)$. Hence, as expected, $H^*(K(Z_2, 1); Z_2) \cong Z_2[x_0]$ where $x_0 = \mathrm{Sq}^0 \iota = \iota$. Now suppose $n = 2$. $\mathrm{ex}(I) = 1$ can only happen if $I = (2^k, 2^{k-1}, \ldots, 1)$. Thus $H^*(K(Z_2, 2); Z_2) = Z_2[\iota, x_0, x_1, \ldots]$ where $x_k = \mathrm{Sq}^{2^k} \cdots \mathrm{Sq}^1 \iota$.

In fact the proof of this is quite easy and is typical of the inductive step. $H^*(K(Z_2, 1); Z_2)$ has a simple system of generators, $x^{2^k} = \mathrm{Sq}^{2^{k-1}} \mathrm{Sq}^{2^{k-2}} \cdots \mathrm{Sq}^1 x$. Now if $\iota \in H^2(K(Z_2, 2); Z_2)$ is the generator, $\Sigma\iota = x$. Thus $\Sigma(\mathrm{Sq}^{2^{k-1}} \cdots$

$Sq^1\iota) = x^{2^k}$, since $\Sigma = \sigma^{-1}f^*$ commutes with the action of Steenrod operations. Hence 28.6 follows in the case $n = 2$ by applying Borel's theorem.

Proof of 28.6 Suppose now that $H^*(K(Z_2, n); Z_2)$ is as stated above. Then a simple system of generators is given by $(Sq^I \iota)^{2^s}$ for $s \geq 0$ and I admissible with $ex(I) < n$. Suppose dim $Sq^I\iota = n + i_1 + \cdots + i_k = m$. Then

$$(Sq^I\iota)^{2^s} = Sq^{2^{s-1}m} \cdots Sq^m Sq^{i_k} \cdots Sq^{i_1} \quad \text{if} \quad s > 0.$$

The sequence $(2^{s-1}m, \ldots, m, i_k, \ldots, i_1)$ is clearly admissible and has excess n. Furthermore, every sequence I of excess n can be written uniquely in this form. Thus $H^*(K(Z_2, n); Z_2)$ has a simple system of generators of the form Sq^I for I admissible and $ex(I) \leq n$. 28.6 in the case $n + 1$ now follows from Borel's theorem since $Sq^J(\Sigma\iota) = \Sigma(Sq^J\iota)$ for any sequence J. ∎

Corollary 28.7 $\mathcal{A}(2)$ has as a basis the admissible mononomials Sq^I.

Proof A basis in dimension k is given by $H^{2k+1}(K(Z_2, k + 1); Z_2)$ by 27.5. This is generated by all admissible monomials of excess $\leq k$ and dimension k. However dim $I = k$ implies $ex(I) \leq k$. Hence the admissible monomials Sq^I of dimension k form a basis in dimension k. ∎

We list a basis in dimension k for $k \leq 10$.

k						
0	1					
1	Sq^1					
2	Sq^2					
3	Sq^3	$Sq^{(2,1)}$				
4	Sq^4	$Sq^{(3,1)}$				
5	Sq^5	$Sq^{(4,1)}$				
6	Sq^6	$Sq^{(5,1)}$	$Sq^{(4,2)}$			
7	Sq^7	$Sq^{(6,1)}$	$Sq^{(5,2)}$	$Sq^{(4,2,1)}$		
8	Sq^8	$Sq^{(7,1)}$	$Sq^{(6,2)}$	$Sq^{(5,2,1)}$		
9	Sq^9	$Sq^{(8,1)}$	$Sq^{(7,2)}$	$Sq^{(6,3)}$	$Sq^{(6,2,1)}$	
10	Sq^{10}	$Sq^{(9,1)}$	$Sq^{(8,2)}$	$Sq^{(7,3)}$	$Sq^{(7,2,1)}$	$Sq^{(6,3,1)}$

Thus the algebra $\mathcal{A}(2)$ is generated by Steenrod operations and is called the Steenrod algebra.

It is quite clear that there must be some relations among the Steenrod operations. For example, Sq^1Sq^2 has dimension 3 and hence $Sq^1Sq^2 = \lambda Sq^3 + \mu Sq^{(2,1)}$. Applying this equation to $x \in H^1(RP^\infty; Z_2)$, one concludes that $\mu = 0$. Applying it to $x^3 \in H^3(RP^\infty; Z_2)$ yields $\lambda = 1$ so $Sq^1Sq^2 = Sq^3$.

In general, if $a < 2b$, 28.7 implies that there is a formula $Sq^aSq^b = \Sigma\lambda_I Sq_I$, where I runs over all admissible sequences of dimension $a + b$ and $\lambda_I \in Z_2$. Using the spaces $RP^\infty \times \cdots \times RP^\infty$, one can calculate the coefficients λ_I and prove

Theorem 28.8 (*The Adem Relations*) If $a < 2b$,

$$\mathrm{Sq}^a\mathrm{Sq}^b = \sum_{j=a-b+1}^{j=[a/2]} \binom{b-j-1}{a-2j} \mathrm{Sq}^{a+b-j}\mathrm{Sq}^j.$$

We will prove 28.8 by applying a little more theory and a little less calculation than the above outlined method. Our method is due to Kristensen [38; 39].

By the Cartan formula and induction, we have for any sequence $I = (i_k, \ldots, i_1)$, and any two cohomology classes x and y

$$\mathrm{Sq}^I(x \cdot y) = \sum \mathrm{Sq}^J x \cdot \mathrm{Sq}^{I-J} y$$

where the sum is taken over all sequences $J = (j_k, \ldots, j_1)$ with $j_s \leq i_s$, and $I - J = (i_k - j_k, \ldots, i_1 - j_1)$. We define a linear map

$$\varphi\colon \mathcal{A}(2) \to \mathcal{A}(2) \otimes \mathcal{A}(2)$$

by

$$\varphi(\mathrm{Sq}^I) = \sum \mathrm{Sq}^J \otimes \mathrm{Sq}^{I-J}$$

for I admissible.

Lemma 28.9 Suppose there are operations β, β_i, and β_i' such that for all x and y

$$\beta(x \cdot y) = \sum_i \beta_i(x) \cdot \beta_i'(y).$$

Then $\varphi(\beta) = \sum \beta_i \otimes \beta_i'$.

Proof Let $\varphi(\beta) - \sum \beta_i \otimes \beta_i' = \sum_i \alpha_i \otimes \alpha_i' \in \mathcal{A}(2) \otimes \mathcal{A}(2)$. $\varphi(\beta)(x \otimes y) = \beta(x \cdot y)$ since this is clearly true when β is an admissible monomial. Consequently,

$$\sum \alpha_i(x) \cdot \alpha_i'(y) = 0.$$

Let $k > \dim \sum \alpha_i \otimes \alpha_i'$ and $\iota \in H^k(K(Z_2, k); Z_2)$. Let $x = p_1{}^*(\iota)$ and $y = p_2{}^*(\iota)$ where p_1 and p_2 are the projections $K(Z_2, k) \times K(Z_2, k) \to K(Z_2, k)$. Now

$$\begin{aligned} 0 &= \sum \alpha_i(x) \cdot \alpha_i'(y) \\ &= \sum \alpha_i(p_1{}^*(\iota)) \cdot \alpha_i'(p_2{}^*(\iota)) \\ &= \sum p_1{}^*(\alpha_i(\iota)) \cdot p_2{}^*(\alpha_i'(\iota)) \\ &= \sum \alpha_i(\iota) \mathbin{\overline{\times}} \alpha_i'(i). \end{aligned}$$

Now 25.15 implies that $\sum \alpha_i(\iota) \otimes \alpha_i'(\iota) = 0$. However the mapping

$$\mathcal{A}(2) \otimes \mathcal{A}(2) \to H^*(K(Z_2, k); Z_2) \otimes H^*(K(Z_2, k); Z_2)$$

is an isomorphism in this dimension, so $\sum \alpha_i \otimes \alpha_i' = 0$. ∎

Corollary 28.10 Suppose $\varphi(\alpha) = \sum \alpha_i \otimes \alpha_i'$ and $\varphi(\beta) = \sum \beta_j \otimes \beta_j'$. Then

$$\varphi(\alpha\beta) = \sum \alpha_i \beta_j \otimes \alpha_i' \beta_j'.$$

Proof $\alpha\beta(x \cdot y) = \alpha(\sum \beta_j x \cdot \beta_j' y) = \sum \alpha_i \beta_j x \cdot \alpha_i' \beta_j' y$. The result follows from 28.9. ∎

Proposition 28.11 (*Milnor*) The diagrams

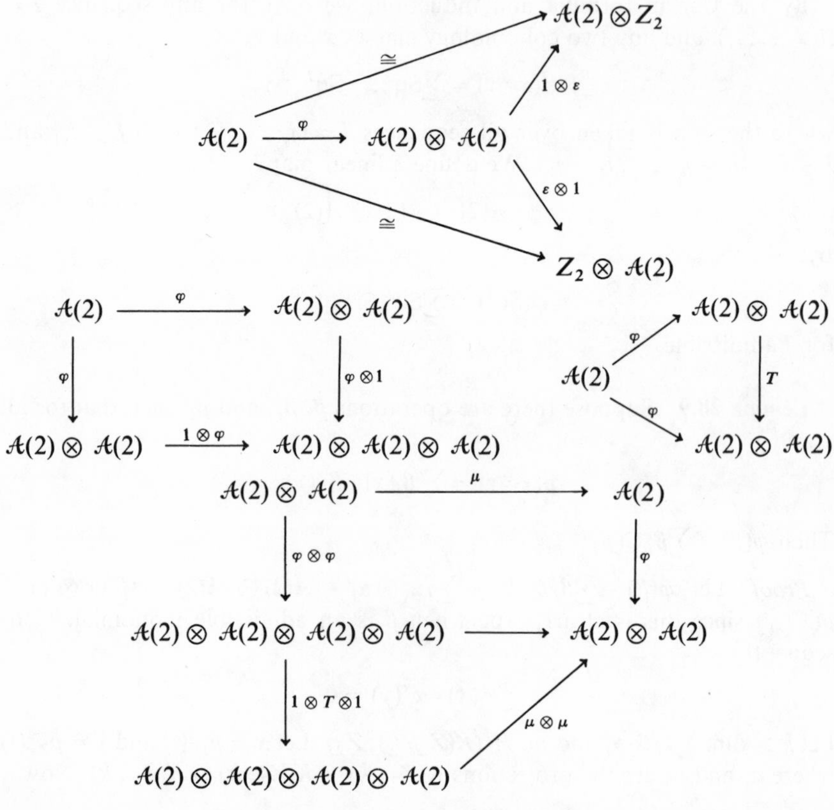

commute, where μ is composition, T is transposition, and ε: $\mathcal{A}(2) \to Z_2$ is given by

$$\varepsilon(x) = \begin{cases} 0, & \dim x > 0 \\ x, & \dim x = 0. \end{cases}$$

Proof All of the diagrams except the last one follow from the definition. The last diagram is 28.10. ∎

The mapping φ is called a comultiplication and ε is called a counit. The first three diagrams express that $\mathcal{A}(2)$ is a coassociative cocommutive coalgebra. These diagrams are duals to the diagrams one ordinarily has for algebras. The last condition says that φ is an algebra homomorphism where the multiplication in $\mathcal{A}(2) \otimes \mathcal{A}(2)$ is given by $(\mu \otimes \mu)(1 \otimes T \otimes 1)$. These conditions (except for cocommutativity) are usually expressed by saying that $\mathcal{A}(2)$ is a graded Hopf algebra.

Let I be an admissible sequence.

Definition 28.12 Define $\kappa_I \colon \mathcal{A}(2)_n \to \mathcal{A}(2)_{n-\dim I}$

$$\varphi(x) = \sum \kappa_I(x) \otimes \mathrm{Sq}^I$$

These operations were first defined by Kristensen [38].

By 28.7 we can write equations

$$\mathrm{Sq}^I \mathrm{Sq}^J = \sum \lambda_M^{I,J} \mathrm{Sq}^M, \qquad \mathrm{Sq}^T = \sum \alpha_M{}^T \mathrm{Sq}^M$$

where the sums are taken over all admissible mononomials Sq^M and the coefficients $\lambda_M^{I,J}$, $\alpha_M{}^T$ are 0 or 1.

Proposition 28.13

$$\kappa_M(xy) = \sum \lambda_M^{I,J} \kappa_I(x) \kappa_J(y);$$
$$\kappa_M(\kappa_L(x)) = \sum \alpha_M{}^J \alpha_L^{I-J} \kappa_I(x).$$

The first sum is taken over all admissible sequences I, J with $\dim I + \dim J = \dim M$. The second sum is taken over all admissible sequences I and all sequences J with $J \leq I$.

Proof The first formula follows from 28.10 and the second from the coassociativity of φ, i.e., $(\varphi \otimes 1)\varphi = (1 \otimes \varphi)\varphi$. We will do only the first as the proofs are similar and easy.

By 28.10,

$$\begin{aligned}\sum \kappa_M(xy) \otimes \mathrm{Sq}^M &= \left(\sum \kappa_I(x) \otimes \mathrm{Sq}^I\right)\left(\sum \kappa_J(y) \otimes \mathrm{Sq}^J\right)\\ &= \sum \kappa_I(x)\kappa_J(y) \otimes \mathrm{Sq}^I \mathrm{Sq}^J\\ &= \sum \lambda_M^{I,J} \kappa_I(x) \kappa_J(y) \otimes \mathrm{Sq}^M.\end{aligned}$$

The sums here are taken over all I, J, and M that are admissible. ∎

Corollary 28.14 Let $\kappa = \kappa_{(1)}$ and $\kappa' = \kappa_{(2,1)}$. Then

$$\kappa(xy) = \kappa(x)y + x\kappa(y).$$
$$\kappa'(xy) = \kappa'(x)y + x\kappa'(y) + \kappa(\kappa(x))\kappa(y).$$
$$\kappa(\mathrm{Sq}^n) = \mathrm{Sq}^{n-1}, \qquad \kappa'(\mathrm{Sq}^n) = 0.$$

Proof If $\lambda_{(1)}^{I,J} \neq 0$, dim I + dim $J = 1$. Hence we must have $I = (1)$ and $J = (0)$ or $I = (0)$ and $J = (1)$. In these cases $\lambda_{(1)}^{I,J} = 1$ so 28.13 yields

$$\begin{aligned}\kappa(xy) &= \kappa_{(0)}(x)\kappa(y) + \kappa(x)\kappa_{(0)}(y)\\ &= x\kappa(y) + \kappa(x)y.\end{aligned}$$

Suppose now that $\lambda_{(2,1)}^{I,J} \neq 0$. Then we have the following possibilities:

I	J
3	0
(2,1)	0
2	1
1	2
0	3
0	(2,1)

However $\mathrm{Sq}^1\mathrm{Sq}^2 = \mathrm{Sq}^3$. Hence the only cases in which $\lambda_{(2,1)}^{I,J} \neq 0$ are

I	J
(2,1)	0
2	1
0	(2,1)

Thus we have

$$\kappa'(xy) = \kappa'(x)y + \kappa_{(2)}(x)\kappa(y) + x\kappa'(y).$$

Now $\kappa(\kappa(x)) = \kappa_{(2)}(x)$ by the second part of 28.13. This proves the second formula. The formulas for $\kappa(\mathrm{Sq}^n)$ and $\kappa'(\mathrm{Sq}^n)$ follow from the Cartan formula. ▮

Proposition 28.15 $\mathrm{Sq}^{2n+1}\,\mathrm{Sq}^{n+1} = 0.$

Proof By 28.7 we have

$$\mathrm{Sq}^{2n+1}\mathrm{Sq}^{n+1} = \sum \lambda_I \mathrm{Sq}^I.$$

We first prove that $\lambda_I = 0$ if I has length greater than 2. To prove this we apply this equation to $x \mathbin{\bar{\times}} \iota_k \in H^{k+1}(RP^\infty \times K(Z_2, k); Z_2)$:

$$\begin{aligned}\mathrm{Sq}^{n+1}(x \mathbin{\bar{\times}} \iota_k) &= x \mathbin{\bar{\times}} \mathrm{Sq}^{n+1}\iota_k + x^2 \mathbin{\bar{\times}} \mathrm{Sq}^n\iota_k\\ \mathrm{Sq}^{2n+1}\mathrm{Sq}^{n+1}(x \mathbin{\bar{\times}} \iota_k) &= x \mathbin{\bar{\times}} \mathrm{Sq}^{2n+1}\mathrm{Sq}^{n+1}\iota_k + x^2 \mathbin{\bar{\times}} \mathrm{Sq}^{2n}\mathrm{Sq}^{n+1}\iota_k\\ &\quad + x^2 \mathbin{\bar{\times}} \mathrm{Sq}^{2n+1}\mathrm{Sq}^n\iota_k + x^4 \mathbin{\bar{\times}} \mathrm{Sq}^{2n-1}\mathrm{Sq}^n\iota_k.\end{aligned}$$

By Exercise 7, Section 27

$$\mathrm{Sq}^J x = \begin{cases} 0, & J \neq (2^s, 2^{s-1}, \ldots, 1)\\ x^{2^{s+1}}, & J \neq (2^s, 2^{s-1}, \ldots, 1).\end{cases}$$

Hence

$$\mathrm{Sq}^I(x \,\bar{\times}\, \iota_k) = \sum \mathrm{Sq}^J x \,\bar{\times}\, \mathrm{Sq}^{I-J}\iota_k$$
$$= \sum x^{2^{s+1}} \,\bar{\times}\, \mathrm{Sq}^{I-I_s}\iota_k$$

where $I_s = (2^s, 2^{s-1}, \ldots, 1)$. Now if I is admissible, $I - I_s$ is admissible. The equation implies that $\lambda_I \mathrm{Sq}^{I-I_s}\iota_k = 0$ for $s \geq 2$. If k is sufficiently large, this implies that $\lambda_I = 0$ if I has length >2. ▮

Now we write

$$\mathrm{Sq}^{2n+1}\mathrm{Sq}^{n+1} = \sum_{s=0}^{n} \lambda_s \mathrm{Sq}^{3n-s+2}\mathrm{Sq}^s.$$

We prove 28.15 by induction on n. Since $\mathrm{Sq}^1\mathrm{Sq}^2 = \mathrm{Sq}^3$, we have

$$\mathrm{Sq}^2 = \kappa(\mathrm{Sq}^3) = \kappa(\mathrm{Sq}^1\mathrm{Sq}^2) = \mathrm{Sq}^2 + \mathrm{Sq}^1\mathrm{Sq}^1.$$

Hence $\mathrm{Sq}^1\mathrm{Sq}^1 = 0$. This is the case $n = 0$. Observe that $\kappa'(\mathrm{Sq}^a\mathrm{Sq}^b) = \mathrm{Sq}^{a-2}\,\mathrm{Sq}^{b-1}$ by 28.14. Assume $\mathrm{Sq}^{2n-1}\mathrm{Sq}^n = 0$. Then

$$0 = \mathrm{Sq}^{2n-1}\mathrm{Sq}^n = \kappa'(\mathrm{Sq}^{2n+1}\mathrm{Sq}^{n+1})$$
$$= \sum_{s=0}^{n} \lambda_s \kappa'(\mathrm{Sq}^{3n-s+2}\mathrm{Sq}^s)$$
$$= \sum_{s=0}^{n} \lambda_s \mathrm{Sq}^{3n-s}\mathrm{Sq}^{s-1}.$$

$\mathrm{Sq}^{3n-s}\mathrm{Sq}^{s-1}$ is admissible so we conclude that $\lambda_s = 0$ for each s and hence $\mathrm{Sq}^{2n+1}\mathrm{Sq}^{n+1} = 0$. ▮

The equations $\mathrm{Sq}^{2n+1}\mathrm{Sq}^{n+1} = 0$ together with the derivation κ yield more relations. Applying κ we have

$$0 = \kappa(\mathrm{Sq}^{2n+1}\mathrm{Sq}^{n+1}) = \mathrm{Sq}^{2n}\mathrm{Sq}^{n+1} + \mathrm{Sq}^{2n+1}\mathrm{Sq}^n;$$

since $\mathrm{Sq}^{2n+1}\mathrm{Sq}^n$ is admissible, we have

$$\mathrm{Sq}^{2n}\mathrm{Sq}^{n+1} = \mathrm{Sq}^{2n+1}\mathrm{Sq}^n;$$

applying κ again we get

$$\mathrm{Sq}^{2n-1}\mathrm{Sq}^{n+1} + \mathrm{Sq}^{2n}\mathrm{Sq}^n = \mathrm{Sq}^{2n}\mathrm{Sq}^n + \mathrm{Sq}^{2n+1}\mathrm{Sq}^{n-1}$$

or

$$\mathrm{Sq}^{2n-1}\mathrm{Sq}^{n+1} = \mathrm{Sq}^{2n+1}\mathrm{Sq}^{n-1}.$$

Proceeding in this way we obtain relations

$$\mathrm{Sq}^{2n-2}\mathrm{Sq}^{n+1} + \mathrm{Sq}^{2n-1}\mathrm{Sq}^n = \mathrm{Sq}^{2n}\mathrm{Sq}^{n-1} + \mathrm{Sq}^{2n+1}\mathrm{Sq}^{n-2}$$

or

$$\mathrm{Sq}^{2n-2}\mathrm{Sq}^{n+1} = \mathrm{Sq}^{2n}\mathrm{Sq}^{n-1} + \mathrm{Sq}^{2n+1}\mathrm{Sq}^{n-2}.$$

If we apply this process r times, we get a formula for $\mathrm{Sq}^{2n+1-r}\mathrm{Sq}^{n+1}$ in terms of admissible monomials. This will be our proof of 28.8.

Proof of 28.8 Observe that if $j < a - b + 1$ or $j > [a/2]$, $\binom{b-j-1}{a-2j} = 0$. Hence we can consider the sum over all j. Let $r = 2b - a$. We will prove 28.8 by induction on r. If $r = 1$, the left-hand side is $\mathrm{Sq}^{2b-1}\mathrm{Sq}^{b} = 0$. The right-hand side is

$$\sum\binom{b-j-1}{2b-2j-1}\mathrm{Sq}^{a+b-j}\mathrm{Sq}^{j}.$$

This can only be nonzero if $2b - 2j - 1 \le b - j - 1$, and $b - j - 1 \ge 0$. Thus all the terms are 0. This completes the proof in the case $r = 1$. Suppose the formula is proved if $a = 2b - r$:

$$\mathrm{Sq}^{2b-r}\mathrm{Sq}^{b} = \sum\binom{b-j-1}{2b-2j-r}\mathrm{Sq}^{3b-j-r}\mathrm{Sq}^{j}.$$

Applying κ we get

$$\begin{aligned}&\mathrm{Sq}^{2b-r-1}\mathrm{Sq}^{b} + \mathrm{Sq}^{2b-r}\mathrm{Sq}^{b-1}\\ &\qquad = \sum\binom{b-j-1}{2b-2j-r}\{\mathrm{Sq}^{3b-j-r-1}\mathrm{Sq}^{j} + \mathrm{Sq}^{3b-j-r}\mathrm{Sq}^{j-1}\},\end{aligned}$$

$$\mathrm{Sq}^{2b-r}\mathrm{Sq}^{b-1} = \mathrm{Sq}^{2(b-1)-(r-2)}\mathrm{Sq}^{b-1},$$

so by induction, we have

$$\mathrm{Sq}^{2b-r}\mathrm{Sq}^{b-1} = \sum\binom{b-j-2}{2b-2j-r}\mathrm{Sq}^{3b-r-1-j}\mathrm{Sq}^{j}.$$

Hence

$$\begin{aligned}&\mathrm{Sq}^{2b-(r+1)}\mathrm{Sq}^{b}\\ &\quad = \sum\left\{\binom{b-j-2}{2b-2j-r} + \binom{b-j-1}{2b-2j-r} + \binom{b-j-2}{2b-2j-r-2}\right\}\mathrm{Sq}^{3b-r-j-1}\mathrm{Sq}^{j}\\ &\quad = \sum\left\{\binom{b-j-2}{2b-2j-r-1} + \binom{b-j-2}{2b-2j-r-2}\right\}\mathrm{Sq}^{3b-r-j-1}\mathrm{Sq}^{j}\\ &\quad = \sum\binom{b-j-1}{2b-2j-r-1}\mathrm{Sq}^{3b-r-j-1}\mathrm{Sq}^{j}. \quad \blacksquare\end{aligned}$$

Although 28.8 expresses the Adem relations in a compact formula, it is very complicated and sometimes inconvenient. If one wishes to calculate all the Adem relations in a given range of dimensions, it is much easier to begin with 28.15 and apply κ. The expression for $\mathrm{Sq}^a\mathrm{Sq}^b$ is then calculated by induction on $k = 2b - a$, using the formula $\mathrm{Sq}^a\mathrm{Sq}^b = \mathrm{Sq}^{a+1}\mathrm{Sq}^{b-1} + \kappa(\mathrm{Sq}^{a+1}\mathrm{Sq}^b)$. Using induction, the right-hand side of this equation is already tabulated in terms of admissible monomials.

Table 1 (pages 358–359) gives the expressions for $\mathrm{Sq}^a\mathrm{Sq}^b$ for $a < 2b \leq 12$. We now show that there are no other relations in $\mathcal{A}(2)$.

Corollary 28.16 Let $\mathcal{A}$ be the algebra over Z_2 generated by Sq^i subject to the Adem relations (28.8) Then $\mathcal{A} \cong \mathcal{A}(2)$.

Proof One can easily define an algebra homomorphism $\lambda\colon \mathcal{A} \to \mathcal{A}(2)$ by 28.8. λ is onto and it suffices to show that λ is 1–1. To do this it is sufficient to show that the admissible monomials in $\mathcal{A}$ generate $\mathcal{A}$ as a vector space.

For any sequence $I = (i_1, \ldots, i_k)$, we define the moment of I by $m(I) = \sum_{s=1}^k s i_s$. $m(\mathrm{Sq}^0) = 0$, $m(\mathrm{Sq}^1) = 1$, and all other monomials have moment greater than 1. We prove that Sq^I is a sum of admissible monomials by induction on $m(I)$. Suppose $i_s < 2i_{s+1}$. Then

$$\begin{aligned}\mathrm{Sq}^I &= \mathrm{Sq}^{(i_1, \ldots, i_{s-1})}\mathrm{Sq}^{i_s}\mathrm{Sq}^{i_{s+1}}\mathrm{Sq}^{(i_{s+2}, \ldots, i_k)}\\ &= \sum \lambda_j \mathrm{Sq}^{(i_1, \ldots, i_{s-1})}\mathrm{Sq}^{i_s+i_{s+1}-j}\mathrm{Sq}^j\mathrm{Sq}^{(i_{s+2}, \ldots, i_k)}\\ &= \sum \lambda_j \mathrm{Sq}^{I_j}.\end{aligned}$$

Now

$$\begin{aligned}m(I_j) &= m(I) + s(i_{s+1} - j) + (s+1)(j - i_{s+1})\\ &= m(I) + (j - i_{s+1}) < m(I),\end{aligned}$$

since $2(j - i_{s+1}) \leq i_s - 2i_{s+1} < 0$. By induction Sq^{I_j} can be written as a sum of admissible monomials and hence Sq^I can as well. ∎

An element $\alpha \in \mathcal{A}(2)$ is called indecomposable if it cannot be written in the form $\sum \alpha_i \alpha_i'$ with $\dim \alpha_i > 0$ and $\dim \alpha_i' > 0$.

Proposition 28.17 The indecomposable elements of $\mathcal{A}(2)$ are the elements Sq^{2^n} for $n \geq 0$.

Proof Clearly the only elements of $\mathcal{A}(2)$ that can possibly be indecomposable are the elements Sq^k for $k > 0$. Now we let $x \in H^1(RP^\infty; Z_2)$ be a generator. Then

$$\mathrm{Sq}^s x^{2^n} = \begin{cases} 0, & s \neq 2^n \\ x^{2^{n+1}}, & s = 2^n. \end{cases}$$

If Sq^{2^n} were decomposable, we would have $x^{2^{n+1}} = \mathrm{Sq}^{2^n}x^{2^n} = \sum \alpha_i \alpha_i' x^{2^n} = 0$. Thus Sq^{2^n} is indecomposable. Suppose on the other hand that $s \neq 2^n$. Then $2^k < s < 2^{k+1}$ for some k and

$$\mathrm{Sq}^{s-2^k}\mathrm{Sq}^{2^k} = \binom{2^k - 1}{s - 2^k}\mathrm{Sq}^s + \sum_{j>0} \lambda_j \mathrm{Sq}^{n-j}\mathrm{Sq}^j.$$

By 28.20 (in the Appendix),

$$\binom{2^k - 1}{s} \equiv 1 \pmod 2$$

for $0 \leq s \leq 2^k - 1$. Thus Sq^s occurs with a nonzero coefficient in the formula and so it is decomposable. ∎

One of the simplest applications of the Adem relations is to the Hopf invariant problem. Let $\alpha: S^{2n-1} \to S^n$. Then $\tilde{H}^*(S^n \cup_\alpha e^{2n})$ has generators ι_1 and ι_2 in dimensions n and $2n$. Hence $\iota_1{}^2 = k\iota_2$ for some integer k. k depends on the choice of generators up to sign, but otherwise depends only on the homotopy type of $S^n \cup_\alpha e^{2n}$ and hence only on the homotopy class of α (by Exercise 22, Section 14). Thus we can define a transformation $H: \pi_{2n-1}(S^n) \to Z$. H is called the Hopf invariant. $H(\eta) = H(2\iota) = H(\nu) = 1$ by 26.35 and Exercise 5, Section 14.

Proposition 28.18 If there exists an element $\alpha \in \pi_{2n-1}(S^n)$ with $H(\alpha)$ odd, $n = 2^s$.

Proof Let $c_\rho: \tilde{H}^*(S^n \cup_\alpha e^{2n}) \to \tilde{H}^*(S^n \cup_\alpha e^{2n}; Z_2)$ be the coefficient transformation induced by the epimorphism $\rho: Z \to Z_2$. By Exercise 18, Section 23, c_ρ is a ring homomorphism and hence $c_\rho(\iota_1)^2 = c_\rho(\iota_2)$. Thus $\mathrm{Sq}^n c_\rho(\iota_1) = c_\rho(\iota_2)$. Since $H^{n+1}(S^n \cup_\alpha e^{2n}; Z_2) = 0$ for $0 < i < n$, $\mathrm{Sq}^i c_\rho(\iota_1) = 0$ for $0 < i < n$. Hence 28.17 implies that $\mathrm{Sq}^n c_\rho(\iota_1) = 0$ unless $n = 2^s$. ∎

In fact this phenomenon occurs iff $n = 1, 2, 4$, or 8 (see 29.19). The cases $n = 1, 2$, and 4 correspond to the maps $2\iota: S^1 \to S^1$, $\eta: S^3 \to S^2$, and $\nu: S^7 \to S^4$. A map $\sigma: S^{15} \to S^8$ with $H(\sigma) = 1$ can also be constructed using the Cayley number multiplication. (The construction in Section 7 does not work for the Cayley numbers because they are not associative and hence $\sim$ is not an equivalence relation.)

As a final application of 28.8, we prove:

Proposition 28.19 Let $\eta \in \pi_3(S^2)$ be the Hopf map. Then $\{S^n(S\eta \circ \eta)\} \neq 0$ for each $n \geq 0$.

Proof Let us write η_k for $S^{k-2}\eta\colon S^{k+1} \to S^k$. Our object is to prove that $\{\eta_k \circ \eta_{k+1}\} \neq 0$ for each k. Supposing $\eta_k \circ \eta_{k+1} \sim *$, we can construct an extension

$$S^{k+1} \cup_{\eta_{k+1}} e^{k+3} \xrightarrow{\theta} S^k$$

of η_k. Let X be the mapping cone of θ. Then X has cells in dimensions 0, k, $k+2$, and $k+4$:

$$X = S^k \cup_\theta C(S^{k+1} \cup_{\eta_{k+1}} e^{k+3}).$$

$$X^{k+2} \equiv S^k \cup_{\eta_k} e^{k+2} \equiv S^{k-2}CP^2 \qquad \text{and} \qquad X/S^k \equiv S^{k+2} \cup_{\eta_{k+2}} e^{k+4} \equiv S^k CP^2$$

Choose generators $x_n \in H^n(X; Z_2)$ for $n = k$, $k+2$, and $k+4$. Since $X^{k+4} \equiv S^{k-2}CP^2$, $\mathrm{Sq}^2 x_k = x_{k+2}$. The mapping $\pi\colon X \to S^k CP^2$ induces a homomorphism in cohomology and $\pi^*(\sigma^k y) = x_{k+2}$, $\pi^*(\sigma^k y^2) = x_{k+4}$ where $y \in H^2(CP^2; Z_2)$ is a generator. Hence $\mathrm{Sq}^2 x_{k+2} = x_{k+4}$.

Now by 28.8, $\mathrm{Sq}^2\mathrm{Sq}^2 = \mathrm{Sq}^3\mathrm{Sq}^1$. Hence

$$x_{k+4} = \mathrm{Sq}^2 x_{k+2} = \mathrm{Sq}^2\mathrm{Sq}^2 x_k = \mathrm{Sq}^3\mathrm{Sq}^1 x_k = 0,$$

since X has no cells in dimension $k+1$. This contradiction implies that such a space X cannot exist and hence that $\eta_k \circ \eta_{k+1} \not\sim *$. ∎

Appendix

In calculating Adem relations it is often useful to have an algorithm for calculating $\binom{a}{b}$ modulo 2.

Proposition 28.20 Let $a = \sum_{i=0}^m a_i 2^i$ and $b = \sum_{i=0}^m b_i 2^i$ be binary expansions. Then

$$\binom{a}{b} \equiv \prod_{i=0}^{m} \binom{a_i}{b_i} (\text{mod } 2).$$

Proof $(1+x)^2 \equiv 1 + x^2$ (mod 2), so by induction, $(1+x)^{2^n} \equiv 1 + x^{2^n}$. Hence

$$(1+x)^a \equiv (1+x)^{\Sigma a_i 2^i} \equiv \prod_{i=0}^{m} (1 + x^{2^i})^{a_i} \equiv \prod_{i=0}^{m} \left(\sum \binom{a_i}{s} x^{s2^i}\right).$$

The coefficient of x^b in this product is

$$\prod_{i=0}^{m} \binom{a_i}{b_i}. \quad ∎$$

Seminar Problem

The structure maps of 28.11 make $\mathcal{A}(2)$ into a Hopf algebra. Since this structure is self-dual, and dim $\mathcal{A}(2)_n$ is finite, $\mathcal{A}(2)^*$ is also a Hopf algebra. Prove the Milnor theorem [3].

Theorem 28.21 (*Milnor*)

$$\mathcal{A}(2)^* \cong Z_2[\xi_1, \xi_2, \ldots]$$

as algebras, where $Z_2[\xi_1, \xi_2, \ldots]$ is a polynomial algebra with generators ξ_i in dimension $2^i - 1$. The diagonal in $\mathcal{A}(2)^*$ is given by

$$\mu^*(\xi_r) = \sum_{i=0}^{r} \xi_{r-1}^{2^i} \otimes \xi_i .$$

ξ_i is defined by $\phi(x) = \langle \phi, \xi_i \rangle x^{2^i}$ for $\phi \in \mathcal{A}(2)$ of dimension $2^i - 1$ and $x \in H^1(RP^\infty \cdot Z_2)$ the nonzero element.

Z_p Cohomology Operations

There are Adem relations for the operations P^r analogous to those in 28.8. These can be proven by the same method as in this section. One proves a version of the Borel theorem for Z_p cohomology (this theorem is much more complicated in statement than the Z_2 version). From this one can calculate $H^*(K(Z_p, n); Z_p)$. This information can then be utilized in determining the relations among Z_p cohomology operations. Since P^r raises dimension by $2r(p-1)$, the Bockstein β_p associated with the sequence $0 \to Z_p \to Z_{p^2} \to Z_p \to 0$ is not in the algebra generated by the operations P^r. We define a Steenrod operation with Z_p coefficients to be any operation in the algebra generated by β_p and the operations P^r. Then every stable Z_p cohomology operation is a Steenrod operation and the relations take on the form

Theorem 28.22 (*Adem Relations*) If $a < pb$,

$$P^a P^b = \sum_{t=0}^{[a/p]} (-1)^{a+t} \binom{(p-1)(b-t)-1}{a-pt} P^{a+b-t} P^t.$$

If $a \le pb$,

$$P^a \beta_p P^b = \sum_{t=0}^{[a/p]} (-1)^{a+t} \binom{(p-1)(b-t)}{a-pt} \beta_p P^{a+b-t} P^t + \sum_{t=0}^{[(a-1)/p]} (-1)^{a+t-1} \binom{(p-1)(b-t)-1}{a-pt-1} P^{a+b-t} P^t.$$

(See [67] for details.)

Exercises

1. Use the method of 28.19 to show that $S^n(2\nu) \nsim *$ in $\pi_{n+7}(S^{n+4})$ and $S^n(\nu \circ S^3\nu) \nsim *$ in $\pi_{n+10}(S^{n+4})$.

2. Show that $H^*(K(Z, n); Z_2)$ is a polynomial ring with one generator $x_I = \mathrm{Sq}^I\, \iota$ for each admissible sequence $I = (i_k, \ldots, i_1)$ with $\mathrm{ex}(I) < n$ and $i_1 > 1$ for $n \geq 2$.

3. Prove that

$$\mathrm{Sq}^1\mathrm{Sq}^n = \begin{cases} \mathrm{Sq}^{n+1} & \text{if } n \text{ is even} \\ 0 & \text{if } n \text{ is odd.} \end{cases}$$

4. Prove that if X is a homotopy associative H space and k is a field, $H^*(X; k)$ is a graded Hopf algebra.

5. If E and F are spectra, define $\tilde{E}_k(F)$ as $\varinjlim \tilde{E}_{k+n}(F_n)$ where the homomorphisms are

$$\tilde{E}_{k+n}(F_n) \xrightarrow[\cong]{\sigma} \tilde{E}_{k+n+1}(SF_n) \xrightarrow{(f_n)_*} \tilde{E}_{k+n+1}(F_{n+1})$$

Show that if $\underline{X}$ is the suspension spectrum of X, $\tilde{E}_k(\underline{X}) \cong \tilde{E}_k(X)$. Prove that $\tilde{E}_k(F) \cong \tilde{F}_k(E)$. (See Exercise 16, Section 15.)

6. Show that $H_{m+k}(K(\pi, m); \rho) \cong H_{m+k}(K(\rho, m); \pi)$ for $k < m$. (Hint: Use Exercise 5).

7. Prove that if E_n is a ring spectrum, $F_*(E)$ is a commutative ring.

8. A group is called a 2-group if every element has order 2^s for some $s \geq 0$. Show that if π is a finite 2-group, $H_{m+k}(K(\pi, m))$ and $H^{m+k}(K(\pi, m))$ are finite 2-groups for $k < m$. (Use Exercise 6.)

9. Show that the map $RP^\infty \wedge \cdots \wedge RP^\infty \to K(Z_2, n)$ induces a monomorphism in Z_2 cohomology in dimensions less than $2n$. (Hint: It is only necessary to show that the admissible monomials Sq^I take linearly independent values on the $\iota \,\overline{\wedge} \cdots \overline{\wedge}\, \iota$.) (30.12)

29

K-Theories

Historically, the first examples of "extraordinary" cohomology theories are the *K*-theories that arise from the study of vector bundles. It is the purpose of this section to give an exposition (rather than a development) of *K*-theory. We do not, therefore, discuss the most general *K*-theories, nor prove the Bott periodicity theorems which give *K*-theory its power. We sketch two important applications of *K*-theory: the solutions of the Hopf invariant problem and the vector field problem.

All spaces in this section will be Hausdorff. Let k be one of the division rings R, C, or H.

Definition 29.1 A k vector bundle is a locally trivial bundle $\xi = \{E, \pi, B\}$ such that each fiber $\pi^{-1}(b)$ has the structure of a vector space over k, and there exist coordinate transformations $\varphi_\alpha \colon U_\alpha \times k^n \to \pi^{-1}(U_\alpha)$ that are linear over each point $b \in B$ (i.e., $\varphi_\alpha|_{b \times k^n} \colon k^n \to \pi^{-1}(b)$ is linear). ξ is called an n-plane bundle or an n-dimensional vector bundle if $\pi^{-1}(b)$ has dimension n for each $b \in B$.

Not every vector bundle has a dimension. However it is easy to see that over each component of the base, $\pi^{-1}(b)$ has constant dimension. Hence each vector bundle with a connected base has a dimension.

It is possible to put more restrictions on the φ_α than that they are linear (such as orientation preserving), but we will not consider any such refinements.

Vector bundles occur readily in geometric situations. We list some important examples.

Examples

1. The bundle $k^{n+1} - \{0\} \to kP^n$ given in Exercise 12, Section 11 is a k-line bundle. For each $x \in kP^n$, $\pi^{-1}(x)$ is a one-dimensional k-linear subspace of k^{n+1} and the coordinate transformations are k-linear (see also Example 4, Section 11).

2. Consider the case $k = R$ and $n = 1$ in the above example. This is the Möbius band projected onto a central circle. (See Section 26. We must, of course, transpose a vector space structure onto (0, 1) by a homeomorphism $R^1 \equiv (0, 1)$.) This is a nontrivial vector bundle that is easy to visualize, and is a good picture to keep in mind. The twisting that occurs in this bundle is in some sense typical of the complications that distinguish a general vector bundle from a product bundle: $k^n \times B \to B$.

3. The tangent bundle to a differential manifold $\tau(M) \to M$ is a real n-plane bundle [49, Chapter 2]. This is most easily defined if M is differentiably imbedded in R^{n+k}. Let $M^n \subset R^{n+k}$ be a C^1 imbedding; i.e., M^n is covered by coordinate systems $\{U_\alpha, \varphi_\alpha\}$ with $\varphi_\alpha\colon R^n \to U_\alpha \subset M^n \subset R^{n+k}$ differentiable and such that the Jacobian of φ_α has rank n. We then define $\tau(M)$ as follows. For each $x \in M$, let T_x be the tangent space to M at x. Then the total space of $\tau(M)$ is $\{(x, y) \in M \times R^{n+k} \mid x + y \in T_x\}$ and $\pi\colon \tau(M) \to M$ is given by $\pi(x, y) = x$. The coordinate functions $\tilde{\varphi}_\alpha\colon U_\alpha \times R^n \to \tau(M)$ are given by

$$\varphi_\alpha(u, t_1, \ldots, t_n) = \left(\varphi_\alpha(u), \sum \frac{\partial \varphi_\alpha}{\partial u_i} t_i\right).$$

This depends on the imbedding $M \subset R^{n+k}$, but one can show that different imbeddings give equivalent vector bundles (in the sense of 29.2).

Definition 29.2 Let $\xi = \{E, \pi, B\}$ and $\xi' = \{E', \pi', B'\}$ be vector bundles. A bundle map $f\colon \xi \to \xi'$ is a pair (f_E, f_B) of maps $f_E\colon E \to E'$ and $f_B\colon B \to B'$ such that

(a) $\pi' f_E = f_B \pi$;

(b) $f_E|_{\pi^{-1}(x)}$ is an isomorphism for each $x \in B$.

f is called an equivalence if $B = B'$ and $f_B = 1$. In this case we write $\xi \cong \xi'$.

Lemma 29.3 Equivalence of bundles over B is an equivalence relation.

Proof Reflexivity and transitivity are immediate. To prove symmetry, let $f\colon \xi \to \xi'$ be an equivalence. Then $f_E\colon E \to E'$ is a continuous 1–1 correspondence. To see that f_E is open it suffices to show that $f_E|_{\pi^{-1}(U_\alpha)}$ is open. In terms of local coordinates this is given by $(x, v) \to (x, A_x v)$ where A_x is a nonsingular linear transformation depending continuously on x. This map has a continuous inverse since matrix inversion is continuous. Hence $f_E|_{\pi^{-1}(U_\alpha)}$ is a homeomorphism and f_E is open. ∎

A bundle will be called trivial if it is equivalent to a product bundle $B \times k^n \to B$.

We will write $\mathrm{Vect}_n(X)$, for the set of equivalence classes of n-dimensional k-vector bundles over X. This is in fact a contravariant functor. For each map $f: X' \to X$ and vector bundle ξ over X, there is a vector bundle $f^*(\xi)$ over X' (see Exercise 5, Section 11). One defines a bundle map (π_2, f) where $\pi_2: f^*(E) \to E$ is the projection. Then $f^*(\xi) = \{f^*(E), \pi_1, X'\}$ is a vector bundle with coordinate functions $\tilde{\varphi}_\alpha: U_\alpha \times k^n \to f^*(E)$ given by $\tilde{\varphi}_\alpha(u, x) = (u, \varphi_\alpha(f(u), x))$. It is easy to see that (π_2, f) is bundle map from $f^*(\xi)$ to ξ, and that if $\xi \cong \xi'$, $f^*(\xi) \cong f^*(\xi')$. Consequently the transformation $\xi \to f^*(\xi)$ makes the set $\mathrm{Vect}_n(X)$ into a cotravariant functor.

Finally we observe that if $f: \xi \to \xi'$ is a bundle map, $\xi \cong f^*(\xi')$. In fact the map f induces a bundle map $e: \xi \to f^*(\xi')$ so that $\xi \to f^*(\xi') \to \xi'$ is the bundle map f. Since $e_B = 1$, e is an equivalence.

Let $k^\infty = \bigcup_{n=1}^{\infty} k^n$ with the weak topology.

Definition 29.4 A Gauss map for a k-vector bundle ξ is a continuous map $F: E \to k^m$ for some m, $1 \le m \le \infty$ such that for each $x \in B$, $F|_{\pi^{-1}(x)}$ is a linear monomorphism.

Under mild restrictions, a Gauss map always exists. In Example 1, the inclusion $E = k^{n+1} - \{0\} \subset k^{n+1}$ is a Gauss map. In the case of a differential manifold $M^n \subset R^{n+k}$ (differentiably imbedded) the mapping $F: \tau(M) \to R^{n+k}$ given by $F(x, y) = y$ is a Gauss map.

Proposition 29.5 If $\xi = \{E, \pi, B\}$ is a vector bundle and B is paracompact, a Gauss map exists for ξ.

In the case that B is compact, one can find a Gauss map $F: E \to k^m$ with $m < \infty$. We will prove 29.5 in this case. The general case is a little more complicated [48, Theorem 7; 34, 13, 5.5].

Proof in this case Choose a finite collection of coordinate neighborhoods $U_{\alpha_1}, \ldots, U_{\alpha_m}$ that cover B. Choose an associated partition of unity f_{α_i}. Define $F_{\alpha_i}: E \to k^n$ by

$$F_{\alpha_i}(e) = \begin{cases} f_{\alpha_i}(\pi(e)) \cdot \pi_2 \varphi_{\alpha_i}^{-1}(e), & \pi(e) \in U_{\alpha_i} \\ 0, & \pi(e) \notin U_{\alpha_i}. \end{cases}$$

Thus F_{α_i} is a linear monomorphism on $\pi^{-1}(x)$ if $f_{\alpha_i}(x) \neq 0$. A Gauss map $F: E \to k^{mn}$ is thus defined by

$$F(x) = (F_{\alpha_i}(x), \ldots, F_{\alpha_m}(x)). \quad ∎$$

Let V be a vector space over k. We will describe an n-plane bundle $\gamma^n(V)$ which is universal in the sense that if ξ is an n-plane bundle and $F\colon E(\xi) \to V$ is a Gauss map, there is a bundle map $f\colon \xi \to \gamma^n(V)$ (and hence $\xi \cong f^*(\gamma^n(V))$.

The description of $\gamma^n(V)$ is easy enough. As base space we take $G_n(V)$—the set of all n-dimensional subspaces of V. ($G_n(V)$ is called the Grassmanian on V.) As total space we take $E_n(V)$ to be the set of all pairs $(x, M) \in V \times G_n(V)$ with $x \in M$. We define $\pi_n\colon E_n(V) \to G_n(V)$ by $\pi_n(x, M) = M$. Then $\gamma^n(V) = \{E_n(V), \pi_n, G_n(V)\}$. Now suppose we are given an n-plane bundle ξ and a Gauss map $F\colon E(\xi) \to V$. We define $f\colon \xi \to \gamma^n(V)$ by

$$f_B(x) = \{F(\pi^{-1}(x))\}, \qquad f_E(e) = (F(e), f_B(\pi(e))).$$

One easily checks that $\pi_n f_E = f_B \pi$ and that f_E is an isomorphism on each fiber. We have carefully sidestepped the question of how we will topologize $G_n(V)$. Continuing in this vein, we will describe the local product structure in $\gamma^n(V)$. Choose a continuous inner product in V. For each $M \in G_n(V)$ let $P(M)$ be the orthogonal projection onto M. (This is well defined even if V is infinite dimensional since M is finite dimensional.) Let $U(N) = \{M \in G_n(V) \mid P(M)|_N$ has rank $n\}$. $N \in U(N)$ so $\{U(N)\}$ is an (open) cover of $G_n(V)$. We define $\varphi_N\colon U(N) \times N \to \pi^{-1}(U(N))$ by $\varphi_N(M, x) = (M, P(M)(x))$. This is a 1–1 correspondence which is linear on each fiber.

We must find a topology in which f_B and φ_N are continuous and $U(N)$ is open. The details of this are a little delicate and often neglected. To make f_B continuous it is sufficient for $f_B|_{U_\alpha}$ to be continuous for each coordinate neighborhood U_α. For $x \in U_\alpha$, $f_B(x) = \{F\varphi_\alpha(x \times k^n)\}$. Let $L_n \subset V^n$ be the set of linearly independent n-tuples of vectors in V. Give L_n the induced topology. Let $\rho\colon L_n \to G_n(V)$ be the natural map which assigns to each n-tuple its span. ρ is onto and we give $G_n(V)$ the quotient topology. $f_B|_{U_\alpha}$ factors through L_n and the map $U_\alpha \to L_n$ is given by $x \to (F\varphi_\alpha(x \times v_1), \ldots, F\varphi_\alpha(x \times v_n))$ where $v_1, \ldots, v_n$ is a basis for k^n. Thus f_B is continuous. To ensure that φ_N is continuous and $U(N)$ is open it is only necessary to check that $M \to P(M)$ is continuous, where $P(M) \in V^V$ and V^V has the function space topolgy. Let $V_n \subset L_n$ be the subset of orthogonal n-tuples. Then $\rho(V_n) = G_n$ and V_n is a closed subset of L_n. Thus G_n has the quotient topology on V_n and it is only necessary to verify that $V_n \subset L_n \xrightarrow{\rho} G_n(V) \to V^V$ is continuous. This composite is given by $(x_1, \ldots, x_n) \to f$ where $f(x) = \sum (x, x_i)x_i$. Since the adjoint $V_n \times V \to V$ is continuous, we are done.

Since $G_n(V) \to V^V$ is continuous and 1–1, $G_n(V)$ is Hausdorff. In fact, if V has dimension $n + p < \infty$, $G_n(V)$ is a differential manifold of dimension np [49]. We conclude:

Proposition 29.6 For each n-plane bundle ξ with paracompact base, there is a bundle map $f\colon \xi \to \gamma^n(k^\infty)$ and hence $\xi \cong f^*(\gamma^n(k^\infty))$. ∎

The mapping f depends on the choice of a Gauss map. One can show that any two Gauss maps are homotopic through Gauss maps. One first uses a linear homotopy to put one Gauss map in even dimensions and the other in odd dimensions and then takes a linear homotopy between them. (For details see [34, I3, 6.2].) This induces a homotopy between the respective bundle maps. Thus an equivalence class of bundles determines a homotopy class of maps from B to $G_n(k^\infty)$. Conversely a map $f\colon B \to G_n(k^\infty)$ induces a vector bundle over B and one can show that the equivalence class of this bundle only depends on the homotopy class of f [34, I3, 4.7; 65, 11.5; 9, 1.4.3]. Hence

Theorem 29.7 (*Classification Theorem*) Let B be paracompact. Then the transformation $f \to f^*(\gamma^n(k^\infty))$ induces a 1–1 correspondence

$$[B, G_n(k^\infty)] \to \mathrm{Vect}_n(B).$$

We define some notation in universal use. $BO(n) = G_n(R^\infty)$, $BU(n) = G_n(C^\infty)$, and $BSp(n) = G_n(H^\infty)$. Here $O(n)$, $U(n)$, and $Sp(n)$ are the orthogonal, unitary, and symplectic groups of $n \times n$ matrices, and our baptism is based on homotopy equivalences $O(n) \simeq \Omega BO(n)$, $U(n) \simeq \Omega BU(n)$, and $Sp(n) \simeq \Omega BSp(n)$. (See Exercise 8.)

We describe now the Whitney sum of two vector bundles. This is a generalization of the notion of direct sum of vector spaces to vector bundles. If ξ and η are vector bundles over B, $\xi \oplus \eta$ will be a bundle over B such that the fibers in $\xi \oplus \eta$ will be the direct sum of the fibers in ξ and η. To construct such a bundle we consider the vector bundle $\xi \times \eta = \{E \times E', \pi \times \pi', B \times B'\}$. Let $d\colon B \to B \times B$ and define $\xi \oplus \eta = d^*(\xi \times \eta)$. Thus a point in $E(\xi \oplus \eta)$ is a pair $(e, e') \in E \times E'$ with $\pi(e) = \pi'(e')$.

One immediately checks that there are equivalences

$$\xi \oplus \eta \cong \eta \oplus \xi$$
$$\xi \oplus (\eta \oplus \lambda) \cong (\xi \oplus \eta) \oplus \lambda$$
$$0 \oplus \xi \cong \xi$$

where 0 is the 0-plane bundle. Finally, if $\xi \cong \xi'$ and $\eta = \eta'$, $\xi \oplus \eta = \xi' \oplus \eta'$. Thus the set of equivalence classes of vector bundles over B is a semigroup with the operation $\oplus$.

There is a natural way of producing from this semigroup an abelian group. The construction, called the Grothendieck construction, is as follows. Let $F(X)$ be the free group generated by isomorphism classes of vector bundles over X. Given a vector bundle ξ we write $[\xi]$ for the corresponding element of $F(X)$. Let R be the subgroup generated by all elements of the form $[\xi \oplus \eta] - [\xi] - [\eta]$, and let $K_k(X)$ be the quotient group. The functor $K_k(X)$ is called K-theory.

In the case $k = R$ or C it is possible but a little more complicated to extend some other functors from vector spaces to vector bundles. Of particular interest is the tensor product $\xi \otimes \eta$ of two vector bundles, and the exterior power [17, AIII, §7] of a vector bundle $\lambda^i(\xi)$. These operations satisfy the laws

$$\xi \otimes \eta \cong \eta \otimes \xi$$
$$(\xi \otimes \eta) \otimes \zeta \cong \xi \otimes (\eta \otimes \zeta)$$
$$\xi \otimes (\eta_1 \oplus \eta_2) \cong (\xi \otimes \eta_1) \oplus (\xi \otimes \eta_2)$$
$$\xi \otimes 1 \cong \xi$$

where 1 is a trivial line bundle, and

$$\lambda^0(\xi) \cong 1$$
$$\lambda^1(\xi) \cong \xi$$
$$\lambda^k(\xi \oplus \eta) \cong \sum_{i+j=k} \lambda^i(\xi) \otimes \lambda^j(\eta)$$

If $\xi \cong \xi'$ and $\eta = \eta'$, then $\xi \otimes \eta = \xi' \otimes \eta'$ and $\lambda^i(\xi) \cong \lambda^i(\xi')$. Details of these constructions can be found in [34, I5, Section 6; 9, §1.2].

Since the tensor product distributes over the Whitney sum, it induces a natural commutative ring structure on $K_k(X)$ and the exterior power operations define natural transformations $\lambda^i\colon K_k(X) \to K_k(X)$ such that

$$\lambda^0(x) = 1, \qquad \lambda^1(x) = x,$$
$$\lambda^k(x + y) = \sum_{i+j=k} \lambda^i(x)\lambda^j(y).$$

It is easy to see that $K_k(P)$ is isomorphic to the integers if P is a one-point space, and the isomorphism is a ring isomorphism (in case $k = R$ or C). The operations are given by $\lambda^i(n) = \binom{n}{i}$.

We now define reduced K-theory by $\tilde{K}_k(X) \cong \operatorname{coker}\{K_k(*) \to K_k(X)\}$. $\tilde{K}_k$ is thus a functor from $\mathcal{CG}$ to $\mathcal{M}_Z$. If we choose a base point $* \in X$, this induces a splitting

$$K_k(X) \cong \tilde{K}_k(X) \oplus Z.$$

It is easy to see that $\tilde{K}_k(X) \cong \ker\{K_k(X) \to K_k(*)\}$ where the later group depends on the choice of a base point. The advantage of the later group is that a ring structure is transferred onto $\tilde{K}_k(X)$ since the kernel in question is an ideal. This ring structure conceivably depends on the choice of $*$.

We give an alternative description of reduced K-theory when X is compact.

Proposition 29.8 Let X be compact and ξ be a vector bundle over X. Then there is a bundle ξ' over X with $\xi \oplus \xi'$ trivial.

Proof Let $F: E(\xi) \to k^m$ be a Gauss map with $m < \infty$. Choose an inner product in k^m and let $E' = \{(u, x) \in k^m \times X \mid u \cdot y = 0 \text{ for all } y \in F(\pi^{-1}(x))\}$. Thus E' is the orthogonal complement of $F(E(\xi))$ in k^m. (A good picture to look at here is the Möbius band imbedded in $D^2 \times S^1$.) We claim that $\xi' = \{E', \pi_2, X\}$ is a vector bundle and $\xi \oplus \xi' = \{k^m \times X, \pi_2, X\}$. For details see [49, 2.20]. (In the case of the Möbius band μ over S^1, one easily sees that $\mu' \cong \mu$.) ▮

We will think of each integer n as an n-dimensional trivial bundle.

Definition 29.9 Two vector bundles ξ and η over X are stably equivalent (written $\xi \cong_s \eta$) if there are trivial bundles n and m such that $\xi \oplus n \cong \eta \oplus m$.

This is clearly an equivalence relation. We define a function φ from the equivalence classes to $\tilde{K}_k(X)$ by $\varphi(\{\xi\}) = [\xi] \pmod{Z}$. This is well defined and 1–1.

Proposition 29.10 If X is compact, φ is a 1–1 correspondence.

Proof Let $x \in K_k(X)$. We can write $x = [\xi] - [\eta]$, and since X is compact, $x = [\xi \oplus \eta'] - m$ by 29.8. Hence $x \equiv [\xi \oplus \eta'] \pmod{Z}$ and thus φ is onto. ▮

We define mappings $\text{Vect}_n(X) \to \text{Vect}_{n+1}(X)$ by $\xi \to 1 \oplus \xi$. Corresponding to this there is a continuous mapping $\iota: G_n(k^\infty) \to G_{n+1}(k^\infty)$ given by $M \to k \oplus M \subset k \oplus k^\infty \cong k^\infty$. Then $\iota^*(\gamma^{n+1}(k^\infty)) \cong 1 \oplus \gamma^n(k^\infty)$ so we have a commutative diagram

$$\begin{array}{ccc} \text{Vect}_n(X) & \longrightarrow & \text{Vect}_{n+1}(X) \\ \updownarrow & & \updownarrow \\ [X, G_n(k^\infty)] & \xrightarrow{\iota_*} & [X, G_{n+1}(k^\infty)] \end{array}$$

We define $G(k^\infty) = \bigcup_{n=1}^{\infty} G_n(k^\infty)$ with the weak topology.

Proposition 29.11 If X is compact and connected, there is a 1–1 correspondence

$$[X, G(k^\infty)] \leftrightarrow \tilde{K}_k(X)$$

Proof There is a 1–1 correspondence

$$\varinjlim \text{Vect}_n(X) \leftrightarrow \tilde{K}_k(X)$$

given by $\xi \to [\xi] - [\dim \xi]$. (Every vector bundle ξ over a connected space has a constant dimension, written $\dim \xi$.) This is clearly well defined, 1–1, and onto. There is also a 1–1 correspondence

$$\varinjlim \text{Vect}_n(X) \leftrightarrow \varinjlim\, [X, G_n(k^\infty)];$$

since X is compact and $G(k^\infty)$ has the weak topology, the conclusion follows by the method in 15.10. ∎

$G_n(k^\infty)$ is classically written BO, BU, or BSp in the cases $k = R$, C, or H.

Theorem 29.12 (*Bott Periodicity Theorem*)

$$\Omega^2 BU \simeq BU \times Z, \qquad \Omega^4 BSp \simeq BO \times Z, \qquad \Omega^4 BO \simeq BSp \times Z$$

There are basically three methods of proving this theorem, all of which are quite complicated. The first method, which is the original method of Bott [15, 16], uses Morse theory to analyze ΩX for X a Lie group. The best reference for a proof in this spirit is [52] where all the prerequisite Morse theory is developed. Bott's proof was quickly followed by homotopy-theoretic proofs [70] in the case of BU, and [23] in the general case). The third method of proof is to analyze directly a vector bundle over $S^2 X$. A proof in the case of BU by this method was given [10] and this was later generalized [8]. Ideal references are [8; 23].

Theorem 29.13 There are spectra K, KO, and KSp such that:

(a) $\tilde{K}^n(X) \cong \tilde{K}^{n+2}(X)$.
(b) $\widetilde{KO}^n(X) \cong \widetilde{KSp}^{n+4}(X)$.
(c) $\widetilde{KSp}^n(X) \cong \widetilde{KO}^{n+4}(X)$.

Furthermore, if X is compact we have:

(d) $\tilde{K}^0(X) \cong \tilde{K}_C(X)$, $K^0(X) \cong K_C(X)$.
(e) $\widetilde{KO}^0(X) \cong \tilde{K}_R(X)$, $KO^0(X) \cong K_R(X)$.
(f) $\widetilde{KSp}^0(X) \cong \tilde{K}_H(X)$, $KSp^0(X) \cong K_H(X)$.

Proof By 29.12, BO, BU, and BSp are H spaces. In fact they are connected CW complexes [48, VI; 67] so they are WANEs. Thus by Exercise 11, Section 21, we have

$$[X, G(k^\infty)] \leftrightarrow [(X, *), (G(k^\infty), *)]$$

for X paracompact and connected. Thus for arbitrary compact X, we have

$$\tilde{K}_k(X) \leftrightarrow [(X, *), (Z \times G(k^\infty), *)]$$

We define Ω-spectra as follows:

$$K_n = \begin{cases} BU \times Z, & n \text{ even} \\ \Omega BU, & n \text{ odd.} \end{cases}$$

$$KO_n = \Omega^j(BO \times Z), \qquad n = 8r + j, \quad 0 \le j < 8.$$

$$KSp_n = \Omega^4 KO_n.$$

The maps are determined by the homotopy equivalences in 29.12. ∎

In particular KSp is completely determined by KO. Since $KO(X)$ is a ring whereas $KSp(X)$ is not, one usually neglects KSp altogether. $KO(X)$, $K(X)$, and $KSp(X)$ are called real, complex, and symplectic K-theory respectively.

Tensor product of vector bundles determines a map $BO \wedge BO \to BO$, and $BU \wedge BU \to BU$, although care must be taken at this point. $BU \wedge BU$, for example, is not compact and hence a map $BU \wedge BU \to BU$ does not correspond to a vector bundle. However if $X \subset BU$ is a finite subcomplex, the inclusion map determines a bundle ξ_X over X up to stable equivalence. Let π_1 and π_2 be the projections $X \times X \to X$. Then $\eta_X = \pi_1{}^*(\xi_X) \otimes \pi_2{}^*(\xi_X)$ is a well-defined bundle over $X \times X$ and determines a map μ_X: $X \times X \to BU$. The maps μ_X are compatible up to homotopy, and this is enough to define a map $\tilde{\mu}$: $BU \times BU \to BU$ by Exercise 5, Section 15. Let μ: $BU \wedge BU \to BU$ be the map determined by $\tilde{\mu} - \pi_1 - \pi_2$: $BU \times BU \to BU$. One defines a map $(BU \times Z) \wedge (BU \times Z) \to (BU \times Z)$ by μ on the 0-components and $\tilde{\mu}$ on the other components, together with multiplication of integers. Maps $\Omega(BU \times Z) \wedge (BU \times Z) \to \Omega(BU \times Z)$ and $\Omega(BU \times Z) \wedge \Omega(BU \times Z) \to BU \times Z$ are then determined and these maps make K into a ring spectrum. The proof of associativity and commutativity are subtle since the requisite diagrams are only given to commute on each finite subcomplex of the domain space. This is not sufficient in general to prove that they commute, but in the circumstances the difficulties are easily overcome. See [26]. The multiplication induced in $K(X)$ can easily be seen to be the tensor product of vector bundles.

Similar considerations produce a map $BO \wedge BO \to BO$ and give KO the structure of a ring spectrum.

Thus we have

Theorem 29.14 K and KO are ring spectra and if X is compact the induced multiplications $K^0(X)$ and $KO^0(X)$ coincide with the multiplications $K_C(X)$ and $K_R(X)$. ∎

Before discussing applications, we give an account of the coefficient rings $\{K^n(*)\}$ and $\{KO^n(*)\}$. We will write H for the Hopf bundle over S^2 (the complex line bundle of Example 1). Let $b \in K^{-2}(*)$ correspond to $[H] - 1 \in \tilde{K}_C(S^2) \cong \tilde{K}^{-2}(S^0) \cong K^{-2}(*)$.

Proposition 29.15 The periodicity isomorphism

$$\beta_n\colon K^n(X) \to K^{n-2}(X)$$

is given by $\beta_n(x) = x \,\bar{\times}\, b$.

This follows from the proof of 29.12 [8; 23].

In particular, there is a class $c \in K^2(*)$ such that $1 = c \,\bar{\times}\, b$. Since $U(n)$ is connected (see Exercise 5), $BU(n)$ is simply connected for each n and hence $\pi_1(BU) = 0$. Thus $\tilde{K}_C(S^1) = 0$ and we have

Corollary 29.16

$$K^n(*) = \begin{cases} 0, & n \text{ odd} \\ Z, & n \text{ even.} \end{cases}$$

$K^{2n}(*)$ is generated by b^n if $n < 0$ and c^n if $n > 0$. Furthermore, $bc = 1$. ▌

A more compact statement of 29.16 is $K^*(*) \cong Z[b, c]/(bc - 1)$.

Proposition 29.17

$$KO^n(*) \cong \begin{cases} Z_2, & n \equiv -1 \text{ or } -2 \pmod 8 \\ Z, & n \equiv 0 \pmod 4 \\ 0, & \text{otherwise.} \end{cases}$$

See [34, II 8, 5.2; 16] or Exercise 7.

The generator of $KO^{-1}(*) \cong \widetilde{KO}(S^1)$ is $[\mu] - 1$ where μ is the Möbius line bundle (Example 2). The generator of $KO^{-2}(*) \cong \widetilde{KO}(S^2)$ is $([\mu] - 1) \barwedge ([\mu] - 1) = [\mu \otimes \mu] - 1$. Other multiplicative relations are given by

Proposition 29.18 Generators $x \in KO^{-8}(*)$, $y \in KO^8(*)$, $u \in KO^{-4}(*)$ and $v \in KO^4(*)$ may be chosen such that $xy = 1$, $u^2 = 2x$, and

$$\begin{array}{ll} y^r \text{ generates } K^{8r}(*); & \mu^2 y^r \text{ generates } K^{8r-2}(*); \\ \mu y^r \text{ generates } K^{8r-1}(*); & u y^r \text{ generates } K^{8r-4}(*). \end{array}$$

(If $r < 0$, we interpret y^r as x^{-r}.)

We will now indicate two celebrated applications of K-theory to geometric problems.

Recall that we defined a transformation $H\colon \pi_{2n-1}(S^n) \to Z$ in Section 28 and proved that if $H(\alpha)$ is odd, $n = 2^s$ (28.18).

Theorem 29.19 (*Adams*) If $H(\alpha)$ is odd, $n = 1, 2, 4$, or 8.

This was first proved using other methods than the one given here [1]. The version we give is conceptually and technically simpler [6]. The importance of 29.19 is that it is the most difficult step in the cyclic proof of:

Theorem 29.20 The following are equivalent:

(1) $n = 1, 2, 4$, or 8.
(2) R^n has the structure of a normed algebra.
(3) R^n has the structure of a division algebra.
(4) S^{n-1} is parallelizable (i.e., $\tau(S^{n-1})$ is trivial).
(5) S^{n-1} is an H space.
(6) There exists a map $f\colon S^{2n-1} \to S^n$ with $H(f) = 1$.

For a detailed exposition of the proof of 29.20 see [27]. We will indicate the method here. (1) ⇒ (2) ⇒ (3) are pretty easy. Suppose R^n is a division algebra. Choose a basis $e_1, \ldots, e_n$ with $e_1 = 1$, $n-1$ linearly independent vector fields are defined on S^{n-1} by

$$V_i(a) = ae_i - \frac{(a, ae_i)}{\|a\|^2} a, \qquad i = 2, \ldots n.$$

Hence (3) ⇒ (4).

There is a projection $\tau(S^{n-1}) \to S^{n-1}$ given by projecting a vector tangent to S^{n-1} onto S^{n-1} by a line through the origin. If S^{n-1} is parallelizable we thus have a map $R^{n-1} \times S^{n-1} \equiv \tau(S^{n-1}) \to S^{n-1}$. It is easily seen that this map extends over $(R^{n-1})^\infty \times S^{n-1}$ and that the resulting map is a multiplication on S^{n-1} with two sided unit. Thus (4) ⇒ (5).

The proof that (5) ⇒ (6) is more complicated. Let

$$X * Y = X \times I \times Y \Big/ \begin{matrix} (x, 0, y) \sim (x, 0, y') \\ (x, 1, y) \sim (x', 1, y). \end{matrix}$$

If we are given a map $f\colon X \times Y \to Z$, the Hopf construction is the map $h(f)\colon X * Y \to \Sigma Z$ given by $h(f)(x, t, y) = (f(x, y), t)$. Suppose $X = Y = Z = S^{n-1}$ and f is a multiplication with two-sided homotopy unit. There is a homeomorphism $\theta\colon S^{m-1} * S^{n-1} \to S^{m+n-1}$ given by $\theta(u, t, v) = (u \cos(\pi t/2), v \sin(\pi t/2))$. Under this identification, $h(f)\colon S^{2n-1} \to S^n$.

Lemma 29.21 $h(f)$ has Hopf invariant 1.

Proof Let $M = \Sigma Z \cup_{h(f)} C(X * Y) = S^n \cup_{h(f)} e^{2n}$. Let $\beta \in H^n(M)$. We will prove that β^2 is a generator of $H^{2n}(M)$. There is a homeomorphism

$$\varphi\colon X * Y \to CX \times Y \cup X \times CY \subset CX \times CY$$

given by

$$\varphi(u, t, v) = \begin{cases} (u, (v, 1 - 2t)), & t \le \frac{1}{2} \\ ((u, 2t - 1), v), & t \ge \frac{1}{2}. \end{cases}$$

We identify $X * Y$ with $\varphi(X * Y)$ under this homeomorphism. Thus we define a relative homeomorphism

$$L\colon (CX \times CY, X * Y) \to (M, \Sigma Z)$$

by

$$L((x, s), (y, t)) = \begin{cases} (t, ((s, x), y)) \in C(CX \times Y) & \text{if } s \ge t \\ (s, (x, (t, y))) \in C(X \times CY) & \text{if } s \le t. \end{cases}$$

This formula determines a point of $C(X * Y)$ under $C\varphi^{-1}$ and hence a point of M. Choose $x_0 \in X$. Then we can define $a_1: Y \to Z$ and $\iota_1: CY \to CX \times CY$ by $a_1(y) = f(x_0, y)$ and $\iota_1(y, t) = ((x_0, 0), (y, t))$. Then we have a commutative diagram

$$\begin{array}{ccc}
(CX \times CY, X * Y) & \xrightarrow{L} & (M, \Sigma Z) \\
\uparrow & & \uparrow \\
(CX \times CY, CX \times Y) & \xrightarrow{\xi_1} & (M, C^+Z) \\
\uparrow{\scriptstyle \iota_1} & & \uparrow \\
(CY, Y) & \xrightarrow{Ca_1} & (C^-Z, Z)
\end{array}$$

where C^+Z and C^-Z are the cones in ΣZ with $t \geq \frac{1}{2}$ and $t \leq \frac{1}{2}$ respectively. Let $\beta^+ \in H^n(M, C^+Z)$ and $\beta' \in H^n(CY, Y)$ be generators corresponding to $\beta \in H^n(M)$. Since $a_1 \sim 1$, $(\xi_1)^*(\beta^+) = 1 \mathbin{\bar{\times}} \beta'$. Similarly, by choosing $y_0 \in Y$ we can define a similar diagram and $(\xi_2)^*(\beta^-) = \beta' \times 1$ (note: $X = Y$). Now $H^{2n}(CX \times CY, X * Y)$ is generated by

$$\beta' \mathbin{\bar{\times}} \beta' = (\beta' \mathbin{\bar{\times}} 1) \cup (1 \mathbin{\bar{\times}} \beta') = \xi_2{}^*(\beta^-) \cup \xi_1{}^*(\beta^+) = L^*(\beta^- \cup \beta^+).$$

Since L is a relative homeomorphism, $H^{2n}(M, \Sigma Z)$ is generated by $\beta^- \cup \beta^+$. Let $j: (M, \varnothing) \to (M, \Sigma Z)$. Then $\beta^2 = j^*(\beta^- \cup \beta^+)$ is a generator. ∎

To prove that (6) ⇒ (1), let $f: S^{2n-1} \to S^n$ and $M = S^n \cup_f e^{2n}$. Let $\Delta_M: M \to M \wedge M$ be the diagonal map. Since $\Delta_M|_{S^n}$ factors through $\Delta_{S^n}: S^n \to S^n \wedge S^n$ it is nullhomotopic. Thus there is a factorization of Δ_M

$$\begin{array}{ccc}
 & S^{2n} & \\
{\scriptstyle \mu}\nearrow & & \searrow{\scriptstyle \nu} \\
M & \xrightarrow{\Delta_M} & M \wedge M
\end{array}$$

where $\mu: M \to S^{2n}$ is the quotient map. Now $\pi_{2n}(M \wedge M)$ is generated by the inclusion $\gamma: S^{2n} \equiv S^n \wedge S^n \to M \wedge M$, so $\nu = k\gamma$ for some integer k. Since $(\Delta_M) * (a \mathbin{\bar{\times}} b) = a \cup b$, we can evaluate both homomorphisms in cohomology and it follows that $k = H(f)$. Thus $H(f)$ determines the homomorphism $(\Delta_M)^*$ in any cohomology theory. We consider $\tilde{K}^0(M)$. There is an exact sequence

$$\tilde{K}^1(S^{2n}) \leftarrow \tilde{K}^0(S^n) \leftarrow \tilde{K}^0(M) \leftarrow \tilde{K}^0(S^{2n}) \leftarrow \tilde{K}^{-1}(S^n);$$

by 18.9. This reduces to (by 29.16)

$$0 \leftarrow Z \leftarrow \tilde{K}^0(M) \leftarrow Z \leftarrow 0,$$

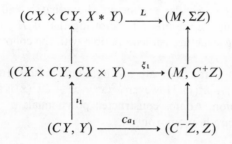

so $\tilde{K}^0(M)$ is free of rank 2. Let $x_n = ([H]-1) \barwedge \cdots \barwedge ([H]-1) \in \tilde{K}(S^{2n})$. Then x_n is a generator. Let $\xi = \mu^*(x_n)$ and choose $\eta \in \tilde{K}^0(M)$ so that $\iota^*(\eta) = x_n/2$ where $\iota\colon S^n \to M$ is the inclusion (n is even by 28.18). Then ξ and η generate $\tilde{K}^0(M)$. Now $\eta^2 = (\Delta_M) * (\eta \barwedge \eta) = \mu^* \nu^* (\eta \barwedge \eta) = \mu^*(H(f) x_n/2 \barwedge x_n/2) = H(f)\mu^*(x_n) = H(f)\xi$. Similar arguments show that $\xi^2 = 0$ and $\xi\eta = 0$ since $\iota^*(\xi) = 0$. We complete the proof by showing that if $H(f)$ is odd, and $n \neq 2$, 4, or 8 this ring structure is incompatible with the cohomology operations λ^i.

The operations λ^i for $i > 1$ are not homomorphisms. In fact $\lambda^2(x+y) = \lambda^2(x) + \lambda^1(x)\lambda^1(y) + \lambda^2(y)$. However $\psi^2(x) = x^2 - 2\lambda^2(x)$ is a homomorphism. In a similar fashion, Adams constructed polynomials ψ^k in the operations $\lambda^1, \ldots, \lambda^k$ which are homomorphisms.

Theorem 29.22 (*Adams*) There are natural transformations $\psi^k\colon K(X) \to K(X)$ satisfying:

(a) $\psi^k(x+y) = \psi^k(x) + \psi^k(y)$.
(b) If x is a line bundle, $\psi^k(x) = x^k$.
(c) $\psi^k(xy) = \psi^k(x)\psi^k(y)$.
(d) $\psi^k(\psi^l(x)) = \psi^{kl}(x)$.
(e) If p is a prime $\psi_p(x) \equiv x^p \pmod{p}$.
(f) If $u \in \tilde{K}(S^{2n})$, $\psi^k(u) = k^n u$.

The best references for a proof of this are [9; 34, II,12].

Let us now apply these operations to $\tilde{K}^0(M)$. By (f),

$$\psi^k(\xi) = \psi^k(\mu^*(x_n)) = \mu^*(\psi^k(x_n)) = k^n\mu^*(x_n) = k^n\xi.$$

Let $\psi^k(\eta) = \alpha_k\eta + \beta_k\xi$. Then

$$\alpha_k x_n/2 = \iota^*(\alpha_k\eta + \beta_k\xi) = \iota^*(\psi^k(\eta)) = \psi^k(\iota^*(\eta)) = \psi^k(x_n/2) = k^{n/2}x_n/2,$$

so $\alpha_k = k^{n/2}$. By (d), $\psi^2(\psi^3(\eta)) = \psi^3(\psi^2(\eta))$. However,

$$\psi^3(\psi^2(\eta)) = \psi^3(2^{n/2}\eta + \beta_2\xi) = 2^{n/2}(3^{n/2}\eta + \beta_3\xi) + 3^n\beta_2\xi,$$

and $\psi^2(\psi^3(\eta)) = 3^{n/2}(2^{n/2}\eta + \beta_2\xi) + 2^n\beta_3\xi$. Thus

$$2^{n/2}\beta_3 + 3^n\beta_2 = 3^{n/2}\beta_2 + 2^n\beta_3, \text{ or } 2^{n/2}(2^{n/2}-1)\beta_3 = 3^{n/2}(3^{n/2}-1)\beta_2.$$

By (e), $\psi^2(\eta) \equiv \eta^2 \equiv \xi \pmod{2}$. Thus β_2 is odd. Consequently $2^{n/2} \mid 3^{n/2} - 1$. It now follows from elementary arithmetical arguments that $n = 2$, 4, or 8. ∎

The second important application of K-theory is to the question of vector fields on spheres. Recall that there is a nonzero vector field on S^{n-1} iff n is even (13.19). Vector fields $V_1, \ldots, V_k$ are said to be linearly independent if for every $x \in S^{n-1}$, $V_1(x), \ldots, V_k(x)$ are linearly independent in $\tau(S^{n-1})_x$. A classical problem is to determine the maximal number of linearly independent vector fields on S^{n-1}. The problem is to find a function $r(n)$ such that S^{n-1} has $r(n) - 1$ linearly independent vector fields but it does not have $r(n)$ linearly independent vector fields. 13.19 can be restated as $r(2n) > 1$ and $r(2n + 1) = 1$.

We first consider the problem of constructing vector fields. It is natural, in view of 3.5 to construct linear vector fields, i.e., linear nonsingular functions $f\colon E^n \to E^n$ with $x \cdot f(x) = 0$. Let $\rho(n) - 1$ be the maximal number of such functions such that for each $x \neq 0, f_1(x), \ldots, f_{\rho(n)-1}(x)$ are linearly independent. Then $r(n) \geq \rho(n)$. The determination of $\rho(n)$ is a problem in linear algebra. As an example we have three linearly independent vector fields on S^3 given by the equations

$$f_1(x_1, x_2, x_3, x_4) = (-x_2, x_1, -x_4, x_3).$$
$$f_2(x_1, x_2, x_3, x_4) = (-x_3, x_4, x_1, -x_2).$$
$$f_3(x_1, x_2, x_3, x_4) = (-x_4, -x_3, x_2, x_1).$$

Theorem 29.23 (*Radon–Hurwitz–Eckmann*) Let $n = (2k + 1)2^{c+4d}$ with $0 \leq c \leq 3$. Then $\rho(n) = 2^c + 8d$.

For a proof see [34; II 11].

We list the first few values of this rather complicated function.

n	2	4	6	8	10	12	14	16	18	20	22
$\rho(n)$	2	4	2	8	2	4	2	9	2	4	2

This very unlikely looking function is in fact a best possible result.

Theorem 29.24 (*Adams*) $\rho(n) = r(n)$.

The proof of 29.24 is complicated and we will give only the briefest outline. First we observe that if there are k linearly independent vector fields on S^n, there are k orthogonal vector fields on S^n. This follows from the Gram–Schmidt orthogonalization process [12]. Consequently the problems of finding linearly independent or orthogonal sets of vector fields are the same. We will concentrate on the later problem. Recall that in the discussion before 29.6 we introduced a space V_k, which we now write as $V_k(R^n)$, consisting of orthogonal k-tuples in R^n. There are continuous maps

$$V_{k-1}(R^{n-1}) \xrightarrow{\iota} V_k(R^n) \xrightarrow{\pi} S^{n-1}$$

given by $\iota(x_1, \ldots, x_{k-1}) = (x_1, \ldots, x_{k-1}, *)$ and $\pi(x_1, \ldots, x_n) = x_n$ where $* = (0, \ldots, 0, 1)$. In fact π is a locally trivial bundle with fiber $V_{k-1}(R^{n-1})$ (compare with Exercise 13, Section 11) [65, 7.8; 34, I7, 3.8]. The spaces V_k (and sometimes L_k) are called Stiefel manifolds. (One can prove that they are manifold by induction on k using this locally trivial bundle.)

Now it is easy to see that a map $f\colon S^{n-1} \to V_k(R^n)$ such that $\pi f = 1$ corresponds to a collection of $k - 1$ orthogonal vector fields on S^{n-1}. Thus we concern ourselves with the lifting problem

$$\begin{array}{ccc} & & V_k(R^n) \\ & \overset{f}{\nearrow} & \Big\downarrow \pi \\ S^{n-1} & \xrightarrow{\;1\;} & S^{n-1} \end{array}$$

Since π is a locally trivial bundle, it has the homotopy lifting property. Thus a lifting f exists iff f exists up to homotopy. At this point we have converted the problem into a homotopy theory problem. The next step is due to James [35].

Proposition 29.25 (*James*) $r(n) \le r(kn)$.

Proof We construct a map

$$V_k(R^m) * V_k(R^n) \xrightarrow{h} V_k(R^{m+n})$$

by $h(u_1, \ldots, u_k, t, v_1, \ldots, v_k) = (w_1, \ldots, w_k)$ where $w_i = (u_i \cos(\pi t/2), v_i \sin(\pi t/2))$. Then the diagram

$$\begin{array}{ccc} V_k(R^m) * V_k(R^n) & \xrightarrow{\;h\;} & V_k(R^{m+n}) \\ \Big\downarrow \pi * \pi & & \Big\downarrow \pi \\ S^{m-1} * S^{n-1} & \underset{\equiv}{\xrightarrow{\;\theta\;}} & S^{m+n-1} \end{array}$$

commutes where $\theta(u, t, v) = (u \cos(\pi t/2), v \sin(\pi t/2))$. Thus the maps $f\colon S^{m-1} \to V_k(R^m)$ and $g\colon S^{n-1} \to V_k(R^n)$ with $\pi f = 1$ and $\pi g = 1$ yield $e = h \circ (f \times g) \circ \theta^{-1}\colon S^{m+n-1} \to V_k(R^{m+n})$ and $\pi e = 1$. Consequently $r(m + n) \ge \min\,(r(m), r(n))$. ∎

Corollary 29.26 Suppose k is odd and $\rho(kn) = r(kn)$. Then $\rho(n) = r(n)$.

Proof $\rho(n) \le r(n) \le r(kn) = \rho(kn) = \rho(n)$. ∎

Our task will then be to show that for each n there is an odd integer k such that $r(kn) \le \rho(kn)$, i.e., there does not exist $f\colon S^{kn-1} \to V_{\rho(n)+1}(R^{kn})$ with $\pi f = 1$.

The next step is to replace $V_k(R^n)$ in our discussion by RP^{n-1}/RP^{n-k-1}. There is a natural map

$$\lambda\colon RP^{n-1}/RP^{n-k-1} \to V_k(R^n)$$

such that $\pi\lambda$ is the map $p_{RP^{n-2}}$. We will define λ as a composition

$$RP^{n-1}/RP^{n-k-1} \xrightarrow{\lambda_1} O(n)/O(n-k) \xrightarrow{\lambda_2} V_k(R^n);$$

λ_1 is induced by a map $\gamma\colon RP^{n-1} \to O(n)$ defined as follows. Let l be a line through the origin in R^n. Then $\gamma(l)$ is the reflection through the hyperplane perpendicular to l. λ_2 is a homeomorphism defined by $\lambda_2(A) = (Ae_{n-k+1}, \ldots, Ae_n)$ for $A \in O(n)$, and $e_1 \ldots, e_n$ an orthogonal basis for R^n.

Proposition 29.27 λ is a $2(n-k)$-equivalence.

Proof If $k = 1$, λ is a homeomorphism. We proceed by induction on k. Suppose $k > 1$ and consider the commutative diagram

$$\begin{array}{ccc} \pi_r(RP^{n-1}/RP^{n-k-1}, RP^{n-2}/RP^{n-k-1}) & & \\ \Big\downarrow \lambda_* & \searrow^{(\pi\lambda)_*} & \\ & & \pi_r(S^{n-1}, *) \\ \pi_r(V_k(R^n), V_{k-1}(R^{n-1})) & \xrightarrow[\pi_*]{\cong} & \end{array}$$

By 16.30, $(\pi\lambda)_*$ is a $(2n-k-2)$-isomorphism, hence λ_* is a $2(n-k)$-isomorphism (since $k > 1$). We complete the inductive step by using the 5-lemma and the exact sequences for the above pairs. ▮

Corollary 29.28 Suppose $\rho(n) \neq r(n)$. Then if $k \geq (2\rho(n)+1)/n$ and k is odd, the map

$$RP^{kn-1}/RP^{kn-\rho(n)-2} \xrightarrow{\pi\lambda} S^{kn-1}$$

has a right homotopy inverse α (i.e., $(\pi\lambda) \circ \alpha \sim 1$).

Proof If there exist $\rho(n)$ linearly independent vector fields on S^{n-1}, there exists $\rho(n) = \rho(kn)$ vector fields on S^{kn-1} for k odd by 29.26. Thus a map $f\colon S^{kn-1} \to V_{\rho(n)+1}(R^{kn})$ exists with $\pi f = 1$. By 29.27 $f \sim \lambda\alpha$ for some map $\alpha\colon S^{kn-1} \to RP^{kn-1}/RP^{kn-\rho(n)-2}$ and $\pi\lambda\alpha \sim \pi f = 1$. ▮

It is in this form that a contradiction is proven. Such a map α would induce a homomorphism

$$K_*(S^{kn-1}) \xrightarrow{\alpha_*} K_*(RP^{kn-1}, RP^{kn-1-\rho(n)-2})$$

compatible with the operations ψ^k. One then needs to calculate these groups, $(\pi\lambda)_*$, and the operations ψ^k to show that such a homomorphism cannot exist. Equivalently, one can consider the dual situation by imbedding all the spaces involved in a large sphere and applying 26.21. This has the advantage that the K-cohomology groups are easier to calculate because of the cup product structure. The details are found in [2; 34].

Exercises

1. Show that the bundle $k^{n+1} - \{0\} \to kP^n$ from Exercise 12, Section 11 is a k-line bundle.

2. Let $A \subset B$ and let $\xi = \{E, \pi, B\}$ be a vector bundle. Define $\xi|_A = \{\pi^{-1}(A), \pi|_{\pi^{-1}(A)}, A\}$. Show that $\xi|_A$ is a vector bundle equivalent to $i^*(\xi)$ where $i : A \to B$ is the inclusion.

3. Show that $G_1(k^{n+1}) \equiv kP^n$. Conclude that $G_1(k^\infty) \equiv kP^\infty$. Let i_n: $kP^n \to kP^\infty$ be the inclusion. Show that $i_n{}^*(\gamma^1(k^\infty))$ is the line bundle of Exercise 12, Section 11. (30.12)

4. Using γ: $(RP^n, RP^{n-1}) \to (O(n+1), O(n))$ and 13.11 prove that there is a commutative diagram

$$\begin{array}{ccc} \pi_{r+1}(S^n) & \xrightarrow{\partial} & \pi_r(O(n)) \\ \uparrow E & & \downarrow \pi_* \\ \pi_r(S^{n-1}) & \xrightarrow{\alpha_*} & \pi_r(S^{n-1}) \end{array}$$

where

$$\deg \alpha = \begin{cases} 2 & \text{if } n \text{ is even} \\ 0 & \text{if } n \text{ is odd.} \end{cases}$$

(Exercise 6)

5. Use the fiberings

$$O(n-1) \to O(n) \to S^{n-1}, \qquad U(n-1) \to U(n) \to S^{2n-1},$$
$$Sp(n-1) \to Sp(n) \to S^{4n-1}$$

to prove that

$\pi_r(O(n)) \to \pi_r(O(n+1))$ is an $(n-1)$-isomorphism.
$\pi_r(U(n)) \to \pi_r(U(n+1))$ is a $2n$-isomorphism.
$\pi_r(Sp(n)) \to \pi_r(Sp(n+1))$ is a $(4n+2)$-isomorphism.

Use the homeomorphisms $U(1) \equiv S^1$ and $Sp(1) \equiv S^3$ and the above to make the calculations

$$\begin{array}{lll} \pi_0(U(n)) = 0. & \pi_3(U(n)) = Z \text{ if } n > 1. & \pi_i(Sp(n)) = 0, \quad i < 3. \\ \pi_1(U(n)) = Z. & \pi_3(U(1)) = 0. & \pi_3(Sp(n)) = Z. \\ \pi_2(U(n)) = 0. & \pi_4(U(2)) = Z_2. & \pi_4(Sp(n)) = Z_2. \end{array}$$

6. By Exercises 4, 5, and 29.27, and the homeomorphism $O(2) \equiv S^1 \amalg S^1$ to prove that

$$\pi_1(O(n)) = Z_2, \quad n > 2. \qquad \pi_3(O(4)) = Z \oplus Z.$$
$$\pi_2(O(n)) = 0. \qquad \pi_3(O(5)) = Z \quad \text{or} \quad Z \oplus Z_2.$$
$$\pi_3(O(3)) = Z.$$

7. Using Exercises 5, 6, and 29.12 to prove

$$\pi_{8n}(O) = Z_2. \qquad \pi_{8n+4}(O) = 0.$$
$$\pi_{8n+1}(O) = Z_2. \qquad \pi_{8n+5}(O) = 0.$$
$$\pi_{8n+2}(O) = 0. \qquad \pi_{8n+6}(O) = 0.$$
$$\pi_{8n+3}(O) = Z \quad \text{or} \quad Z \oplus Z_2. \qquad \pi_{8n+7}(O) = Z. \qquad (29.17)$$

8. Show that there are compatible locally trivial bundles

$$O(k) \to V_k(R^n) \to G_k(R^n), \qquad U(k) \to V_k(C^n) \to G_k(C^n),$$
$$Sp(k) \to V_k(H^n) \to G_k(H^n)$$

for $k \le n \le \infty$. Using the map λ and its complex and quaternionic analogues show that $V_k(R^\infty)$, $V_k(C^\infty)$, and $V_k(H^\infty)$ are contractible. Conclude that there are maps $O(k) \to \Omega BO(k)$, $U(k) \to \Omega BU(k)$, and $Sp(k) \to \Omega BSp(k)$ inducing isomorphisms in homotopy (and hence homotopy equivalences).

9. Find a generalization of 29.21 to determine the multiplicative structure in $E^*(M)$ from the homomorphisms $(a_1)^*$ and $(a_2)^*$ for any ring spectrum E and any map $f\colon X \times Y \to Z$.

10. Calculate $K(CP^n)$ as a ring. (Hint: Use the method employed in 26.35.)

30

Cobordism

This section is intended as an introduction to cobordism theory. There are two aspects to this. The first is the reduction of the geometric problem to one in homotopy theory. We give a brief sketch of this. It involves techniques from differential topology (see [43, 49, 53]) which have little to do with this work. We give a more detailed account of the solution to the homotopy theory problem.

Cobordism was first described by Poincaré [57]. His notion of homology is essentially the same as the modern notion of cobordism. In fact the solution of the unoriented cobordism classification problem leads to a spectrum and the corresponding homology theory has a geometric description very similar to classical singular homology.

For simplicity we study only the unoriented cobordism theory. At the end of the section we will give some indications of other cobordism theories.

At this point we will assume that the reader is familiar with some of the elementary aspects of differential topology. We consider only compact C^∞ manifolds. Two such manifolds without boundary are called cobordant if there is a third manifold whose boundary is their disjoint union. This is an equivalence relation. Write $\mathfrak{N}_n$ for the set of equivalence classes of compact C^∞ n-manifolds.

$\{\mathfrak{N}_n\}$ is in fact a graded ring. The sum is induced by the disjoint union of manifolds and the product by the cartesian product. 0 is represented by the empty manifold and 1 by the one-point manifold. Clearly every element of $\mathfrak{N}_*$ has order 2 since $\partial(M \times I) = (M \cup M) \cup \varnothing$.

Suppose now that M and N are C^∞ manifolds and $f\colon M \to N$ is a C^∞ imbedding. Then the tangent map $df\colon \tau(M) \to \tau(N)$ is a monomorphism of bundles. Hence $\tau(M)$ is a subbundle of $f^*(\tau(N))$. We define the normal bundle ν of the imbedding to be the quotient bundle. Then there is an isomorphism

$$f^*(\tau(N)) \cong \tau(M) \oplus \nu.$$

ν can be visualized as the bundle of vectors in $\tau(N)$ which are orthogonal to M. Thus, for example, the normal bundle of the usual imbedding $S^n \subset R^{n+1}$ is a one-dimensional trivial bundle. As one might expect from this example, we have:

Theorem 30.1 (*Tubular Neighborhood Theorem*) Let $M^n \subset R^{n+k}$ be an imbedding. Then there is a neighborhood of M^n in $E(\nu)$ which is mapped diffeomorphically onto a neighborhood of M^n in R^{n+k}.

This is true in more generality. We may replace R^{n+k} by any manifold N^{n+k}. For a proof, see [49, 3.6].

Let ξ be a vector bundle over a paracompact space X. Choose a Riemannian metric in ξ and let $D(\xi)$ be the subspace of $E(\xi)$ consisting of all vectors v with $\|v\| \leq 1$. Let $S(\xi)$ be the subspace consisting of all vectors v with $\|v\| = 1$. $D(\xi)$ and $S(\xi)$ are called the associated disk and sphere bundles with fibers D^n and S^{n-1} respectively. Furthermore $D(\xi)$ and $S(\xi)$ do not depend on the choice of a metric—up to bundle equivalence.

Definition 30.2 $T(\xi) = D(\xi)/S(\xi)$ is called the Thom space of ξ.

Proposition 30.3 If X is compact, $T(\xi) \equiv E(\xi)^\infty$.

Proof Clearly $D(\xi) - S(\xi) \equiv E(\xi)$. Since $D(\xi)$ is compact and $E(\xi)$ is regular, $E(\xi)^\infty \equiv (D(\xi) - S(\xi))^\infty \equiv D(\xi)/S(\xi)$ by 1.6. ▮

Proposition 30.4 (a) If $A \subset X$ is a closed subspace, $T(\xi|_A) \subset T(\xi)$ as a closed subspace.

(b) If $X = \bigcup_\alpha X_\alpha$ has the weak topology and each X_α is closed, $T(\xi) = \bigcup_\alpha T(\xi|_{X_\alpha})$ with the weak topology.

(c) Given bundles ξ over X and η over Y with $X, Y \in \mathcal{CG}$, $T(\xi \times \eta) \equiv T(\xi) \wedge T(\eta)$.

Proof (a) $D(\xi|_A)$ is a closed subset of $D(\xi)$. Hence the 1–1 continuous map $D(\xi|_A)/S(\xi|_A) \to D(\xi)/S(\xi)$ is closed.

(b) There is a well-defined 1–1 continuous map

$$\bigcup_\alpha T(\xi|_{X_\alpha}) \to T(\xi)$$

which is onto and closed.

(c) Since $X, Y \in \mathcal{CG}$, it is sufficient to prove this for X and Y compact by (b) and 15.14. By Exercise 6, Section 26 and 30.3 we have

$$\begin{aligned} T(\xi \times \eta) &\equiv E(\xi \times \eta)^\infty \\ &\equiv (E(\xi) \times E(\eta))^\infty \equiv E(\xi)^\infty \wedge E(\eta)^\infty \equiv T(\xi) \wedge T(\eta). \quad \blacksquare \end{aligned}$$

Let γ^k be the universal k-plane bundle over $BO(k)$. We will write $MO(k) = T(\gamma^k)$. The inclusion $\iota\colon BO(k) \to BO(k+1)$ defines an isomorphism

$$\iota^*(\gamma^{k+1}) \cong \gamma^k \oplus 1 \equiv \gamma^k \times R^1.$$

Since ι is closed, there is an inclusion

$$MO(k) \wedge S^1 \equiv T(\gamma^k \times R^1) \equiv T(\iota^*(\gamma^{k+1})) \to T(\gamma^{k+1}) \equiv MO(k+1).$$

We designate this inclusion by $mo(k)$. Thus $MO = \{MO(k), mo(k)\}$ is a spectrum.

Next we indicate how a cobordism class of compact n-manifolds determines an element of $\pi_n(MO)$.

We need:

Theorem 30.5 (*Whitney*) Let M^n be a compact C^∞ manifold. Then there is a differentiable imbedding $M^n \subset R^{2n+1}$ as a closed subset.

For a proof, see [49, 1.32].

Let M^n be a compact C^∞ manifold; choose an imbedding $M^n \subset R^{n+k}$ for some k and a tubular neighborhood U of M^n in R^{n+k}. U is diffeomorphic with a neighborhood V of M^n in $E(\nu)$. Choose a Riemannian metric in ν. Since M is compact, we can find a smaller neighborhood U' of M^n in R^{n+k} homeomorphic to the set of all vectors with length $< \varepsilon$ in $E(\nu)$. This is diffeomorphic with $E(\nu)$ and hence there is a map

$$S^{n+k} \equiv (R^{n+k})^\infty \xrightarrow{r} V^\infty \equiv E(\nu)^\infty \equiv T(\nu)$$

since V is open in R^{n+k} (see Exercise 1, Section 26). This does not depend on the choice of ε up to homotopy.

Now choose a map $f\colon M^n \to BO(k)$ which classifies ν. Since M^n is compact, f is closed and hence

$$f_E\colon E(\nu) \to E(\gamma^k)$$

is closed. This consequently induces a map

$$T(\nu) \xrightarrow{T(f_E)} MO(k).$$

The composition $T(f_E)r\colon (S^{n+k}, *) \to (MO(k), *)$ represents an element in $\pi_{n+k}(MO(k), *)$ and thus an element in $\pi_n(MO)$. Conceivably this depends on the choice of an imbedding. However, if k is large, one can show that any two imbeddings are isotopic (i.e., homotopic through imbeddings). This is enough to guarantee that the homotopy class does not depend on the imbedding for k large. By imbedding the cobordism, one can see that it depends only on the cobordism class of M. To see that the element in $\pi_n(MO)$ does not depend on k, consider an inclusion $M^n \subset R^{n+k} \subset R^{n+k+1}$. Let ν' be the normal bundle of the composition. Then $\nu' \cong \nu \oplus 1 \equiv \nu \times R^1$; let

$$r' \colon S^{n+k+1} \to T(\nu') \equiv T(\nu) \wedge S^1.$$

Then $r' \equiv S(r)$. It is easy to verify that the diagram

$$\begin{array}{ccc} T(\nu) \wedge S^1 & \xrightarrow{T(f_E) \wedge 1} & MO(k) \wedge S^1 \\ \equiv & & \\ T(\nu \times R^1) & & \Big\downarrow mo(k) \\ \equiv & & \\ T(\nu') & \xrightarrow{T(f_{E'})} & MO(k+1) \end{array}$$

commutes. Thus $T(f_{E'}) \circ r' = mo(k) \circ S(T(f_E) \circ r)$. Consequently the element $\theta(M) \in \pi_n(MO)$ is well defined.

Theorem 30.6 (*Thom*) $\theta \colon \mathfrak{N}_* \to \pi_*(MO)$ is an isomorphism of graded rings.

For details, see [19, Chapter I; 69, Chapter II; 49, Chapter III; 42, Chapter 5].

We will now show how to calculate $\pi_*(MO)$. We begin by calculating $H^*(BO; Z_2)$ and then $H^*(MO; Z_2)$.

Theorem 30.7 $H^*(BO(n); Z_2) \cong Z_2[w_1, \ldots, w_n]$. The inclusion $BO(n-1) \to BO(n)$ induces a homomorphism ι^* with $\iota^*(w_i) = w_i$ for $i < n$ and $\iota^*(w_n) = 0$. $\dim w_i = i$.

Corollary 30.8 $H^*(BO; Z_2) \cong Z_2[w_1, \ldots, w_n, \ldots]$.

Proof of 30.7 Let $BSO(n)$ be the simply connected covering space of $BO(n)$, and $\theta \colon BO(n) \times BO(1) \to BO(n)$ be the map classifying the tensor product $\pi_1^*(\gamma^n) \otimes \pi_2^*(\gamma^1)$. Then the map $BSO(n) \times RP^\infty \to BO(n) \times BO(1) \to BO(n)$ induces isomorphisms in homotopy groups. Hence $BO(n) \simeq BSO(n) \times RP^\infty$. We show that $H^*(BSO(n); Z_2) \simeq Z_2[w_2, \ldots, w_n]$.

In Section 29 we considered maps $\gamma^n \colon RP^{n-1} \to SO(n)$ such that the diagram

$$\begin{array}{ccc} RP^{n-1} & \xrightarrow{\gamma^n} & SO(n) \\ & \searrow^{p_{RP^{n-2}}} & \Big\downarrow \pi \\ & & S^{n-1} \end{array}$$

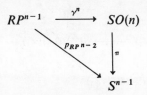

commutes. It follows that $\pi^*;\ H^{n-1}(S^{n-1}; Z_2) \to H^{n-1}(SO(n); Z_2)$ is 1–1. By Exercise 6, Section 21, the map

$$H^r(SO(n); Z_2) \xrightarrow{\iota^*} H^r(SO(n-1); Z_2)$$

is an isomorphism for $r \leq n-2$. We now prove by induction that $H^*(SO(n); Z_2)$ has a simple system of generators $x_1, \ldots, x_{n-1}$ with $\dim x_k = k$, $\iota^*(x_k) = x_k$ if $k < n-1$, and $\iota^*(x_{n-1}) = 0$. This is clear if $n = 2$ since $SO(2) \equiv S^1$. We apply the Leray–Hirsch theorem (Exercise 27, Section 26) to the locally trivial bundle (see Exercise 13, Section 11)

$$SO(n-1) \to SO(n) \to S^{n-1}$$

with $R = Z_2$. Suppose $x_1, \ldots, x_{n-2}$ is a simple system of generators for $H^*(SO(n-1); Z_2)$. Since ι^* is an isomorphism for $r \leq n-2$, there are classes $x_i \in H^*(SO(n); Z_2)$ with $\iota^*(x_i) = x_i$ for $i \leq n-2$. It follows that the products $x_{i_1} \cdots x_{i_s}$ which form a Z_2 basis for $H^*(SO(n-1); Z_2)$ are the image of the corresponding products in $H^*(SO(n); Z_2)$. By the Leray–Hirsch theorem $x_1, \ldots, x_{n-2}$ and $x_{n-1} = \pi^*(e)$ $(e \neq 0)$ form a simple system of generators for $H^*(SO(n); Z_2)$, completing the induction.

We wish to apply the Borel theorem (28.4), so we must find elements $w_i \in H^*(BSO(n); Z_2)$ with $\Sigma w_i = x_{i-1}$ for $2 \leq i \leq n$. (By Exercise 8, Section 29, $SO(n) \simeq \Omega BSO(n)$.) From the commutative diagram

$$\begin{array}{ccccc}
H_k(BO(k), BO(k-1)) & \xleftarrow{\ \Sigma\ } & H_{k-1}(O(k), O(k-1)) & \xrightarrow[\simeq]{\pi_*} & H_{k-1}(S^{k-1}) \\
\cong \uparrow h & & \uparrow h & & \uparrow \cong \\
\pi_k(BO(k), BO(k-1)) & \xleftarrow{\ \cong\ } & \pi_{k-1}(O(k), O(k-1)) & \xrightarrow[\simeq]{\pi_*} & \pi_{k-1}(S^{k-1})
\end{array}$$

we see that Σ is an isomorphism and hence by the universal coefficient theorem (25.16)

$$H^r(BO(k), BO(k-1); Z_2) \xrightarrow{\Sigma} H^{r-1}(O(k), O(k-1); Z_2)$$

is an isomorphism for $r \leq k$. Next consider the ladder

$$\begin{array}{ccccccccc}
\cdots \longrightarrow & H^r(BO(k), BO(k-1); Z_2) & \longrightarrow & H^r(BO(k), Z_2) & \xrightarrow{\iota^*} & H^r(BO(k-1); Z_2) & \longrightarrow \cdots \\
& \downarrow \Sigma & & \downarrow \Sigma & & \downarrow \Sigma & \\
\cdots \longrightarrow & H^{r-1}(O(k), O(k-1); Z_2) & \longrightarrow & H^{r-1}(O(k); Z_2) & \longrightarrow & H^{r-1}(O(k-1); Z_2) & \longrightarrow \cdots \\
& \nwarrow \pi^* & & \nearrow \pi^* & & & \\
& & H^{r-1}(S^{k-1}; Z_2) & & & &
\end{array}$$

ι^* is an isomorphism if $r \leq k-1$ and if $r = k$, ker ι^* consists of an element w with $\Sigma w_k = x_{k-1}$. Consequently it is a simple matter to construct, by induction, classes w_i with the required properties. ▮▮

Theorem 30.9 $\tilde{H}^*(MO; Z_2) \cong H^*(BO; Z_2)$.

Proof We will construct a compatible sequence of isomorphisms

$$\theta_n\colon H^r(BO(n); Z_2) \to \tilde{H}^{n+r}(MO(n); Z_2)$$

The proof will then follow by taking limits. $BO(n) = \bigcup_{k=1}^{\infty} G_n(R^k)$, and there are bundle maps

$$\begin{array}{ccc} E(\gamma^n(R^k)) & \longrightarrow & E(\gamma^n) \\ \downarrow & & \downarrow \\ G_n(R^k) & \longrightarrow & BO(n) \end{array}$$

Since $G_n(R^k)$ is compact, we have $MO(n) = \bigcup_{k=1}^{\infty} E(\gamma^n(R^k))^\infty$. The isomorphism θ_n will in turn be induced by a compatible sequence of isomorphisms

$$\theta_n^{\,k}\colon H^r(G_n(R^k); Z_2) \to \tilde{H}^{n+r}(E(\gamma^n(R^k))^\infty; Z_2).$$

Now $G_n(R^k)$ is a compact $n(n-k)$-manifold. Hence $E(\gamma^n(R^k))$ is an $(n + n(n-k))$-manifold. Since $\pi\colon E(\gamma^n(R^k)) \to G_n(R^k)$ is a homotopy equivalence, 26.24 and 26.28 imply

$$\begin{aligned} H^r(G_n(R^k); Z_2) &\cong H_{n(n-k)-r}(G_n(R^k); Z_2) \\ &\cong H_{n(n-k)-r}(E(\gamma^n(R^k)); Z_2) \\ &\cong \tilde{H}^{n+r}(E(\gamma^n(R^k))^\infty; Z_2). \end{aligned}$$ ▮

The Whitney sum $\pi_1^*(\gamma^n) \oplus \pi_2^*(\gamma^m) \equiv \gamma^n \times \gamma^m$ is classified by a bundle map

$$\gamma^n \times \gamma^m \to \gamma^{n+m}.$$

This induces structure maps

$$MO(n) \wedge MO(m) = T(\gamma^n) \wedge T(\gamma^m) \equiv T(\gamma^n \times \gamma^m) \to T(\gamma^{n+m}) = MO(m+n)$$

The inclusion of a point in $BO(n)$ induces a bundle map

$$R^n \to E(\gamma^n),$$

and hence induces a map $S^n \equiv T(R^n) \to MO(n)$ for each $n \geq 1$. These maps make MO into a ring spectrum.

Lemma 30.10 If E is a properly convergent ring spectrum and k is a field $\tilde{H}^*(E; k)$ is a coalgebra.

Proof The maps

$$\tilde{H}^{n+m+r}(E_{n+m}; k) \to \tilde{H}^{n+m+r}(E_n \wedge E_m; k) \cong \bigoplus_{s+t=r} \tilde{H}^{n+s}(E_n; k) \oplus \tilde{H}^{m+t}(E_m; k)$$

are independent of m and n if $m, n > r$. This yields a diagonal map $\psi: \tilde{H}^*(E; k) \to \tilde{H}^*(E; k) \otimes \tilde{H}^*(E; k)$. Similarly one constructs a counit $1 \in \tilde{H}^0(E; k) \cong \tilde{H}^s(E_s: k)$ and the coassociativity and cocommutativity properties follow from the corresponding properties for E. ∎

We will call a graded coalgebra C connected if $C^n = 0$ for $n < 0$ and C^0 is freely generated over k by the counit.

Lemma 30.11 (*Milnor–Moore*) Let A be a connected Hopf algebra over a field k. Let M be a connected coalgebra over k with counit $1 \in M^0$ and a left module over A such that the diagonal map $\psi: M \to M \otimes M$ is a map of A modules. Suppose the map $v: A \to M$ given by $v(a) = a \cdot 1$ is a monomorphism. Then M is a free left A module.

For a proof, see [54; 69].

Theorem 30.12 $\tilde{H}^*(MO; Z_2)$ is free over $\mathcal{A}(2)$.

Proof All that we need to show is that the nontrivial map $MO \to HZ_2$ induces a monomorphism in cohomology. We will show that the maps

$$MO(n) \to K(Z_2, n)$$

induce monomorphisms in dimensions less than $2n$. This is enough by 27.5 since both MO and HZ_2 are properly convergent.

The Whitney sum map

$$E(\gamma^1) \times \cdots \times E(\gamma^1) \to E(\gamma^n)$$

induces an isomorphism in $H^0(\ ; Z_2)$. By the proof of 30.9, the map on Thom spaces

$$\alpha: MO(1) \wedge \cdots \wedge MO(1) \to MO(n)$$

induces an isomorphism in $\tilde{H}^n(\ ; Z_2)$. Now $MO(1) = D(\gamma^1)/S(\gamma^1)$. By Exercise 3, Section 29, $S(\gamma^1) \equiv S^\infty$ and hence is contractible. Thus $MO(1) \simeq D(\gamma^1) \simeq BO(1) \equiv RP^\infty$. Thus the composition

$$RP^\infty \wedge \cdots \wedge RP^\infty \to MO(n) \to K(Z_2, n)$$

induces an isomorphism in $\tilde{H}^n(\ ; Z_2)$ and is therefore the map considered in Exercise 9, Section 28. The conclusion follows. ∎

Proposition 30.13 Let E be a properly convergent spectrum. Suppose that every element in $\pi_*(E)$ has order p and $\tilde{H}^*(E; Z_p)$ is a free module over $\mathcal{A}(p)$

generated by classes $\{x_\alpha\}$ and that there are only finitely many α in each dimension. Then $\pi_*(E)$ is free over Z_p generated by classes $\{u_\alpha\}$ of the same dimension and $\langle x_\alpha, \rho h(u_\alpha)\rangle = 1$ where $h\colon \pi_*(E) \to \tilde{H}_*(E)$ is the Hurewicz homomorphism, $\rho\colon \tilde{H}_*(E) \to \tilde{H}_*(E; Z_p)$ is the coefficient homomorphism, and $\langle\ ,\ \rangle$ is the Kronecker product. In particular, h is a monomorphism.

Proof Let

$$X_n = \prod_{\dim \alpha < n} K(Z_p, n + \dim x_\alpha).$$

$\tilde{H}^r(E; Z_p) \cong \tilde{H}^{r+n}(E_n; Z_p)$ for $n > r$. If $\dim x_\alpha < n$ choose $\varphi_\alpha\colon E_n \to K(Z_p, n + \dim x_\alpha)$ representing x_α. Let $\varphi\colon E_n \to X_n$ have components $\{\varphi_\alpha\}$. Then $\varphi^*\colon \tilde{H}^r(X_n; Z_p) \to \tilde{H}^r(E_n; Z_p)$ is an isomorphism if $r < 2n$. The statement about $\pi_*(E)$ will be proved in dimensions less than m if we show that $\varphi_*\colon \pi_{n+r}(E_n) \to \pi_{n+r}(X_n)$ is an isomorphism for $r \le m < n$. By 25.20 $\varphi_*\colon \tilde{H}_r(E_n; Z_p) \to \tilde{H}_r(X_n; Z_p)$ is a $(2n-1)$-equivalence. Thus by the generalized Hurewicz theorem (Exercise 11, Section 22), $\varphi_*\colon \pi_r{}^S(E_n; Z_p) \to \pi_r{}^S(X_n; Z_p)$ is a $(2n-1)$-equivalence. We now consider the universal coefficient exact sequences (Exercise 8, Section 25). Note that if every element of π has order p, $\pi \otimes Z_p \cong \pi \cong \mathrm{Tor}(\pi, Z)$. Thus we have a commutative diagram

$$\begin{array}{ccccccccc}
0 & \longrightarrow & \pi_r{}^S(E_n) & \longrightarrow & \pi_r{}^S(E_n; Z_p) & \longrightarrow & \pi_{r-1}^S(E_n) & \longrightarrow & 0 \\
 & & \downarrow{\scriptstyle \varphi_*} & & \downarrow{\scriptstyle \varphi_*} & & \downarrow{\scriptstyle \varphi_*} & & \\
0 & \longrightarrow & \pi_r{}^S(X_n) & \longrightarrow & \pi_r{}^S(X_n; Z_p) & \longrightarrow & \pi_{r-1}^S(X_n) & \longrightarrow & 0
\end{array}$$

which is exact for $r < 2n - 1$. Since φ_* in the middle is an isomorphism, φ_* on the left is a monomorphism and φ_* on the right is an epimorphism. Thus $\varphi_*\colon \pi_r{}^S(E_n) \to \pi_r{}^S(X_n)$ is an isomorphism for $r < 2n - 1$. We have thus calculated $\pi_*(E)$ in dimensions less than $n - 2$; it is free on generators $u_\alpha\colon S^{\dim x_\alpha + n} \to E_n$ with $\varphi u_\alpha \not\sim *$. Let

$$a_\alpha = h(\varphi_*(u_\alpha)) \in H_{n+\dim x_\alpha}(K(Z_p, n + \dim x_\alpha)).$$

Let $\langle i_\alpha, \rho a_\alpha\rangle = 1$. Then $\varphi^*(\imath_\alpha) = x_\alpha$ and

$$\begin{aligned}
\langle x_\alpha, \rho h u_\alpha\rangle &= \langle \varphi^*(\imath_\alpha), \rho h u\rangle = \langle \imath_\alpha, \varphi_*(\rho h(u_\alpha))\rangle \\
&= \langle \imath_\alpha, \rho h(\varphi_*(u_\alpha))\rangle = \langle \imath_\alpha, \rho a_\alpha\rangle = 1. \quad \blacksquare
\end{aligned}$$

Corollary 30.14 $\mathfrak{N}_* = \pi_*(MO)$ is a Z_2 vector space with one generator for each $\mathcal{A}(2)$ free generator in $\tilde{H}^*(MO; Z_2)$.

Proof Clearly MO is properly convergent for

$$\tilde{H}^{r+n+1}(MO(n+1)/SMO(n); Z_2) \cong H^r(BO(n+1), BO(n); Z_2) = 0$$

if $r < n + 1$. By 30.6 every element in $\pi_*(MO)$ has order 2. Thus by 30.12 and 30.13 we are done. ∎

As yet we have not determined the multiplicative structure in $\pi_*(MO)$. To do this one must calculate the multiplicative structure in $\tilde{H}_*(MO; Z_2)$.

Let $U^n \in \tilde{H}^{2n}(MO(n); Z_2)$ be a generator. Then we have a commutative diagram

$$\begin{array}{ccc} SMO(n-1) & \xrightarrow{SU^{n-1}} & SK(Z_2, n-1) \\ \downarrow{\scriptstyle mo(n-1)} & & \downarrow{\scriptstyle (hZ_2)_{n-1}} \\ MO(n) & \xrightarrow{U^n} & K(Z_2, n). \end{array}$$

Hence $mo(n-1)^*(\mathrm{Sq}^n U^n) = 0$. On the other hand, consider

$$\alpha: RP^\infty \wedge \cdots \wedge RP^\infty \to MO(n).$$

Then $\alpha^*(\mathrm{Sq}^n U^n) = (\iota \barwedge \cdots \barwedge \iota)^2 \neq 0$. Since the kernel of the homomorphism $H^n(BO(n); Z_2) \to H^n(BO(n-1); Z_2)$ is generated by w_n, we have $\theta_n(w_n) = \mathrm{Sq}^n U^n$. It follows that $\theta(w_n) = \mathrm{Sq}^n U$ where

$$\theta: H^*(BO; Z_2) \to \tilde{H}^*(MO; Z_2)$$

is the isomorphism from 30.9 and $U = \theta(1)$.

Fix $m, n > k$ and let $\mu: MO(m) \wedge MO(n) \to MO(m+n)$ be the map giving the ring spectra structure. Let $u_k = \mathrm{Sq}^k U^{m+n}$. Then

$$\begin{aligned} \mu^*(u_k) &= \mathrm{Sq}^k \mu^*(U^{m+n}) = \mathrm{Sq}^k(U^m \barwedge U^n) \\ &= \prod_{i+j=k} \mathrm{Sq}^i U^m \barwedge \mathrm{Sq}^j U^n = \sum_{i+j=k} u_i \barwedge u_j. \end{aligned}$$

This determines the coalgebra structure in $\tilde{H}^*(MO; Z_2)$.

Now the maps $\gamma^m \times \gamma^n \to \gamma^{m+n}$ classifying the Whitney sum determine compatible maps $BO(m) \times BO(n) \to BO(m+n)$. Thus BO is an H space. By the definition of θ, it preserves the coalgebra structure. Hence the diagonal in $H^*(BO; Z_2)$ is given by $\psi(w_k) = \sum_{i+j=k} w_i \otimes w_j$.

The multiplication in BO makes $H_*(BO; Z_2)$ into an algebra, whose multiplication is the dual to ψ.

Proposition 30.15 $H_*(BO; Z_2) \cong Z_2[y_1, y_2, \ldots]$.

Proof We first study the map $BO(n) \times BO(1) \xrightarrow{\psi_{n,1}} BO(n+1)$. By the above analysis, $(\psi_{n,1})^*(w_j) = w_j \otimes 1 + w_{j-1} \otimes w_1$ for $j \leq n$ and $(\psi_{n,1})^*(w_{n+1}) = w_n \otimes w_1$. Consider the composition

$$\overbrace{BO(1) \times \cdots \times BO(1)}^{n} \to \cdots \to BO(n-1) \times BO(1) \to BO(n)$$

By induction one sees that $\psi^*(w_j) = j$th elementary symmetric function in the polynomial generators of $H^*(BO(1) \times \cdots \times BO(1); Z_2)$. Thus Im $\bar{\psi}^*$ is the subring of symmetric polynomials and in particular, $\bar{\psi}^*$ is a monomorphism. It follows that $\bar{\psi}_*$ is an epimorphism. Since every element of $H_*(BO(1) \times \cdots \times BO(1); Z_2)$ is a tensor product of elements in $H_*(BO(1); Z_2)$ it follows that the images of these elements generate $H_*(BO; Z_2)$ as a ring. Let $y_n \in H_n(BO; Z_2)$ be the element in dimension n which is in the image of $H_n(BO(1); Z_2)$. Then the elements y_i generate $H_*(BO; Z_2)$. Since the diagonal in $H^*(BO; Z_2)$ is cocommutative, $H_*(BO; Z_2)$ is commutative. Since the rank of $H_*(BO; Z_2)$ is the same as the rank of $Z_2[y_1, \ldots, y_k, \ldots]$ in each dimension, the conclusion follows. ∎

Now one can easily check that the isomorphism $\tilde{H}_r(MO; Z_2) \cong H_r(BO; Z_2)$ is multiplicative by using the diagrams

$$\begin{array}{ccc} D(\gamma^n) \times D(\gamma^m) & \longrightarrow & D(\gamma^{n+m}) \\ \downarrow & & \downarrow \\ BO(n) \times BO(m) & \longrightarrow & BO(n+m) \end{array}$$

Consequently,

Corollary 30.16 $\tilde{H}_*(MO; Z_2) \cong Z_2[y_1, \ldots, y_n, \ldots]$. ∎

Now $H^*(MO; Z_2) \cong \mathcal{A}(2) \otimes C$ for some vector space C. Hence $H_*(MO; Z_2) \cong \mathcal{A}_*(2) \otimes C_*$. Furthermore, $C_* = \text{im } h(\pi_*(MO))$. h is also multiplicative, so $C_* \cong \pi_*(MO)$ as algebras. Since $\mathcal{A}_*(2) \cong Z_2[\xi_1, \xi_2, \ldots]$ with one polynomial generator ξ_i for each $i = 2^n - 1$ by 28.21, we have

Corollary 30.17 (*Thom*) $\mathfrak{N}_* \cong \pi_*(MO) \cong Z_2[x_2, x_4, x_5, \ldots]$ with one generator x_i for each $i \neq 2^n - 1$. $h(x_i) \equiv y_i$ (modulo decomposable elements). ∎

Definition 30.18 If ξ is a vector bundle over X, we will define classes $w_i(\xi) \in H^i(X; Z_2)$ by $w_i(\xi) = f^*(w_i)$ where $f\colon X \to BO$ is a classifying map for ξ. $w_i(\xi)$ is called the ith Stiefel–Whitney class of ξ.

Proposition 30.19 The Stiefel–Whitney classes have the following (characteristic) properties:

(a) If ξ is an n-plane bundle, $w_i(\xi) = 0$ for $i > n$.
(b) If $f\colon W \to X$, $w_i(f^*(\xi)) = f^*(w_i(\xi))$.
(c) (Whitney Formula) $w_k(\xi \oplus \eta) = \Sigma_{i+j=k} w_i(\xi) w_j(\eta)$.
(d) Let H be the canonical line bundle over RP^∞. Then $w_1(H) \neq 0$.

Proof (a), (b), and (d) are obvious. (c) follows from the formula $\psi(w_k) = \Sigma_{i+j=k} w_i \otimes w_j$. ∎

Definition 30.20 For any compact connected manifold M^n and any homogeneous polynomial $p(w_1, \ldots, w_k)$ of degree n, define the normal Stiefel–Whitney number corresponding to p as

$$\langle p(w_1(v), \ldots, w_k(v)), [M]\rangle \in Z_2$$

where v is the stable normal bundle of M.

Since $w_i(v \oplus m) = \Sigma w_j(v) w_{i-j}(m) = w_i(v)$, the cohomology class $p(w_1(v), \ldots, w_k(v))$ depends only on the stable isomorphism class of v.

Proposition 30.21 If $p(w_1, \ldots, w_k) \in H^n(BO; Z_2)$ and M is a compact connected n-manifold, the Stiefel–Whitney number corresponding to p is given by

$$\langle p(w_1, \ldots, w_k), \theta^* h(\{M\})\rangle$$

where $h: \mathfrak{N}_* \to \tilde{H}_*(MO; Z_2)$ is the Hurewicz homomorphism and $\theta^*: \tilde{H}_*(MO; Z_2) \to H_*(BO; Z_2)$ is the dual to the Thom isomorphism (30.9). In particular, two manifolds are cobordant iff all their Stiefel–Whitney numbers are equal.

Proof Let $f: M \to BO(m)$ classify v. Then

$$\langle p(w_1(v), \ldots, w_k(v)), [M]\rangle = \langle p(w_1, \ldots, w_k), f_*([M])\rangle.$$

Now $h\{M\} = T(f)_* r_*(\iota)$ where $\iota \in H_{n+k}(S^{n+k}; Z_2)$ is nonzero, and $r: S^{n+k} \to T(v)$. Thus $\theta^* h\{M\} = f_* \varphi' r_*(\iota)$ where $\varphi': \tilde{H}_{n+k}(T(v); Z_2) \to H_n(M; Z_2)$ is the isomorphism in Exercise 1. Thus we need only show that $\varphi' r_*(\iota) = [M]$. By Exercise 1, it is only necessary to show that the inclusion $E(v) \subset E^{n+k}$ induces an isomorphism in $H_0(\ ; Z_2)$. This follows since M is connected. ∎

By calculating the normal Stiefel–Whitney classes for RP^n, one can prove

Proposition 30.22 $x_{2n} \equiv \{RP^{2n}\}$ (modulo indecomposable elements). That is $\{RP^{2n}\}$ may be taken as a ring generator of $\mathfrak{N}_*$ in dimension $2n$.

For details of the calculation, see [69]. Compare this result with Exercise 15, Section 26.

We give a geometric description of the groups $MO_*(X, A)$ due to Atiyah [7]. One should observe the similarity between this description and the classical definition of singular homology. (See Exercise 9, Section 21.) Simplices are replaced by manifolds.

For a fixed pair (X, A) define a singular manifold in (X, A) to be a mapping $f: (M, \partial M) \to (X, A)$ where M is a compact C^∞ manifold of dimension n.

Such a manifold will be said to bord when there is a compact C^∞ manifold N of dimension $n+1$ and a mapping $F: N \to X$ such that (i) M is a submanifold of ∂N, (ii) $F|_M = f$, and (iii) $F(\partial N - M) \subset A$. Two singular manifolds $(M, \partial M, f)$ and $(M', \partial M', f')$ will be called bordant if their disjoint union $(M \cup M', \partial M \cup \partial M', \partial f \cup \partial f')$ bords. This is an equivalence relation and we write $\mathfrak{N}_n(X, A)$ for the set of equivalence classes. Clearly $\mathfrak{N}_n(*, \phi) = \mathfrak{N}_n$.

Theorem 30.23 (*Atiyah*) If $(X, A) \in \mathfrak{CG}^2$, there is a natural isomorphism

$$\mathfrak{N}_n(X, A) \cong MO_n(X, A).$$

The proof of this will be based on

Lemma 30.24 $\mathfrak{N}_*(X, A)$ is a homology theory with type 1 excision on $\mathfrak{C}^2$.

This is proved in [19, 1, 5.1].

Proof of 30.23 We construct a natural transformation

$$\phi: \mathfrak{N}_n(X, A) \quad \to \quad MO_n(X, A)$$

and prove that it is an isomorphism. Let $f: (M, \partial M) \to (X, A)$ be a singular n-manifold, and $E(v)$ the total space of the normal bundle to $M - \partial M$. A mapping $\alpha: E(v) \to E(v) \times (M - \partial M)$ is defined by $\alpha(x) = (x, \pi(x))$, where $\pi: E(v) \to M - \partial M$ is the projection. This map is proper and hence induces

$$\alpha': T(v) \quad \to \quad T(v) \wedge M^+/\partial M^+,$$

since $M^+/\partial M^+ \equiv (M - \partial M)^\infty$. Thus the composition

$$S^{n+k} \xrightarrow{r} T(v) \xrightarrow{\alpha'} T(v) \wedge M^+/\partial M^+ \xrightarrow{T(f_E)\wedge f} MO(k) \wedge X^+/A^+$$

defines an element of

$$\varinjlim \pi_{n+k}(MO(k) \wedge (X^+/A^+)) = \widetilde{MO}_n(X^+/A^+) = MO_n(X, A).$$

This is well defined by arguments similar to before, and gives the natural transformation φ. In particular, if X is a point, φ is an isomorphism by 30.6. By induction (as in Exercise 10, Section 18) φ is an isomorphism for each CW pair (X, A). The theorem will follow from 21.7 if we show that $\mathfrak{N}_*(S(X)) \cong \mathfrak{N}_*(X)$ where $S(X)$ is the singular complex of X. Let (M, f) represent an element of $\mathfrak{N}_n(X)$. Since M is a compact C^∞ manifold, it is a CW complex [52, I, Part 3]. Hence there exists $g: M \to S(X)$ with $\pi g \sim f$. The homotopy $H: M \times I \to X$ gives a cobordism from (M, f) to $\pi_*(M, g)$. Hence π_* is onto. Suppose $\pi_*(M, f) = 0$ where $f: M \to S(X)$. Since every element of $\mathfrak{N}_*(S(X)$

can be represented by such a pair it is sufficient to show that (M, f) bounds. Suppose $\partial N = M$ and $h: N \to X$ is a map with $h|_M = \pi f$. By 16.17 there is a map $g: N \to S(X)$ with $g|_M = f$. Hence (M, f) bounds and we are done. ∎

The homology and cohomology theories derived from unoriented cobordism are not essentially different from ordinary Z_2 homology and cohomology because of the following.

Proposition 30.25 (a) If $(Y, B) \in \mathcal{CG}^2$,

$$MO_*(Y, B) \cong \mathfrak{N}_* \otimes H_*(Y, B; Z_2)$$

(b) If (Y, B) is a finite relative CW complex[28]

$$MO^*(Y, B) \cong \mathfrak{N}_* \otimes H^*(Y, B; Z_2)$$

Proof In the proof of 30.13, we constructed a spectrum $X = \{X_n, x_n\}$ with $X_n = \prod_{\dim \alpha < n} K(Z_2, n + \dim \alpha)$ and a map $\varphi: MO \to X$. φ induces natural transformations

$$\varphi: MO_*(Y, B) \to X_*(Y, B), \qquad \varphi: MO^*(Y, B) \to X^*(Y, B).$$

We now observe that $\mathfrak{N}_* \otimes H_*(Y, B; Z_2)$ and $\mathfrak{N}_* \otimes H^*(Y, B; Z_2)$ are homology and cohomology theories. The only point needing attention is the exactness axiom, but for Z_2 modules $A \otimes B \cong A \otimes_{Z_2} B$ and since all Z_2 modules are free $\mathfrak{N}_* \otimes_{Z_2}$ is an exact functor. We produce natural transformations

$$X_*(Y, B) \xrightarrow{\psi_1} \mathfrak{N}_* \otimes H_*(Y, B; Z_2)$$

$$X^*(Y, B) \xrightarrow{\psi_2} \mathfrak{N}_* \otimes H^*(Y, B; Z_2)$$

as follows.

Let $\psi_\alpha: X_n \to K(Z_2, n + \dim \alpha)$ be the projection if $\dim \alpha < n$ and otherwise trivial. This defines a map of spectra $\psi_\alpha: X \to E^\alpha$ where $E_n^\alpha = K(Z_2, n + \dim \alpha)$. Clearly $E_m^\alpha(Y, B) = H_{m+\dim \alpha}(Y, B; Z_2)$ and similarly for cohomology. The maps ψ_1 and ψ_2 are given by

$$\psi_1(u) = \sum \alpha \otimes \psi_\alpha(u) \qquad \text{and} \qquad \psi_2(u) = \sum \alpha \otimes \psi_\alpha(u).$$

Thus $\psi\varphi$ is a natural transformation:

$$MO_*(Y, B) \to \mathfrak{N}_* \otimes H_*(Y, B; Z_2), \qquad MO^*(Y, B) \to \mathfrak{N}_* \otimes H^*(Y, B; Z_2)$$

These transformations are isomorphisms if $(Y, B) = (*, \varnothing)$. Hence by Exercise 10, Section 18 and 21.7 the first part of 30.25 is proven. The second part is even easier. ∎

[28] This can be improved. One only needs to assume that Y is compact of finite dimension and B is closed by [24, X, 10.1].

We now define a natural transformation $\mu\colon MO_*(X) \to H_*(X;\ Z_2)$. For each pair (M, f) let $\mu((M, f)) = f_*([M])$. If $M = \partial N$ and $h\colon N \to X$ with $h|_M = f$, we have a commutative diagram

$$\begin{array}{ccc} H_n(M; Z_2) & \xrightarrow{f_*} & H_n(X; Z_2) \\ \Big\downarrow{\scriptstyle \iota_*} & \nearrow & \\ H_n(N; Z_2) & & \end{array}$$

By Exercise 12, Section 26 there is a commutative diagram

$$\begin{array}{ccc} H^0(N; Z_2) & \cong & H_n(N, M; Z_2) \\ \Big\downarrow{\scriptstyle \cong} & & \Big\downarrow{\scriptstyle \partial} \\ H^0(M; Z_2) & \cong & H_{n-1}(M; Z_2) \end{array}$$

from which it follows that ∂ is an isomorphism and hence $\iota_*([M]) = 0$. Thus $f_*([M]) = 0$ and μ is well defined.

Proposition 30.26 (*Thom*) μ is onto.

Remark This says that every Z_2 homology class is represented by a map from a manifold. The corresponding statement for integral homology is false.

Proof Consider the composition

$$\varphi\colon \mathfrak{N}_* \otimes H_*(X; Z_2) \to MO_*(X) \xrightarrow{\mu} H_*(X; Z_2)$$

Define $\varphi_\alpha(x) = \varphi(\alpha \otimes x)$. $\varphi_\alpha\colon H_n(X;\ Z_2) \to H_{n+\dim\alpha}(X;\ Z_2)$, and clearly $\varphi(\Sigma\alpha_i \otimes x_i) = \Sigma\varphi_{\alpha_i}(x_i)$. If $\dim\alpha > 0$, $\varphi_\alpha = 0$ since for each n-dimensional CW complex K, $\varphi_\alpha = 0$. Thus $\varphi(x) = \varphi_0(x_0)$ for some operation $\varphi_0\colon H_n(X; Z_2) \to H_n(X;\ Z_2)$. Since $\mu \neq 0$ when $X = S^n$. $\varphi_0 \neq 0$ in this case. Let K be an n-dimensional CW complex. Then we have a commutative diagram

$$\begin{array}{ccccc} 0 = H_n(K^{n-1}; Z_2) & \longrightarrow & H_n(K; Z_2) & \longrightarrow & H_n(\bigvee S_\beta{}^n; Z_2) \\ \Big\downarrow{\scriptstyle \varphi_0} & & \Big\downarrow{\scriptstyle \varphi_0} & & \Big\downarrow{\scriptstyle \varphi_0 = 1} \\ 0 = H_n(K^{n-1}; Z_2) & \longrightarrow & H_n(K; Z_2) & \longrightarrow & H_n(\bigvee S_\beta{}^n; Z_2) \end{array}$$

Thus $\varphi_0 = 1$ in this case and thus for any CW complex. By 21.7, $\varphi_0 = 1$ and thus $\mu(\Sigma\alpha_i \otimes x_i) = x_0$. μ is clearly onto. ∎

Other cobordism theories yield analogous results, but 30.25 and 30.26 do not hold in general. The simplest generalization is to consider cobordism of oriented manifolds. In this case, two oriented manifolds M and N are cobordant if there is an oriented manifold K with $\partial K = M \cup N$ and such that the orientation induced on ∂K (through the isomorphism $H_{n+1}(K, \partial K) \cong H_n(\partial K)$) agrees with that on M and is opposite to that on N. One writes this as $\partial K = M \cup - N$. Thus it is no longer true that every element has order 2, since $\partial(M \times I) = M \cup - M$. An orientation on M determines an orientation of both the tangent and normal bundles. That is, the linear transformations that occur in comparing coordinate neighborhoods must be orientation preserving (as maps $R^n \to R^n$). Such bundles are classified by maps into $BSO(n)$, and oriented cobordism is classified by $\pi_n(MSO)$ where $MSO = \{MSO(n), mso(n)\}$ is the Thom spectrum obtained from the universal bundle over $BSO(n)$. $H_*(MSO; Z_2) \simeq Z_2[u_2, u_3, \ldots]$. The existence of elements of infinite order in $\pi_*(MSO)$ makes the calculation problem harder. The solution is rather complicated (see [69; 19]).

This is the first example of cobordism of manifolds with "structure." The structure given is an orientation of the stable normal bundle. (This is equivalent to an orientation of the tangent bundle.) Such an orientation corresponds to a choice of a homotopy class of liftings in the double covering:

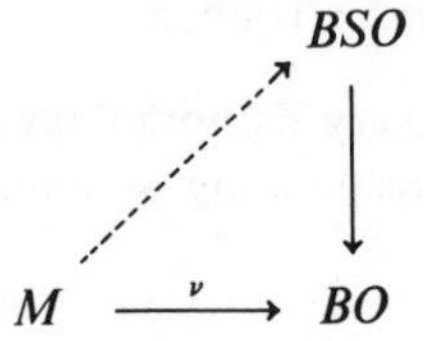

In more generality, we can consider pairs consisting of a manifold M and a lifting λ

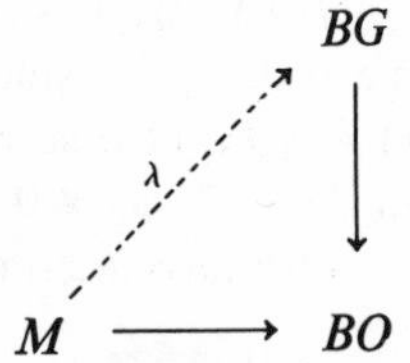

where G is a suitable subgroup of O. (For example, $U = \bigcup_{n=1}^{\infty} U(n)$, $Sp = \bigcup_{n=1}^{\infty} Sp(n)$. Such liftings are determined if M is a complex (simplectic) manifold.)

The generalized Thom theorem says that such cobordism classes are classified by $\pi_*(MG)$. ([69]).

The two cases U and Sp are of special interest. $\pi_*(MU)$ was calculated by

Milnor [51] and has a particularly simple structure. $\pi_*(MU) \cong Z[x_2, x_4, \ldots]$ where $x_{2n} = [CP^n]$. The cohomology theory MU^* has turned out to be useful in many applications to homotopy theory. ([63; 75]).

The case of MSp is at this time a mystery. $\pi_*(MSp)$ is not, as yet, completely known, although much partial information is available [61; 58; 59; 60; 37].

Exercises

1. Let ξ be a vector bundle over a compact manifold X. Then $T(\xi) = E(\xi)^\infty$. Prove that

$$\tilde{H}^{n+r}(T(\xi); Z_2) \cong H^r(M; Z_2)$$
$$\tilde{H}_{n+r}(T(\xi); Z_2) \cong H_r(M; Z_2).$$

This is called the Thom isomorphism theorem and is true without assuming that X is a compact manifold. (30.21).

2. The connected sum of two n-manifolds M and N is defined as follows. Let $U \subset M$ and $V \subset N$ be coordinate neighborhoods. Remove disks D and D' from U and V and attach a tube $S^{n-1} \times I$ to $M - D \cup N - D'$ by connecting one end to the boundary of D and the other end to the boundary of D' by homeomorphisms. Prove that the quotient space $M \# N$ is a manifold. Show that $M \# N$ is cobordant to $M \cup N$. (See Fig. 7.7.)

3. Calculate $H^*(BU)$ and $H^*(BSp)$ as rings.

4. Prove the exactness and homotopy axioms for $\mathfrak{N}_*(X, A)$. (See Exercise 9, Section 20.)

TABLE 1

	$b = 1$	2	3
a	0	3	0
23	0	3.1	5, 4.1
22	23.11	0	5.1
21	23.10	0	5.2
20	23.9, 22.10	21.10	0
19	23.8	21.9	0
18	23.7, 22.8, 21.9	21.8, 20.9	19.9
17	23.6, 21.8	21.7	19.8
16	23.5, 22.6, 20.8	21.6, 20.7, 19.8	19.7, 18.8
15	23.4	21.5, 19.7	19.6
14	23.3, 22.4, 21.5, 19.7	21.4, 20.5, 18.7	19.5, 18.6, 17.7
13	23.2, 21.4, 19.6,	21.3	19.4, 17.6
12	23.1, 22.2, 20.4 19.5, 18.6	21.2, 20.3, 19.4, 17.6	19.3, 18.4, 16.6
11	23, 19.4	21.1, 19.3, 17.5	19.2
10	22, 21.1, 18.4, 17.5	18.3, 17.4, 21, 20.1, 16.5	19.1, 18.2, 17.3, 15.5
9	21, 17.4	17.3	19, 17.2, 15.4
8	20, 16.4	19, 16.3, 15.4	18, 16.2, 15.3, 14.4
7	0	15.3	15.2
6	15.3	14.3	14.2, 13.3
5	15.2	0	13.2
4	15.1, 14.2	13.2	12.2
3	15	13.1	0
2	14, 13.1	13, 12.1	11.1
1	13	0	11
	$b = 12$	11	10

The notation $a.\,b$ represents $\mathrm{Sq}^a\mathrm{Sq}^b$, and the notation x, y represents

ADEM RELATIONS Sq^aSq^b FOR $a < 2b \leq 12$

4	5	6	a
5	0	7	1
6, 5.1	6.1	7.1	2
7	7.1	0	3
7.1, 6.2	9, 8.1, 7.2	10, 8.2	4
7.2	9.1	11, 9.2	5
7.3	9.2, 8.3	11.1, 10.2, 9.3	6
0	9.3	11.2	7
0	9.4	11.3, 10.4	8
17.8	0	11.4	9
17.7	0	11.5	10
17.6, 16.7	15.7	0	11
17.5	15.6	0	13
17.4, 16.5, 15.6	15.5, 14.6	13.6	12
17.3, 15.5	15.4	13.5	11
17.2, 16.3, 14.5	15.3, 14.4, 13.5	13.4, 12.5	10
17.1	15.2, 13.4	13.3	9
17, 16.1, 15.2, 13.4	15.1, 14.2, 12.4	13.2, 12.3, 11.4	8
15.1, 13.3	15	13.1, 11.3	7
14.1, 13.2, 12.3	14, 13.1, 11.3	13, 12.1, 10.3	6
13.1	13, 11.2	0	5
12.1, 11.2	12, 11.1, 10.2	11, 9.2	4
11.1	11	9.1	3
10.1	10, 9.1	9, 8.1	2
0	9	0	1
9	8	7	

$x + y$. Thus for example, 15.5, 14.6 represents $Sq^{15}Sq^5 + Sq^{14}Sq^6$.

References

1. Adams, J. F., On the non-existence of elements of Hopf invariant one, *Ann. of Math.* **72** (1960), 20–104.
2. Adams, J. F., Vector fields on spheres, *Ann. of Math.* **75** (1962), 603–632.
3. Adams, J. F., "Stable Homotopy Theory" (Lecture Notes in Mathematics, Vol. 3). Springer-Verlag, Berlin, 1964.
4. Adams, J. F., "Stable Homotopy and Generalized Homology." Mimeographed notes, University of Chicago (1971).
5. Adams, J. F., "Algebraic Topology: A Student's Guide." Cambridge Univ. Press, London and New York, 1972.
6. Adams, J. F., and Atiyah, M. F., *K*-theory and the Hopf invariant, *Quart. J. Math. Oxford Ser.* **17** (1966), 31–38.
7. Atiyah, M. F., Bordism and cobordism, *Proc. Cambridge Philos. Soc.* **57** (1961), 200–208.
8. Atiyah, M., *K*-Theory and reality, *Quart. J. Math. Oxford Ser.* **17** (1966), 367–386.
9. Atiyah, M., "*K*-Theory" (notes by D. W. Anderson). Benjamin, New York, 1967.
10. Atiyah, M., and Bott, R., On the periodicity theorem for complex vector bundles, *Acta Math.* **112** (1964), 229–247.
11. Barratt, M. G., Track Groups I, *Proc. London Math. Soc.* (1955), 71–106.
12. Birkhoff, G., and MacLane, S., "A Survey of Modern Algebra." Macmillan, New York, 1953.
13. Blakers, A. L., and Massey, W. S., The homotopy groups of a triad II, *Ann. of Math.* **55** (1952), 192–201.
14. Borel, A., La transgression dans les espaces fibrés pincipaux, *C. R. Acad. Sci. Paris Sèr. A-B.* **232** (1951).
15. Bott, R., The stable homotopy of the classical groups, *Proc. Nat. Acad. Sci. U.S.A.* **43** (1957), 933–935.
16. Bott, R., The stable homotopy of the classical groups, *Ann. of Math.* **70** (1959), 313–337.
17. Bourbaki, N., "Éléments de Mathématique Livre II," Chapitre IX. Hermann, Paris, 1959.
18. Cartan, H., and Eilenberg, S., "Homological Algebra." Princeton Univ. Press, Princeton, New Jersey, 1956.
19. Conner, P. E., and Floyd, E. E., "Differentiable Periodic Maps." Academic Press, New York, 1964.

20. Dold, A., "Lectures on Algebraic Topology." Springer-Verlag, Berlin, 1972.
21. Dress, A., Zur Spectralsequenz von Fasserungen, *Invent. Math.* **3** (1967), 172–178.
22. Dugundji, J., "Topology." Allyn and Bacon, Rockleigh, New Jersey, 1966.
23. Dyer, E., and Lashof, R., A topological proof of the Bott periodicity theorems, *Ann. Mat. Pura Appl.* (4), **54** (1961), 231–254.
24. Eilenberg, S., and Steenrod, N., "Foundations of Algebraic Topology." Princeton Univ. Press, Princeton, New Jersey, 1952.
25. Giever, J. B., On the equivalence of two singular homology theories, *Ann. of Math.* **51** (1950), 178–191.
26. Gray, B., On spaces of the same n-type for all n, *Topology* **5** (1966), 241–243.
27. Gray, B., "The Hopf Invariant and Related Problems." Aarhus Universitet, Lecture Notes Series, No. 24, April 1970.
28. Greenberg, M. J., "Lectures on Algebraic Topology." Benjamin, New York, 1967.
29. Hocking, J. G., and Young, G. S., "Topology." Addison-Wesley, Reading, Massachusetts, 1961.
30. Hopf, H., Über die Topologie der Gruppen-Mannigfaltigkeiten und ihre Verallgemeinerungen, *Ann. of Math.* **42** (1941), 22–52.
31. Hu, S. T., "Homotopy Theory." Academic Press, New York, 1959.
32. Huber, P., Homotopical cohomology and Čech cohomology, *Math. Ann.* **144** (1961), 73–76.
33. Hurewicz, W., Beiträge zur Topologie der Deformation II, *Proc. K. Akad. van Wet. Amst.* **38** (1935), 521–528.
34. Husemoller, D., "Fiber Bundles." McGraw-Hill, New York, 1966.
35. James, I., The intrinsic join: A study of the homotopy groups of Stiefel manifolds, *Proc. London Math. Soc.* **8** (1958), 507–535.
36. Kelley, J. L., "General Topology." Van Nostrand, Princeton, New Jersey, 1955.
37. Kochman, S., Lecture at University of Illinois, Chicago Circle, 1972.
38. Kristensen, L., On a Cartan formula for secondary cohomology operations, *Math. Scand.* **16** (1965), 97–115.
39. Kristensen, L., Lectures on cohomology operations. Aarhus Universitet, 1969.
40. Lefschetz, S., "Introduction to Topology." Princeton Univ. Press, Princeton, New Jersey, 1949.
41. Liulevicius, A., "Characteristic Classes and Cobordism, Part I." Aarhus Universitet Lecture Notes Series No. 9, 1967.
42. Liulevicius, A., "Lecture Notes on Characteristic Classes." Nordic Summer School in Mathematics, Aarhus Universitet, 1968.
43. Liulevicius, A., and Lashof, R., "Topology and Geometry of Locally Euclidean Spaces." Mimeographed notes, University of Chicago, 1973.
44. MacLane, S., "Homology." Academic Press, New York, 1963.
45. Massey, W. S., "Algebraic Topology: An Introduction." Harcourt, Brace, Jovanovich, New York, 1967.
46. Maunder, C. R. F., "Algebraic Topology." Van Nostrand, Princeton, New Jersey, 1970.
47. Milgram, J., "The Definition and Properties of Steenrod Squares." Mimeographed notes, Princeton University, 1970.
48. Milnor, J., and Stasheff, J., "Characteristic Classes." Annals of Maths Studies No. 76 (1973).
49. Milnor, J., "Differential Topology." Notes by J. Munkres, mimeographed notes, Princeton University, Fall, 1958.
50. Milnor, J., On spaces having the homotopy type of a CW-complex, *Trans. Amer. Math. Soc.* **90** (1959), 272–280.

51. Milnor, J., On the cobordism ring Ω^* and a complex analogue I, *Amer. J. Math.* **82** (1960), 505–521.
52. Milnor, J., "Morse Theory." Notes by M. Spivak and R. Wells (Annals of Mathematics, Studies No. 51). Princeton Univ. Press, Princeton, New Jersey, 1963.
53. Milnor, J., "Topology from the Differentiable Viewpoint." University Press of Virginia, Charlottesville, Virginia, 1965.
54. Milnor, J., and Moore, J., On the structure of Hopf algebras, *Ann. of Math.* **81** (1965), 211–264.
55. Northcott, D. G., "An Introduction to Homological Algebra." Cambridge Univ. Press, London and New York, 1960.
56. Peterson, F. P., Some non-embedding problems, *Bol. Soc. Mat. Mexicana* **2** (1957), 9–15.
57. Poincaré, H., Analysis Situs, *Journal de l'Ecole Polytechique* **1** (1895), 1–121.
58. Ray, N., Indecomposables in Tors. MS_p, *Topology* **10** (1971), 261–270.
59. Ray, N., The symplectic bordism ring, *Proc. Cambridge Philos. Soc.* **71** (1972), 271–282.
60. Ray, N., Realizing symplectic bordism classes, *Proc. Cambridge Philos. Soc.* **71** (1972), 301–305.
61. Segal, D., On the symplectic cobordism ring, *Comment. Math. Helv.* **45** (1970), 159–169.
62. Serre, J. P., Homologie singulierè des espaces fibrés I. La suite spectrale, *C. R. Acad. Sci. Paris Sèr. A-B* (1950), 1408–1410.
63. Smith, L., On the complex Bordism of Finite Complexes. Proceedings of the Advanced Study Institute on Algebraic Topology, August 1970, Aarhus Universitet, Various Publications, Series No. 13, pp. 513–566.
64. Spanier, E. H., "Algebraic Topology." McGraw-Hill, New York, 1966.
65. Steenrod, N., "The Topology of Fiber Bundles." Princeton Univ. Press, Princeton, New Jersey, 1951.
66. Steenrod, N. E., Cyclic reduced powers of cohomology classes, *Proc. Nat. Acad. Sci. U.S.A.* **39** (1953), 217–223.
67. Steenrod, N., "Cohomology Operations." Written and revised by D. B. A. Epstein (Annals of Mathematics, Studies No. 49). Princeton Univ. Press, Princeton, New Jersey, 1962.
68. Steenrod, N. E., A convenient category of topological spaces, *Michigan Math. Journal*, **14** (1967).
69. Stong, R., "Notes on Cobordism Theory." Princeton Univ. Press, Princeton, New Jersey, 1968.
70. Toda, H., A topological proof of theorems of Bott and Borel–Hirzebruch for homotopy groups of unitary groups, *Mem. Coll. Sci. Univ., Kyoto Ser. A. Math.* **32** (1959), 103–119.
71. Vick, J. W., "Homology Theory: An introduction to algebraic topology." Academic Press, New York, 1973.
72. Waley, A., (transl.), "*Tao Tê Ching*" (from "The Way and its power, a study of the *Tao Tê Ching* and its place in Chinese Thought"). Grove Press, New York, 1958.
73. Whitehead, J. H. C., Combinatorial Homotopy I, *Bull. Amer. Math. Soc.* **55** (1949), 213–245.
74. Whitehead, J. H. C., On a certain exact sequence, *Ann. of Math.* **52** (1950), 51–110.
75. Zahler, R., Existence of the Stable Homotopy Family $\{\gamma_t\}$, *Bull. Amer. Math. Soc.* **79** (1973), 787–789.
76. Zeeman, E. C., A proof of the comparison theorem for spectral sequences, *Proc. Cambridge Philos. Soc.* **53** (1957), 57–62.

Index

An n following a page number indicates a topic mentioned in a footnote.

A

Absolute homotopy extension
  property (AHEP), 117
Abstract semisimplicial complex, 155
Abstract simplicial complex, 155
Action of the fundamental group, 89
Acyclic, 191
Adams operations, 336
Adem relations, 313, 322, 359
Adjointness, 56–60
Admissible monomial, 311
Affine subspace (plane), 92
AHEP, 117
Alexander duality theorem, 283
Alexander cohomology, 212
Amalgamated product, 40
Antipodal map, 14
Arc component, 15n
Arc(-wise) connected, 15n
Associated compactly generated space, 52
Attaching map, 99
Axioms for chain complexes, 191
  for homology and cohomology, 186

B

Ball, 4
Barratt theorem, 149
Barratt–Whitehead lemma, 202
Barycenter, 94
Barycentric coordinates, 90
Barycentric subdivision, 94, 147
Base point, 20
Betti number, 192
Binomial coefficients mod 2, 321
Blakers–Massey theorem, 103, 143
Bockstein, 268
Bockstein homomorphism, 182
Bockstein sequence, 182
Borel's theorem, 311
Borsuk–Ulam theorem, 311
Bott periodicity theorem, 331
Boundary, 191
Brouwer degree (theorem), 220
Brouwer fixed point theorem, 108
Bundle map, 325

C

Cap product, 242
Cartan formula, 301
Cartesian product topology, 53
Category, 17
Čech cohomology, 212
Cell, 8
  relative, 99
Cell complex, 113
Cellular, 115, 133
  approximation, 133
  homotopy, 134
Chain, 189
  complex, 191
  map, 191
Characteristic class, 351
Characteristic map, 100, 113

Closure finite, 115
Coalgebra, 314
Coassociative, 314
Cobordant, 342
Cobordism, 342
Coboundary, 191
Cochain, 190
  complex, 190
  map, 191
Cocommutative, 314
Cocycle, 191
Coefficient groups, 177
Coefficient operations, 294
Cohomology operation, 180
  of type (E, m, F, n), 295
Cohomology theory
  reduced, 170
  unreduced, 183
Cohomology with compact supports, 284
Commute, 9
Compact open topology, 55
Compactly generated, 50
Complex, 91
  cell, 113
Complex K-theory, 332
Composite, 17
Composition (of paths), 72
Comultiplication, 314
Cone
  reduced, 63
  unreduced, 68
(n-)Connected, 130
Connected sum, 45, 357
Connective, 180
Connective covering space, 164
Continuity, 209
Contractible, 31
Convex, 93
Convex hull, 98
Coordinate neighborhood, 35, 77
Counit, 314
Covering space, 35
  n-fold, 36, 48
  existence, 47
Cube, 4
Cup product, 236, 238
CW complex, 115
  regular, 147
  semisimplicial, 146
  simplicial, 147
Cycle, 191

D

Degree, 220
  of a map of spectra, 180
Dimension, 91, 113
Dimension axiom, 186
Direct limit, 123
Directed set 122
  system, 122
Disk bundle, 343
Domain, 17
Dual pairing, 285
Duality theorems, 283, 291

E

Eilenberg–MacLane space, 158
Eilenberg–Steenrod aixoms, 186
Equivalence,
  in a category, 21
  homotopy, 30
  k-equivalence, 135
  of vector bundles, 325
Euclidean space, 4
Euler characteristic, 199, 283, 287, 291
Evaluation mapping, 55
Exact sequence, 74
  of a fibering, 84
  of a pair, 74
  of a triad, 88
Exactness axiom, 184
Excess, 311
Exicision axiom, 184, 209
Excisive, 40
Exterior algebra, 311
Exterior power operations, 329
External product, 233
Extraordinary cohomology, 168n
Extraordinary homology, 168n

F

Face, 91
Fiber, 77, 86
Fiber bundle, 77
Fibration
  Hurewicz, 86, 89
  Serre, 79
Five lemma, 142

Fixed point, 1
Fixed point theorem
  Brouwer, 108
  Lefschetz, 288
Folding map, 64
Forgetful functor, 19
Freudenthal suspension theorem, 108, 145
Freyd conjecture, 145
Function space topology (in $\mathcal{CG}$), 55
Functor
  covariant, 18
  contravariant, 19
Fundamental category, 27
Fundamental class, 271
Fundamental group, 23
Fundamental theorem of algebra, 38

G

Gauss map, 326
Geometric finite simplicial complex, 91
Giever–Whitehead theorem, 149
Graded abelian group, 169
Graded commutative, 235
Graded module, 235
Graded ring, 235
Grassmanian, 327
Grothendieck construction, 328
Group ring, 89

H

H space, 64
Ham sandwich theorem, 287
Handle, 48
HLP, 79
Homologous, 191
Homology
  of CW complexes, 187
  of simplicial complexes, 196–197
  singular, 214
Homology operation, 180
Homology theory
  ordinary, 186
  reduced, 170
  nureduced, 183
Homomorphism (of graded abelian groups), 169
Homotopic, 13, 15, 21, 30
  relative to *A*, 22
Homotopy, 13, 15, 21, 30
  relative, 22
Homotopy associative, 67
Homotopy axiom, 184
Homotopy class, 14, 16, 21
Homotopy commutative, 68
Homotopy equivalence, 30
Homotopy group, 64
  relative, 70
  of a spectrum, 177
Homotopy inverse, 67
Homotopy lifting property (HLP), 79
Homotopy type, 30
Hopf algebra, 315
Hopf construction, 334
Hopf invariant, 320, 334
Hopf map, 44, 78
Hopf theorem, 167
Hurewicz fibering, 86, 89
Hurewicz homomorphism, 216, 217
Hurewicz theorem, 216, 217

I

Identity functor, 19
Inessential, 69
Intersection pairing, 292
Invariance of domain, 285
Inverse system, 122
Isomorphic, 21
(*k*-)Isomorphism, 135
Isotopic, 345

J

Join, 93, 334
Jordan separation theorem, 284

K

K-theory, 328
Kill, 160
Klein bottle, 48
Knot, 46
Knot equivalence, 46
Kronecker product, 242
Künneth theorems, 261–263, 267, 268

L

Ladder, 76, 176
Lebesgue's covering lemma, 3
Lefschetz fixed point theorem, 288
Lens space, 49
Leray–Hirsch theorem, 293
Limit
  direct, 123
  inverse, 128
  right, 123
Linear, 92
Linear dimension, 92
Linearly independent (vector fields), 337
Locally trivial bundle, 77
Long exact sequence, 175, 186
Loop, 22
Loop space, 65

M

Manifold, 269
  with boundary, 290
Map
  of a category, 18
  of spectra, 180
Mapping cone, 118
Mapping cylinder, 134
Mayer–Vietoris sequence, 214
Mesh, 96
Milnor–Moore theorem, 348
Module spectrum, 225
Möbius band, 49, 270
Monodromy theorem, 36
Moore space, 159
Morphism, 17
Multiplication, 64

N

Natural transformation, 19
Neighborhood extension property, 209
Nondegenerate base point, 144
Normal bundle, 343
Null homotopic, 68

O

Object, 17
Obstruction set, 166
Obstruction theory, 165–166
Operation (homology or cohomology), 180
Ordinary cohomology, 166, 186
  spectral reduced, 170
Ordinary homology, 186
Orientable, 271
(E-)orientable, 271
Orientation, 269
(E-)orientation, 271
Oriented along $\{X\}$, 271
Oriented cobordism, 356
Orthogonal group, 89

P

Pairing, 224
Path, 22
  based, 22
Path component, 15n
Path connected, 15n
Path lifting property, 35
Peterson imbedding theorem, 307
Phantom map, 165
Poincaré duality, 283
Pointed category, 20
Polyhedron, 91
Positively graded, 169
Postnikov
  section, 163
  system, 163
  tower, 163
Principal ideal domain (module theory), 256
Product bundle, 78
Product topology in $\mathcal{CG}$, 53
Projective space, 44
  homology of, 195, 252
  sphere bundles over, 78
Properly convergent, 182

Q

Quadratic construction, 297
Quaternions, 46

R

Range, 17
Real K-theory, 332
Reduced cone, 63
Reduced cohomology theory, 170
Reduced homology theory, 170
Reduced K-theory, 329
Reduced suspension, 62
Regular CW complex, 147
Relative cell complex, 120
Relative (n-)cell, 99
Relative CW complex, 120
Relative homeomorphism, 209
Relative homotopy group, 70
Resolution, 134, 137, 257
Right limit, 123
Ring spectrum, 225

S

Scalar product, 9
Semisimplicial CW complex, 146
  abstract, 155
Serre exact sequence, 213, 214
Serre fibering, 79
Serre's theorem, 311
Simple system (of generators), 310
Simplex, 4
Simplicial approximation theorem, 97
Simplicial complex, 91
Simplicial CW complex, 147
  abstract, 155
Simplicial homology, 196–197
Simplicial map, 97, 149
Simply connected, 32
Singular chain complex, 214
Singular cohomology, 214
Singular complex
  ad hoc, 134, 145
  functorial, 146
Singular extraordinary cohomology and homology, 200
Singular homology, 214
Singular manifold, 352
Skeleton, 91, 113
Slant product, 233
Smash product, 58
Spectral cohomology (reduced), 170
Spectral homology (reduced), 170
Spectrum, 169
  suspension, 169
  Ω-, 169
Sphere, 4
Sphere bundle, 343
Spherical, 277
  stably, 277
Split, 85
Stable cohomology operation, 180
Stable cohomotopy, 179
Stable equivalence (of vector bundles), 330
Stable homology operation, 180
Stable homotopy group, 145
Stably spherical, 277
Standard simplex, 90
Steenrod algebra, 312
Steenrod operation, 301
Steenrod pth power operation, 308
Steenrod square, 296, 301
Stiefel manifold, 337
Stiefel–Whitney classes, 351
Stiefel–Whitney number, 352
Strong deformation retract, 32
Subcomplex, 91, 114, 120
Surface, 45
Suspension, 311
  functor, 107, 112
  reduced, 62
  theorem, 109
  unreduced, 69
Suspension spectrum, 169
Symplectic K-theory, 332

T

Tangent, bundle, 78, 325
Taosim, 17n
Tensor product
  of chain complexes, 247
  of graded groups, 245
  of graded homomorphisms, 248n
  of vector bundles, 329
Thom imbedding theorem, 307
Thom space, 343
Torsion coefficients, 192
Trace, 288

Track groups, 62
Trefoil, 46
Triad, 88
Triangulation, 91
(k-) Trivial, 165
Trivial map, 64
Trivial vector bundle, 326
Tubular neighborhood theorem, 343
Type 1 excision, 184
Type 2 excision, 184

U

Universal coefficient theorems, 262–264, 266, 268
Universal property (covering spaces), 47
Unreduced cohomology theory, 183
Unreduced cone, 68
Unreduced homology theory, 183
Unreduced suspension, 69

V

Van Kampen theorem, 40
Vector bundle, 324
Vector field, 10
Vector field problem, 338
Vertices, 90

W

Weak absolute neighborhood extensor (WANE), 204
Weak homotopy equivalence, 133, 137
Weak topology, 115
Wedge axiom, 179
Wedge product, 58
Well-pointed, 144
Whitehead theorem, 139, 217
Whitney formula, 351
Whitney imbedding theorem, 344
Whitney sum, 328

A 5
B 6
C 7
D 8
E 9
F 0
G 1
H 2
I 3
J 4